Food, Pharmaceutical and Bioengineering Division 2022

Held at the 2022 AIChE Annual Meeting

Phoenix, Arizona, USA
13-18 November 2022

ISBN: 978-1-7138-7910-7

Printed from e-media with permission by:

Curran Associates, Inc.
57 Morehouse Lane
Red Hook, NY 12571

Some format issues inherent in the e-media version may also appear in this print version.

Printed with permission by Curran Associates, Inc. (2023)

For permission requests, please contact AIChE
at the address below.

AIChE
120 Wall Street, FL 23
New York, NY 10005-4020

Phone: (800) 242-4363
Fax: (203) 775-5177

www.aiche.org

Additional copies of this publication are available from:

Curran Associates, Inc.
57 Morehouse Lane
Red Hook, NY 12571 USA
Phone: 845-758-0400
Fax: 845-758-2633
Email: curran@proceedings.com
Web: www.proceedings.com

TABLE OF CONTENTS

DIVISION PLENARY: FOOD, PHARMACEUTICAL, AND BIOENGINEERING DIVISION (INVITED TALKS)

FERMENTATION AND PROCESS ENGINEERING IN FOOD AND BIOPROCESS INDUSTRIES

METABOLIC ENGINEERING FOR FOOD, FEED, AND BIOPRODUCTS

MICROBIOME AND NATURAL PRODUCTS IN FOOD, HEALTH, AND BIOPROCESSING

POSTER SESSION: BIOENGINEERING

POSTER SESSION: ENGINEERING FUNDAMENTALS IN LIFE SCIENCE

POSTER SESSION: FOOD AND BIOPROCESS ENGINEERING

ADVANCES IN BIOCATALYSTS AND BIOCATALYTIC PROCESSES

FOOD, PHARMACEUTICAL AND BIOENGINEERING FACULTY CANDIDATES SESSION I

MOLECULAR AND CELLULAR SENSING TECHNOLOGIES

CELL CULTURE ENGINEERING AND BIOPHARMACEUTICAL MANUFACTURING

NEW TOOLS AND STRATEGIES FOR METABOLIC ENGINEERING

ENGINEERING PROTEIN THERAPEUTICS

ENGINEERING BACTERIA FOR NOVEL CHEMISTRY AND INTERACTIONS

SYSTEMS AND QUANTITATIVE BIOLOGY: DISEASE MECHANISMS AND THERAPIES

BIOMOLECULAR ENGINEERING III

STEM CELLS AND TISSUE ENGINEERING

CELLS, ORGANS, AND LABS ON A CHIP

DRUG DELIVERY

MICRO- AND NANO-SCALE TECHNOLOGIES IN LIFE SCIENCES I

SYSTEMS BIOLOGY OF DEVELOPMENT AND CANCER

MEMBRANE AND PROCESS TECHNOLOGIES IN FOOD AND BIOPROCESS ENGINEERING

BIOTECHNOLOGICAL PRODUCTION OF FOOD AND FEED INGREDIENTS

BIOBASED AND BIODERIVED PROCESSES FOR VALUE ADDED CHEMICALS AND ADVANCED MATERIALS

COMPUTATIONAL, STRUCTURE, BIOPHYSICAL PROTEIN ENGINEERING

NEW METHODS IN PROTEIN ENGINEERING

BIOCATALYSIS AND ENZYME ENGINEERING

PROTEIN ASSEMBLIES AND AGGREGATES

COMPUTATIONAL AND SYSTEMS BIOLOGY TOOLS FOR METABOLIC ENGINEERING AND CELL CHARACTERIZATION

METABOLIC PLATFORM DEVELOPMENT- NON-CONVENTIONAL SPECIES AND SYSTEMS

ADVANCES IN METABOLIC ENGINEERING- EUKARYOTIC ORGANISMS

ADVANCES IN METABOLIC ENGINEERING- PROKARYOTIC ORGANISMS

SYNTHETIC BIOLOGY OF UNDERUTILIZED ORGANISMS WITH UNIQUE PHENOTYPES

NEW APPROACHES FOR GENE AND PROTEIN REGULATION

DIAGNOSTIC TECHNOLOGIES FOR CLINICAL APPLICATIONS

DRUG DELIVERY STRATEGIES FOR IMMUNOMODULATION

DRUG DELIVERY FOR APPLICATIONS IN THE BRAIN, NERVOUS SYSTEM, AND MUSCULOSKELETAL SYSTEM

MICRO- AND NANO-SCALE TECHNOLOGIES IN LIFE SCIENCES II

SYSTEMS BIOLOGY FOR ENGINEERING METABOLISM

SYSTEMS BIOLOGY FOR ENGINEERING MICROBES

CELL AND TISSUE ENGINEERING: MECHANICAL CUES AND CELL BEHAVIOR

CELL AND TISSUE ENGINEERING: ENGINEERING THE IMMUNE RESPONSE

CELL AND TISSUE ENGINEERING: ENGINEERING 3D TISSUES TO MODEL DISEASE AND DEVELOPMENT

CELL AND TISSUE ENGINEERING: ENGINEERING IN AGING AND AGING ASSOCIATED DISEASES

GENERAL TOPICS IN SYNTHETIC BIOLOGY AND METABOLIC ENGINEERING

BIOMOLECULAR ENGINEERING AND THE IMMUNE SYSTEM

BIOMOLECULAR ENGINEERING II

BIOMOLECULAR ENGINEERING I

FOOD, PHARMACEUTICAL AND BIOENGINEERING FACULTY CANDIDATES SESSION II

394a Division Plenary: Synthetic Biology: Putting Synthesis into Biology

Huimin Zhao

Synthetic biology is the design of novel or improved biological systems using engineering principles. It is a rapidly growing area with broad applications in medical, chemical, food, and agricultural industries. In this talk, I will highlight our recent work in the development and application of novel foundational synthetic biology tools. Specifically, I will introduce four interrelated stories, including: (1) development of the Illinois Biological Foundry for Advanced Biomanufacturing (iBioFAB) for next-generation synthetic biology applications; (2) development of new strategies and tools for discovery of novel natural products from genomes and metagenomes; (3) development of genome-scale engineering tools for rapid metabolic engineering applications; and (4) development of new genome engineering tools for human gene therapy and fundamental studies of cell biology.

394b Early Career Award: Discovery, Domestication, and Engineering of Diverse Microbes for a Circular Economy

Mark Blenner

Microorganisms have always been integral to natural circular processes in the environment. They serve an important role in converting complex biological molecules into central building blocks that can be converted back into complex products. As a society, we now seek similar circularity in the semi-synthetic world that we've constructed through over a century of synthetic organic chemistry. Recent advances in genomics and synthetic biology have enabled exploitation of a wide array of microbes engineered to produce value-added products from biomass-derived and synthetically derived substrates. Our group has mainly focused on oleaginous yeast – that use a variety of low-value and waste feedstocks and can naturally accumulate a large amount of lipids. We have engineered multiple genetic tools to rapidly engineer Yarrowia lipolytica, and used these tools to elucidate cryptic metabolic pathways, and to produce oleochemicals and secondary metabolites by taking advantage of naturally high flux pathways. We will also describe our work with emerging microbes that have more favorable properties for utilizing low-value and waste substrates, including lignin derived aromatics and plastic waste.

394c Area 15a Plenary: Opportunities for Cell Engineering for Cellular Agriculture

David L. Kaplan

Mixing the right cells, support materials, and nutrients/growth factors to guide cells toward a programmed tissue (food) outcome ex vivo is the goal for cellular agriculture, with a focus on muscle and fat. A tissue engineering approach using cells from livestock and other animals, offers an alternative opportunity to generate foods of the future, including new routes to protein-enriched or meat-like foods. At the same time, bioengineering approaches can be utilized to selectively enhance features of these cells to improve food quality and nutrition. Some of our strategies will be described, with a focus on moving cells cultivated in vitro for food to the next generation of control related to nutrition, health, stability and function.

394d Area 15c Plenary: Outsmarting pathogens with protein engineering: lessons from pertussis, cytomegalovirus and coronavirus.

Jennifer Maynard

There is growing interest in development of antibodies to provide instant immunity that protects susceptible individuals against infectious diseases, especially pathogens for which no vaccine is yet available. However, many of these pathogens, especially those that manifest as highly transmissible or latent infections, express complex arrays of virulence proteins and are adept at avoiding the immune system. Some organisms have developed strategies to selectively destroy anti-pathogen antibodies, while others create decoy epitopes that trick the host immune system into generating antibodies that are at best non-protective and, at worst, enhance pathogenesis. Design of pathogen-resistant antibodies can represent novel therapies and in turn guide development of protective vaccines for these complex pathogens. This talk will provide an overview of our progress in engineering antibodies and antigens resistant to specific immune defense mechanisms used by pathogens including *B. pertussis,* human cytomegalovirus, and SARS-CoV-2.

There is growing interest in development of antibodies to provide instant immunity that protects susceptible individuals against infectious diseases, especially pathogens for which no vaccine is yet available. However, many of these pathogens, especially those that manifest as highly transmissible or latent infections, express complex arrays of virulence proteins and are adept at avoiding the immune system. Some organisms have developed strategies to selectively destroy anti-pathogen antibodies, while others create decoy epitopes that trick the host immune system into generating antibodies that are at best non-protective and, at worst, enhance pathogenesis. Design of pathogen-resistant antibodies can represent novel therapies and in turn guide development of protective vaccines for these complex pathogens. This talk will provide an overview of our progress in engineering antibodies and antigens resistant to specific immune defense mechanisms used by pathogens including *B. pertussis,* human cytomegalovirus, and SARS-CoV-2.

394e Area 15d/e Plenary: Tissue-Inspired Synthetic Biomaterials

Shelly Peyton

Improved experimental model systems are critically needed to better understand cancer progression and bridge the gap between lab bench proof-of-concept studies, validation in animal models, and eventual clinical application. Many methods exist to create biomaterials, including hydrogels, which we use to study cells in contexts more akin to what they experience in the human body. Our lab has multiple approaches to create such biomaterials, based on combinations of poly(ethylene glycol) (PEG) with peptides and zwitterions. In this presentation, I will discuss our synthetic approaches to building life-like materials, how we use these systems to grow cells and understand how a cell's environment, particularly the extracellular matrix regulates cancer cell growth, dormancy, and drug sensitivity.

448a Relationship of Different Cultivation Conditions to the Efficacy of D-Lactic Acid Fermentation By *Sporolactobacillus Terrae*sbt-1

Nichakorn Amornbunchai, Sitanan Thitiprasert, Nuttha Thongchul

Lactic acid is a versatile organic acid which contains the chiral properties, including L-isomer and D-isomer. Recently, lactic acid is emerged in the variety of applications; in particular, its demand is driven by rapid expansion of poly(lactic acid) (PLA) market. PLA is synthesized by polymerization of lactic acid and employed for various applications including packaging, cosmetic, and pharmaceutical industries due to its biodegradability nature. The lactic acid molecule has a chiral carbon that confers optical purity. It can be found in two different optical isomers including L(+)-lactic acid and D(-)-lactic acid. Highly optically pure lactic acid is required for the synthesis of high-quality plastic, and poly-L-lactic acid (PLLA) and poly-D-lactic acid (PDLA) are produced from L-lactic acid and D-lactic acid, respectively. Microbial fermentation has been extensively used as the major route for the synthesis of optically pure lactic acid. In comparison of L-lactic acid, the production of optically pure D-lactic acid has attracted to develop due to the demand for producing thermostable PLA, but large-scale production of D-lactic acid has not yet been commercialized. The main obstacle to produce optically pure D-lactic acid is the market price that their cost should be competitive. Several limitations, including the use of alternative substrates, sensitivity to decrease lactic acid yield and productivity, and decreased optical purity of lactic acid have been being discussed. Furthermore, more cost-effective production of optically pure lactic acid is needed to meet the current huge demand. In addition, fermentation conditions might influence the optical purity of lactic acid. Two optically pure isomers of lactate were produced from pyruvate through reactions catalyzed by either the chiral-specific D-lactate dehydrogenase (D-LDH) or L-lactate dehydrogenase (L-LDH). However, studies on the associations of L- and D-LDH with the optical purity of lactate are limited. In this study aims to investigate the associations of L- and D-lactate dehydrogenase with the optical purity of lactate under different cultivation conditions. Comparison of D-lactic acid fermentation performance by *S. terrae* SBT-1 was carried out in flask culture. Different nitrogen sources, including yeast extract and corn steep liquor (CSL) could influence the optical purity of lactic acid as well as enzymes responsible for L- and D-lactic acid production. Decreasing optical purity of D-lactic acid was observed in fermentation medium containing CSL as the nitrogen source. Consistently, lower L-LDH and D-LDH were also found. This indicated that the proper nitrogen source could impact the correlation between the optical purity of D-lactate and enzyme activities in *S. terrae* SBT-1.

448b Multi-Information Source / Objective Bayesian Optimization of Serum-Free Cell Culture Media for Cellular Agriculture

Zachary Cosenza, David E. Block

The emerging field of cellular agriculture, in which bioreactors are used to grow mammalian cells for human consumption in leu of carbon-intensive animal agriculture, is predicated on use of inexpensive and serum-free (no animal components) culture media. Designing such a media is a difficult task due to the number of metabolites in use, nonlinear component-wise interactions, and the difficulty in making measurements that reflect the complex dynamics of cell growth and survival. We address these issues by using a Bayesian optimization algorithm in which a multi-information source Gaussian process model explores the media design space while maximizing a multi-objective (expected hypervolume improvement) acquisition function to find optimal media component concentrations. Using this acquisition function the uncertainty in the Gaussian process model us iteratively reduced while the cell growth metric and component cost are maximized and minimized respect. Multiple cell growth assays are used in concert, increasing the quality of the model predictions. The result was a methodology that discovered a high-growth / minimized cost medium in few experiments while exploring the pareto trade-off curve between cell growth and medium cost.

448c Fermentation Process Optimization for Enhancing D-Arabitol Production Performance in a Novel Strain of *Wickerharmomyces Anomalus* BKK11-4

Nuttha Thongchul, Sitanan Thitiprasert, Jesnipit Tammakes

D-Arabitol, also known as arabinitol, is an important pentitol compound. It has been listed in the top 12 biobased building block chemicals in polyols group compounds that targeted further development of industrial biotechnology and research. D-Arabitol is widely used in many consumer products as an ideal sugar substitute for diabetes patients, including artificial sweeteners, and oral healthcare products, herbicides, and antipathogenic disease medicines. D-Arabitol can be also used as an intermediate for chemical synthesis such as xylitol, propylene glycol, and ethylene glycol. To date, arabitol is commercially produced by the chemical reduction of arabinonic acid lactones. The drawbacks of the chemical synthesis include consumption of costly catalysts, operation under high temperature (100 °C), and difficulty in product separation and purification in downstream recovery. Arabitol production via biosynthesis platform has therefore attracted more interest as the alternative production route. The challenge for biotechnologists is to understand the fundamental knowledges on sugar metabolism by microbes for enhancing arabitol production. Yeasts have been selected as the potential microbes for biosynthesizing sugar alcohols including arabitol. Here, we present our in-house osmotolerant yeast isolate *Wickerharmomyces anomalus* BKK11-4 as the biofactory to produce D-arabitol. *W. anomalus* BKK11-4 could ferment high glucose concentration of 200 g/L for D-arabitol under aerobic conditions. Mixing and aeration were found to influence fermentation productivity and yield. Complete sugar consumption was observed at high agitation and aeration rates. While organic nitrogen source like yeast extract mainly controlled fermentation productivity and carbon flux ratio to D-arabitol.

448d Exploring the Potential of NON-Growth Associated POLY-3-Hydroxybutyrate Production with Recombinant *Escherichia coli* through Dynamic Flux Balance Analysis

Willians Santos, Galo Carrillo le Roux, José Gregório Cabrera Gomez

For the past decades there is a push for the development of sustainable processes due to increasing concerns with environmental impacts. Plastic disposal, for instance, has become a major problem due to how long conventional oil-derived plastics take to degrade when disposed, and because of problems with toxicity and bioaccumulation (ANDREESSEN; TAYLOR; STEINBÜCHEL, 2014; BRADFORD, 2017; PARKER, 2018). In this scenario, bioplastics such as poly-3-hydroxybutyrate (PHB) are seem as a possible and more environmentally friendly alternative for certain applications (ANDREESSEN; TAYLOR; STEINBÜCHEL, 2014; CHEN, 2009; KESHAVARZ; ROY, 2010; ZHENG et al., 2020). But PHB production costs remain a barrier for its broader commercialization (GODINHO, 2005; ZHANG et al., 2014; ZHENG et al., 2020). Many researches therefore were focused on trying to improve the PHB production processes through the use of metabolic engineering strategies. Recombinant microorganisms such as *Escherichia coli* harboring the PHB synthesis pathway from *Cupriavidus necator* can often offer advantages over naturally producing microorganisms such as a higher growth rate and easier polymer extraction (GODINHO, 2005). While the literature points that recombinant *E. coli* produces PHB even during growth (CARLSON; WLASCHIN; SRIENC, 2005), a dedicated growth phase followed by a dedicated production phase can lead to higher productivities, which theoretically can be achieved with strategies such as genetic toggle-switches, where after reaching a predetermined cell concentration, a trigger shifts carbon flux from biomass to product formation (ANESIADIS; CLUETT; MAHADEVAN, 2008; ZHUANG et al., 2013). This scenario leads to an optimization problem in order to find how much of a given total glucose should be used for the growth phase, with the remaining glucose being used for the PHB production phase. With the use of Dynamic Flux Balance Analyses simulations and an estimation of the monthly gross profit for each scenario, this work explores the two-phase PHB production potential of recombinant *E. coli* for both aerobic and anaerobic batch cultures. The best result for the two-phase simulations under aerobic conditions is achieved when using 20%(mol) of the available glucose on the growth phase and the remaining 80% in the PHB production phase, leading to a final yield of 0.47 g PHB/g glu, titer of 11.73 g PHB/L, and productivity of 1.51 g PHB/L.h, which results in a monthly gross profit of USD 5460.09, in the conditions simulated. For the anaerobic two-phase simulations, the best performance within a realistic PHB content is achieved when 30%(mol) of the available glucose is used on the growth phase and the remaining 70% in the PHB production phase, leading to a final yield of 0.32 g PHB/g glu, titer of 7.99 g PHB/L, and productivity of 0.85 g PHB/L.h, which results in a monthly gross profit of USD 758.02. These results show how DFBA can be used to identify the approaches that lead to the most profitable scenarios, setting therefore more precise targets for future metabolic engineering strategies.

References

ANDREESSEN, B.; TAYLOR, N.; STEINBÜCHEL, A. Poly(3-Hydroxypropionate): a Promising Alternative to Fossil Fuel-Based Materials. **Applied and Environmental Microbiology**, v. 80, n. 21, p. 6574–6582, 1 nov. 2014.

ANESIADIS, N.; CLUETT, W. R.; MAHADEVAN, R. Dynamic metabolic engineering for increasing bioprocess productivity. **Metabolic Engineering**, v. 10, n. 5, p. 255–266, 2008.

BRADFORD, A. **What is a Polymer?** Disponível em: <https://www.livescience.com/60682-polymers.html>. Acesso em: 14 jan. 2018.

CARLSON, R.; WLASCHIN, A.; SRIENC, F. Kinetic Studies and Biochemical Pathway Analysis of Anaerobic Poly-(R)-3-Hydroxybutyric Acid Synthesis in Escherichia coli. **Applied and Environmental Microbiology**, v. 71, n. 2, p. 713–720, fev. 2005.

CHEN, G.-Q. A microbial polyhydroxyalkanoates (PHA) based bio- and materials industry. **Chemical Society Reviews**, v. 38, n. 8, p. 2434, 2009.

GODINHO, D. **Produção de Poli[3-hidroxibutirato-co-3- hidroxivalerato] por Escherichia coli recombinante a partir de Glicose e Ácido Propiônico**. [s.l.] Universidade Federal de Santa Catarina, 2005.

KESHAVARZ, T.; ROY, I. Polyhydroxyalkanoates: bioplastics with a green agenda. **Current Opinion in Microbiology**, v. 13, n. 3, p. 321–326, jun. 2010.

PARKER, L. **We made plastic. We depend on it. Now we're drowning in it.** Disponível em: <https://www.nationalgeographic.com/magazine/2018/06/plastic-planet-waste-pollution-trash-crisis/>. Acesso em: 13 jan. 2018.

ZHANG, Y. et al. Engineering of Serine-Deamination pathway, Entner-Doudoroff pathway and pyruvate dehydrogenase complex to improve poly(3-hydroxybutyrate) production in Escherichia coli. **Microbial Cell Factories**, v. 13, n. 1, p. 172, 16 dez. 2014.

ZHENG, Y. et al. Engineering biosynthesis of polyhydroxyalkanoates (PHA) for diversity and cost reduction. **Metabolic Engineering**, v. 58, n. June 2019, p. 82–93, mar. 2020.

ZHUANG, K. et al. Dynamic strain scanning optimization: an efficient strain design strategy for balanced yield, titer, and productivity. DySScO strategy for strain design. **BMC Biotechnology**, v. 13, n. 1, p. 8, 6 dez. 2013.

Acknowledgments

We gratefully acknowledge support of the RCGI – Research Centre for Greenhouse Gas Innovation, hosted by the University of São Paulo (USP) and sponsored by FAPESP – São Paulo Research Foundation (2014/50279-4 and 2020/15230-5) and Shell Brasil, and the strategic importance of the support given by ANP (Brazil's National Oil, Natural Gas and Biofuels Agency) through the Randamp;D levy regulation.

448e Production of High-Value Methylxanthines Using Engineered Whole-Cell Biocatalysts

Meredith Mock, Ryan Summers

Methylxanthines are natural products with many applications in the food, pharmaceutical, and cosmetic industries. While some methylxanthines, such as caffeine (1,3,7-trimethylxanthine) and theophylline (1,3-dimethylxanthine), are commonly found in nature, other compounds are not generated in sufficient quantities for economical production *via* extraction from natural sources. Instead, these rare compounds are produced synthetically using harsh conditions that result in racemic mixtures of products and low yields. Paraxanthine (1,7-dimethylxanthine) has potential as a treatment for Parkinson's disease, while 7-methylxanthine can serve to reduce progression of myopia, but retail price of these compounds is over 31,000- and 13,000-fold higher, respectively, than caffeine. The purpose of this project is to develop microbial strains and processes for the biocatalytic production of paraxanthine and 7-methylxanthine.

We have constructed 19 strains of *Escherichia coli* capable of producing paraxanthine from caffeine using a mutant *N*-demethylase (NdmA4) from the soil bacterium *Pseudomonas putida* CBB5. Strains were grown, harvested by centrifugation, and used as whole-cell biocatalysts in a resting cell reaction. The strain with the highest activity, MBM019, was used in a bench-scale reaction to produce 0.98 mM paraxanthine from 2.6 mM caffeine, with subsequent purification of paraxanthine, resulting in 104 mg of powdered product. We also demonstrate a newly-discovered promiscuous activity of NdmA4 toward paraxanthine, resulting in generation of 7-methylxanthine when paraxanthine concentration reaches approximately 1 mM. Taking advantage of the promiscuity of NdmA4, we carried out a series of four reactions by swapping spent cells with fresh cells to produce 2.6 mM 7-methylxanthine from 4.7 mM caffeine and isolated 177 mg pure 7-methylxanthine powder.

To improve 7-methylxanthine production, we have developed a co-culture reaction system using two distinct "specialist" strains. Strain MBM020 produced theobromine (3,7-dimethylxanthine) from caffeine using the wild-type N_1-demethylase NdmA, while strain MBM021 converted theobromine to 7-methylxanthine using NdmB, the wild-type N_3-demethylase. When the strains are mixed at a 1:1 ratio, we observed rapid and efficient conversion of caffeine to 7-methylxathine in a single reaction. The mixed-culture reaction system produced 2.23 mM 7-methylxanthine from 2.29 mM caffeine over 5 h, resulting in 153 mg pure 7-methylxanthine isolated from a 500 mL reaction. In summary, we have developed strains and reaction processes to produce the high-value biochemicals paraxanthine and 7-methylxanthine using caffeine as a feedstock, which will enable efficient production of these compounds and facilitate their use as pharmaceutical components.

448f An Integrated Biological Process for the Valorization of Lignin-Derived Aromatics to Poly(4-vinylphenol)

William Ederer, Lakshmi Meghana Bathineni, Apurv Mhatre, Hemant Choudhary, John M. Gladden, Alberto Rodriguez, Arul Mozhy Varman

Phenolic polymers such as polyphenols and polyphenylenes are generated industrially for several applications but are typically associated with harsh reaction conditions and environmentally hazardous chemicals, such as formaldehyde. On the other hand, hydroxycinnamic acids, such as p-coumaric acid (CA), are found in high concentrations in the underutilized lignin-derived hydrolysates and represent a renewable and sustainable feedstock for the production of various aromatics and phenolics. To that end, recently a strain of *Corynebacterium glutamicum* has been developed by the Joint Bioenergy Institute to express a Phenolic Acid Decarboxylase (PAD) gene, which can convert CA into 4-vinylphenol (4VP). 4VP is cytotoxic but can be polymerized by ligninolytic enzymes such as laccases or peroxidases into less-toxic poly(4-vinylphenol) (PVP). This work investigates the potential of polymerizing 4VP *in situ* by adding ligninolytic enzymes into the fermentation media to polymerize 4VP into PVP as it is produced, while reducing cellular toxicity to aid in chemical conversion.

The engineered *C. glutamicum* strain was cultured in the presence of CA to produce 4VP, with a maximum yield of 80.75%. Simultaneously, two ligninolytic enzymes, laccase and horseradish peroxidase (HRP), were explored in an *in vitro* experiment for their ability to polymerize 4VP, with laccase achieving full polymerization within 45 minutes and HRP able to polymerize 54.06% of 4VP in 24 hours. The resulting polymers were further analyzed by using nuclear magnetic resonance and gas permeation chromatography, validating the synthesis of PVP from 4VP with the addition of laccase or HRP. Finally, the *C. glutamicum* strain was evaluated for its ability to grow in the presence of hydrogen peroxide, which is a necessary reagent for HRP functionality, and it was able to reach an optical density of 3.69 within 24 hours. These findings suggest that *in situ* polymerization may be possible. Further work is underway to explore the enzyme kinetics at different pH, validate the potential of polymerization *in situ,* study the fermentative benefits associated with *in situ* polymerization. This will be followed by additional analytical studies to characterize the resulting PVP.

448g [Keynote] Enhancing Prehydrolysates Fermentability By Adding Nucleophilic Amino Acids and Proteins in Biomass Pretreatment

Yequan Sheng, Yu Zhang, Maobing Tu

Dilute acid pretreatment produced a considerable amount of carbonyl compounds in the biomass prehydrolysates, which significantly inhibited the sequential microbial fermentation. To reduce the release of carbonyl inhibitors, a new approach of pretreatment with amino acids and proteins has been developed to improve the fermentability of prehydrolysates. Cysteine (Cys), histidine (His), soy protein isolate (SPI) and bovine serum albumin (BSA) were added into dilute acid pretreatment of aspen (DAPA). The prehydrolysates were fermented by *Saccharomyces cerevisiae*, the glucose consumption rate in the prehydrolysates was increased from 0.32 g/L/h to 1.35, 3.22, 1.02 and 1.61 g/L/h, respectively. The substrates remaining after pretreatment were applied to enzymatic hydrolysis. Unexpectedly, it was observed that 72 h hydrolysis yields of DAPA-Cys and DAPA-His decreased from 71.35% (DAPA) to 63.93% and 28.11%, respectively. While, the 72 h hydrolysis yields of DAPA-SPI and DAPA-BSA increased to 75.04%, and 72.38%, respectively. The results showed that BSA was the most effective additives to enhance the prehydrolysate fermentability and enzymatic hydrolysis. It increased the ethanol productivity of prehydrolysates from 0.33 g/L/h (without addition) to 1.54 g/L/h. The final yield was promoted from 0.05 to 0.44 g/g glucose. In addition, total 22 potential inhibitors in prehydrolysates have been identified and quantified by gas chromatography-mass spectrometry (GC/MS). The results showed the addition of histidine in pretreatment reduced inhibitors much more than the addition of cysteine, SPI and BSA.

499a Screening of a Combinatorial Pathway Library to Improve Isobutanol Production in *Saccharomyces Cerevisiae*

Joshua Dietrich, Francesca Gambacorta, Brian F. Pfleger

The yeast *Saccharomyces cerevisiae* is a well-studied and tractable microbe with considerable potential for use in production of next-generation biofuels such as isobutanol. *S. cerevisiae* is also tolerant to a variety of stressors present in lignocellulosic hydrolysate, a feedstock that does not compete with food resources. To maximize isobutanol production from hydrolysate with *S. cerevisiae*, we built a platform strain of *S. cerevisiae* that can ferment both glucose and xylose, in which we additionally deleted several genes to prevent formation of unwanted byproducts and increase flux towards isobutanol. In parallel, we sought to optimize the isobutanol biosynthesis pathway by screening a combinatorial library of diverse homologs of the five isobutanol pathway genes (>15 homologs per gene), each driven by either a strong, medium, or weak promoter. This variation of homologs as well as promoter strength, while increasing the library complexity to a final size of $\sim 10^9$, allows for the simultaneous identification of active homologs and optimal expression levels. Preliminary screening of this library relied on the ability of an active isobutanol pathway to restore growth in a fermentation-deficient strain and has already provided insights into pathway bottlenecks and enabled production of isobutanol at yields close to the highest reported in academic literature. Additionally, the presence of several false-positives in our growth selection revealed promiscuous homologs of 2-ketoisovalerate decarboxylase that act on pyruvate and restore ethanol fermentation, deepening our understanding of substrate specificity of these enzymes. Further screening of the library will give us a more complete picture of the combinatorial space, and this data can be used to build machine learning models to elucidate trends in the most active isobutanol cassettes and predict designs that maximize isobutanol production.

499b Metabolic Engineering of *Paraburkholderia Sacchari* for Improved Bioplastic Co-Polymer Production.

Dianna Morris, Cheryl Immethun, Mark Wilkins, Rajib Saha

Recent research endeavors have turned to generating useful chemicals from biological platforms as an environmentally responsible alternative to non-sustainable sources. Here we share development of the non-model bacterium, *Paraburkholderia sacchari* (hereafter *P. sacchari*), as a microbial factory for production of polyhydroxyalkanoates (PHAs), which serve as precursors for biodegradable plastic. The particular PHAs of interest produced by *P. sacchari* include poly(3-hydroxybutyrate) (hereafter PHB) and the co-polymer produced by the combination of PHB and 3-hydroxyvalerate (hereafter 3HV) called poly(3-hydroxybutyrate-co-3-hydroxyvalerate) (hereafter PHBV). PHBV is particularly interesting due to its improved toughness and elasticity over other PHAs, making it more competitive with petroleum-based plastics. While growth studies have shown high PHB production by *P. sacchari* in the presence of single carbon sources, PHBV production can only be achieved with additional co-fed substrates such as proprionate. When proprionate is added to the media, *P. sacchari* produces PHBV with a 3HV fraction of 28%, however, this is also accompanied by a significant drop in overall PHA accumulation (from 75% of cell dry weight to 45%) and inhibition to growth. Strategies to either engineer an increase in the efficiency of converting proprionate to 3HV could produce PHBV with less growth inhibition thus resulting in a more industrially useful strain. To address this short-coming, we have improved incorporation of proprionate by engineering *P. sacchari* to upregulate propionyl-CoA transferase, the enzyme which converts proprionate and acetyl-CoA into propionyl-CoA. We have done so using the propionyl-CoA transferase gene (*pct*) from *Cupriavidus necator* H16. This strategy in combination with overexpression of acetyl-CoA reductase along the PHB production pathway (using the *phaB* gene from *C. necator*) has shown a drastic increase in PHBV production over wild type. Through these methods, we have shown promising progress toward an industrially useful strain of *P. sacchari* for PHBV production.

499c Engineering a Mixotrophic Clostridium Consortium Directly Exchanging Cellular Materials Via Cell Fusion for Biochemical Production with CO_2 Fixation.

Hyeongmin Seo, Jonathan Otten, Noah Willis, John Hill, Eleftherios Papoutsakis

Engineering microbial consortia provides insights into syntrophic microbial interactions and expands metabolic capability for complicated functions that individual population is unable to perform. Here, we present a synthetic Clostridium consortium that enables biosynthesis of biofuels, chemicals, and materials with transformative yields from sugar catabolism. As a case study, we targeted production of isopropanol, which serves as biofuel component or a valuable chemical with a market size of $2.65B (2019) andamp; a CAGR of 8.2% (2020-27). We engineered a non-sporulating *Clostridium acetobutylicum* M5 strain which lacks the megaplasmid pSOL1 that encodes for the solventogenic pathways (e.g., acetone, butanol) in wild type *C. acetobutylicum*. Several genes of *C. acetobutylicum* M5 were successfully deleted using CRISPR/Cas9 and mazF toxin-antitoxin system, minimizing undesirable byproduct formation. To optimize acetone production, a key precursor for isopropanol, acetone biosynthesis pathway was redesigned and constructed under the control of strong and constitutive promoters and computationally designed ribosome binding sites. Engineered *C. acetobutylicum* M5 chassis expressing the synthetic acetone pathway produced acetone without butyrate formation. The acetone producing *C. acetobutylicum* M5 strain was paired with *C. ljungdalhii*, which consumes CO_2 and H_2 and converts acetone to isopropanol using endogenous dehydrogenases. Through heterologous cell fusion, the Clostridium strains efficiently exchanged cellular materials, enabling isopropanol production with supra-theoretical yields. We examined the cell fusion between the engineered *C. acetobutylicum* M5 and *C. ljungdalhii* using anaerobic fluorescence reporters, visualizing exchange of cellular materials between the two strains. The Clostridium consortium platform represents an engineered syntrophic microbial consortium for carbon neutral or negative production of biofuels, chemicals, and materials.

499d Metabolic Engineering of *Clostridium Tyrobutyricum* for Increased Intracellular NADH Pool and High-Yield *N*-Butanol Production

Jun Feng, Jialei Hu, Shang-Tian Yang

Clostridium tyrobutyricum is a native butyric acid productor. Due to its high metabolic flux toward butyryl-CoA, *C. tyrobutyricum* has also been engineered to overexpress a bifunctional aldehyde/alcohol dehydrogenase (ALD/ADH) for n-butanol production. However, the mutant *C. tyrobutyricum* OE::*adh*E2 (Ct-*adh*E2) also produced large amounts of butyrate, acetate, and ethanol, which not only reduced the butanol yield but also increased the downstream processing cost for product purification. Compared to butyrate and ethanol, butanol biosynthesis requires two more NADH. The limited intracellular NADH pools is limiting butanol production in the fermentation. In this study, we aimed to increase the reducing equivalent (NADH) for butanol biosynthesis in *C. tyrobutyricum*, which should increase butanol selectivity over ethanol and acids. Formate dehydrogenase (FDH) could facilitate the regeneration of NADH from NAD^+ with format as the electron donor. Ferredoxin-$NAD(P)^+$ oxidoreductase (FNR) is involved in electron transfer from the reduced ferredoxin to $NAD(P)^+$. Thus, we hypothesized that overexpressing FNR (encoded by *fnr* and *nfn*AB) or FDH (encoded by *fdh*) could increase intracellular NAD(P)H pool and butanol biosynthesis. *C. tyrobutyricum* mutant strains co-expressing *adh*E2 with *fdh*, *fnr*, or *nfn*AB were constructed and evaluated for their performance in batch fermentation. In general, mutants co-expressing *fdh*, *fnr* or *nfn*AB with *adh*E2 showed increases in C4/C2, alcohol/acid, and butanol/ethanol ratios, confirming their benefits on increasing intracellular NADH pool to favor butanol biosynthesis. In batch fermentation kinetics studies carried out in serum bottles, one engineered *C. tyrobutyricum* strain produced butanol at 0.3 g/g glucose, more than 50% increase compared to the strain expressing only *adh*E2. Metabolic engineering strategies for further increasing butanol production by enhancing metabolic flux toward butyryl-CoA via overexpressing heterologous thiolase (*thl*) and 3-hydroxybutyryl-CoA dehydrogenase (*hbd*) and knockout of the CoA transferase (*cat*1) responsible for the biosynthesis of butyrate in *C. tyrobutyricum* will also be discussed in the presentation.

499f Dynamic Regulation of 4-Hydroxycoumarin Synthesis in Engineered *Escherichia coli*

Yusong Zou, Jianli Zhang, Jian Wang, Yajun Yan

4-Hydroxycoumarin (4-HC) is an important pharmacophore for manufacturing of oral anticoagulants. The *de novo* biosynthetic pathway of 4-HC has been established by condensation of salicoyl-CoA and malonyl-CoA in *Escherichia coli*. However, the imbalance between the co-precursors, salicylate and malonyl-CoA, would hinder the production of 4-HC. To address the imbalance, we recruited an engineered MarR-based salicylate biosensor system to dynamically control the 4-HC pathway. In this dynamic control system, the buildup of salicylate will activate the downstream 4-HC pathway under control of the engineered PmarO promoter and repress the major PEP consumption pathway via salicylate-inducible CRISPRi, redirecting the carbon flux to 4-HC production. The pyruvate released from the 4-HC production will be taken up for TCA cycle and malonyl-CoA biosynthesis. Further CRISPRi knockdown of the fatty acid pathway increased the availability of malonyl-CoA for 4-HC production. Compared with the parental strain, the engineered strain with the dynamic control system significantly increased 4-HC production from glycerol. This study demonstrated the over-production of 4-HC by integrating a bifunctional dynamic control system, which could serve as a new strategy for intelligent metabolic rewiring and balance in biosynthetic pathways.

499g [Keynote] Systems and Synthetic Biology: Constructing Smart and Programmable Microbes to Address Global Problems

Tae Seok Moon

The past decade has witnessed the tremendous power of systems and synthetic biology in the creation of genetic parts, devices, and systems, which helps understand complex biological systems. However, its potential for real-world applications has not been fully exploited. One of its promising applications is the construction of programmable cells that are able to integrate multiple environmental signals and to implement synthetic control over biological processes. My research interests are focused on developing microbes that are able to process multiple input signals and to generate user-defined outputs. Specifically, I aim to build genetic programs in order to control various bacterial processes such as gene expression, chemical reactions, and evolution. I will present unpublished results of my research projects by discussing the potential and challenges of systems and synthetic biology to address global problems, including plastic and agricultural waste issues, non-invasive diagnostics and disease treatment using smart probiotics, sustainable bioproduction, and biocontainment of engineered organisms.

22a Inducible Directed Evolution of an Anticancer Terpene Biosynthetic Pathway in *E. coli*

Zidan Li, Catherine Odhiambo, Ibrahim Al'Abri, Gavin Williams, Nathan Crook

A central challenge in metabolic engineering lies in identifying the correct expression levels and activities of enzymes that lead to the production of a desired product. While a plethora of methods enable efficient exploration of expression space for entire metabolic pathways, enzymatic activity improvements are primarily achieved through directed evolution of single enzymes. We wished to understand whether directed evolution of more complete metabolic pathways would lead to greater productivities than evolution of single enzymes alone. To do this, we took production of (S)-(-)-perillyl alcohol (POH), a putative anticancer molecule, as a case study. Studies have shown that POH and its derivatives can pass the blood-brain barrier and treat glioblastoma with few side effects. The low amounts that can be isolated from plant sources, as well as the complexity and high cost of chemical synthesis, has made microbial production an ideal platform for POH synthesis. We have developed a minimal, highly efficient enzymatic pathway for POH biosynthesis in *E. coli* and a transcription factor-based biosensor to detect POH production *in situ*. However, further improvement in productivity has been hampered by the difficulty of evolving the large POH biosynthetic pathway (~ 10 kb). To solve this issue, we used Inducible Directed Evolution (IDE), which is capable of evolving DNA sequences up to 36 kb in length. Specifically, we used a mutagenesis plasmid to introduce random mutations into the POH biosynthetic pathway, followed by transfer into a biosensor-containing strain via a temperate bacteriophage. Variants with high cellular fluorescence, corresponding to high POH productivity, were recovered by multiple rounds of fluorescence-activated cell sorting spanning over 10^6 variants, and the spectrum of mutations present during each round were determined by next generation sequencing. By synthetically introducing combinations of key mutations, we investigated whether synergies were occurring between mutations in different enzymes. These results demonstrate the utility of inducible directed evolution for biosensor-enabled screens and point to a pathway-wide framework for performing directed evolution on multigene metabolic pathways.

22b Cas12a-Assisted Precise Targeted Cloning Using In Vivo Cre-*Lox* Recombination

Behnam Enghiad, Chunshuai Huang, Fang Guo, Guangde Jiang, Bin Wang, Huimin Zhao

Direct cloning represents the most efficient strategy to access the vast number of uncharacterized natural product biosynthetic gene clusters (BGCs) for the discovery of novel bioactive compounds. However, due to their large size, repetitive nature, or high GC-content, large-scale cloning of these BGCs remains an overwhelming challenge. We report a scalable direct cloning method named Cas12a-assisted precise targeted cloning using in vivo Cre-*lox* recombination (CAPTURE) which consists of Cas12a digestion, a DNA assembly approach termed T4 polymerase exo + fill-in DNA assembly, and Cre-*lox* in vivo DNA circularization. We applied this method to clone 47 BGCs ranging from 10 to 113 kb from both Actinomycetes and Bacilli with ~100% efficiency. Heterologous expression of cloned BGCs led to the discovery of 15 previously uncharacterized natural products including six cyclic head-to-tail heterodimers with a unique 5/6/6/6/5 pentacyclic carbon skeleton, designated as bipentaromycins A–F. Four of the bipentaromycins show strong antimicrobial activity towards both Gram-positive and Gram-negative bacteria such as methicillin-resistant *Staphylococcus aureus*, vancomycinresistant *Enterococcus faecium*, and bioweapon *Bacillus anthracis*. Due to its robustness and efficiency, CAPTURE coupled with heterologous expression provides an effective strategy for large-scale discovery of novel natural products.

22c The Role of Sigb on the (stress) Response of the Pathogen *Listeria Monocytogenes* to Novel Sustainable Processing Technologies.

Melina Kitsiou, Lisa Purk, Jorge Gutierrez-Merino, Kimon-Andreas Karatzas, Oleksiy Klymenko, Eirini Velliou

INTRODUCTION

Consumers, in the recent years, are seeking minimally processed foods that do not contain chemical preservatives or antibiotics and are produced in a more sustainable way. Novel processing technologies i.e. ultrasound, cold atmospheric plasma (CAP), high hydrostatic pressure and natural antimicrobials are such alternative approaches that can be used as mild decontamination methods. However, as these treatments are less harsh than classic heat sterilisation, the desirable microbial inactivation (5 log CFU/ml) might not occur, leading to contaminated and unsafe products. This is due to the nature of those mild/alternative technologies, i.e., due to their mode of action, they cause less damage and death to bacteria as compared to classic treatments, e.g., sterilisation[1]. For this reason, novel combinatory treatments (the so-called hurdle approach) and their microbial stress response should be investigated[2].

The principle of the hurdle technology approach on microbial safety, is the utilization of two or more methods/approaches/processes to enable microbial inactivation and consequently to ensure food safety[3]. A combination of them could lead to synergistic and/or complementary actions against bacteria. For example, one technology/treatment could be used to damage or increase the permeability of the bacterial cell wall and another one to interfere with the intracellular metabolism[4]. To develop novel hurdle technology approaches, it is important to understand the mode of action and the microbial stress response for each treatment. For example, it is known that the gene regulator, SigB, plays a major role in the environmental stress adaptation of *L. monocytogenes* for multiple type of environmental stresses such as thermal, acid, or osmotic stress. A recent study has found that SigB can be overexpressed as a protective measure, transforming the cells to hyperestistant but susceptible to oxidative stress[5]. *The aim of our work is to develop novel green hurdle technology approaches, targeting the weakness of L. monocytogenes stress response.* More specifically, acid stress, Cold Atmospheric Plasma (CAP) and the grape seed extracts (GSE-which have antimicrobial properties) were assessed individually and/or combined.

METHODS

L. monocytogenes and its mutant (ΔSigB) was cultured in Tryptone Soy Broth supplemented with 0.6% of yeast extract (TSBYE) until mid-exponential phase of growth (10^5 CFU/ml). Thereafter, mid-exponential cells were treated with different stress factors, i.e., acid, grape seed extracts or CAP. For the *mild acid stress treatment*, the cells (cell pellets) were exposed in TSBYE with 0.35ml hydrochloric acid (HCl, 1M) at pH 4.5. The cells were challenged for 30min in this mild acid condition. For the *grape seed extract (GSE)* treatment the cells were challenged for 2h in TSBYE with 1% GSE. For the *CAP treatment* 300μl of each tube (10^5 CFU/ml) were placed in a 12 well plate and CAP treatment was applied using flow rate of 4ml/min for 1 to 3 minutes. Two different combinatory treatments were applied: 1) hydrochloric acid and CAP treatment and 2) GSE and CAP treatment. The viable and sub-lethally damaged cells were enumerated (using the spread plate technique in TSAYE and PALCAM agar) after each individual and combinatory treatment.

RESULTS

Both sets of combined treatments induced a reduction in the cell concentration. In contrast, individual treatments did not affect, at all or as much, the cell survival. The most effective hurdle approach was the combination of GSE and CAP treatment. This is due to the increased numbers of sub-lethal damaged cells caused by the individual GSE treatment. The increased sensitivity of *L. monocytogenes* to oxidative stress was clearer for 3 minutes of CAP treatment, when the combined treatments were applied, as cells post-treatment were undetectable (<1 log CFU/ml). Lower synergistic effect of the treatments was observed in the mutant strain (SigB deletion) validating the initial hypothesis of the presence of SigB linked with *L. monocytogenes* susceptibility to oxidative stress.

CONCLUSIONS

This work shows the importance of SigB on *L. monocytogenes* to novel sustainable processing approaches and their combination. This insight to the microbial stress response is essential for the development of novel and robust green/sustainable hurdle technology approaches, i.e., application of GSE coupled with CAP treatment in a hurdle approach.

ACKNOWLEDGEMENTS

This work was supported by the Doctoral College and the Department of Chemical and Process Engineering of the University of Surrey, United Kingdom. E.V. is grateful to the Royal Academy of Engineering for an Industrial Fellowship and to the Medical Research Council UK for a New Investigator Research Grant (MR/V028553/1).

References:

1. Bahrami A, Moaddabdoost Baboli Z, Schimmel K, Jafari SM, Williams L. Efficiency of novel processing technologies for the control of Listeria monocytogenes in food products. *Trends Food Sci Technol.* 2020;96(May 2019):61-78. doi:10.1016/j.tifs.2019.12.009
2. Velliou EG, Van Derlinden E, Cappuyns AM, Goossens J, Geeraerd AH, Devlieghere F, Van Impe JF. Heat adaptation of Escherichia coli K12: Effect of acid and glucose. *Procedia Food Sci.* 2011;1:987-993. doi:10.1016/j.profoo.2011.09.148
3. Khan I, Tango CN, Miskeen S, Lee BH, Oh DH. Hurdle technology: A novel approach for enhanced food quality and safety – A review. *Food Control.* 2017;73:1426-1444. doi:10.1016/j.foodcont.2016.11.010
4. Costello KM, Smet C, Gutierrez-Merino J, Bussemaker M, Van Impe JF, Velliou EG. The impact of food model system structure on the inactivation of Listeria innocua by cold atmospheric plasma and nisin combined treatments. *Int J Food Microbiol.* 2021;337(August 2020):108948. doi:10.1016/j.ijfoodmicro.2020.108948
5. Boura M, Keating C, Royet K, Paudyal R, O'Donoghue B, O'Byrne CP, Karatzas KA. Loss of sigb in listeria monocytogenes strains egd-e and 10403s confers hyperresistance to hydrogen peroxide in stationary phase under aerobic conditions. *Appl Environ Microbiol.* 2016;82(15):4584-4591. doi:10.1128/AEM.00709-16

22d Characterization of Microbial Interactions in Stressed Environments for Construction of Synthetic Communities

Bradley Biggs, Markus de Raad, Trent Northen, Adam P. Arkin

Microbial communities play vital roles in areas ranging from the environment (e.g., rhizosphere) to human health (e.g., gut microbiome). Understanding the key interactions among microbial species in these communities stands to benefit engineers who desire to leverage the greater functional capacity of the community, compared to individual species, toward biotechnological goals. Here, we use paired synthetic microbial communities to study the interactions of bacterial strains isolated from the Oak Ridge Reservation, a site with high metal and acid contamination. By identifying various mechanisms of cooperation and competitions, we hope to extend our learnings and construct higher order synthetic communities robust to toxic environments with applications in remediation and waste upgrading.

22e Screening the Rhizobiome for Plant Growth Promoting Bacteria That Improve Maize Growth and Development

Ryan Hansen, Niloy Barua, Kayla Clouse, Maggie Wagner, Thomas Platt

Plant growth promoting bacteria (PGPB) are important for sustainable food production and have the potential to alleviate some of the environmental and health-related drawbacks of chemical fertilizers. However, many PGPB are limited in application because they are unreliable, inefficient, and have high host specificity, presenting considerable risks to farmers and producers. This is due in part to the poor survival of PGPB inoculants in plant rhizosphere. In this presentation, we report on the discovery and application of new, synergistic PGPB consortia that generate enhancements in corn growth metrics relative to single-strain inoculations by improving PGPB survival in the rhizosphere. Corn roots were sampled from a farmer's field in central Kansas and washed to remove the microbial communities from the rhizosphere. These microbial community members were then combined with *Azospirillum brasilense*, a well-known nitrogen fixing PGPB, in a novel, high-throughput microfluidic screening system. The device rapidly screened thousands of unique consortia consisting of *A. brasilense* combined with a small number of rhizosphere isolates to identify strain combinations that maximized *A. brasilense* growth and survival. Co-inoculation of *A. brasilense* onto corn seedlings with and without the synergistic isolate collections was then characterized in plant growth chambers and compared to single strain inoculations. Corn seedlings inoculated in consortia of *A. brasilense* and synergistic isolates showed higher growth rates, and in some cases, higher root and shoot mass and enhanced root architectures. Root colonization assays revealed that the synergistic consortia enabled *A. brasilense* to colonize at higher levels in the rhizosphere in order to enhance plant growth. These studies present a new approach for bioprospecting the rhizosphere of agriculturally-relevant crops to discover new and novel bacterial isolates for use in PGPB formulations, and suggests that combining well-characterized, nitrogen-fixing rhizobacteria with synergistic isolates in consortia has potential to improve the efficiency of PGPB biofertilizers.

22f Tools and Methods for Engineering Colonization of Plant Microbiomes

John Van Schaik, Akansha Pandey, Nathan Crook

Enhancing crop production is a necessity with increasing demand for staple crops and utilization of carbon/nitrogen. Plant-associated microbiomes offers one avenue for addressing this demand. While large advancements have been gained recently due to advancements in high-throughput technology, building enhanced or new functionality into existing microbiomes has remained a challenge due to sparse materials and methods along with poor colonization of the desired trait. In the work discussed here, we detail the development of genetic tools and utilization of these tools for elucidating colonization factors into the Maize microbiome using a model synthetic community (SynCom). Utilizing this simple yet representative SynCom, we have been able to develop highly efficient and standard transformation methods along with standardized genetic expression and transformation vectors. Furthermore, we show how these standardized tools and methods can be used to rapidly characterize endogenous promoter sequences and identify functions to enhance member colonization of the plant host.

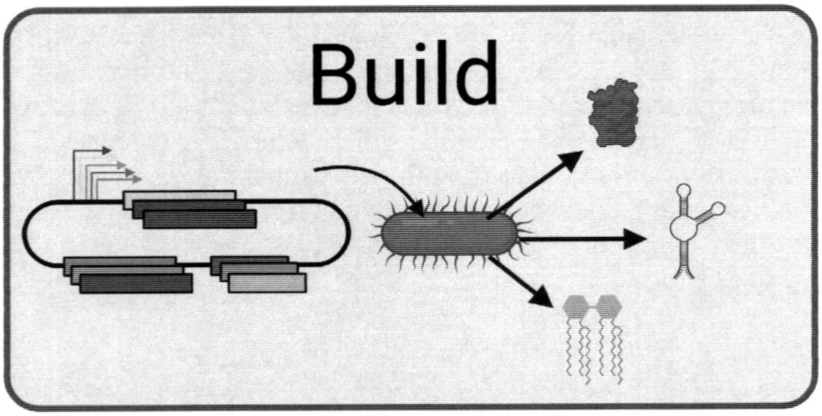

22g [Keynote] Engineering Bacteria from the Plant Root Microbiome to Explore and Manipulate Plant-Microbe Interactions

Jonathan M. Conway

Chemical signaling in the plant microbiome can have drastic effects on microbial community structure, and on plant host growth and development. Relatively few plant-microbe interactions are mechanistically understood and filling this knowledge gap is necessary to manipulate and engineer these interactions for the development of new technologies for the bio-agriculture, bio-fuel, and bio-chemical industries. Through genetic engineering of non-model bacteria from the plant root microbiome and utilization of engineered plants, we can probe these interactions, define them mechanistically, and manipulate these bacteria, plants, and their interactions for application. Here, I will present past and ongoing work to probe plant-microbe interactions involving 1) microbial evasion and suppression of the plant immune system and 2) chemical communication in the root microbiome involving auxin plant hormones. Through genetic engineering of plant-associated bacteria, we have identified bacterial genes that play essential roles in suppression of the flg22-mediated plant immune response and auxin hormone degradation. This fundamental genetic understanding will enable us to engineer microbes with improved plant colonization and the ability to improve plant host growth and health.

664b Sampling of Contact Lens Saline for Sars-Cov-2 Virus Detection Applications

Hazim Aljewari, Shadrach Ibinola, Kevin Clark, Vicki Thompson, Audie Thompson, Shannon Servoss, Robert Beitle

Since a contact lens adsorb biological materials such as, proteins, microbes, small molecules, and even viruses, it was hypothesized that the solution used to clean the lens overnight could be used as a source of a biological sample. This hypothesis was tested for SARS-CoV-2 (covid-19) detection using loop-mediated isothermal amplification (LAMP) PCR. LAMP PCR primers were designed to detect a unique sequence of the RNA dependent RNA polymerase gene of SARS-CoV-2 virus. Results showed that LAMP PCR could successfully be used to detect heat inactivated virus in as little as thirty minutes. This presentation will discuss the motivation and rationale of the use of a contact lens and cleaning solution as a passive sorbent and recovery solution for biological materials, provide a short overview of LAMP PCR and primer design, and discuss results that indicate a viable path forward for this concept.

164a A Polyketide Synthase-Based Platform for Making Plastic Monomers of Polydiketoenamines

Zilong Wang, Seokjung Cheong, Robert W. Haushalter, Jay Keasling

Plastics in use today are predominantly single-use and rarely recycled. The linearity of their life cycles is not only wasteful from a resource and energy perspective, it has also resulted in environmental stresses with >6 billion metric tons of plastic waste. This study provides a platform to biologically produce a specific type of lactones, the important molecules for synthesis of a new class of dynamic covalent polymer networks, known as vitrimers, which combine the processing and recycling ease of thermoplastics with the performance advantages of thermosets. Regarding circularity, most vitrimers are differentiated from classical thermosets in that they can be chemically de-polymerized, typically into small molecules or short oligomers, including lactones and diacids. For decades, microbial production of commodity chemicals is limited in the diversity of the molecules produced by natural or modified enzymes. Our technology of recombining the Type I polyketide synthases (PKSs) and other enzymes demonstrates a promising strategy for the synthesis of diversified lactones for making a variety of vitrimers with different properties.

164c Thermoacidophilic Archaea Enhance Bioleaching of Chalcopyrite for Copper Recovery

Mohamad Javad Haghighat Manesh, Dan Willard, April Lewis, Robert M. Kelly

As the availability of critical metals (e.g., copper, nickel, uranium) declines, there is a need for recovery processes that are efficient and environmentally friendly. Bioleaching is one option and now accounts for over 15% of the copper recovered worldwide, but processes are rudimentary and slow. Pilot-scale data show that extremely thermoacidophilic microorganisms from the domain Archaea can leach copper at much higher rates and yields than mesophilic bacteria, although the reasons for this enhancement are not known[1]. Among these thermoacidophiles, members of the order Sulfolobales have attracted the most attention. Key to the effectiveness of these archaea is the function of the *fox* cluster, which encodes proteins that catalyze iron oxidation[2], their ability to use reduced inorganic sulfur compounds as an energy source[3], the formation of biofilms on mineral surfaces[4], and the regulation of biotic and abiotic reactions to release copper from ores[5].

Chalcopyrite ($CuFeS_2$) is the most abundant source of copper. Copper extraction in chalcopyrite involves the oxidation of the mineral initially, releasing copper, sulfur, and ferric iron. If ferric iron is oxidized to ferrous iron, it can also lead to chalcopyrite oxidation, leading to more copper release from the ores. However, this process is slow. In addition, formation of sulfur or sulfur compounds on the surface of the mineral might lead to passivation, limiting access to the surface of the mineral and thus, hindering the leaching process. The characteristics of thermoacidophiles enhance bioleaching, since several members are known to oxidize iron or sulfur. Furthermore, the acidic environment, and the high temperature in which these microorganisms grow, have shown to further increase the leaching rates. Among these thermoacidophiles is *Sulfolobus acidocaldarius* (*Saci*), not known to be a good sulfur oxidizer, *but* has available genetic tools for metabolic engineering via an uracil-auxotrophy. Efforts are underway to engineer this archaeon to be a bioleacher[6].

Here, thermoacidophiles from the order Sulfolobales were examined for their potential for iron and chalcopyrite oxidation. Furthermore, the bioleaching process is optimized using the galvanic interactions between pyrite and chalcopyrite in a reaction containing the mixture. Biofilm formation of the microorganisms are investigated on the surface of the mineral. Finally, metabolic engineering tools of *Saci* are being used to modify its metabolism to enable it to bioleach chalcopyrite via enhanced iron and sulfur oxidation.

References

[1] D. W. Dew Van Buuren, C. ,. McEwan, K. andamp;. Bowker C., "Bioleaching of base metal sulphide concentrates : A comparison of high and low temperature bioleaching," *J. South. Afr. Inst. Min. Metall.*, vol. 100, no. 7, pp. 409–413, Nov. 2000, doi: 10.10520/AJA0038223X_2619.

[2] J. A. Counts, N. P. Vitko, and R. M. Kelly, "Fox Cluster Determinants for Iron Biooxidation in the Extremely Thermoacidophilic Sulfolobaceae," *Environ. Microbiol.*.

[3] K. S. Auernik and R. M. Kelly, "Identification of components of electron transport chains in the extremely thermoacidophilic crenarchaeon Metallosphaera sedula through iron and sulfur compound oxidation transcriptomes," *Appl. Environ. Microbiol.*, vol. 74, no. 24, pp. 7723–7732, 2008.

[4] A. M. Lewis *et al.*, "The biology of thermoacidophilic archaea from the order Sulfolobales,"

FEMS Microbiol. Rev., no. fuaa063, Jan. 2021, doi: 10.1093/femsre/fuaa063.

[5] K. S. Auernik and R. M. Kelly, "Impact of Molecular Hydrogen on Chalcopyrite Bioleaching by the Extremely Thermoacidophilic Archaeon <em>Metallosphaera sedula</em>," *Appl. Environ. Microbiol.*, vol. 76, no. 8, p. 2668, Apr. 2010, doi: 10.1128/AEM.02016-09.

[6] B. M. Zeldes *et al.*, "Determinants of sulphur chemolithoautotrophy in the extremely thermoacidophilic Sulfolobales," *Environ. Microbiol.*, vol. 21, no. 10, pp. 3696–3710, 2019.

250b Enzymatic Reactive Extraction of Fermentation Products for the Production of Short-Chain Esters

Danika Kartchner, Aditya Sarnaik, Dylan Ellis, Apurv Mhatre, Ryan Davis, Somnath Shinde, Arul Mozhy Varman

Esters such as ethyl acetate and ethyl lactate are biodegradable solvents with applications across multiple industries including adhesives, food, and pharmaceuticals. However, the conventional process for producing esters is not eco-friendly because they are largely derived from crude oil. More recently, metabolic engineering has been explored as an alternative route for the production of esters from renewable feedstocks, but the resulting ester titers were fairly low. A major factor for this low titer is that the esterification reaction is not thermodynamically favorable in the aqueous phase as water is a byproduct of the reaction. We propose that by performing reactive extraction this limitation can be overcome as the reaction would mainly happen in the interface and the product would be continuously extracted to the organic phase. In this work, we have identified the optimal conditions required for the production of esters and modeled the enzymatic kinetics.

Our results showed that, for production of ethyl acetate through this reactive extraction approach, Novozyme 435 exhibited significant improvements in esterification with chloroform as the extractant, resulting in an almost 85% conversion efficiency. Further optimizations with phase ratios, pH, and incubation time were performed. It was demonstrated that the pH 6.0 was the most optimum where ethyl acetate titer was found to improve 10 times than that at pH 7.0 with the phase ratio of 1:1. Kinetic studies further demonstrated that reactive extraction at 37oC provides maximum productivity for ethyl acetate synthesis. After initial optimization studies, reactive extraction was performed on cell broth resulting from fermentation that accumulated ethanol and acetate. However, we observed a ~7.5X decrease in ethyl acetate production in the cell broth versus simulated samples that had the same concentration of reactants.

As a last step, a mathematical model was developed for the reactive extraction which will be useful for scaling up studies. The mathematical model employed assumes that enzyme kinetics rather than diffusion was the rate limiting step and that the concentrations of reactants at the interface are equivalent to the initial concentration of reactants. Vmax was found to be 18.5 mmol min-1g-1 of catalyst used, and the Michaelis constants(Km) were 0.957 M and 0.00557 M for acetic acid and ethanol respectively. Overall, we hope that by optimizing the reactive extraction process for ester production, biorefineries could become more competitive and economically feasible for numerous applications.

164d Engineering Human Mesenchymal Bodies for Exosome Secretion in a Novel 3-D Printed Microchannel Bioreactor.

Richard Jeske, Xingchi Chen, Logan Mulderrig, Wenhao Cheng, Changchun Zeng, Jingjiao Guan, Daniel T. Hallinan Jr., Yan Li

Human Mesenchymal Stem Cells (hMSCs) and their derived products hold great potential in tissue engineering and as therapeutics in a wide range of diseases. While traditionally expanded in 2D on plastic culture dishes, hMSCs possess the ability to aggregate into "spheroids" which are commonly used as a preconditioning technique to enhance their therapeutic potential by upregulating stemness, immunomodulatory capacity, and anti-inflammatory and pro-angiogenic secretome. While there are a number of methods to form hMSC spheroids, few studies have investigated the impact on aggregate properties stemming from dynamic and static aggregation techniques. Furthermore, recent studies have indicated that the main mechanistic mode of hMSC action occurs through their secretome, including extracellular vesicles/exosomes, which contain therapeutically relevant proteins and nucleic acids. In this study, a novel 3D-printed microchannel bioreactor was developed to dynamically form hMSC spheroids under wave motion in order to assess how dynamic microenvironment conditions alter hMSC properties in relation to statically cultured aggregates, in particular the biogenesis of extracellular vesicles/exosomes. Our study found that dynamic aggregation has impacts on hMSC proliferation, morphology, and exosome production. This study advances our knowledge on a commonly used preconditioning technique that could be beneficial in wound healing, neural tissue regeneration, and autoimmune disorders.

164e Leveraging Advances in Proteomics to Mathematically Model Cell Signaling Processes: A Case Study on TGFβ Signaling in Valve Interstitial Cells

Daniel P. Howsmon, Toni M. West, Michael S. Sacks

Background: With no pharmaceutical treatments that prevent or delay the onset of calcific aortic valve disease (CAVD), treatment options consist of interventional repair or replacement. Valve interstitial cells (VICs) are the predominant cell type contributing to increased extracellular matrix (ECM) deposition and remodeling as well as calcification in CAVD. A detailed understanding of the signaling network driving VIC activation and expression of ECM-related genes would facilitate the rational discovery of suitable drug targets for mitigating, preventing, and/or reversing CAVD.

Methods: Herein, we stimulated VIC activation with transforming growth factor beta 1 (TGFβ1) to induce their phenotypic transition to myofibroblasts. VIC lysates were collected at various time points and subjected to bottom-up (phospho-)proteomics analysis based on a data independent acquisition (DIA) strategy. We use a small sample of the VIC lysate prior to phosphoenrichment to determine the protein copies/cell and to normalize the phospho-enriched samples. The time course of a subset of the discovered phosphorylations were modeled as a dynamic reaction network.

Results: A suitable model of TGFβ1-induced VIC activation was identified from phosphoproteomics data. In addition to canonical signaling through SMAD2/3, we have incorporated numerous non-canonical mechanisms. Sensitivity analysis enables model simplification and the identification of key reactions that drive this phenotypic transition in response to TGFβ1.

Implications: By coupling (phospho-)proteomics experiments with scientific computing, we can develop large, predictive models that enable the rational identification of suitable drug targets for combatting CAVD. These models can be extended in the future with other stimuli and further validated with knockout/knockdown studies to enhance our understanding of the mechanisms of VIC phenotypic transitions in CAVD.

164f Amyloid Aggregation of Enzymes Under Non-Denaturing Conditions

Timothy Charlton, Emre Erkanli, Jin Ryoun Kim

Aberrant associations into amyloid aggregates can occur from unfolded states of proteins and are associated with various diseases. Recent evidence demonstrates that amyloid aggregation of proteins can also occur from conformational states directly accessible from native states, without substantial unfolding. The presence of aggregation-prone regions (APRs) near the active sites of many enzymes and the conformation dynamics, required for the functional roles, contribute to the high risk of amyloid aggregation even under non-denaturing conditions. These aberrant intermolecular associations occur through exposures of APRs. While the evolutionary pressures have optimized folded states of proteins for serving biological functions without the aggregation, conditions that enzymes face in industrial settings are different from where enzymes have evolved. While amyloid aggregation of an unfolded protein has been extensively studied, little is known about amyloid aggregation of a folded protein and a strategy for prevention thereof.

In our study presented herein, a folded enzyme, *Bacillus circulans* xylanase (BCX), aggregates into amyloid fibrils during incubation under a non-denaturing condition. We also found that the aggregation is concomitant with the loss of enzymatic activity and can be reduced by peptides derived from other amyloid proteins. Moreover, we found that different enzymes, including those with high conformational stability, also aggregates into amyloid fibrils under non-denaturing conditions, further supporting that amyloid aggregation of enzymes under non-denaturing conditions is more common than what has been reported. Collectively, our study enhances an understanding of amyloid aggregation of enzymes under non-denaturing conditions toward advances in industrial biochemical processes.

164g Peptide Aggregation Induced Immunogenic Rupture (PAIIR) on Breast Cancer Spheroid Model

Gokhan Gunay

Peptide Aggregation Induced Immunogenic Rupture (PAIIR) on Breast Cancer Spheroid Model

Gokhan Gunay M.Sc., PhD Candidate: Gokhan.Gunay-1@ou.edu

Inducing cell death was the primary goal of treatment strategies against cancer for decades. However, emerging results are showing that the type of cell death dictates the progression of the tumor. Induction of apoptosis within the tumor micro environment (TME) leads to suppression of the immune system, increased proliferation of cancer cells and reduced inflammation, all of which increase the progression of cancer [1]. Immunogenic cell death (ICD) on the other hand results in secretion, release or surface expression of damage-associated molecular patterns (DAMPs), natural adjuvants capable of engaging immune system[2]. Current ICD inducing strategies involve chemotherapeutics, radiation therapy and photothermal therapy. However, their effects on inducing ICD and DAMP release have been predominantly studied in classical monolayer cultures. 3D tumor spheroid models offer a bridge to in vivo studies and reflect physiological relevance to in vivo tumors in terms of drug resistance and zone formation due to reduced nutrient and oxygen diffusion [3]. Therefore, utilization of such models for understanding ICD and DAMP release will serve as a more physiologically relevant and high-throughput system for not just understanding the effects of current strategies but also engineer new materials.

Previously we engineered a co-assembly of oppositely charged peptide (CoOP) platform[4] which leverages the characteristics of amino acids such as hydrophobicity to induce immunogenic cell death[5], which is called peptide aggregation induced immunogenic rupture (PAIIR). Here, we applied this newly developed technology into a breast cancer spheroid model by using EMT6 cells. We compared PAIIR technology with a known ICD inducer mitoxantrone and a known non-ICD inducer cisplatin. We showed that PAIIR treatment is superior in terms of DAMP release from spheroids and the released DAMPs can engage dendritic cells into a pro-inflammatory phenotype. Engineering new materials that can induce ICD within TME hold great potential for evoking adaptive immunity. Spheroids offer a bridge between in vitro and in vivo studies through reflecting certain physiological features of in vivo tumors.

Figure 1 Peptide aggregation induced immunogenic rupture (PAIIR) for induction of ICD in breast cancer spheroid model. A. Spheroid formation technique. B. Extracellular ATP release. C. Design of bone marrow dendritic cells (BMDCs) activation. D. IL-1b release from BMDCs

References

1. Gadiyar, V. *et al.* Cell Death in the Tumor Microenvironment: Implications for Cancer Immunotherapy. *Cells* **9**, 2207 (2020).
2. Krysko, D. V. *et al.* Immunogenic cell death and DAMPs in cancer therapy. *Nat Rev Cancer* **12**, 860–875 (2012).
3. Yan, F. *et al.* Characterization and quantification of necrotic tissues and morphology in multicellular ovarian cancer tumor spheroids using optical coherence tomography. *Biomed. Opt. Express, BOE* **12**, 3352–3371 (2021).
4. Hamsici, S., White, A. D. andamp; Acar, H. Peptide framework for screening the effects of amino acids on assembly. *Science Advances* (2022) doi:10.1126/sciadv.abj0305.
5. Gunay, G. *et al.* Peptide Aggregation Induced Immunogenic Rupture (PAIIR). *BioRxiv* **472230**, (2021).

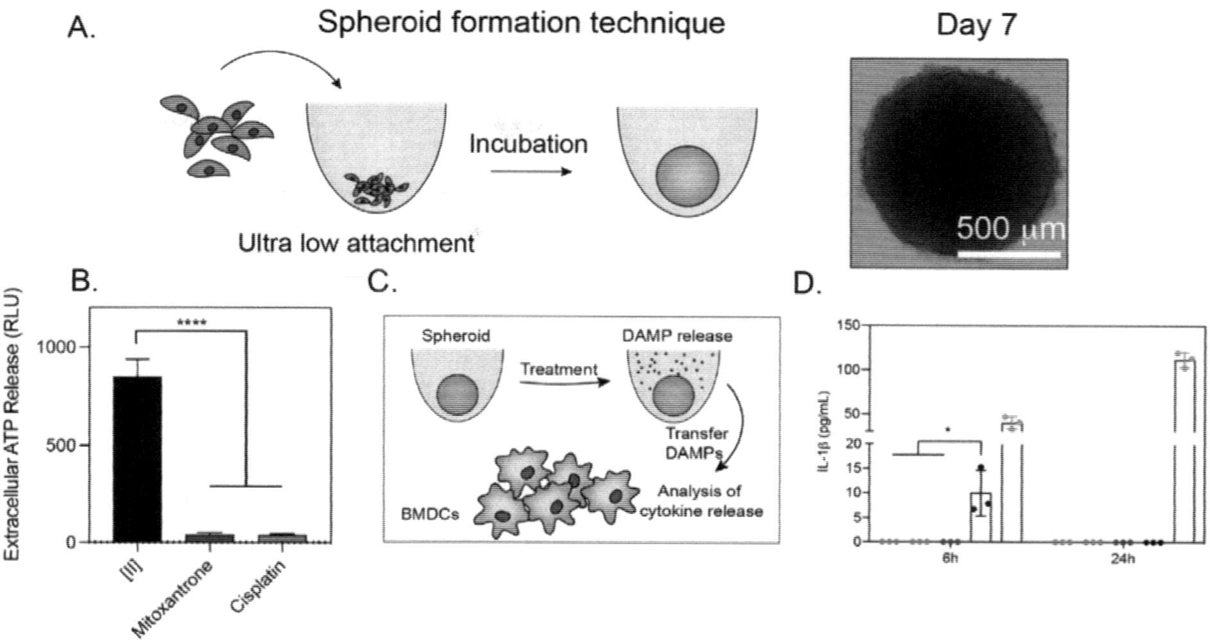
A.
Spheroid formation technique
Day 7
Incubation
Ultra low attachment
500 µm
B.
Extracellular ATP Release (RLU)
1000
500
0
****
[II]
Mitoxantrone
Cisplatin
C.
Spheroid
Treatment
DAMP release
Transfer DAMPs
BMDCs
Analysis of cytokine release
D.
IL-1β (pg/mL)
150
100
50
20
15
10
5
0
*
6h
24h

164h Systematic Evaluation of Protein-Small Molecule Conjugates on Yeast Surface

Manjie Huang, Abbigael Harthorn, Benjamin J. Hackel, James Van Deventer

Proteins made of canonical amino acids have limited chemical diversity, which constrains their interactions with biological targets, including in enzyme active sites. Noncanonical amino acids (ncAAs) and efficient conjugation reactions such as "click" chemistry offer powerful strategies to add small molecule functionalities to protein-based therapeutics. Here, we demonstrate our ability to construct and characterize diverse conjugates on the yeast surface during our systematic evaluation of fibronectin-based protein-small molecule "hybrid" conjugates. Starting from 6 recently described fibronectin-maleimide-sulfonamide conjugates, we systematically varied amino acid attachment points and small molecule warheads to generate a combinatorial set of hybrids. Our flow cytometric analyses identified new hybrid structures that show isoform-specific carbonic anhydrase binding with double-digit nanomolar affinities despite drastic linker and small molecule warhead alterations from the original conjugates. We observed that amino acid side chains can greatly change the hybrid affinity even when a strong warhead like acetazolamide is used. In addition, we found new hybrids containing ncAAs and a lower potency small molecule warhead that showed high affinity to the target enzyme. Biochemical characterization of a promising hybrid in solution indicates that observations made on the yeast surface are consistent with observations made with the same hybrid in soluble form. These findings highlight promising opportunities for discovering protein-small molecule hybrids on the yeast surface by integrating a broad chemical toolkit with powerful yeast display screening and analysis platforms.

164i Modulated Cell-Free Protein Synthesis System for Biomanufacturing of Multiple Disulfide Bond Containing Therapeutic Proteins

Claire Lanclos, Parker Hannan, Yongchan Kwon

The cell-free system (CFS) activates complex biological processes without using intact living cells. Unlike the living system, CFS utilizes the cell extract retaining cellular components that allows direct access to biocatalytic enzymes, easy substrate addition and product removal, and rapid sampling. The distinctive perspective of the CFS brings the technological advantages over traditional fermentation methods such as decoupling of protein synthesis from cell growth and reproduction, which allows engineers to solely utilize the catalysts from cells, enabling the study of the protein synthesis process by examining the role of the supplements such as chaperones, elongation factors, ribosomes in the CFS. In this work, we studied *E. coli* crude cell extracts that facilitate multiple disulfide bonds containing therapeutic protein production. We established a strategy for modulating multiple cell extracts harboring periplasmic chaperone proteins and assembled them in optimal conditions. To achieve the overarching goal, we first utilized Gaussia luciferase for tuning the modulated CFS and then applied the strategy for the biomanufacturing of different human therapeutic proteins containing multiple disulfide bonds. We expect that the modulated cell-free biomanufacturing platform for disulfide bond-containing protein will promote high-yielding therapeutic protein and increase opportunities for discovering new therapeutic proteins.

164j Engineer Novel Functional Proteins in Plant Cell Culture for Industrial and Biomedical Applications

Jianfeng Xu

Genetic engineering of plants and plant cells has been used for generating transgenic plants with new traits or for "molecular farming" to produce valuable recombinant proteins. We aim to leverage two unique post-translational modifications – "glycosylphosphatidylinositol (GPI) anchor" and "hydroxyproline (Hyp)-*O*-glycosylation" – to strategically design and engineer functional recombinant proteins in plants and plant cells for industrial and biomedical applications. GPI anchoring leads to the attachment of glycolipid anchor to the C-terminus of a protein, which was found to dramatically promote the intracellular transport of anchored proteins in plant cells and display the proteins on the cell surface. GPI anchoring modification also facilitated N-linked and Hyp-*O*-linked glycosylation, resulting in more homogeneous glycosylation of the anchored proteins. Hyp-*O*-glycosylation is a plant-specific posttranslational modification that involves the addition of the arabino-oligosaccharides or arabinogalactan polysaccharides to the residues. Hyp-*O*-glycosylation was found to facilitate extracellular secretion of associated proteins and protect proteins from proteolytic degradation. For an industrial application, the GPI anchor and a Hyp-*O*-glycosylation module $(SP)_{20}$ consisting of 20 repeats of the "Ser-Pro" motif was engineered to the thermophilic E1 endoglucanase (cellulase) in tobacco plants to reconstruct the plant cell wall for enhanced processability of the cell wall biomass, leading to more than 50% increase of the saccharification efficiency. The two functional modules were also engineered to an anti-TNFα molecule (single-chain fragment variable of the human anti-TNFα antibody) in tobacco BY-2 cells to create a new type of oral biologic to treat inflammatory bowel disease (IBD). Our preliminary data indicate that the plant cells produced oral anti-TNFα biologic was able to ameliorate the disease in the dextran sulfate sodium (DSS)-induced colitis mouse model and induce a reduction of histopathological markers.

164k Acoustic Force Spectroscopy Enables Multiplexed Single-Molecule Characterization of Protein-Carbohydrate Binding

Markus Hackl, Shishir Chundawat

To rationally engineer more efficient enzymes for cellulosic biomass deconstruction for industrial biotechnology applications or develop protein binders targeting extracellular matrix enriched in carbohydrates for therapeutic applications, it is necessary to better understand the complex protein-carbohydrate interfacial interactions. Carbohydrate binding module (CBM) proteins or lectins are often associated with driving such surface-tethered protein-ligand interfacial interactions. However, it is not well known how proteins recognize, bind, and dissociate from polysaccharide surfaces due to the lack of suitable toolkit to study CBM-substrate binding interactions. Here, we outline a general approach to methodically study the unbinding behavior of CBMs from model polysaccharide surfaces using a highly multiplexed single-molecule force spectroscopy assay technique called acoustic force spectroscopy. We use this technique to study how CBM dissociate from cellulose surface to infer the molecular mechanism governing substrate recognition and dissociation. Our approach can be broadly applied to study multivalent protein-polysaccharide binding interactions relevant to other carbohydrates such as starch, chitin, or hyaluronan to engineer more efficient ligand binders or biocatalysts.

164l Designing Versatile Beta Roll Peptide Scaffolds

Matthew Lucia, Devin Golla, Farid Khoury, Scott Banta

The Block V RTX "beta roll" peptide domain, located in the CyaA enzyme of *B. pertussis*, has been recognized in the past as a potential platform for ion-responsive protein capture, due to its ability to reversibly fold in millimolar concentrations of calcium ions. While the beta roll is a good candidate for selective binding, the applicability of the wildtype beta roll domain as a scaffold is narrow due to the limited number of designable positions on its binding face that can form protein-biomolecule interactions. We approached the challenge of designing novel lengthened and widened Block V domains by setting structural homology constraints and computationally predicting the thermodynamic conformations of residues. Utilizing this strategy, the most unfavorable residues in the new beta roll structures were identified and the surrounding areas were designed for optimal thermodynamic stability. Fluorescence resonance energy transfer (FRET) experiments have shown that computationally designed beta rolls with one and two additional N-terminal consensus loops fold in the presence of calcium ions. Concurrent results indicate that the conformation of residues in the additional N-terminal loops maintain an internal hydrophobic continuum within the structure. As has been shown for the recently characterized Block IV-Block V region of the RTX peptide, maintaining this region may help stabilize the scaffold and propagate the folding of additional loops. Widening the beta roll contributes similarly to an increase in thermodynamic stability. Computational models and crystallography data has indicated that greater thermodynamic stability of the structure in the absence of calcium is correlated to an increase in calcium binding affinity. The lengthening and widening processes can be conducted simultaneously to assemble structures that have the thermodynamic benefits corresponding to both structures. A larger binding face can partake in a greater diversity of protein-biomolecule interactions, which can be evolved to bind to the new scaffold via phage-assisted continuous evolution (PACE). We plan to express and crystallize novel lengthened and widened beta rolls to further validate these computational results, and to leverage this directed evolution pathway to improve the scaffold's ability to bind biomolecular targets.

164m Thermostable Enzymes for Error Correction of DNA

Rushant Sabnis, Madeline Jordan, Paul de Figueiredo, Qing Sun

The ability to synthesize accurate DNA is important for advancements in synthetic biology and biotechnology. Contrary to DNA replication *in vivo, in vitro* DNA synthesis lacks DNA repair mechanism and pathways to detect and repair mismatches. Enzymatic error correction for DNA synthesis is a cost-effective method of reducing errors in synthetically generate DNA. DNA-based error correction enzymes are known to detect specific base-pair mismatches and work by cleaving both DNA strands at 3' end of the mismatched site. Currently, these error correction enzymes are highly accurate but lack thermal stability and demonstrate poor enzyme kinetics. Thermophiles were found to have lower mutation rate as compared to mesophiles and therefore, surveyor endonucleases in thermophilic bacteria were explored and characterized for functioning error correction activity and higher temperature stabilities. This specific study focuses on an error correction enzyme from *Thermococcus Barophilus* which is abbreviated as enzyme TbaMP. This enzyme was sequence optimized, cloned, expressed in *Escherichia coli*, and purified for activity and thermostability tests. Activity assay demonstrated that the enzyme TbaMP had superior G/C mismatch detection and correction ability as compared to the commercial standard. Ultimately, this data supports the efficacy of thermophilic error correction enzymes and supports further investigation and application of this enzyme in DNA error correction.

164n Exploring Structure-Function Relationships Governing Activity of the Cyanobacterial Bicarbonate Transporter Sbta

Sydney Parrish, Christopher Jones, Amanda Godar, David Nielsen, Brent L. Nannenga

Cyanobacteria play a large and essential role in the global carbon cycle, collectively contributing to nearly 30% of net primary inorganic carbon (C_i) fixation. In addition, they also represent programmable platforms for the sustainable bioproduction of fuels and chemicals directly from sunlight and carbon dioxide (CO_2). Photosynthetic carbon fixation, however, depends on the ability to cyanobacteria to efficiently uptake dissolved C_i which, under physiological conditions, exists mainly in the form of bicarbonate (HCO_3^-). SbtA is one of two main HCO_3^- transporters in cyanobacteria, and it is highly conserved across most species. That said, past reports have shown that individual SbtA homologs can display diverse flux vs. affinity behaviors, suggesting that small sequence differences can impart diverse changes in function. Accordingly, our group is interested in enhancing our structure-function understanding of SbtA and using these insights to engineer improved HCO_3^- transporters. To this end, several approaches are being applied in parallel, including i) structural modeling to identify features of putative importance to controlling HCO_3^- uptake rates among SbtA homologs, ii) structure determination of purified SbtA expressed in both *Synechococcus* sp. PCC 7002 and *Escherichia coli,* and iii) functional characterization of SbtA variants in C_i-uptake deficient PCC 7002 mutants as well as *E. coli* strains lacking carbonic anhydrase (CA).

Recently, the structure of SbtA from *Synechocystis* sp. PCC 6803 (expressed in *E. coli*) was reported and has been used to develop Robetta models of SbtA's from PCC 7002, PCC 6803, and *Cyanobium* sp. PCC 7001; three homologs displaying vastly different kinetics. We are still using these models to learn about inherent structure-function relationships and generate new hypotheses regarding HCO_3^- transport behaviors displayed by different SbtA homologs.

In support of attaining higher resolution structures, we have developed a platform for *sbtA* expression and purification from heterologous *E. coli* that yields sufficient levels of SbtA for single particle cryo-EM. Furthermore, we are working to advance our structural objectives by developing protocols for *sbtA* expression and purification from native, cyanobacterial hosts using a newly developed T7-RNA polymerase expression system that has been shown to enable very high protein production (up to two thirds of total cellular protein content in PCC 7002).

Finally, we have developed a series of Ci-uptake deficient mutants of PCC 7002 to serve as facile genetic screens in which alterations in transporter function then correlate directly with cell growth. Meanwhile, to provide an additional, heterotrophic platform for SbtA functional screening, we have also developed a CA-free strain of *E. coli* that is also optimized for succinate fermentation; thereby providing a strong C_i sink. Ultimately, these collectively efforts will further our understanding of key structural features that influence transport activity (i.e., flux and/or affinity) by SbtA, as well as enhance our ability to engineer faster growing and more productive cell factories.

164o A Novel Thiolase with Enhanced Activity of ATP-Independent Triacetic Acid Lactone (TAL) Production

Seokjung Cheong, Zilong Wang, Jose Henrique Pereira, Andy Degiovanni, Carolina Araujo Barcelos, Robert W. Haushalter, Paul D. Adams, Jay Keasling

Thiolase is the enzyme that catalyzes the non-decarboxylative Claisen condensation reaction forming a carbon-carbon bond between the carbonyl carbon of an acyl-CoA serving as the primer and alpha carbon of another acyl-CoA serving as the extension unit. This type of carbon bond forming reaction does not require the beta-carboxylation of extension unit acyl-CoA and uses acetyl-CoA instead of malonyl-CoA as extension unit in most cases, hence allowing more flexibility in substrate structure and bypassing the ATP-consuming acetyl-CoA carboxylation step to supply malonyl-CoA. Therefore, many researchers are exploiting thiolase for carbon-chain elongation in bioproduction of chemicals. In one recent exemplary report (Tan *et al.* 2020. *Nature Catalysis*, 3, 593-603), researchers discovered that *Cupriavidus necator* thiolase BktB is able to condense between acetoacetyl-CoA primer and acetyl-CoA extension unit to generate triacetic acid lactone (TAL), a platform chemical for productions of valuable chemicals and a potential monomer of recyclable polymers, and *Escherichia coli* strains expressing BktB exhibited higher titer than strains expressing 2-pyrone synthase, a native and commonly used polyketide synthase for TAL production, due to higher ATP energy efficiency.

While *C. necator* BktB is already an energy-efficient TAL producing enzyme, through screening selected BktB homologs with different phylogenetic distances, this research discovered a new thiolase with ~30 times higher *in vitro* TAL production activity and ~10 times higher TAL titer from *E. coli* in *in vivo* small-scale plate fermentation compared to *C. necator* BktB. When growing in a larger scale fermenter, *E. coli* strain expressing this thiolase produced ~ 3 g/L of TAL, higher than the highest reported TAL titer by *E. coli* despite the lethality of TAL upon *E. coli*. Successful crystallization and subsequent substrate docking further helped explain the reaction mechanism and reveal the contributing residues for enhanced TAL synthesis activity of this enzyme.

This new discovered thiolase with much higher ATP-independent TAL production activity could benefit the biorenewable TAL production industry in microbial platforms like *E. coli* or other organisms with higher TAL tolerance, and its structural analysis results could aid the future rational designs for activity improvement or expansion of functional and structural range of substrates hence expanding the potential product diversity of biorefinery.

164p Engineering Antibody Fusion Proteins for Targeted Intracellular Therapeutic Delivery

Kyle Kaeo, Seth Ludwig, Jamie Spangler

Intracellular proteins make up a significant portion of the proteosome, leading to an interest in achieving immune modulation via intracellularly targeted therapeutics. Most intracellular protein-targeted drugs are small molecules, which can easily pass through the cell membrane but are hindered by a lack of specificity and short half-life. Antibodies, thanks to their greater contact surface area, can have superior target specificity, and persist longer in the blood. and can target specific receptors and cell types. Fusing a therapeutic to an antibody presents a promising solution, but antibodies lack an efficient path into the cell.

We recently discovered a multiparatopic antibody that binds multiple epitopes on the immune checkpoint protein programed death-ligand 1 (PD-L1) to induce its downregulation. This platform promotes endocytosis through clustering receptors together on the surface and enhancing internalization and degradation. We sought to exploit this mechanism for intracellular delivery by introducing endosomalytic peptides to induce escape of the multiparatopic antibody into the cytosol. We inserted two different endosomalytic peptides into the downregulating antibody, one before and after the hinge region. We found that these peptides do not hinder downregulation and we observed decreased lysosomal accumulation of the antibodies containing these peptides.

To determine the efficacy of this modified antibody as a vehicle for an intracellular therapeutic, we fused Cas9 to the C-terminus of the heavy chain (HC) of our anti-PD-L1 multiparatopic antibody. We further demonstrated robust PD-L1 downregulation and endosomal escape for these fusion proteins. These results offer proof-of-concept establishing a promising new vehicle for intracellular delivery, with wide-ranging applications including gene therapy, signaling modulation, and genetic engineering.

164q A Novel Thermophilic Enzyme for DNA Based Error Correction

Rushant Sabnis, Madeline Jordan, Paul de Figueiredo, Qing Sun

The ability to synthesize accurate DNA is important for advancements in synthetic biology and biotechnology. Contrary to DNA replication *in vivo, in vitro* DNA synthesis lacks DNA repair mechanism and pathways to detect and repair mismatches. Enzymatic error correction has emerged as a cost effective and convenient method for error mitigation that are embedded in synthetically generated DNA fragments. Several commercial enzymes have shown to reduce error rate of synthetic genes *in vitro*. However, due to limitations related to enzyme kinetics and thermal stability, the error correction efficiencies are inadequate for downstream applications. Prior research has reported that despite environments more favorable to mutations, the mutation frequency in hyperthermophiles is lower than their mesophilic bacteria counterparts. We hypothesize that enzymes obtained from thermophiles will demonstrate higher robustness and better error correction efficiencies. We identified, expressed, purified, and characterized one error correction enzyme from thermophilic archaeon *Thermococcus Barophilus*. We tested our hypothesis by comparing the enzyme activity and found that the thermophilic enzyme has better performance as compared to the commercially available error correction enzyme in detecting and correcting G/C mismatches. Furthermore, we carried out kinetic and activity assay of the purified enzyme. We observed that the thermophilic enzymes exhibit lower unwanted exonuclease activity that is responsible for error free DNA template degradation.

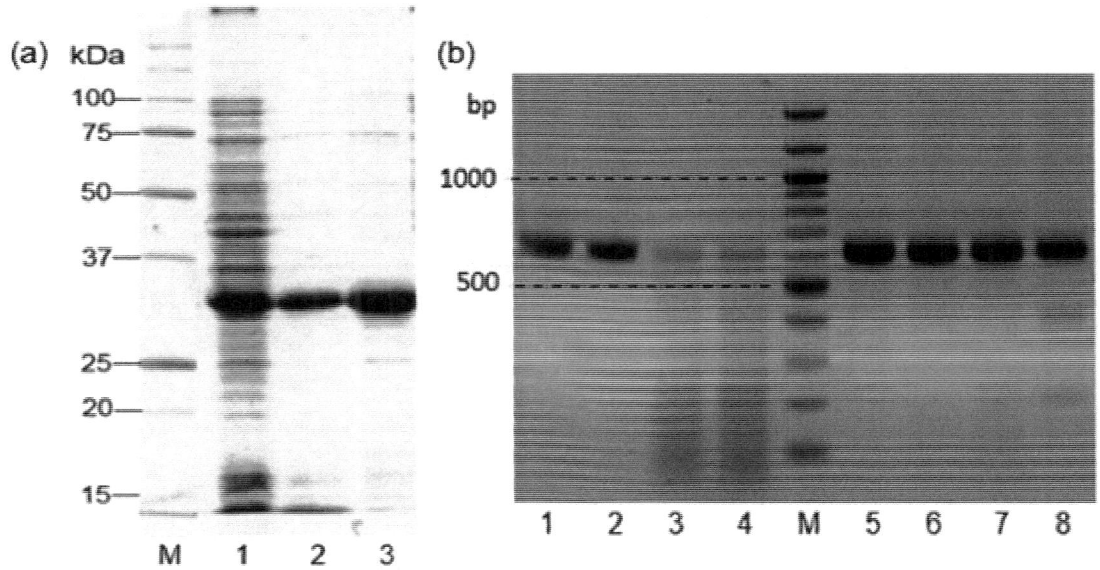

Figure 1: Purification, Expression and Activity comparison of thermophilic enzyme with commercial Enzyme. (a) Protein expression and purification; Lane 1: cell lysate post sonication; Lane 2: His-tag purified protein Lane 3: Heat purified protein.**(b): Activity assay of with normalized concentration His-Tag purified enzymes and comparison with commercial enzyme.** Lane 1: Homoduplex with no enzyme; Lane 2: Heteroduplex with no enzyme; Lane 3: Homoduplex with commercial enzyme; Lane 4: Heteroduplex with commercial enzyme; Lane M: Marker; Lane 5: Homoduplex without thermophilic enzyme; Lane 6: Heteroduplex without thermophilic enzyme; Lane 7: Homoduplex with thermophilic enzyme; Lane 8: Heteroduplex with thermophilic enzyme

164r The Influence of Electrostatic Distribution on Small Molecule Binding of Coiled-Coil Protein Microfibers

Julia Monkovic, Dustin Britton, Sihan Jia, Chengliang Liu, Jin Kim Montclare

The ability to engineer the solvent-exposed surface of self-assembling coiled-coils allows one to achieve higher order hierarchical assembly such as nano or microfibers. Currently these materials are being developed for a range of biomedical applications, including drug delivery systems, however ways to mechanistically optimize the coiled-coil structure for drug binding has yet to be explored. Our laboratory has previously leveraged the functional properties of the naturally occurring cartilage oligomeric matrix protein coiled coil (COMPcc), not only for its favorable motif but also for the presence of a hydrophobic pore to allow for small molecule binding. Here, we present a small library of protein microfibers derived from the parent sequence of COMPcc with various electrostatic potentials, with the aim to investigate the influence of higher order assembly on candidate small molecule, curcumin, drug delivery. The impact of drug binding on coiled-coil mutants was characterized by fluorescence spectroscopy and confocal microscopy. Structural characterization was also performed via circular dichroism and attenuated response Fourier transform infrared spectroscopy. Our findings demonstrate that redistribution of the surface charge impacts not only its supramolecular assembly but also its thermal stability after drug binding. We elucidate this relationship with the support of Rosetta docking simulations, revealing the impact of surface charge on side-chain packing in the pore. For all variants, protein melting temperature saw a modest to extreme improvement upon small molecule encapsulation. By continuing to investigate these structure-property-function relationships, we can construct electrostatically optimized protein fibers capable of efficiently encapsulating and stabilizing small molecules at a higher rate than the current state-of-the-art coiled coil drug delivery system.

164s Synthesis and Evaluation of Megamolecule Dendrimers in Cancer Therapy

Blaise Kimmel, Milan Mrksich, John Wilson

Antibody therapies have the potential to improve treatment outcomes for patients with solid tumors by allowing for selective binding of overexpressed receptors, largely sparing healthy cells. However, we lack the ability to readily tailor the structure of these therapeutics, which contributes to factors such as off-target immunogenic responses, inability to penetrate solid tumors, and both high dosage and infusion requirements. Recent work from the Mrksich group demonstrated a rapid synthesis approach to construct modular protein scaffolds (termed megamolecules), confirming the programmability of these therapeutics. It is believed that tuning the topology of megamolecule nanostructures will enable the generation of a new class of cell-specific biologics. In this talk, I will explore our approach to address current limitations in precise protein engineering and targeted therapy design. Here, we leverage the megamolecule strategy to construct of multi-valent megamolecules with hypervalent affinity for HER2+ tissue. In order to accomplish this, we first evaluate a panel of single domain antibody fragments (nanobodies) that have varied affinity for HER2+ receptors. Programming these affinity domains into multimeric structures – up to the 2nd generation – revealed atomically precise near megadalton protein dendrimers with a controlled number of anti-HER2 domains. We evaluate these structures *in vitro* for their selective binding affinity to HER2+ tissue and related avid affinity of the dendrimers to structural architecture. Further, in breast cancer tumor mice models, we explored the in vivo effects of our hypervalent megamolecules for improved on-target binding and efficacy through biodistribution studies. The resulting innovation from this study has the potential to enable a shift in our approach to construct high efficacy, programmable nanotherapeutics that results in minimizing off-tumor cytotoxic effects and improves tumor reduction *in vivo.*

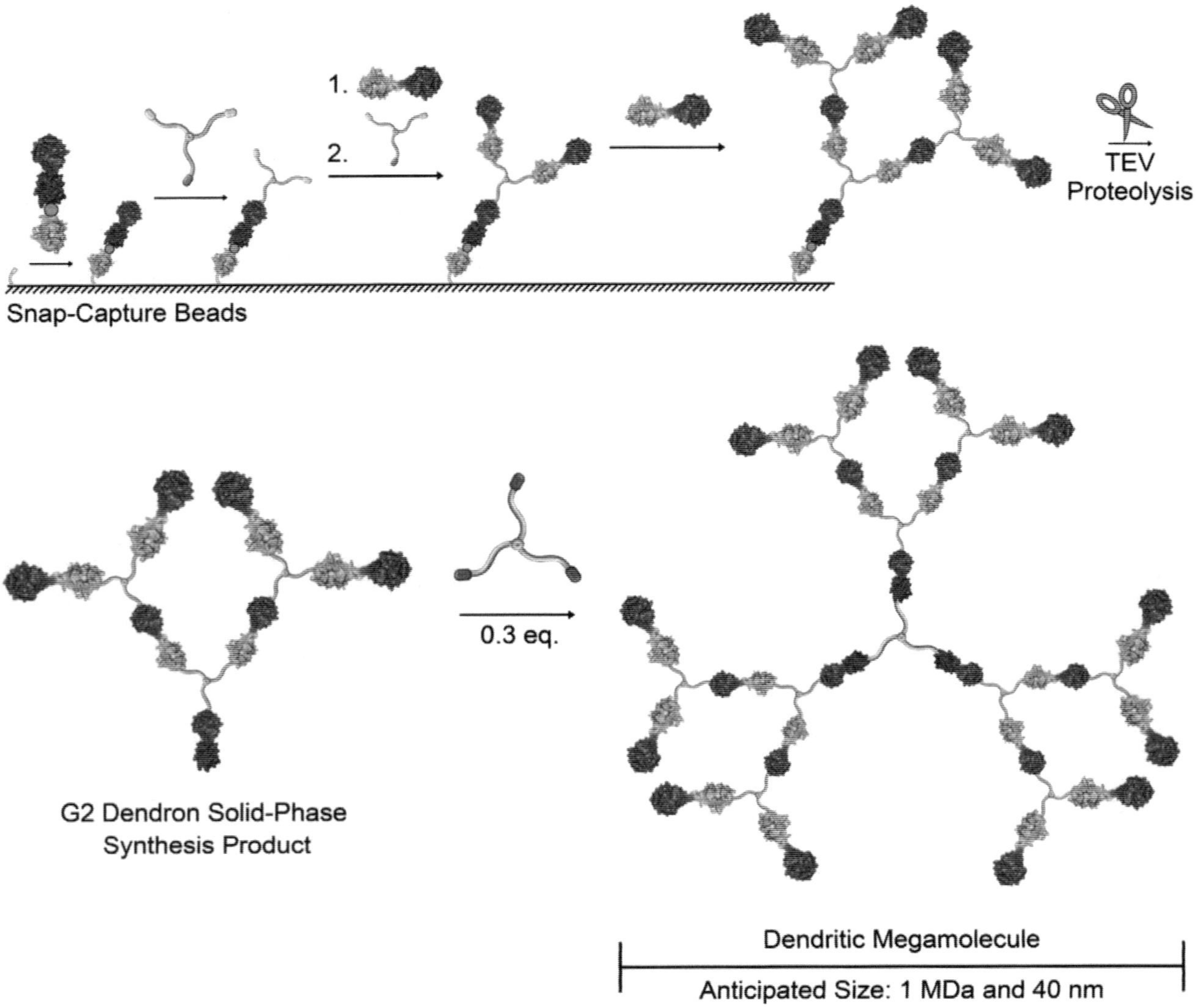
1.
2.
TEV
Proteolysis
Snap-Capture Beads
0.3 eq.
G2 Dendron Solid-Phase
Synthesis Product
Dendritic Megamolecule
Anticipated Size: 1 MDa and 40 nm

164t Leveraging Synthetic Biology and Gut-on-a-Chip Systems to Investigate the Mechanistic Role of H_2s in the Gut

Justin Hayes, Benjamin Woolston, Ryan Koppes

The human gut microbiota is a complex biological system composed of trillions of bacteria and their interactions with the host. Hydrogen sulfide (H_2S), a metabolite of various gut microbes, has been correlated with diseases such as IBD and colorectal cancer. However, H_2S has also been shown to protect tissue from oxidative stress and act as a signaling molecule. Past studies have investigated H_2S effects on gut epithelial cells but were done in static cultures and H_2S was not microbially derived. Traditionally, static epithelial cultures and animal models are used to study the gut but are plagued by low transcriptomic similarity, high drug translation failure rates, and ethical issues. In vitro flow systems, such as the gut-on-a-chip (GoC), designed with human-derived tissue are easier to probe than animal models and more genetically representative than static cultures. We hypothesized the effects of H_2S on gut epithelial cells could be concentration-dependent and more accurately investigated in a GoC system. Researchers estimate H_2S levels in the human gut vary from 0.2 to 3 mM. To investigate our hypothesis, we engineered the bacterium *Escherichia coli* for tunable production of H_2S via titratable expression of l-cysteine desulfidase, which catalyzes the conversion of l-cysteine to H_2S. Since *E. coli* has various native enzymes with l-cysteine desulfidase activity which could limit our ability to explore lower biologically relevant H_2S levels, we first targeted four native *E. coli* genes for deletion via the CRISPR-Cas9 system, resulting in a strain with very low background sulfide production. We then assessed desulfidases from *Fusobacterium nucleatum* and the native *E. coli* yhaOM operon which we expressed in plasmids with varying promoter strengths. Experiments in Hungate tubes confirmed our ability to vary H_2S levels from 0.025 to 1 mM. Next, we introduced our strains into a GoC system and showed we can tune H_2S levels from 0.2 to 1.5 mM for up to 72 hours. Studies are underway using this system to investigate how different concentrations of H_2S affect gut permeability, monolayer integrity, and genetic expression. It is postulated H_2S induces cell cycle arrest and DNA damage which we will investigate with qRT-PCR of the epithelial cells. Through our work, we aim to answer fundamental questions about H_2S role in the human gut and showcase a platform for integrating synthetic biology and GoC systems to study gut metabolites with ambiguous roles in disease.

164u Novel Magnetic Biosensor for COVID-19 Surveillance through Wastewater-Based Epidemiology

Stefano Ciannella, Sowrav Barua, Jenifer Gomez Pastora

Since the first case of human infection by the novel coronavirus disease (COVID-19) was identified in December 2019, the severe acute respiratory syndrome coronavirus 2 (SARS-CoV-2) has caused around 6 million deaths worldwide according to recent estimates [1]. Despite the preventative efforts to avoid transmission, healthcare systems have been continuously overwhelmed over the last two years. Since the virus is shed in feces and urine [2], wastewater-based epidemiology (WBE) is considered a promising alternative to estimate the regional infection rates of COVID-19, by measuring the levels of SARS-CoV-2 RNA in the water, which could provide a faster and more economical estimate than individual testing approaches [3]. In this work, a magnetic biosensor is proposed to detect and quantify SARS-CoV-2 RNA in wastewater. Our sensor integrates magnetic nanoparticles within an immiscible filtration assisted by surface tension (IFAST) device. First, the superparamagnetic iron oxide nanoparticles (SPIONS) will be synthesized to target specific SARS-CoV-2 proteins (S protein), whose binding capability will be assessed through ELISA tests. Second, the miniaturized IFAST-liked device will be fabricated, and the viral load will be quantified through fluorescence measurements. Our system efficiency will be also compared to the conventional RT-qPCR assays. The proposed device can be fabricated at a relatively low cost and has the potential to be easily operated at the point-of-care for COVID-19 monitoring, for example, in high-risk settings (nursing homes and hospitals) and low-resource areas (where pit toilets and septic tanks are mostly used). Moreover, this research will generate useful information about the relationship between SARS-CoV-2 RNA concentration in wastewater and the number of reported cases in a specific setting, which foments further research on the prediction power of WBE for COVID-19.

REFERENCES

1. John Hopkins University. Johns Hopkins Coronavirus Resource Center [Internet]. Global cases [cited 2022 Apr 10]. Available from: https://coronavirus.jhu.edu/map.html
2. Patel M, Chaubey AK, Pittman CU, Mlsna T, Mohan D. Coronavirus (SARS-CoV-2) in the environment: Occurrence, persistence, analysis in aquatic systems and possible management. *Total Environ*. 765:142698.
3. Lu D, Zhu DZ, Gan H, Yao Z, Fu Q, Zhang X. Prospects and challenges of using electrochemical immunosensors as an alternative detection method for SARS-CoV-2 wastewater-based epidemiology. *Total Environ.* 777:146239.

164w Yeast Surface Display of Sars-Cov-2 Receptor Binding Domain (RBD) for Diagnostic Purposes

Shadrach Ibinola, Hazim Aljewari, Kaylin Hicks, Sebastian Freeman, Robert Beitle, Shannon Servoss, Kaiming Ye, Ahmed El Masheiti, Kevin Clark, Vicki Thompson, Beth Cobb

Although the COVID-19 pandemic engendered global comradery for vaccine and anti-viral production, it exposed some of the challenges associated with the fast-track of the design of accurate diagnostics. We proposed that the inclusion of biosafety level 1 (BSL-1) capabilities in the arsenal to fight COVID-19 could include yeast surface display technology. First, *Saccharomyces cerevisiae* EBY100 was modified for surface display of SARS-CoV-2 receptor binding domain (RBD). Batch bioreactor culture was performed with selection media to produce the recombinant yeast, and simple centrifugation in contrast to centrifugation, lysis, and fractionation resulted in recovery of the potential diagnostic kit component. Confocal microscopy confirmed the decoration of the surface with RBD (and Green Fluorescent Protein, GFP, in separate cultures as a control). Experiments with lateral flow assays and dot blot assays demonstrated the decorated yeast served as positive and negative controls for the detection of SARS-CoV-2.

164x Engineering Ligand-Specific Biosensors for Aromatic Amino Acids, Neurochemicals, and Other Structurally Similar Compounds

Chenggang Xi, Austin Rottinghaus, Matthew Amrofell, Tae Seok Moon

Microbial biosensors have diverse applications in metabolic engineering and medicine (1, 2). Specific and accurate quantification of chemical concentrations allows for adaptive regulation of enzymatic pathways and temporally-precise expression of diagnostic reporters. Although biosensors should differentiate structurally similar ligands with distinct biological functions, such specific sensors are rarely found in nature and challenging to create. Using E. coli Nissle 1917, a generally regarded as safe microbe, we characterized two biosensor systems that promiscuously recognize aromatic amino acids or neurochemicals. To improve the sensors' selectivity and sensitivity, we applied rational protein engineering by identifying and mutagenizing amino acid residues and successfully demonstrated the first ligand-specific biosensors for phenylalanine, tyrosine, phenylethylamine, and tyramine. Additionally, our approach revealed insights into the uncharacterized structure of the FeaR regulator, including critical residues in ligand binding. These results lay the groundwork for developing kinetically-adaptive microbes for diverse applications (3). In this talk, we will discuss our recent efforts to develop other structurally similar compounds.

1. Amrofell, Rottinghaus, and Moon, Curr Opin Biotechnol, 2020. **66**: p. 11-17.
2. Rottinghaus, Amrofell, and Moon, Biotechnology Journal, 2020. **15**: p. 1900319.
3. Rottinghaus, Xi et al. Cell Systems, 2022. **13**: p204-214.e4.

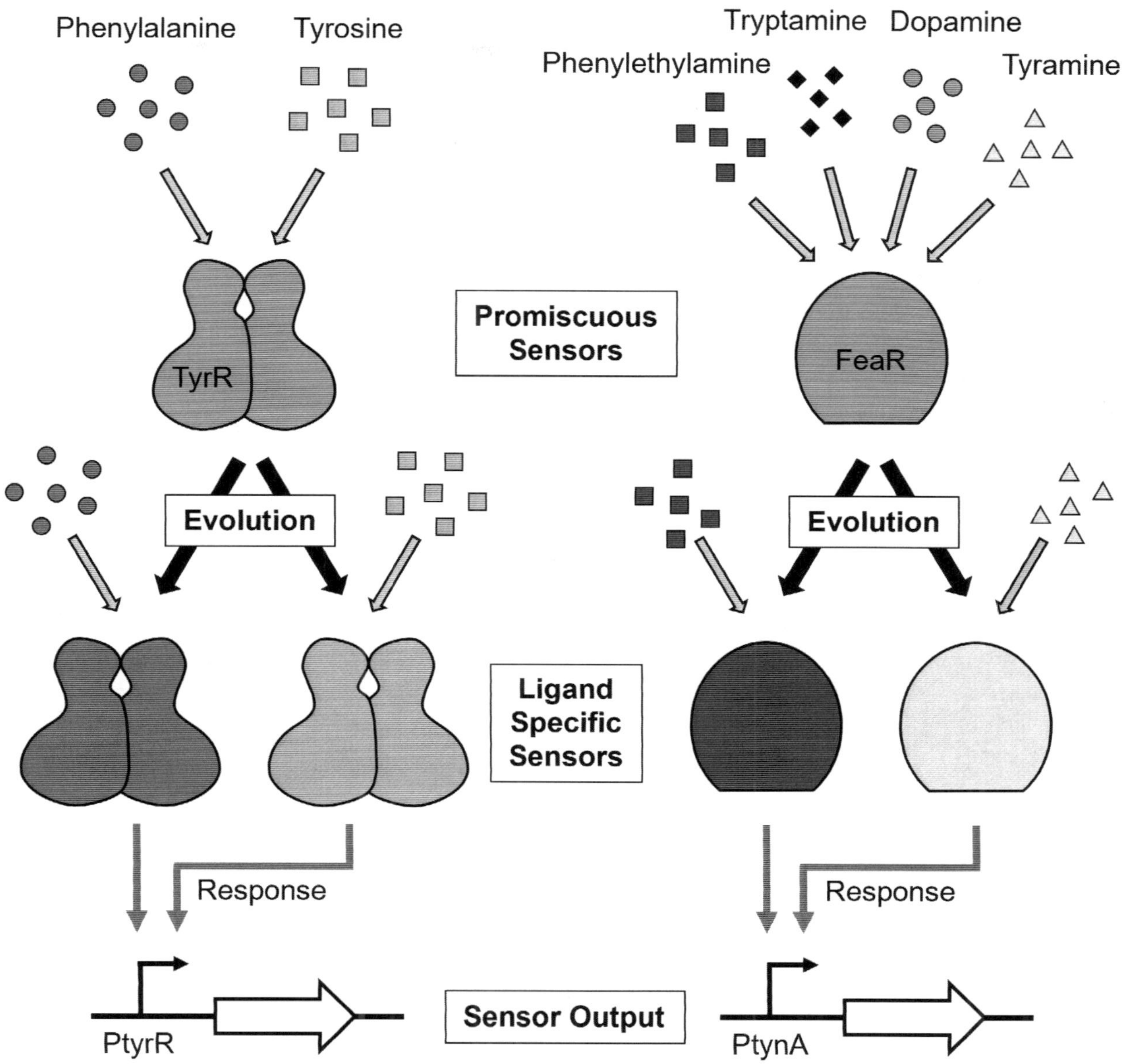

Phenylalanine
Tyrosine
Tryptamine
Dopamine
Phenylethylamine
Tyramine
TyrR
Promiscuous Sensors
FeaR
Evolution
Evolution
Ligand Specific Sensors
Response
Response
PtyrR
Sensor Output
PtynA

164y The Change of Magnetic Susceptibility of Red Blood Cells Depending on the Density.

Hyeon Choe, Mitchell Weigand, Xian Wu, Jacob Strayer, Jenifer Gomez Pastora, Jeffrey Chalmers

Currently, clinical blood test focuses on the measuring the number or concentration of hemoglobin (Hb) in blood. However, the main function of Hb is combination of iron ion with oxygens for gas exchange in cells. Subsequently, the combination ability of Hb can be a criterion for biological problem. The change of magnetic susceptibility of Hb shows the red blood cells have different mobility in magnetic field depending on the combination with oxygen. In the case of combination with oxygen, the iron ions have covalent bond and diamagnetic property, which moves opposite direction from magnetic field. Otherwise, the iron ions have unpaired electrons and paramagnetic property, which moves the same direction following magnetic force. To examine the magnetic susceptibilities of red blood cells, the cell tracking velocimetry (CTV) was used. Dragged by magnetic and gravitational force, the red blood cells show different mobility depending on the shape of cells or the concentration of Hb cells by cells during the CTV analyze. Moreover, the density can be used as a measure of disease diagnosis of red blood cells. For example, the blood cell from sickle cell disease has larger densities than normal blood cell. Also, normal red blood cells have a narrow density distribution whereas sickle cells have a relatively wider density distribution. Therefore, the change in the magnetic susceptibility of red blood cells according to the density is investigated, and the probability of using this property for disease diagnosis will be discussed.

164aa Human Forebrain Organoid-Derived Extracellular Vesicle Labelling with Iron Oxides for in Vitro Magnetic Resonance Imaging

Chang Liu, Mark Marzano, Shannon Helsper, Xingchi Chen, Laurie Muok, Changchun Zeng, Li Sun, Samuel C. Grant, Yan Li

The significant roles of extracellular vesicles (EVs) as intracellular mediators, disease biomarkers, and therapeutic agents, make them as a scientific hotspot. In particular, EVs secreted by human stem cells show significance in treating neurological disorders such as Alzheimer's disease and ischemic stroke. However, the clinical applications of EVs are limited due to their poor targeting capabilities and low therapeutic efficacies after intravenous administration. Superparamagnetic iron oxide (SPIO) nanoparticles are biocompatible and have been shown to improve the targeting ability of EVs. In particular, ultrasmall SPIO (USPIO, <50 nm) is more suitable for labelling nanoscale EVs due to its small size. In this study, induced forebrain neural progenitor cortical organoids (iNPCo) were differentiated from human induced pluripotent stem cells (iPSCs), and the iNPCo expressed FOXG1, Nkx2.1, α-catenin, as well as β-tubulin III. EVs were isolated from iNPCo media, then were loaded with USPIOs by sonication for 30 seconds, followed by incubation at 37°C for one hour. Size and concentration of EV particles were measured by nanoparticle tracking analysis, and no significant changes were observed before and after sonication. miR-155 and miR-221 were decreased after sonication. In vitro magnetic resonance imaging (MRI) revealed the contrast of USPIOs labeled EVs while the unlabeled EVs did not have the contrast. Besides, the labeled EVs showed shorter T1, T2, T2* echo times compared with the gel, while the unlabeled EVs showed longer echo times. Our study demonstrated that USPIO labeling was a feasible approach for in vitro tracking of EVs, which paves the way for further in vivo examination.

164ab Red and White Blood Cell Magnetic Analysis and Separation: Potential for Hematologic Disease Diagnosis

Sowrav Barua, Stefano Ciannella, Jeffrey Chalmers, Jenifer Gomez Pastora

Separation of red and white blood cells from the blood is of paramount importance for medical research; the diagnosis of different pathologies and the evolution of an individual health status are primarily assessed by performing a blood test [1]. Multiple technologies have been developed for the fast and economical separation of blood cells, from them, label-free techniques are interesting because of their several advantages [2]. Since some types of blood cells (erythrocytes and monocytes) have a high iron content [3], magnetophoresis can be performed to isolate these cells without the need of magnetic labels (or magnetic beads). Nevertheless, information about the iron distribution in RBC and monocyte populations in healthy and disease blood needs to be available before any magnetophoretic device can be commercialized for blood separations. In this work, the magnetic mobility, susceptibility, and iron/hemoglobin concentration of red blood cells (RBCs) and monocytes obtained from healthy donors and patients with hematologic diseases (anemias) will be reported. We hypothesize that whereas the magnetic character of RBCs in anemic patients will be abnormal and will show more diamagnetism, the percentage of magnetic monocytes will be higher and might be an indicator of disease. The relationship between the number of magnetic white blood cells (monocytes), RBCs and the presence of hematologic diseases has never been investigated before. Thus, this work will provide important insight into the iron recycling mechanisms in disease and could open the door to the use of simple and economical magnetic devices to diagnose hematologic disorders.

REFERENCES

1. Green, R., andamp; Wachsmann-Hogiu, S. (2015). Development, History, and Future of Automated Cell Counters. *Clinics in Laboratory Medicine*, *35*(1), 1–10. https://doi.org/10.1016/j.cll.2014.11.003
2. Munaz, A., Shiddiky, M. J. A., andamp; Nguyen, N.-T. (2018). Recent advances and current challenges in magnetophoresis based micro magnetofluidics. *Biomicrofluidics*, *12*(3), 031501. https://doi.org/10.1063/1.5035388
3. Kim, J., Gómez-Pastora, J., Weigand, M., Potgieter, M., A. Walters, N., Reátegui, E., F. Palmer, A., Yazer, M., Zborowski, M., andamp; Chalmers, J. J. (2019). A Subpopulation of Monocytes in Normal Human Blood Has Significant Magnetic Susceptibility: Quantification and Potential Implications. *Cytometry Part A*, *95*(5), 478–487. https://doi.org/10.1002/cyto.a.23755

164ac Monitoring the Activity of Optogenetically Engineered Cardiomyocytes Using Microelectrode Arrays Chips

Olurotimi A. Bolonduro, Zijing Chen, Emmanuel Tzanakakis, Brian P. Timko

We report an integrated optogenetic and bioelectronic platform for stable and long-term modulation and monitoring of cardiomyocyte function in vitro. Optogenetic inputs were achieved through expression of a photoactivatable adenylyl cyclase (bPAC), which when activated by blue light caused a dose-dependent and time-limited increase in autonomous cardiomyocyte beat rate. Bioelectronic readouts were achieved through an integrated planar multi-electrode array (MEA) that provided real-time readouts of electrophysiological activity from 32 spatially-distinct locations.

Optogenetic modulation of cellular processes can be achieved in a reversible fashion, with high specificity, and few – if any – side effects compared to pharmacological treatments. The secondary messenger cyclic adenosine monophosphate (cAMP) plays a central role in the control of cardiomyocyte force of contraction and rate of relaxation. Increased cAMP activates the protein kinase A (PKA), which phosphorylates and enhances the function of voltage-dependent calcium channels by inducing Ca^{2+} influx and membrane depolarization.

Through the expression of bPAC, intracellular cyclic AMP ([cAMP]i) can be upregulated leading to an increase in heart beating rate. Rat ventricular cardiomyocytes were transduced with an adenoviral vector, AdbPAC, carrying a cassette with the bPAC gene (featuring a cMyc tag for immunodetection) and the mCherry fluorescent reporter co-expressed through the CMV promoter. Cell transduction was performed as a multiplicity of infection (MOI) ranging between 50 – 1000. The expression of mCherry was detected by fluorescence microscopy and quantified by flow cytometry. The fraction of mCherry-positive cells was 10.2% at MOI of 50, and markedly increased to 52.5% at MOI of 500 with no difference for cells cultured at MOI of 1000. However, cell viability was decreased from 85.0% to 55.3% for MOI of 500 and 1000, respectively. Expression of the bPAC protein was confirmed by western blotting. After 30 minutes of blue light illumination, intracellular cAMP was 2.5- fold higher with the expression of bPAC, as determined by ELISA. No difference was observed between cells infected with the empty adenoviral AdeGFP vector (control group), and cells transduced with AdbPAC but no exposure to blue light. The expression and pattern of distinct cardiomyocyte proteins such as α-actinin and connexin-43 were not altered despite the expression of bPAC as revealed by confocal microscopy.

The integrated MEA consisted of circular 30-micron diameter gold recording elements and SU-8-passivated interconnects fabricated on a silicon / silicon oxide substate. Following fabrication the recording elements were electrochemically coated with porous platinum black which reduced their impedance from ca. 260 to 8.0 kΩ. Neonatal rat cardiomyocytes were seeded onto the MEA and transfected on DIV 3 (days *in vitro*). Bioelectronic readouts revealed spontaneous beating behavior by DIV 6 with signal-to-noise $\geq$ 200 dB.

MEAs provided real-time readouts of cardiomyocyte behavior in response to optical modulation. Irradiation at 24 $\mu W/mm^2$ resulted in a ca. 17% increase in beat rate within 20-25 minutes of irradiation. This elevated beat rate remained stable for at least 2 hours. In addition, bPAC activation could be cycled through repeated "on" and "off" states via time-limited irradiation. Multiplexed readouts revealed that wavefront propagation rates throughout the monolayer remained constant throughout both "on" and "off" states, demonstrating that optical modulation did not affect intercellular coupling. Cardiomyocytes could be modulated reproducibly over at least four days, demonstrating that bPAC expression was stable throughout that period.

Taken together, our results demonstrate the potential of optogenetic control for rational modulation of cardiac function. Real-time readouts enable assessments of cardiac function that could be further integrated into closed feedback loops. These technologies could have clinical applications for cardiac regulation including arrythmia diagnosis and intervention.

164ad Characterization of Mneongreen in the Cell-Free Protein Synthesis System and Optimizations of Signal Output for Biosensor Application

Caroline Copeland, Jeehye Kim, Pearce Copeland, Chloe Heitmeier, Yongchan Kwon

The cell-free system (CFS) has allowed for a new wave of innovative biosensors to emerge. Many traditional biosensors currently used suffer from inaccuracies largely due to the analyte in the sample being removed and placed in a foreign environment. The CFS allows the user to harness the checks and balances that come with the cellular environment these analytes are used to while being able to program complicated circuits to detect it. This allows the user to move away from faulty immediate signal amplification steps that result in high background noise and focus more on accurately detecting the biomarker of interest.

In this study, we increase the performance of the *E. coli*-based CFS to maximize the output protein signal to achieve a lower limit of detection. We first debut a new fluorescent protein, mNeonGreen (mNG), to the CFS and present characterization of its kinetics. With systemic optimization of the DNA sequence, cell-free reaction conditions, environmental contaminants, and storage additives, we could achieve a significantly brighter signal from mNG than commonly used fluorescent proteins (deGFP, x7.75; eGFP, x26.81; sfGFP, x2.64). We anticipate that our work will promote more sensitive cell-free sensor development and applied research for synthetic biology and biotechnology.

164ae Biosensor Engineering for Reliable Production Detection

Jennifer Kaczmarek, Kristala Prather

Metabolic engineering has shown that it is possible to improve microbial cell factories for higher productivities, titers, and yields of useful chemical products. However, there exists a finite knowledge of metabolism, meaning that there is only so much rational improvement to be made to organisms and pathways. To bridge this gap, the use of directed evolution can push both the host organism and the desired enzymatic pathways within these hosts to yield industrially relevant levels. Directed evolution relies on the ability to create large and diverse genetic libraries, and traditional quantification methods fail to achieve the throughput necessary to search through a significant amount of the library diversity to find improved genetic variants. To reach the throughputs needed to access this library sequence space, it is possible to use transcription-factor-based biosensors. These genetic tools allow for the transformation of the relative levels of desired metabolite into a proportional and detectable output, such as fluorescence. Here, we have developed a biosensor for application in the directed evolution of *myo*-inositol oxygenase for improved titers of glucaric acid from glucose. In response to glucuronic acid, the precursor to glucaric acid and the product of the *myo*-inositol enzyme, the biosensor outputs GFP, allowing for the use of fluorescence-activated cell sorting to detect cells with improved enzyme variants. In such screens there is a challenge to detect true improved variants over false positives, making a reliable biosensor the key to a successful screening campaign. Thus, to develop this biosensor for application in screening we have focused on a two-pronged tuning of our biosensor characteristics; both developing the sensor in global and specific contexts.

Desirable global characteristics for biosensors are: dynamic range, sensitivity, and operating range of the sensor. We tuned these characteristics by modifying the promoter and verified these changes for *in vivo* detection of product. The specific characteristics we chose were those that were screen dependent. In a system seeking improved producers, the product transport was a key element to tune. By preventing the importation of desired product while maintaining product export, we were able to obtain a system that was sensitive only to intracellular production. Another screen-specific challenge to the system was the lack of orthogonality between the production and biosensor vectors which impacted fluorescence. By creating a single plasmid system, we were able to see that the fluorescence output was more consistent and reliable for system application. By focusing on both global and specific contexts of the biosensor we were able to create a system that has qualities that increase the likelihood to detect true improved mutants amidst the false positives.

164af Combining the Use of Predictive Modelling Tools and Experimental Data to Optimize Formulation Design for Novel Modalities in Biologics

Julia E. Vela Ramirez, Marco Blanco, Suzanne M. D'Addio, Heidi Ferguson, Suman Luthra

The recent diversification of biologic molecules in Discovery and Development in the pharmaceutical industry showcases the exciting possibilities of new therapeutics. While protein engineering has enabled the synthesis of complex structures that can be tailored to fit the necessary efficacy, safety, shelf-life, and physicochemical profiles, they also pose new challenges for the developability risk assessment of these constructs utilizing workflows established for more mature modalities (e.g., mAbs). While the decision to progress a molecule through Development involves considering technical and commercial challenges, a key aspect of this decision resides in rapidly and accurately assessing critical liabilities that might present downstream and having in place mitigation strategies to minimize their impact.

The use of predictive computational modeling tools to identify critical quality attributes, mimic stress conditions, and optimize experimental design in early and late Discovery, is one approach that can help teams make data-driven decisions in biologic programs with novel modalities and minimize resources. However, these approaches also need to be adapted to accurately simulate the behavior of these constructs. Therefore, before they can be implemented in standard workflows, they must be benchmarked against appropriate experimental information.

In this work, MOE (Molecular Operating Environment) was used to predict potential liabilities in a fusion protein and compared with experimental results. This novel methodology required that each of the construct subunits were simulated separately. To achieve this, a homology model was built and molecular attributes, post-translational modifications (PTM) and intermolecular interactions were evaluated under several buffer and stress conditions. The results of these computational studies were then used for comparison with experimental findings. The data package included outputs from several characterization techniques that evaluated aggregation, PTM generation, colloidal, physical, and chemical stability. This comprehensive analysis allowed the benchmarking of identified molecule liabilities, highlighted trends, and confirmed the accuracy and value of the predictive modeling. The use of these modeling tools for assessing drug candidates based on novel modalities will optimize the selection of experimental conditions and improve resource allocation to swiftly identify critical liabilities and enable molecule progression.

164ag Overexpression of Native Gene Presumably Encoding for D-Xylose Reductase in *Escherichia coli* leads to High Xylitol Production

Angelo Banares, Grace Nisola, Won-Keun Lee, Wook-Jin Chung

Xylitol has a high global demand due to its wide-range application primarily in the food and pharmaceutical industries. Xylitol can also be utilized as a building block molecule for high-value chemical production. Currently, industrial xylitol production is through chemical synthesis involving catalytic hydrogenation of xylose. This process is expensive, requiring high energy expenditure and a series of purification steps thus affecting the market value of xylitol. Hence, biotechnological xylitol production emerges as a promising alternative to chemical synthesis. Bio-production of xylitol from D-xylose heavily relies on expression of heterologous xylose reductase (XR) that predominantly originates from yeasts, and some from filamentous fungi and bacteria into the host organisms. So far, there is no known native XR in *Escherichia coli* that catalyzes D-xylose to xylitol. Thus, it requires expression of heterologous XR from other organisms to produce xylitol. Surprisingly, plasmid-based overexpression of endogenous *yahK* gene in *E. coli* W3110 devoid of D-xylose isomerase pathway resulted in the production of xylitol as confirmed by GC-MSD and HPLC analysis. Using the said strain, high xylitol titer of 20.77 g/L (83% of theoretical yield) was produced via large-scale fermentation from D-xylose with glycerol as a co-substrate. Subsequently, an even higher yield of 30 g/L xylitol (100% of theoretical yield) was attained through whole-cell biocatalysis. So far, this study is the first to demonstrate xylitol production in *E. coli* using its native protein (YahK) suggesting the possibility of the presence of a silent xylose-oxidoreductase pathway (XORP) in *E. coli.* This work was supported by NRF funded by The Ministry of Science and ICT (2021R1F1A1045612), Basic Science Research Program through the Ministry of Education (2020R1A6A1A03038817).

164ah Leveraging Propionate-Induced Growth Inhibition in *Corynebacterium Glutamicum* to Evolve Improved Methylmalonyl-CoA-Dependent Polyketide Synthases

Chunjun Zhan, Robert W. Haushalter, Jay Keasling

Corynebacterium glutamicum is a promising yet largely unexplored host for the production of polyketide products. Here we evaluate the feasibility of using *C. glutamicum* as a host for engineered polyketide production. We introduced module-based strategies aimed at decreasing carbon loss and increasing the intracellular concentration of methylmalonyl-coenzyme A (a key precursor for polyketide production) to express heterologous polyketide biosynthetic pathways that use methylmalonyl-coA as a substrate. Based on the elucidation of the mechanism of propionate inhibition of growth, growth-coupled evolution strategy was introduced to rescue cell growth and evolve polyketide synthase in *C. glutamicum*. Results showed that introduction methylmalonyl-CoA dependent polyketide production pathway would relieve this propionate-induced growth inhibition, opening the door to leverage this fitness advantage for selection of evolved polyketide synthase. These experiments demonstrate that we can engineer this industrial host to express PKS enzymes heterologously and use them to condense methylmalonyl-coA into more complex products.

164ai Metabolic Engineering of Four Host Microbes for the Production of Green Leaf Volatiles and Precursor Molecules

Qingyun Dan, Jing Huang, Jay Keasling

Green leaf volatiles (GLVs) are six carbon aldehydes, alcohols and esters that are released by plant upon physical damage, which are vital to plant defense mechanism. They also serve as plant hormones. Due to their strong grassy odor, they are important aroma compounds in flavors and fragrances. To this date, no microbial production of such compounds using glucose as the carbon source have been reported. We have successfully engineered *Saccharomyces cerevisiae* and *Yarrowia lipolytica* to produce mg/L level of trans-2-hexenal (leaf aldehyde) and cis-3-hexenol (leaf alcohol), via genetic integration of plant GLV biosynthetic pathway genes and other auxiliary genes. We have also investigated the GLV degradation pathway in *Yarrowia lipolytica* and identified candidate enzymes. Furthermore, we will address our current progress on producing GLVs and their precursor molecules, hexenoic acids, in *Streptomyces albus* and *Streptomyces coelicolor* by heterologous expression of type I modular polyketide synthases (PKSs). This de novo approach utilizes a rationally designed chimeric PKS that involves an efficient butyryl-CoA/ethylmalonyl-CoA loading module, a FabA-type dehydratase from *Shawenella*, and a hypothetical terminal reductase domain from *Dictyostelium*.

164aj Introduction of Abiological Reactions into Biosynthesis

Jing Huang, Aindrila Mukhopadhyay, Jay Keasling

Reactions in chemistry are much more abundant and diverse than in biology. If those reactions in chemistry can be brought into microorganisms, it will broaden the types of compounds that can be biosynthesized inside the cell and offer the possibility for diversifying valuable natural compounds in vivo, such as medicine precursors. Carbene-mediated reactions have wide applications in chemical synthesis, but occur rarely in biology. Here we report the introduction of a carbene-mediated reaction with a biosynthesized carbene precursor into a microbe and integration of it with a natural biosynthesis pathway to generate new compounds. The biosynthesis pathway for the carbene precursor was identified, cloned and heterologously expressed to produce the precursor. Natural P450s were screened and improved by directed evolution to catalyze the reaction between the biosynthesized carbene precursor and the other substrate, which is also biosynthesized by introducing the corresponding natural pathway, with moderate yield and excellent diastereoselectivity. In this case, all the components involved in the abiological reaction, the carbene precursor, the other substrate and the enzyme catalyzing the reaction, are biosynthesized. This work sets a platform for introducing more abiological carbene mediated or related reactions into microbes for bio-manufacturing and contributes to green chemistry.

164ak Functional Characterization of paclitaxel Transporters in *Taxus Cuspidata* to Enhance Secretion

Md Tahsin Rahi, Cassandra Brzycki Newton, Alexandra Harrison, Lauren Revene, Susan Roberts

Paclitaxel is a diterpenoid secondary metabolite naturally produced in *Taxus* species. The inherent medicinal activity of this drug has already been recognized and it has been widely used to treat various types of cancer. The advancement of plant cell culture technology has enabled industries to commercially produce paclitaxel to meet global demand. However, limited product storage capacity and the possibility of product degradation within plant cells limits overall yields. Manipulation of the paclitaxel biosynthetic and related pathways can be used to increase production. However, consideration of both transport and degradation can lead to superior strategies for yield improvements. Although the ABC transporter family is commonly responsible for exporting various plant secondary metabolites across the plasma membrane, no paclitaxel-specific transporters have yet been identified and characterized. In this study, putative paclitaxel transporter genes were identified by screening transcriptomic data for genes co-regulated with taxane biosynthesis and further comparing the homology of those genes with ABC transporters in other plants using BLASTP. From these data, we identified four uncharacterized proteins that display 70-80% homology for ABC-B or ABC-G transporters in related plant species. To characterize functional activity of these putative transporters, a knockout strain of *S. cerevisiae*, (AD1-8) which lacks all the essential ABC transporters in its genome was transformed with putative transporter genes following PEG-lithium-acetate based yeast transformation. The transgenic AD1-8 strain was then studied through the application of exogenous paclitaxel over time where extracellular and cell-associated paclitaxel levels were compared with that of the wild type (control) strain. A significant reduction in cell-associated paclitaxel concentration was observed in AD 1-8 expressing two of the four transporter genes, suggesting paclitaxel-specific transport activity of those putative genes. Efforts are now underway to overexpress relevant transporter genes in *T. cuspidata* cell lines to improve paclitaxel synthesis in culture through this novel strategy.

164al Characterizing a cAMP/Crp Mediated Metabolic Futile Cycle in Bacterial Persisters

Hanny Ngo, Sayed Golam Mohiuddin, Mehmet Orman

Persister cells are a small subpopulation of phenotypic variants in a bacterial cell population that can temporarily tolerate a high concentration of antibiotics (1). Studying these phenotypes is essential because they are known to cause recurrent infections (2, 3) and serve as a reservoir for the emergence of drug-resistant mutants. Here, we focus on type I persisters, which are non-growing, multi-drug-tolerant cells; are majorly formed via passage through the stationary phase (1); and have a high level of redox activities maintained by RNA and endogenous protein degradation (4). Our approach attempts to describe the metabolic mechanism that facilitates the formation of type I persisters through the CRP/cAMP-mediated metabolic futile cycle in the stationary phase. Understanding the persister metabolism will enable the development of novel antipersister therapeutics.

In this study, the deletion of *crp* (encoding a global transcriptional regulatory protein) and *cyaA* (encoding adenylate cyclase that synthesizes cAMP) in *Escherichia coli* reduce type I persister levels significantly compared to wild-type parental strain (Fig. 1A-B). Using an inducible green fluorescent protein (GFP) expression system and protein dilution method, we demonstrated that the *crp* and *cyaA* deletions reduce the non-growing cell formation during the stationary phase (Fig. 1C). CRP/cAMP is the DNA-binding transcriptional dual regulator, regulating many genes associated with the central metabolism, so we also performed untargeted mass spectrometry analysis to quantify the metabolites in the wild-type and the mutant *crp* strain in the stationary phase. The results revealed that metabolites associated with energy metabolism are significantly down regulated in the mutant strain compared to the wild-type (Fig. 1D). When we knocked out some of the genes associated with energy metabolism, we found deletions of these genes also reduce the persistence in *E. coli*. Overall, with the data presented here, we expect to generate a comprehensive map of the metabolic mechanisms facilitating type I persister cell formation in *E. coli*.

References

1. Balaban, N.Q., Merrin, J., Chait, R., Kowalik, L. and Leibler, S., 2004. Bacterial persistence as a phenotypic switch. Science, 305 (5690), pp.1622-1625.
2. Lewis, K. (2007). Persister cells, dormancy and infectious disease. Nature Reviews Microbiology, 5(1), 48-56.
3. Lewis, K. (2010). Persister cells. Annual review of microbiology, 64, 357-372.
4. Orman, M.A. and Brynildsen, M.P., 2015. Inhibition of stationary phase respiration impairs persister formation in E. coli. Nature communications, 6, p.7983.

Fig. 1: CRP/cAMP-mediated metabolic futile cycle facilitates persister formation in stationary-phase in *E. coli* (A-B) Persister levels (log-scale) of wild-type (WT) and mutant strains. Late stationary phase cells (i.e., cell populations cultured for 24h) were transferred to fresh LB and treated with ampicillin and ofloxacin to enumerate persisters, as measured by colony-forming units (CFUs). WT persister levels at the late stationary phase were significantly higher than those of mutant strains (N=4). **(C)** Non-growing cell levels of WT and mutant strains. GFP-positive late stationary-phase cells (harboring an inducible GFP expression system, pQE-80L-*gfp*) were transferred to fresh media (without inducer) to monitor their proliferation by a fluorescent protein-dilution method. Non-growing cells that did not resuscitate retained their high GFP levels (t=2.5h). A representative biological replicate was shown here. All 3 biological replicates consistently resulted in similar trends. **(D)** Metabolomics data of the WT and the mutant strain in the late stationary-phase. Red and blue dots represent the metabolites that are significantly upregulated or downregulated in the mutant strain compared to WT, respectively (P-value< 0.05, ANOVA test).

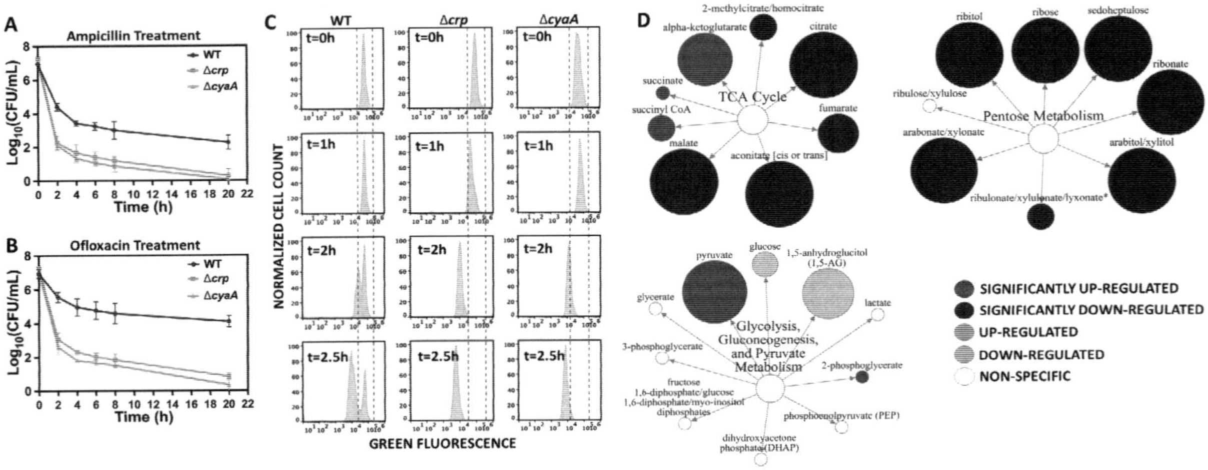

A
Ampicillin Treatment
WT
Δcrp
ΔcyaA
Log10(CFU/mL)
Time (h)
B
Ofloxacin Treatment
WT
Δcrp
ΔcyaA
Log10(CFU/mL)
Time (h)
C
WT
Δcrp
ΔcyaA
t=0h
t=1h
t=2h
t=2.5h
NORMALIZED CELL COUNT
GREEN FLUORESCENCE
D
TCA Cycle
2-methylcitrate/homocitrate
alpha-ketoglutarate
citrate
succinate
succinyl CoA
fumarate
malate
aconitate [cis or trans]
Pentose Metabolism
ribitol
ribose
sedoheptulose
ribonate
ribulose/xylulose
arabonate/xylonate
arabitol/xylitol
ribulonate/xylulonate/lyxonate*
Glycolysis, Gluconeogenesis, and Pyruvate Metabolism
pyruvate
glucose
1,5-anhydroglucitol (1,5-AG)
glycerate
lactate
3-phosphoglycerate
2-phosphoglycerate
fructose 1,6-diphosphate/glucose 1,6-diphosphate/myo-inositol diphosphates
phosphoenolpyruvate (PEP)
dihydroxyacetone phosphate (DHAP)
SIGNIFICANTLY UP-REGULATED
SIGNIFICANTLY DOWN-REGULATED
UP-REGULATED
DOWN-REGULATED
NON-SPECIFIC

164am Environment Constrains Fitness Advantages of Division of Labor in Microbial Consortia Engineered for Metabolite Push or Pull Interactions

Ashley Beck, Alissa Bleem, Martina Du, William Harcombe, Hans C. Bernstein, Jeff Heys, Tomas Gedeon, Ross P. Carlson

Fitness benefits from division of labor in microbial consortia are well documented, but the dependency of the benefits on environmental context is poorly understood. Two synthetic *Escherichia coli* consortia were built to test the relationships between exchanged organic acid, local environment, and opportunity costs of different metabolic strategies. Opportunity costs quantify benefits not realized due to selecting one phenotype over another. The consortia catabolized glucose and exchanged either acetic or lactic acid to create producer-consumer food webs. The organic acids had different inhibitory properties and different opportunity costs associated with their positions in central metabolism. The exchanged metabolites modulated different consortial dynamics. The acetic acid-exchanging (AAE) consortium had a 'push' interaction motif where acetic acid was secreted faster by the producer than the consumer imported it, while the lactic acid-exchanging (LAE) consortium had a 'pull' interaction motif where the consumer imported lactic acid at a comparable rate to its production. The LAE consortium outperformed wild type (WT) batch cultures under the environmental context of weakly buffered conditions, achieving a 55% increase in biomass titer, a 51% increase in biomass per proton yield, an 86% increase in substrate conversion, and the complete elimination of byproduct accumulation all relative to the WT. However, the LAE consortium had the tradeoff of a 42% lower specific growth rate. The AAE consortium did not outperform the WT in any considered performance metric. Performance advantages of the LAE consortium were sensitive to environment; increasing the medium buffering capacity negated the performance advantages compared to WT.

164an Engineering *Y. Lipolytica* for the Biosynthesis of Geraniol

Ayushi Agrawal, Zhiliang Yang, Mark Blenner

Geraniol is a monoterpene with wide applications in food, cosmetics, and pharmaceutical industries. Moreover, geraniol is the primary precursor of strictosidine which can be produced by incorporating a 12-step pathway in a geraniol producing strain. Strictosidine serves as the central intermediate for 3,000 monoterpene indole alkaloids with anticancer and other pharmacological properties. Traditionally, the monoterpenes are produced in their native plant sources in minute amounts. Microbial production of monoterpenes provides an alternative to traditional plant-based methods which require laborious extraction and separation methods and result in overharvesting of medicinal plants. Hence, efforts have been focused on metabolic engineering of microbes like *E. coli* and *S. cerevisiae* for the biosynthesis of natural products. However, *E. coli* lacks inner organelles that are present in yeast and cannot perform post-translational modifications making it an unsuitable host. Additionally, *S. cerevisiae* has low natural flux towards natural product synthesis making it an undesirable host. Alternatively, *Yarrowia lipolytica* can serve as an ideal host for natural product biosynthesis because of the availability of genetic engineering and synthetic biology tools.

This project aims to produce geraniol in metabolically engineered *Yarrowia lipolytica*. First, two plant-derived geraniol synthases (GES) from *Catharanthus roseus* (Cr) and *Valeriana officinalis* (Vo) were screened based on literature. Both wild type and truncated mutants of GES (without signal peptide targeting chloroplast) were examined by transforming yeast. It was found that the truncated CrGES produced the most geraniol and thus was used for further experimentation. The starting strain was obtained by overexpression of the truncated HMG1, IDI and truncated CrGES. Enhancing the precursor pool of acetyl-CoA was performed by overexpression of the MVA pathway genes such as ERG10, HMGS or MVK, PMK. Effect of copy number of truncated CrGES was also investigated. The final strain overexpressing 3 copies of truncated CrGES and ERG10, HMGS, tHMG1, IDI was able to produce around 700 mg/L of geraniol in shake flask cultivation. The optimization of fermentation conditions (carbon to nitrogen ratio) led the titer to increase around 1 g/L in shake flask cultivation. The bioreactor experiment is underway to boost the geraniol titers. The future work includes integrating the downstream pathway genes from geraniol to strictosidine to engineer a strictosidine producing strain.

164ao Understanding Strain Instability in Yarrowia Lipolytica

Philip Gitman, Alyssa M. Worland, Yinjie Tang, Mark Blenner

The model oleaginous yeast species, *Yarrowia lipolytica,* is a promising biomanufacturing chassis well suited for production of oleochemicals, terpenoids, and other valuable chemicals of the carotenoid family. Scaleup of these processes, however, has been marred by phenotypic changes, such as loss of titer (or titer instability). The titer instability problem is exemplified by an engineered β-carotene producing *Y. lipolytica* strain that was developed by enhancing flux from acetyl-CoA to terpene precursors and overexpressing *carB* and *carRP* enzymes. This engineered strain produced ~4 g/L β-carotene titer using benchtop bioreactors. However, when moving this strain to a larger bioreactor, cell performance significantly dropped during 60 generations of cell growth. We hypothesized that loss of productivity is due to accumulation of mutations in genes for β-carotene production during prolonged generations of culture growth. To test this hypothesis, we collected samples from shake flask cultures under optimal continuous fermentation conditions (50 mL working volume) and performed Single Molecule, Real-Time (SMRT) HiFi resequencing using PacBio's Sequel II system. The mutation rate and locations of the β-carotene producing strain were determined at 2 different time points in culture growth, associated with 50% (generation 30) and 0% (generation 60) β-carotene titer, respectively. Similarly, RNAseq, targeted metabolomics, and targeted proteomics of the time course samples will be used to identify correlations between gene expression rate and mutation rate/locations, and key genes or enzymes of frequent mutations causing production loss or metabolic shifts. The goal of this work is to investigate the nature of titer instability in this β-carotene strain as a model to understand more broadly the factors that lead to cellular heterogeneity during cell line development and scaleup.

164aq Effect of Alternative Sigma Factors on Metabolic Activity in *E. coli*

Ryan Armstrong, Elizabeth Kornreich, Michael Anderson, Jason Boock

Microbial bioproduction offers several advantages compared to conventional chemical synthesis including use of renewable substrates, catalysis by highly specific enzymes, and processing under labile conditions. To be efficient, bioproduction requires a balance of energy between production of the desired metabolite and general maintenance for cellular growth and function. We hypothesize that alternative sigma factors, which are naturally used by cells to shift metabolic state, can be used similarly to direct expression and resources towards heterologous pathways. Sigma factors control the start of transcription by binding to RNA polymerase (RNAP) and subsequent promoter interaction. Competition for RNAP is thought to occur when multiple sigma factors are present, permitting rerouting of transcriptional resources. We are investigating this competition through controlled and induced expression of non-native, alternative sigma factors B, F, and G from *B. subtilis* in *E. coli*. We have shown that these sigma factors are functional in *E. coli*, directing expression of a GFP reporter when paired with their cognate promoter. Upon sigma factor production, we observe a decrease in cellular growth rate which we believe is due to sigma factor competition for RNAP resources. We are currently confirming our hypothesis using qRT-PCR to show the downregulation of native transcription when non-native sigma factors are present as well as apply the sigma factors to direct expression of a heterologous pathway.

164ar Developing a Novel Microbial Host and Synthetic Biology Tools for Valorizing Waste Polyethylene Terephthalate and Lignin-Derived Compounds

Jinjin Diao, Yifeng Hu, Tae Seok Moon

Polyethylene terephthalate (PET) represents 8% (by weight) of global solid waste. PET chemical recycling has been an option to solve this global problem, but it suffers from its relatively high process cost and the extremely low price of virgin PET. One solution is to upcycle waste PET rather than recycle it to generate the same PET typically with low quality. PET upcycling can be achieved by depolymerizing PET into terephthalic acid (TPA) and ethylene glycol (EG) and biologically converting these monomers into value-added products. However, there are only a handful of reports demonstrating microbial strains capable of growing on both TPA and EG generated from PET as sole carbon sources. To overcome this critical challenge, we have performed strain screening to discover a *Rhodococcus* strain (named RPET) that can grow well on the alkaline hydrolysis products of PET as the sole carbon source without any purification step. Notably, this strain was able to tolerate and grow on a mixture of TPA and EG at extremely high concentrations (up to 0.3M each, total 0.6M) and high osmolarity resulting from alkaline hydrolysis and pH neutralization. The resultant pH neutralized media supported RPET's growth (up to 0.4 g dry cell weight per g PET) without any purification and sterilization step except for their dilution to make up to 0.6M of monomer concentrations.

Adipic acid, a monomer for nylon production, is currently produced from petroleum derivatives, requiring an alternative process for its sustainable "green" production. Muconate can be converted into various chemicals, including adipic acid. Using non-model organisms, multiple labs have demonstrated muconate production from lignin-derived aromatic compounds, with glucose used as a growth substrate. *Rhodococcus opacus* is well suited for valorizing lignin [1-8], but developing this promising chassis had been challenging due to limited genetic engineering tools. To address this issue, we have developed various synthetic biology tools [9-12], including a gene repression system based on CRISPR interference (CRISPRi) and a knockout method. Notably, many synthetic biology tools, developed for *Rhodococcus opacus* [12], were functional in different related species and strains such as RPET.

In this presentation, we will discuss our CRISPRi tool's utility for waste valorization. We have developed and optimized the CRISPRi system, which uses a T7 RNA polymerase system to express a small guide RNA, demonstrating the maximum repression efficiency up to 85% [13]. We also provide a cloning strategy that enables constructing multiple CRISPRi plasmids without any PCR step, facilitating this GC-rich organism's engineering. Using the optimized CRISPRi system, we confirmed the annotated roles of four metabolic pathway genes related to the consumption of benzoate, vanillate, catechol, and acetate. Additionally, we showed our tool's utility by demonstrating the inducible accumulation of muconate. While the maximum muconate yield obtained using our tool was 30% of the yield obtained using gene knockout, our tool showed its inducibility and partial repressibility. Finally, we will discuss our effort for PET conversion into carotenoids and muconate as two demonstration products [14]. This work represents the promise for valorizing waste PET and lignin-derived compounds using *Rhodococcus*.

[1] Chatterjee, A., DeLorenzo, D. M., Carr, R., Moon, T. S., Bioconversion of renewable feedstocks by Rhodococcus opacus. Current opinion in biotechnology 2020, 64, 10-16. [2] Davis, K., Moon, T. S., Tailoring microbes to upgrade lignin. Current opinion in chemical biology 2020, 59, 23-29. [3] Henson, W. R., Campbell, T., DeLorenzo, D. M., Gao, Y., et al., Multi-omic elucidation of aromatic catabolism in adaptively evolved Rhodococcus opacus. Metabolic engineering 2018, 49, 69-83. [4] Yoneda, A., Henson, W. R., Goldner, N. K., Park, K., et al., Comparative transcriptomics elucidates adaptive phenol tolerance and utilization in lipid-accumulating Rhodococcus opacus PD630. Nucleic Acids Res. 2016, 44, 2240–2254. [5] Henson, W. R., Hsu, F.-F., Dantas, G., Moon, T. S., Foston, M., Lipid metabolism of phenol-tolerant Rhodococcus opacus strains for lignin bioconversion. Biotechnology for biofuels 2018, 11, 339. [6] Roell, G. W., Carr, R. R., Campbell, T., Shang, Z., et al., A concerted systems biology analysis of phenol metabolism in Rhodococcus opacus PD630. Metabolic engineering 2019, 55, 120-130. [7] Hollinshead, W. D., Henson, W. R., Abernathy, M., Moon, T. S., Tang, Y. J., Rapid metabolic analysis of

Rhodococcus opacus PD630 via parallel C-metabolite fingerprinting. Biotechnology and bioengineering 2016, 113, 91-100. [8] Anthony, W. E., Carr, R. R., DeLorenzo, D. M., Campbell, T. P., et al., Development of Rhodococcus opacus as a chassis for lignin valorization and bioproduction of high-value compounds. Biotechnology for biofuels 2019, 12, 192. [9] DeLorenzo, D. M., Henson, W. R., Moon, T. S., Development of Chemical and Metabolite Sensors for Rhodococcus opacus PD630. ACS Synthetic Biology 2017, 6, 1973–1978. [10] DeLorenzo, D. M., Moon, T. S., Selection of stable reference genes for RT-qPCR in Rhodococcus opacus PD630. Scientific reports 2018, 8, 6019. [11] DeLorenzo, D. M., Moon, T. S., Construction of Genetic Logic Gates Based on the T7 RNA Polymerase Expression System in Rhodococcus opacus PD630. ACS Synthetic Biology 2019, 8, 1921-1930. [12] DeLorenzo, D. M., Rottinghaus, A. G., Henson, W. R., Moon, T. S., Molecular Toolkit for Gene Expression Control and Genome Modification in Rhodococcus opacus PD630. ACS Synthetic Biology 2018, 7, 727-738. [13] DeLorenzo, D.M., Diao, J., Carr, R., Hu, Y., Moon, T.S., An Improved CRISPR Interference Tool to Engineer Rhodococcus opacus. ACS Synthetic Biology. doi.org/10.1021/acssynbio.0c00591. [14] Moon, T.S. et al. U.S. Provisional Patent Application

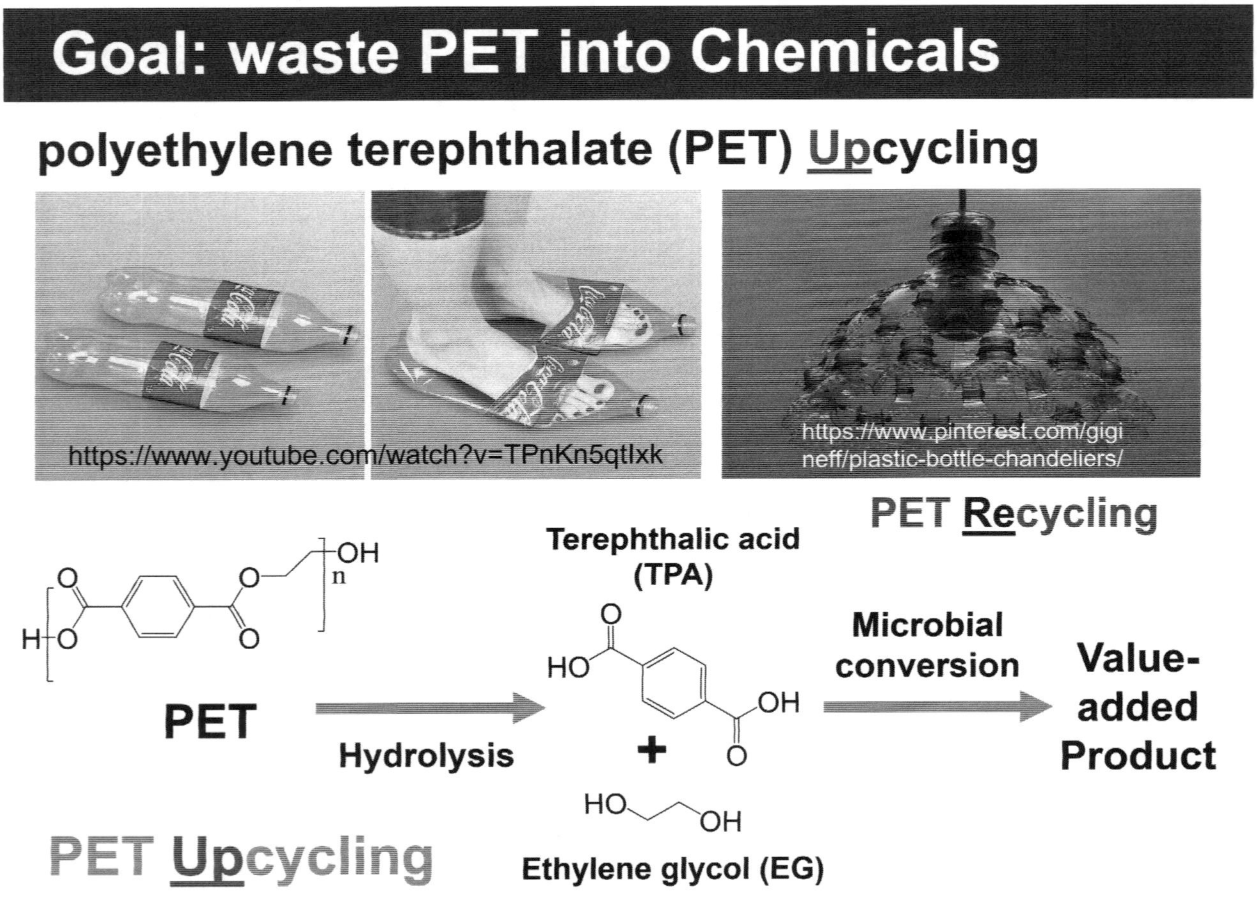

164as Plastics Degradation By the Yellow Mealworm Gut Microbiota and Associated Enzyme Studies

Ross Klauer, Lummy Monteiro, Jyoti Singh, Kevin Solomon, Mark Blenner

Plastics waste is accumulation is a worsening problem due to global reliance on low cost and high durability plastics. Over 7 billion tons of plastics waste have accumulated since 1950. Roughly nine percent of plastics are recycled globally, with economic incentive being a key inhibitor to increasing recycling throughput. Current plastics waste management techniques (incineration, mechanical recycling, and chemical recycling) lack economic viability due to the need for expensive catalysts, high operating conditions, pretreatment, and mechanical sorting. Biodegradation presents a sustainable and economic upcycling solution by eliminating the mentioned cost barriers. The microbiota of insect larvae, namely the yellow mealworm (*T. molitor*), have recently been identified as a platform for plastics biodegradation by demonstrating improved plastics consumption rate over reported microorganisms, which insinuates microbial community synergies.

Only a small fraction of the total number of microorganisms from the guts of plastic-fed mealworms have been isolated and shown to degrade plastics. There exists a knowledge gap and an opportunity to discover plastics-degrading organisms and associated enzymes from the yellow mealworm gut microbiome. In this work, we aim to address this knowledge gap by isolating plastic-degrading organisms from yellow mealworms, by using bioinformatic analyses to determine microbial species and genes prevalent in plastic-fed mealworms, and by doing early work in extracting plastic-degrading enzymes.

Isolates were extracted from dissected mealworm guts and screened for growth with polyethylene as the sole carbon source in order to discover microbes capable of degrading plastics. A subset of unique isolates was discovered, and plastics degradation is indicated by growth on plastic-containing media and scanning electron microscopy (SEM). Metagenomic data provided by the Joint Genome Institute (JGI) from the gut microbiota of worms fed Polystyrene (PS), Low-Density and High-Density Polyethylene (LDPE, HDPE), and Polypropylene (PP) were analyzed to determine microbial taxa and associated genes prevalent to the degradation of each plastic.

Early-stage enzyme studies of supernatant and lysis solutions from microbial isolate cultures were performed on fluorescent LDPE particles to demonstrate in vitro activity. Shotgun proteomics coupled with mass spectrometry and fast protein liquid chromatography (FLPC) were used to identify proteins important for plastics degradation. The results from these studies indicate the presence of plastics degrading enzymes from the plastics-degrading isolates.

The findings detailed in this work will allow for further studies elucidating the degradation pathways of plastics-degrading organisms. Discovering these pathways will allow for the isolation and purification of key enzymes in plastics degradation processes and provide the potential to engineer these pathways for enhanced plastics degradation and upcycling. In parallel, synthetic microbial consortia or synthetic enzyme cascades can be developed in order to mirror the plastics degradation present in the gut of *T. molitor* outside of the worm. The work in these studies and subsequent future studies can have the potential to be scaled for large-scale plastics upcycling in a bioreactor with conditions emulating and improving upon those present in the gut of the yellow mealworm.

164au Integrating Metabolomics and Fluxomics to Study Cancer Metabolism in Low Glucose Environments

Aliya Lakhani

Metabolomics and fluxomics are integral tools for quantitative metabolic analysis. By combining mass spectrometry, isotope tracing, and mathematical modeling, we can quantify metabolite concentrations and fluxes in cancer cells. Using these tools, we investigated the coordination of bioenergetics, biosynthesis, and redox metabolism of cancer cells in the tumor microenvironment. We hypothesized that the tumor microenvironment plays a significant role in dictating cancer cell survival and proliferation. To shed light on how cancer cells use the tumor microenvironment, we assessed their adaptability to high levels of lactate. While cancer cells did not have a growth advantage over healthy cells in the high-lactate environment, we found that an intermittent supply of glucose allowed cancer cells to proliferate faster than healthy cells. By tracing stable isotopes from lactate and glutamine across metabolism, we elucidated the role of lactate in cancer cell survival via gluconeogenesis and energy metabolism.

164av Human Liver Diverticulum Triggers Liver Organogenesis

Daniel Guiggey, Ogechi Ogoke, Natesh Parashurama

Many patients suffer from chronic hepatocellular injury, leading to fibrosis and cirrhosis, which ultimately overwhelms the adult liver's regenerative capacities. Unfortunately, these chronic diseases are escalating globally, in both developing and developed countries, and currently affect more than 800 million people worldwide. The only accepted treatment is orthotopic liver transplantation, but there are several potential therapeutic solutions within the field of liver regenerative medicine. Of these, liver organogenesis has great potential, and has evolved from original studies in mice. Liver organogenesis begins with the liver diverticulum (LD), an out-pocketing of gut that is surrounded by mesoderm (MES) tissue. The role of the LD in liver organogenesis has been largely ignored in current mouse and human studies. We hypothesize that the LD and surrounding MES form a functional tissue complex (LD-MESC) that triggers organogenesis. To elucidate these triggers and better understand the dynamics of global transcriptomic changes in early liver organogenesis, we performed an in-depth bioinformatics analysis, which demonstrated that the murine LD-MESC integrates diverse signals to drive transcription and migration. Our analysis utilized single-cell transcriptomic data from fetal mice, which allowed us to identify and compare several unique cell types. We focused primarily on the transition from definite endoderm (DE, E75), gut tube endoderm (GT, E8.5), migrating hepatoblasts (MH, 9.5), hepatoblasts (HB, E10.5), and hepatomesenchyme (HM, E10.5). After performing quality control and normalization, we were able to analyze the change in gene expression throughout this time period. The data showed several thousand differentially expressed genes between the different cell types, particularly between MH and GT and HB (2102 differentially expressed genes). These differentially expressed gene lists were then able to be further analyzed with genomics tools from DAVID, REACTOME, ENRICHR, as well as our own scoring system. The several analyses used confirmed the significant enrichment of migration gene sets in the MH population, including cell migration (FDR=0.0032), cell adhesion (FDR=6.1E-19), and covalent chromatin modification (FDR=4E-20). Interestingly, we found significant enrichment of important signaling pathways in the MH compared to GT and HB, including Hippo (FDR=0.0039), FoxO (FDR=0.0092), pluripotency signaling (FDR=0.0073), Wnt (FDR=0.051), and TGF-beta (FDR=0.11), indicating a possible role of signaling pathways in migration. We further found significant changes in several metabolic pathways throughout differentiation. Oxidative phosphorylation was found to be a significant downregulated (FDR=1.4E-35) pathway compared to both GT and HB. This same result was found to be significant ($p<0.001$) when utilizing our scoring system, which considered all genes within the pathway. Many of these same gene dynamics were also observed and confirmed by analyzing additional transcriptomic data sets. A similar cell type at E9.5 was found in the Mu et al. dataset with both migration gene enrichment and signaling pathway enrichment. A few of these gene sets include, cell migration (FDR=0.0005), cell adhesion (1.7E-10), hippo signaling (FDR=3.9E-6), and wnt signaling (FDR=0.00076). Taken together, these studies elucidate details of genes expressed during differentiation and a role of signaling from the LD-MESC in triggering early liver organogenesis. These results will have broad implications in liver regenerative medicine.

164aw Spatiotemporal Imaging and Analysis of Mouse and Human Liver Bud Morphogenesis

Daniel Guiggey, Ogechi Ogoke, Natesh Parashurama

The process of liver organogenesis has thus far served as a paradigm for organ formation. However, there still remains a lack of understanding regarding early mouse and human liver bud morphogenesis and early liver volumetric growth. Elucidating dynamic changes in liver volumes is critical for understanding organ development, implementing toxicological studies, and for modeling hPSC-derived liver organoid growth. New visualization, analysis, and experimental techniques are desperately needed. Here, we combine observational data with digital resources, new 3D imaging approaches, retrospective analysis of liver volume data, mathematical modeling, and experiments with hPSC-derived liver organoids. Image slices were downloaded from the 3D Atlas of Human Embryology, eMouseAtlas, and Toronto Centre for Chemogenomic Mouse Imaging Centre. ImageJ was used to trace the liver bud, the STM, and the gut tube for each aligned section, and these traces were combined with the 3D Viewer plugin to form a 3D model of the liver. Using images from fetal mouse and human from several imaging modalities (MRI, micro-CT, optical projection, Handamp;E stain), we were able to demonstrate distinct spatial features and complete full growth curves, which we modeled using a Gompertz curve. Hepatic cords were able to be visualized at E9.0-E9.5 in mouse and day 25 in human as well as evidence of mesenchymal-liver epithelial interactions. The growth curve models were able to be further analyzed and verified with liver weight, cell count data, as well as partial hepatectomy growth. There was an overall 8.74×10^4-fold-increase in volume between E8.5 to E18. However, 7.5×10^3 of this fold change occurs just between E9 and E14. This growth is much more exponential when compared to regeneration growth. Interestingly, the cell count data increased at a similar rate as volumetric growth from E10.5 onward, indicating that proliferation is the main source of growth rather than hypertrophy. Visualization of liver-epithelial and septum transversum mesenchyme (STM) interactions suggests extended interactions, which together with new spatial features, may be responsible for extensive exponential growth. These STM interactions are modeled with a novel in vitro human pluripotent stem cell (hPSC)-derived hepatic organoid system that exhibits cell migration. Our methods enhance our understanding of liver organogenesis, with new 3D visualization, analysis, mathematical modeling, and in vitro models with hPSC. Our approach highlights mouse and human differences and provides potential hypothesis for further investigation in vitro and in vivo.

164ax FOXA1/2 Depletion Drives Global Reprogramming of Differentiation State and Metabolism in a Human Hepatic Cell Line and Human Stem Cell-Derived Hepatic Progenitor Cells

Iyan Warren, Daniel Guiggey, Natesh Parashurama

FOXA factors are critical members of the developmental gene regulatory network (GRN) composed of master transcription factors (TF) that regulates murine cell fate and metabolism in the gut and liver. Nonetheless, how FOXA factors individually and collectively (through changes in interconnected GRN TF) dictate cell fate during human gut and liver differentiation and simultaneously regulate metabolic pathways, is poorly understood. If these molecular mechanisms can be elucidated, FOXA factors could be targeted for human liver diseases, which exhibit altered metabolism and differentiation, like cancer and fibrosis, and for controlling key gene regulatory pathways during human stem cell differentiation, maturation, and organogenesis. In this study, we aimed to determine the role of FOXA2 (and FOXA1 which is believed to compensate for FOXA2) in the regulation of hepatic differentiation and cell metabolism in a human hepatic cell line and during human stem cell differentiation. We employed (siRNA and shRNA), qRT-PCR, Western Blot, RNA-seq and bioinformatics, metabolic pathway maps, and functional metabolic analysis (Seahorse XF Analyzer, lipid accumulation) to determine the effects of FOXA1/2 on liver differentiation and metabolism. Depletion of both FOXA1 and FOXA2 via RNAi in human hepatic (HepG2) cells and during hepatic differentiation significantly downregulated albumin ($p < 0.05$) and GRN TF gene expression (HNF4A, HEX, HNF1B, TBX3) ($p<0.05$) and significantly upregulated endoderm/gut/hepatic endoderm markers (goosecoid (GSC), FOXA3, and GATA4), gut TF (CDX2), pluripotent TF (NANOG), and neuroectodermal TF (PAX6)($p<0.05$), all consistent with a partial reprogramming of cell fate towards earlier and alternate lineages. Conditional, dox-sensitive shFOXA1/2 lines resulted in similar findings and demonstrated evidence of reversibility. RNA-seq and bioinformatic analysis of shFOXA1/2 knockdown in human liver (HepG2) cell line including principal component analysis, volcano plot, and heatmap analysis demonstrated distinct control (scrambled-shRNA) and shFOXA1/2 knockdown cell populations, and 235 significant downregulated genes and 448 upregulated genes. Gene set enrichment analysis employing normalized enrichment scores and false discovery rate cut offs demonstrated hundreds of downregulated genes associated with differentiation, and surprisingly, several hundreds of genes upregulated that are associated with alternate germ layers lineages (cardiac, endothelial, muscle) and neurectoderm (eye, neural) (FDR-qval<0.1). Interestingly, the same bioinformatic analysis approach demonstrated downregulation of genes associated with mitochondria, and hundreds of genes associated with metabolic reactions which were visualized with metabolic pathway maps, including widespread downregulation of glycolysis, citric acid cycle, mitochondrial genes, and alterations in lipid metabolism, pentose phosphate pathway, and ketogenesis (FDR-qval<0.1). Functional metabolic analysis agreed with these findings, demonstrating significantly diminished glycolysis and mitochondrial respiration, and accumulation of lipid droplets. These data demonstrate that FOXA1/2 depletion results impartial de-differentiation and activation of alternate lineages, and disruption of glycolysis, mitochondrial respiration, lipid metabolism, and other key metabolic pathways in human liver (HepG2) cells, together with evidence of de-differentiation of human pluripotent-derived hepatic progenitor cells.

165a Mutation of a Conserved, Hydrophobic, Cryptic Epitope Improves Manufacturability and Immunogenicity of the Sars-Cov-2 RBD

Sergio Rodriguez Aponte, Neil C. Dalvie, J. Christopher Love

The supply of COVID-19 vaccine doses still lags behind the global demand for first time vaccination and booster doses. Distribution of vaccine doses has been far from equitable across the world given the steep prices and logistical challenges that low- and middle-income countries face. Subunit protein vaccine candidates have now been shown to elicit protective responses against SARS-CoV-2 infection, while providing additional benefits for manufacturing capability and stability requirements compared to many currently approved vaccines. Here we report a second-generation engineered RBD sequence variant with enhanced manufacturability and immunogenicity over the wild-type ancestral RBD and a first-generation engineered variant (RBD-L452K-F490W (RBD-J)). Introducing two additional mutations, S383D and L518D, to a hydrophobic cryptic epitope in the RBD core improved expression titers and biophysical stability compared to RBD-J. These two additional mutations in RBD-S383D-L452K-F490W-L518D (RBD-J6) ablated the interaction of two neutralizing antibodies, CR3022 and EY6A, targeting the class 4 epitope on the RBD core, but the protein is still bound by human convalescent sera. Mice immunized with a Beta sequence variant of RBD-J and RBD-J6 displayed on a virus-like particle were protected against challenges with Alpha and Beta variants of SARS-CoV-2. Sera from mice immunized with three doses of a RBD-J6 β – VLP showed comparable neutralizing activity to several variants of concern compared to two doses of Comirnaty.

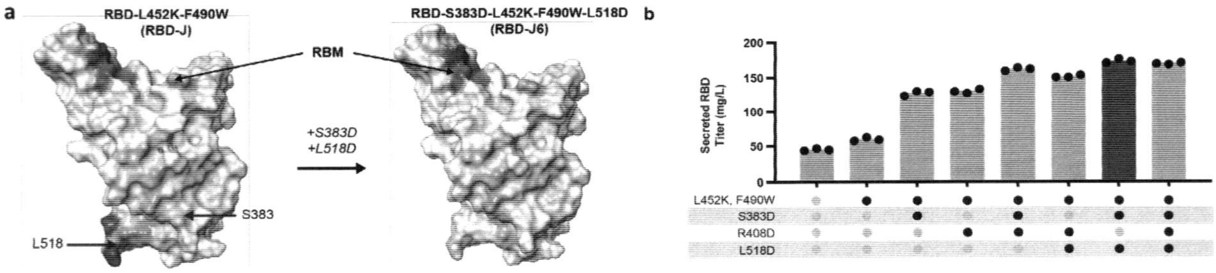

165b Tailoring the Size of Biodegradable Poly (sulfobetaine) Hydrogels for IgG Pulmonary Delivery

Songpei Xie, Saeed Manouchehri, Joshua Ramsey, Clint Aichele

Zwitterionic polymers are biocompatible materials that have shown promise in the area of drug delivery. The dual charges in the molecule help retain protein structure and minimize the immune response against the polymer. Through the use of emulsion polymerization, the size of the hydrogel bead can be controlled to yield an inhalable zwitterionic polymer particle for pulmonary drug delivery. In our research, we developed a pathway to synthesize inhalable, swellable poly (sulfobetaine) (pSB) particles used for pulmonary drug delivery with diameter approximately equal to 1 μm. We characterized the degradation of biodegradable pSB and the associated cytotoxicity. To evaluate the efficacy of biodegradable pSB hydrogels for protein delivery, we used confocal microscopy coupled with UV-vis to quantify loading and release of IgG.

165c Sprayable, Antimicrobial Hydrogels to Improve Wound Infection Treatment

Riannon Smith, Nicole Brogden, Jennifer Fiegel

Burn wounds have high infection rates and pain associated with current treatments; therefore, we aim to develop sprayable, antimicrobial hydrogels to reduce application pain and improve infection treatment. Thermoreversible polymers loaded with antimicrobial agents can be sprayed onto the skin as a liquid at colder temperatures then transition to a gel as it warms on the skin while providing prolonged drug release in the gel state. Poloxamer hydrogels loaded with ciprofloxacin hydrochloride have been developed and evaluated for desirable physical, antimicrobial, and sprayable properties. Many developed formulations have gelation temperatures above room and below skin temperature (21 - 34°C), which is desirable in order to spray while the formulation is a cold liquid and gel once on the warm skin. Using in-line diffusion cells, 50% of the drug is released between 4 and 6 hours. From the drug release experiments, we are also maintaining drug concentrations above the minimum inhibitory concentrations for four common wound pathogens for 18 to 21 hours. The antimicrobial-loaded hydrogels have shown larger zones of bacterial inhibition compared to common topical treatments of burn and chronic wound infections. Through spray pattern analysis, we observed that the type of nozzle, formulation, and distance to a target largely influence the size and shape of the spray pattern, which is important to determine proper dosing of these topical formulations. These results will help guide future development of sprayable, antimicrobial hydrogel treatments to improve current wound infection treatments.

165d Constructing Zonular Articular Cartilage Using a Varied Shear Bioreactor

Terreill Robertson, Bryn Matheson, Bernard Van Wie, Arda Gozen, Ryan Driskell, Lawrence J. Bonassar, Wenji Dong, Iwona Driskell

Osteoarthritis is a critical disease affecting 33 million Americans, causing them to suffer from cartilage degradation. Current practices are unable to restore the native properties of the tissue. The success of growing articular cartilage (AC) *in vitro* relies heavily on developing bioreactors that mimics an environment present in native cartilage. However, current literature does not provide adequate strategies for developing zonally distinct AC. In this work, I aimed to produce functional tissue by developing a shear bioreactor containing a tapered wall to impart shear stress variations at the walls of cell-laden scaffolds. Human articular chondrocytes (Chs) were encapsulated in a porous alginate hydrogel and cultured in a shear bioreactor imparting a fluid flowrate of 5 mL/min. Culture duration occurred for 7 and 14 days to observe the temporal effects of the system. A static culture of Chs served as the control and a static culture of a cell-laden hydrogel served as the negative control. After the culture duration, samples were sectioned into 6 pieces to understand the location dependency of shear effects. Samples were tested for protein secretion, viability, and mRNA expression. It was hypothesized that there will be differences in protein production and gene expression in the different locations due to shear variation. Additionally, a longer culture duration would demonstrate these differences in samples.

165e A Comparative Assessment of the Response of Primary and Metastatic Ovarian Cancer Cells to Cisplatin in 3D Models of Various Structural and Biochemical Configurations

Priyanka Gupta, Melina Kitsiou, Kavitha Madhuri-Thumuluru, Aline Miller, Adedamola Olayanju, Eirini Velliou

INTRODUCTION: Termed as the 'silent killer', epithelial ovarian cancer (EOC) earns its nickname due to its high mortality rate, with 5-year survival rates of only 46%, 46.5% and 38% in UK, USA andamp; Europe respectively (www.cancerresearchuk.org, www.ocrahope.org). This has been attributed to some extent to the EOC's high recurrence rate and resistance to currently available platinum based chemotherapeutic treatment methods. Hence, society, researchers and the clinical community are in dire need for a more in-depth study of the EOC microenvironment, to design better patient specific treatments and to advance current therapeutic methods. Multiple groups have studied and reported the effect of chemotherapeutic agents on various EOC 3D *in vitro* models (spheroids and hydrogels) [1,2,3]. However, there are very few studies wherein a direct comparative study has been carried out between the different *in vitro* 3D models of EOC and the effect of chemotherapy within them. Such comparative evaluation is of particular importance, especially as different 3D in vitro systems capture different structural, biochemical and biomechanical aspects of the tissue microenvironment.

Herein, we report, for the first time, a comprehensive comparative study of three different 3D in vitro platforms, namely (i) spheroids, (ii) synthetic hydrogels of various chemical configurations and (iii) polymeric scaffolds of various Extracellular Matrix (ECM) coating on the cell growth and response to the chemotherapeutic (Cisplatin) for ovary derived (A2780) and metastatic (SK-OV-3) EOC cell lines.

METHODS: Spheroids of A2780 (primary ovarian cancer) and SK-OV-3 (metastatic ovarian cancer) cell lines were fabricated using specialized 96 well round bottom plates, provided by faCellitate (Manheim, Germany). Synthetic PeptiGels (Manchester BIOGEL, UK) were used as per manufacturer's instructions for the hydrogel based culture while polyurethane (PU) scaffold assisted 3D cultures was carried out as per our previously published protocol[4,5,6,7]. The 3D cultures were maintained and monitored for different time periods (10–28 days) depending on the culture platform type. Feasibility of using these models for assessment of chemotherapeutic agent (Cisplatin) was also carried out. Various in situ assays for monitoring the cell viability, cellular apoptosis and spatial organisation were performed using advanced CLSM microscopy

RESULTS andamp; CONCLUSION: We report that all three 3D models can support the growth of EOC, but for different time periods (varying from 7 days to 4 weeks). We have seen that chemoresistance to Cisplatin, *in vitro,* observed especially for metastatic EOC cells, is platform dependent both in terms of structural as well as in terms of biochemical composition of the model/platform. Our study highlights the selection of the appropriate/best 3D *in vitro* model depends on the cell type, experimental time period and experimental question being asked as different models are appropriate even when studying. Our study highlights that the selection of a 3D *in vitro* platform depends on (i) the planned experimental/assessment time period, (ii) the type of cell to be studied, (iii) the site of cell origin in vivo and (iv) the question that needs to be answered [8].

ACKNOWLEDGMENT: The project was supported financially by the 3DBioNet (UKRI). E.V is grateful to the Royal Academy of Engineering for an Industrial Fellowship and to the Medical

Research Council UK for a New Investigator Research Grant (MR/V028553/1), which also financially supports P.G.

REFERENCE:

1. Heredia-Soto, V. *et al. Oncotarget* 9, 21893 (2018)
2. Raghavan, S. *et al. Gynecologic oncology* 138, 181-189 (2015).
3. Loessner, D. *et al. Biomaterials* **31**, 8494-8506 (2010).
4. Totti, S. *et al.* RSC Advances. 8(37): 20928-20940. (2018)
5. Gupta, P. et al. RSC Advances. 9 (71): 41649-41663 (2019)
6. Velliou, E.G *et al.*Bioprocess Biosyst. Eng. 38, 1589–1600 (2015)
7. Wishart , G. *et al.*Cancers 13(23):6080 (2021)
8. Gupta, P. *et al.* Cancers 14 (5): 1274 (2022)

165g Mouse Obesity Increases Peg-Based Micelle Liver Uptake and Decreases Lipid Nanoparticle Liver Uptake through the Modified Activity of Kupffer Cells and Lsecs

Mitch Raith, Uche Anozie, Paul Dalhaimer

Understanding the mammalian response to nanoparticles (NPs) in an obesity environment is an unmet crucial need in nanomedicine since more than two-thirds of adults in the U.S. are overweight to obese. Thus, patients that will be treated with nanoparticles (NPs) will have non-alcoholic fatty liver disease (NAFLD) and chronic inflammation. Yet, NP pharmacokinetics (PKs) and toxicity (TOX) are determined in lean rodents with healthy livers and low inflammation. Thus, we have little understanding of how NPs will behave in a metabolically realistic environment. To bridge this knowledge gap, we determined the PK and TOX of PEG-based cylindrical nanoparticles (CNPs) and lipid-based nanoparticles (LNPs) as a function of diet- and genetic-induced obesity in *ob/ob* and wild-type mice. In PK studies, CNPs localized to fatty livers in *ob/ob* mice on a high-fat diet (HFD) more quickly than to healthy livers in *ob/ob* mice on a restricted low-fat diet (LFD). The PKs of LNP liver signal are greatly affected by mouse weight. LNP liver signal peaks 6 hours post-injection in lean mice, 24 hours post-injection in heavy mice, and 48 hours post-injection in obese mice. We next aimed to determine the reasons for these observations. In lean mice, most NPs rapidly localize to Kupffer cells (KCs) and liver sinusoidal endothelial cells (LSECs) after intravenous injection. But the distribution of NPs between these liver cells is unknown in both lean and obese mice. In lean mice, clodronate liposome-induced-depletion of KCs decreased both CNP and LNP liver localization, with a much stronger reduction in CNP localization. MCT-induced depletion of LSECs increased CNP liver localization and decreased LNP liver localization. This suggests that CNPs interact more strongly with KCs than LSECs and that LNPs interact with both KCs and LSECs. These experiments were not possible in obese mice because of lethality. To determine why CNP localization to fatty livers is higher than to lean livers, we focused on the expression of SR-BI, which binds CNPs and causes CNP cellular uptake. SR-BI liver expression was increased in *ob/ob* mice fed a HFD. SR-BI was also increased in RAW 264.7 murine macrophages cultured in HFD conditions in vitro. On the other hand, in vitro experiments showed that SR-BI does not appear to be important for CNP uptake by LSECs. Also, oleic acid (a model fatty acid) decreased the uptake up of CNPs by LSECs *in vitro* by 100-fold. Conversely, oleic acid increased the uptake up of CNPs by macrophages *in vitro*. To determine if obesity-driven inflammation played an additional role in CNP and LNP liver localization, we administered the leaky-gut endotoxin lipopolysaccharide (LPS) to wild-type mice on a LFD. LPS triggers inflammation in obese mammals. LPS increased the liver localization of CNPs but decreased the liver localization of LNPs. Therefore, LPS-induced inflammation plays a role in the increased CNP localization to the liver. CNPs and LNPs reduced most cytokine and chemokine levels that were elevated in obese *ob/ob* mice. Interestingly, CNPs and LNPs greatly increased IL-5 levels which promotes the survival, differentiation, and chemotaxis of eosinophils.

165h Pulmonary Functionality Changes Following Titanium Dioxide Particulate Exposure and Antioxidant Protection

Jordan Hoops, Timothy Brenza

Fine particulate matter (PM2.5) is a major health concern, impacting the respiratory system through impaired lung function, infection, cancer, and contributing to 48,000 deaths in the United States every year. Exposure to PM2.5 causes oxidative stress through the direct introduction of exogenous ROS and compounds that drive free radical reactions, or indirectly through the recruitment and activation of inflammatory cells which release free radicals. The large surface area of the lungs and direct contact with inhaled air make the respiratory system highly susceptible to oxidative stress related injury. Individuals with immunodeficiencies and pulmonary diseases including asthma and COP are especially susceptible as these conditions have been correlated with decreased antioxidant status in lung lining fluids and tissues.

Titanium dioxide nanoparticles are widely used in industrial and consumer products including paints, plastics, pharmaceuticals, cosmetics, and food products. Large production of this nanoparticle and other engineered nanoparticles are a cause for concern of occupational exposure through inhalation and dermal contact. TiO_2 nanoparticle toxicity has been studied extensively in dermal models, however, respiratory effects and oxidative damage pathways remain a concern.

This study aims to characterize oxidative stress response in human pulmonary cell line A549 following isolated and repeated exposures to titanium dioxide nanoparticles. Size, charge, and morphology of TiO_2 nanoparticles were characterized using DLS, zeta potential, and SEM prior to exposure. Post-exposure investigation included cell viability by MTT assay and physiological response parameters including IL-8 secretion, apoptosis execution determined by caspase 3 activation, ratio of reduced and oxidized glutathione, and quantification of reactive oxygen species. Further, antioxidant pretreatment is investigated to supplement decreased antioxidant capacity and reduce extent of functionality impairment.

165i In Vitro Human Spine Model Manufacturing and Applications to Intrathecal Drug Delivery

Ayankola Ayansiji, Daniel Gehrke, Meenesh Singh, Andreas Linninger

Intrathecal drug (IT) delivery can efficiently target regions in the central nervous system without hindrance of the blood brain barrier. This delivery mode is especially promising for novel therapeutics such as enzyme replacement, gene, or antisense oligonucleotide (ASO) therapies for the treatment of neurodegenerative diseases including Alzheimer's, Huntington's disease, and stroke. Due to the high risk to patients and technical limitations of acquiring high speed CSF flow and bio-distribution data with existing imaging modalities (e.g., MRI, PET), in human in vivo study cannot be performed. Novel, fast, relatively cheap, and subject specific in-vitro method to study the drug bio-dispersion mechanism and optimize the intrathecal drug therapies for individual patients is in great need. Here, we present model design and additive manufacturing process for producing a subject specific spine which replicated the interaction of the human spine with the cerebrospinal fluid (CSF). The model was manufactured using the 3-dimensional (3D) printing technology or additive manufacturing. The printing was done with the used of imaging data set which helps in acquiring the geometry through which 3D mesh was created with the use of computer-aided design (CAD). The developed phantom is watertight, anatomically accurate, physiological functional, deformable - which allows it to produce pulsation as experienced in the real human spine, and transparent – which allows it the be suitable for intrathecal (IT) tracer dispersion experiment as it is very easy to track, in real time, the spread of the tracer in the phantom. The manufactured phantom was used for IT tracer dispersion study using different parameters such as infusion rate, and frequency of pulsation.

165j Non-V600E Braf Mutations in Melanoma and Their Response to Different Clinically Approved Braf Inhibitors

Erik Laurini, Maria Russi, Domenico Marson, Suzana Aulic, Sabrina Pricl

Melanoma is one of the deadliest skin malignancies, and its prevalence is increasing globally. Formerly thought to be a drug-resistant illness, melanoma has seen a dramatic transformation in the treatment landscape since 2011. Indeed, advances in our understanding of the immune system and its interactions with tumors, as well as ever-increasing molecular characterization of this cancer, have paved the way for immunotherapy on the one hand, and molecular target therapies on the other. The rising availability of more powerful technologies, such as Next-Generation Sequencing (NGS), as well as the growing availability of huge genetic panels, has enabled the discovery of a number of prospective therapeutic targets and the development of relevant treatments. These new therapeutic approaches improved melanoma prognosis, resulting in a 5-year survival rate of 34–43%. However, mainly because of primary and acquired drug resistance, the majority of patients will ultimately relapse, and only patients harboring a BRAF mutation, observed in about 50% of cutaneous melanoma, can receive a targeted treatment with BRAF and MEK inhibitors.

Activating mutations in the BRAF gene occur in 40-60% of melanomas with the majority of mutations resulting in V600E/K. Previous cohort studies have identified rates of non-V600E/K BRAF mutations to occur in 5-12% of patients. Despite this, there remains limited evidence characterizing the disease characteristics and outcomes of patients who harbor the non-classical BRAF mutation. In our cohort of 1000 melanoma patients, we identified three groups of patients with respect disease characteristics: a) BRAF wild type; b) BRAF V600E/K positive; and c) BRAF non-V600E/K positive. Among the last group, the most frequent BRAF mutations identified were L597S/Q, A598V, T599Dup, and K601E.

Accordingly, in this work we present the full set of results obtained from a combined computational/experimental approach for the in silico/in vitro characterization of the interaction of different clinically approved BRAF inhibitors (Dabrafenib, Vemurafenib, Encorafenib) and other investigations drugs (AZ628, PLX8394) with all these BRAF mutant isoforms. Remarkably, among all these the activating BRAF V598A variant - also found in papillary thyroid carcinoma - can be effectively inhibited by all considered inhibitors, even in the presence of activating NRAS mutations (Q61R).

165k Biomimetic Virus Nanoparticles for Oral Vaccine Delivery across Gut Organoid Mucosal Models

Qun Wang

Biomimetic Virus Nanoparticles for Oral Vaccine Delivery across Gut Organoid Mucosal Models

Tianjian Tong[1], Yijun Qi[1], Luke D. Bussiere[3], Debarpan Dhar[3], Cathy L Miller[3], Chenxu Yu[2], Qun Wang[1]*

[1]Department of Chemical andamp; Biological Engineering, [2]Department of Agricultural Biosystem and Engineering, [3] Department of Vet Microbiology andamp; Preventive Medicine, Iowa State University

Statement of Purpose: In the age of the COVID-19 pandemic, the significance of effective oral vaccines is as high as it has ever been. Compared to the intramuscular/ intravenous injection as the delivery pathway for vaccines, oral delivery through gastrointestinal mucosa has its unique advantages of high patient acceptance (e.g., no pain), easy administration, and high efficacy via triggering mucosal immunological responses. The effective oral vaccine requires high delivery efficiency across the gastrointestinal epithelia and protection of medically effective payloads (i.e., immunogens) against gastric damage [1]. In this study, hollowed artificial virus nanoparticles (AVNs: silica nanospheres and gold nanocages) with poly-l-lysine (PLL) coating and mammalian orthoreovirus cell attachment protein σ1 functionalization (NC-PLL-σ1) were explored as oral vaccine delivery vehicles (OVDVs).

Methods: To validate the effectiveness of the OVDVs as transporters and to evaluate the differences in payload protection and release with and without the PLL coating, we used the same intestinal monolayer system we reported before [2] mimic to the gastrointestinal epithelia in Figure 1. Utilizing this M cell monolayer system, we aimed to compare the capacity of both complete OVDVs (e.g., HSS-PLL-σ1) and uncoated OVDVs (e.g., HSS-σ1) to transport model payload (e.g., R6G) across the monolayer via M cell-mediated transcytosis pathway. This system serves as an excellent simulation to evaluate the potential effectiveness of the OVDVs at transporting vaccines to reach mucosal-associated lymphoid tissues (MALTs). In addition, via control of the organoid engineering process (different Rank L levels at 200 ng/mL, 100 ng/mL, and 0 ng/mL, respectively), organoid monolayers with different M cell presence were created, as shown in Figure 2. Their effects on the transport of OVDVs and delivery of the model payload were characterized.

Results: The transport of these OVDVs to mucosal lymphoid tissues could be facilitated by microfold (M-cells) mediated transcytosis (via σ1-α2–3-linked sialic acids adherence) across gastrointestinal epithelia. A model payload, PLL coating provided protection and slow-release control of rhodamine 6 G. The transport effectiveness of these OVDVs was tested on intestinal organoid mucosal models *ex vivo*. As shown in Figure 3, when compared with other experimental groups, the complete OVDV system (with PLL-σ1) demonstrated two significant advantages: a significantly higher transport efficiency (198% over control at 48 hrs); and protection of payloads which led to both better transport efficiency and extended-release of payloads (61% over no-coating control at 48 hrs) which could enhance vaccine efficacy. In addition, it was shown that the M cell presence in intestinal organoid monolayers (modulated by Rank L levels) was a determining factor in the transport efficiency of the OVDVs. The complete OVDVs showed great potential as effective oral vaccine and drug delivery systems.

Figure 1. Scheme: T1L σ1 functionalized OVDVs exploit the M cell transcytosis pathway. (a) Model of MRV infection of MALT. (b) Schematic of the transport of R6G loaded OVDVs nanocarriers through M cells incorporated into intestinal organoid monolayers.

Figure 2. Detection of microfold cells in 3D mouse organoids and 2D organoid-derived monolayers.

Figure 3. GNC- PLL-σ, HSS-PLL-σ1 transport total through intestinal monolayers with different Rank L levels (200 ng/mL, 100 ng/mL).

Conclusions: We report a universal setup of oral drug/vaccine delivery vehicles utilizing the σ1-M cell-mediated transcytosis pathway for payload transport across gastrointestinal epithelia, which could effectively improve oral vaccine efficacy.

References:

[1]Davoudi Z, et al. ***Marine Drugs*** 2021, 19(5): 282-298.

[2]Tong T, et al. ***Nanoscale*** 2020, 12(30): 16339-16347.

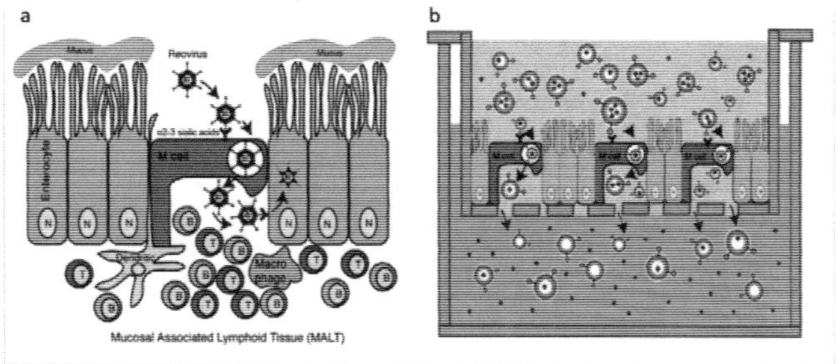

Figure 1. Scheme: T1L σ1 functionalized OVDVs exploit the M cell transcytosis pathway. (a) Model of MRV infection of MALT. (b) Schematic of the transport of R6G loaded OVDVs nanocarriers through M cells incorporated into intestinal organoid monolayers.

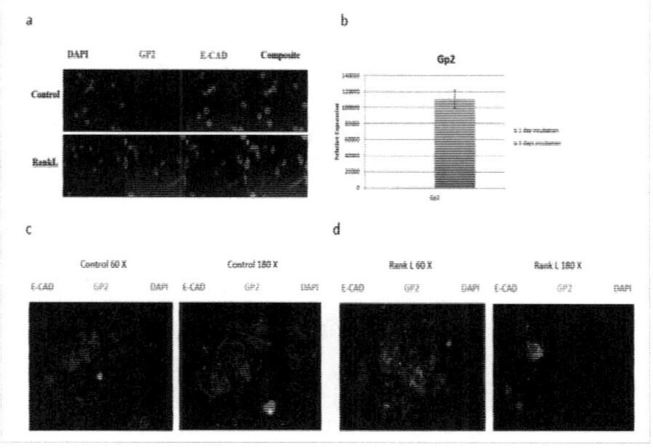

Figure 2. Detection of microfold cells in 3D mouse organoids and 2D organoid-derived monolayers.

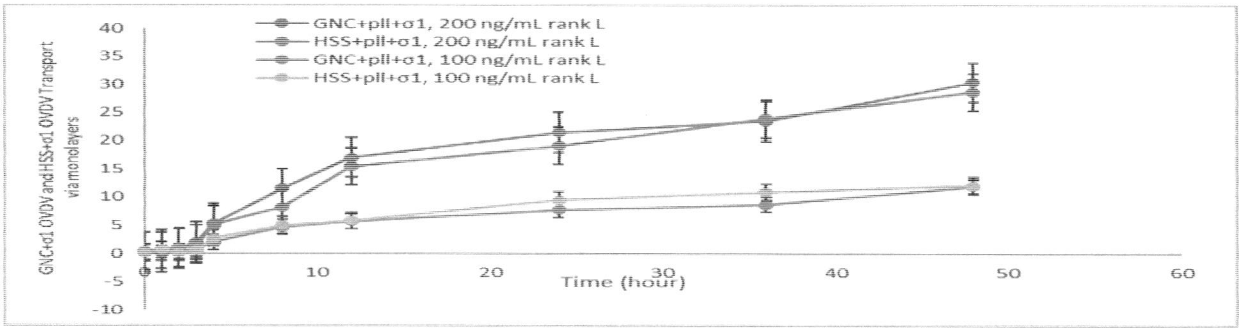

Figure 3. GNC- PLL-σ, HSS-PLL-σ1 transport total through intestinal monolayers with different Rank L levels (200 ng/mL, 100 ng/mL).

165l Triggered Release of Light-Responsive Model Prodrugs from Polymeric Core-Shell Nanocarriers *Via* UV Irradiation

Amy Moreno, Jennifer Vilnot, Stefan Chassaing, Nathalie M. Pinkerton

Encapsulation of therapeutic agents into nanocarriers has the potential to improve the spatial and temporal control required for local administration of cancer drug delivery. To control local administration, many researchers have developed polymeric core-shell nanocarriers responsive to physiological triggers, such as pH, oxidation potential, or enzymatic degradation. Lately, nanocarriers responsive to external stimuli are becoming increasingly popular as they offer a higher degree of control. Of the external modalities, light is the least invasive due to its remote modulation and easy manipulation by controlling wavelength, intensity, and duration. Light-responsive nanocarriers are typically fitted with light sensitive moieties that respond to wavelengths within the UV range. Release of core material from constructs happens by disruption of nanocarriers through a variety of light-induced mechanisms. However, one avenue yet to be explored is the encapsulation of light-cleavable prodrug constructs within nanocarriers. Here, we present a novel polymeric core-shell nanocarrier drug delivery vehicle and show the triggered release of model drug constructs *via* UV irradiation. Prodrug constructs are based around the o-hydroxycinnamic acid platform because of its intrinsic nonfluorescence and capability of absorbing within the UV range to release small alcohols and fluorescent co-product. To show the versatility of the platform, we covalently link a variety of chemistries to form model prodrug constructs. Subsequent UV irradiation of the model prodrugs show the release of model drug, and fluorescent co-product used as a quantification tool for drug release. Lastly, we encapsulated the model prodrugs in polymeric nanocarriers and illuminated *via* UV irradiation to demonstrate controlled light-triggered release of fluorescent reporter and model drug out of the nanocarrier.

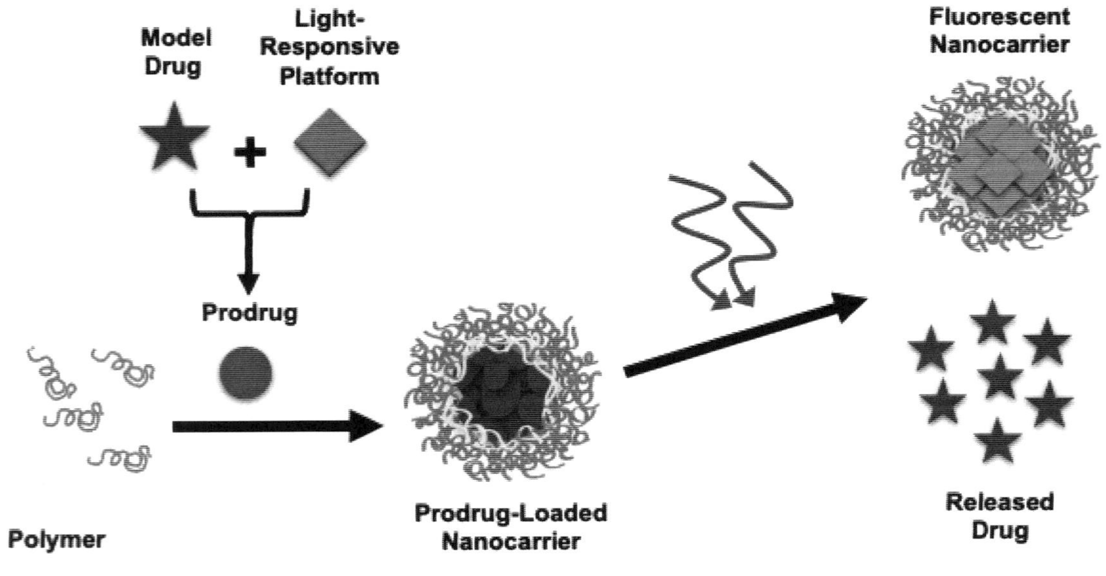

165m Label-Free Optical Electrophysiology Harnessing Bio-Electrochromic Materials Interface

Yuecheng Peter Zhou

Understanding how a network of neurons receive, store and process information in the human brain is a grand scientific challenge of our time. Neurons encode and communicate information through electrical signals. Traditional electrode-based recording methods such as patch clamp or multielectrode arrays are either highly invasive to cells or inflexible to sense electrical fields of cells at user-selected spatial positions with limited recording capacity. Existing optical methods, on the other hand, rely heavily on inserting voltage-sensitive fluorescent molecules into the cell membrane, which suffer from cell phototoxicity, limited recording duration and signal-to-noise ratio due to photobleaching.

Here, I will introduce ElectroChromic Optical REcording (ECORE), a new electrophysiology method that optically reads out the electrical activity of electrogenic cells in a label-free, parallel, and non-perturbative manner harnessing the unique properties of electrochromic polymers. The ECORE method fundamentally differs from any existing voltage recording approaches in that cell electrical signals are read out optically through voltage-sensitive optical absorption of electrochromic polymer thin films. Changes in optical reflection of the film, rather than fluorescence, are detected to probe cell electrical activities through a home-engineered optical system with a high signal-to-noise ratio. In this way, ECORE does not perturb the physiology of cells without the insertion of any molecular probes. It is also not limited by photobleaching or phototoxicity, making it a suitable tool for long-term recording of cell electrical signals over weeks or months. I will take about single color, dual color, and multi-channel optical recording of electrical signals from heart muscle cells, heart tissue, neurons, and brain slices interfaced with electrochromic polymer thin films with high sensitivity and throughput.

My future research will gear towards three directions. **Research area 1:** Developing a cutting-edge label-free optical electrophysiology tool using electrochromic polymers. **Research area 2:** Exploring chemical and biological interactions at the bio-electrochromic materials interface. **Research area 3:** Understanding the bioelectrical, chemical and mechanical cues in brain injury and diseases using the tools developed.

References (9/15):

9. **Y. Zhou** and B. Cui et al. ***J. Amer. Chem. Soc.*** (2021)

8. F. S. Alfonso, **Y. Zhou** and B. Cui et al., ***PNAS*** (2020)

7. **Y. Zhou** and C. M. Schroeder et al., ***Entropy*** (2022)

6. **Y. Zhou** and C. M. Schroeder et al., ***J. Rheol.*** (2021)

5. **Y. Zhou** and C. M. Schroeder et al., ***Nat. Comm.*** (2019)

4. **Y. Zhou** and C. M. Schroeder, ***Phys. Rev. Lett.*** (2018)

3. **Y. Zhou** and C. M. Schroeder et al., ***ACS Cent. Sci.*** (2017)

2. **Y. Zhou** and C. M. Schroeder, ***Macromolecules***, **49**, 8018 (2016)

1. **Y. Zhou** and C. M. Schroeder, ***Phys. Rev. Fluids*** (2016)

165n Characteristics of Electrically Responsive Ferrocene-Conjugated Chitosan/Alginate Hydrogel for Biomedical Application

Prakriti Dhungana, Victoria Messuri, Kyle Duke, Asma Allababdeh, Byung-Wook Park

Pathogenic infection as the most frequent complication of chronic wounds has become a global healthcare challenge because of the impairment of the healing process and the risks for amputations and mortality. The spatiotemporal release of therapeutics from wound dressings is highly desirable in order to manage wound treatment. The aim of this study is to obtain and characterize an electro-responsive ferrocene-chitosan/alginate polyelectrolyte complex (PEC) hydrogel that can be used as a smart wound dressing. First, chitosan/alginate PEC hydrogel was obtained as a control and characterized in terms of chemical properties and drug release kinetics. Natural chitosan (CHI) was chemically conjugated with ferrocene (Fc) moieties to create Fc-CHI. The Fc-CHI was interacted with alginate (ALG) to form Fc-CHI/ALG PEC through electrostatic interaction. The turbidity test was performed to find the optimum ratio between the Fc-CHI and ALG, thus the stoichiometric PEC hydrogel. The PEC hydrogel was characterized by Attenuated Total Reflection Fourier Transform Infrared Spectroscopy (ATR-FTIR), Scanning Electron Microscopy (SEM), Energy Dispersive X-ray Spectrometer (EDS), in addition to the swelling behavior and gel content tests. Comparative analysis of the ATR-FTIR spectra of CHI, Fc-CHI, ALG, and their mixtures indicated the formation of a polyelectrolyte complex. The SEM images showed the porosity of the PEC. The EDS analysis proved the incorporation of the Fc into the CHI by the appearance of the Fc peaks in the analysis. The PEC hydrogel showed a comparative swelling percentage to be 4400% and also showed excellent stability. To demonstrate the drug delivery potential of the developed PEC-based wound dressing, fluorescence (FITC) and FITC-Dextran were used as model drugs. First, the drug loading and release kinetics of the PEC were studied in solution. In 3 days, about 83% and 61% were released of the FITC, and FITC-Dextran, respectively in PBS solution. Secondly, the drug release properties on the phantom skin surface (agarose gel) were investigated using a custom-made electrical stimulus setup incorporated with fluorescence microscopy. The release of the model drugs on the surface was tested in passive (no electrical stimulus) and active (with electrical stimulus) manners. The PEC hydrogels showed an electro responsivity represented by increasing the intensity of the model drugs that diffused through the agarose gel under the electric stimulus. The diffusion coefficients for different drug models were estimated by analyzing images obtained after a time series of acquisition. Based on the results, the developed Fc-CHI/ALG PEC hydrogel is prone to be an enhanced drug release upon the electrical stimulus, compared to the CHI/ALG PEC hydrogel, indicating the presence of Fc may be able to increase the electro-osmosis pressure followed by the spontaneous drug release. The developed PEC hydrogel may be integrated with the bioelectronic devices.

166a Computational Protein Analysis and Design of Aldehyde and Alcohol Dehydrogenases for Enhanced Butanol Biosynthesis in Solventogenic Fermentation

Curtis Moore, Shang-Tian Yang

Biobutanol derived from acetone-butanol-ethanol (ABE) fermentation has potential to replace gasoline as fuel for automobiles powered by internal combustion engines. The roadblocks preventing this transition are the inherent limitations of the biobutanol fermentation process, such as butanol toxicity, expensive feedstocks, and high energy input requirements for butanol recovery. The high cost of biobutanol can be lessened by increasing the product yield via increasing the selectivity of the enzymes aldehyde and alcohol dehydrogenases (ALD and ADH), which convert acetyl-CoA and butyryl-CoA to corresponding aldehydes (acetaldehyde and butyraldehyde) and alcohols (ethanol and butanol), respectively, in the fermentation. Our goal is to discover novel ALD and ADH enzymes capable of performing our reaction of interest using *in-silico* enzyme screening, then design these enzymes for higher selectivity towards our product of choice, butanol, using computational enzyme design. First, bioprospecting or mining protein sequence databases was performed to discover genes/enzymes capable of catalyzing the reactions. Alphafold2 is then used to generate protein structures from mined sequences passing a series of filtering criteria. Ligand Docking in Rosetta is then performed using the generated models and the reaction intermediate of the enzyme's respective reaction. The ligand/protein interface score from these docking studies can be used to evaluate each protein's ability for performing their respective reaction. This procedure for enzyme mining and ligand docking requires insight into the enzymatic mechanism of the reaction in question. The most promising protein sequences will then be expressed in *E. coli*, and their enzymatic activity will be determined via *in vitro* assay. The results can be used in computational protein design to improve the selectivity of the enzymes displaying the highest activity. For this procedure Rosetta enzyme design can be used to mutate residues in the active site of these enzymes to enhance enzyme selectivity.

166b Are Plant-Based Meat Alternatives More Sustainable ?

Nicholas Schneider, Luis Huezo, Rui Shi

Through the use of synthetic biology and fermentation, microbes are turned into tiny factories that mass produce the desired protein—which is then used as a food ingredient or spun into fiber. In recent years, there has been growing interests in making plant-based alternative meats through synthesizing plant-based heme, a protein originally derived from soy plant roots that gives pseudo-meat flavor, color, and texture. In Randamp;D, scientists have been able to identify fast-growing, high-protein microorganisms to produce large quantities of protein in bioreactors. Impossible Foods claims its plant-based burger with heme protein result in significant reduction in energy requirements, greenhouse gas emissions, and overall environmental impacts. There is a tremendous need to elucidate the potential impacts of large-scale production of plant-based ingredients and evaluate the implications for industrial-scale fermentation and down-stream processes. In this study, we have identified different options for the production and purification of plant-based proteins, developed conceptual processes, and performed life cycle assessment to characterize and potential environmental impacts of large scale engineered heme protein production as a food ingredient.

166c Investigation of Antimicrobial Resistance in Listeria Monocytogenes from 2010 through 2021

Robert Hanes Jr., Zuyi Huang

Antimicrobial resistance (AMR) is a serious public health issue. Due to resistance to current antibiotics and a low rate of development of new classes of antimicrobials, AMR is a leading cause of death worldwide. Listeria monocytogenes is a deadly foodborne pathogen that causes listeriosis for the immuno-compromised, the elderly, and pregnant women. Unfortunately, antimicrobial resistance has been reported in L. monocytogenes. This study conducted the first comprehensive statistical analysis of L. monocytogenes isolate data from the National Pathogen Detection Isolate Browser (NPDIB) to identify the genes and trends for AMR in L. monocytogenes.

Principal component analysis was firstly used to project the multi-dimensional data into two dimensions. Hierarchical clustering was then used to identify the significant AMR genes found in L. monocytogenes samples and to assess changes during the period from 2010 through 2021. Statistical analysis of the data identified fosX, lin, abc-f, and tet(M) as the four most common AMR genes found in L. monocytogenes. It was determined that there was not an increase in AMR genes during the studied time period. It was also observed that the number of isolates decreased from 2016 to 2020. This study establishes a baseline for ongoing monitoring of L. monocytogenes for AMR.

References

FDA. Common Foodborne Disease Causes. Available online: https://www.fda.gov/files/food/published/Most-Common-Foodborne-Illnesses-%28PDF%29.pdf (accessed on 13 February 2022)

Lotfollahi, L. Prevalence and antimicrobial resistance profiles of Listeria monocytogenes in spontaneous abortions in humans. Afr. J. Microbiol. Res. 2011, 5, 1990–1993. https://doi.org/10.5897/ajmr11.498

Charpentier, E.; Courvalin, P. Antibiotic resistance in Listeria spp. Antimicrob. Agents Chem. 1999, 43, 2103–2108. https://doi.org/10.1128/aac.43.9.2103

Morvan, A.; Moubareck, C.; Leclercq, A.; Hervé-Bazin, M.; Bremont, S.; Lecuit, M.; Courvalin, P.; Le Monnier, A. Antimicrobial resistance of Listeria monocytogenes strains isolated from humans in France. Antimicrob. Agents Chem. 2010, 54, 2728–2731. https://doi.org/10.1128/AAC.01557-09

Hailu, W.; Helmy, Y.A.; Carney-Knisely, G.; Kauffman, M.; Fraga, D.; Rajashekara, G. Prevalence and antimicrobial resistance profiles of foodborne pathogens isolated from dairy cattle and poultry manure amended farms in northeastern Ohio, the United States. Antibiotics 2021, 10. https://doi.org/10.3390/antibiotics10121450

Pathogen Detection. Available online: https://www.ncbi.nlm.nih.gov/pathogens/ (accessed on 16 January 2022).

166d Effect of Processing Conditions and Investigating Vitamin B-12 Retention in Fortified Soymilk Powder

Priya Singh, Kiruba Krishnaswamy, chung-Ho Lin

Effect of Processing Conditions and Investigating Vitamin B-12 Retention in Fortified Soymilk Powder

Singh, P[1]*, Lin, C.H[3], and Krishnaswamy, K[1,2]

[1]Department of Biomedical, Biological and Chemical Engineering,

[2]Division of Food, Nutrition and Exercise Science,

[3]Center for Agroforestry, School of Natural Resources,

University of Missouri, Columbia, MO, USA

Abstract:

Introduction:

The future food trend especially among Gen Z population in noted to gravitate towards plant based diet.. Various reasons for this shift could be due to health concerns, ethical beliefs, environmental effect, and the idea of plant-based food sources are more sustainable. However, plant-based foods contain insufficient amounts of vitamin B-12, which is essential to maintain neurological function, metabolism, and blood cell count inside our systems. Vitamin B-12 deficiency can cause anemia, central nervous system demyelination, stomach neoplastic tumors, and other diseases. Vitamin B-12 deficiency affects around 10-40% of total global pollution. Soybeans are a highly nutritious crop plant that is grown solely over the world, and there are a number of soy-based food products available on the market, with soymilk being one of the most popular among plant-based diet consumers. Due to high moisture content, soymilk has a limited shelf life and is susceptible to microbial contamination when environmental conditions change.

Method:

In the current study, soymilk was fortified with different concentrations of vitamin B-12 (1.2mg, 2.4mg, and 3.6mg) and spray dried. Spray drying settings were determined based on an optimization study conducted early in the project (inlet temperature 140 °C, aspirator 90%, and feed rate 15%). A comparative study of vitamin B-12 retention in soymilk (liquid medium) and spray dried soymilk powder was conducted. Fortified soymilk and spray dried soymilk powder were stored for 15 days at refrigerator (4°C) and room temperature (25°C), followed by vitamin B-12 quantification to evaluate retention at both temperatures using HPLC-MS/MS. In addition, the physiochemical characteristics, reconstitution, rheological, particle, and thermal properties of vitamin B-12 fortified spray dried soymilk were investigated in order to determine the powder quality and customer acceptability. Furthermore, the in vitro release profile of optimized concentration of vitamin B-12 (2.4mg) fortified soymilk powder was examined by in- vitro digestion stimulating mouth, stomach, and intestinal media. The JMP 14.0.0 software was used to evaluate the fit model and compute statistical statistics for powder characterization parameters.

Result:

The experimental data from all combinations demonstrate that high amount of vitamin B-12 retention in fortified spray dried soymilk powder. Vitamin B-12 retention increases with increase concentration of vitamin (1.2mg, 2.4mg andamp; 3.6mg) in the spray dried soymilk powder. After storing the powder at 4°C and 25°C, no notable changes were observed in vitamin B-12 retention. Soymilk, on the other hand, retained a large level of vitamin B-12 regardless of storage temperature (4°C andamp; 25°C) or duration (15 days). However, the quality of soymilk begins to deteriorate after 4-5 days at refrigerated temperature and after 2 days at room temperature. Furthermore, the results of the powder quality measurement show that as the concentration of vitamin B-12 increased, the color changed from light to intense pinkish in hue, the powder flowing property were in the excellent to good flowing range (excellent to good), the moisture content was less than 5%, the dispersibility was 80-90%, the water solubility was 60-70%, and the hygroscopicity was 4-5% range. Advanced powder properties, such as viscosity, thermal degradation behavior, particle size distribution, particle size, shape, and morphology, are also fall under acceptable range. Based on in-vitro release profile experimental data, the highest amount of vitamin B-12 release was recorded in the intestine.

Significance:

The retention of vitamin B-12 in spray dried soymilk powder may help researchers better understand the impact of high temperature processing on vitamin B-12 activity. Because vitamin B-12 is an essential ingredient that must be consumed on a regular basis, soymilk enriched with the vitamin could be a smart way to protect the population from vitamin B-12 deficient disorders. Fortified soymilk powder could be used for a variety of purposes, including developing reconstituted beverages, infant formulation, and dietary supplements. Therefore, this study has laid the foundation to open opportunities and further commercialize the vitamin B-12 fortified soymilk powder in the plant-based food sector.

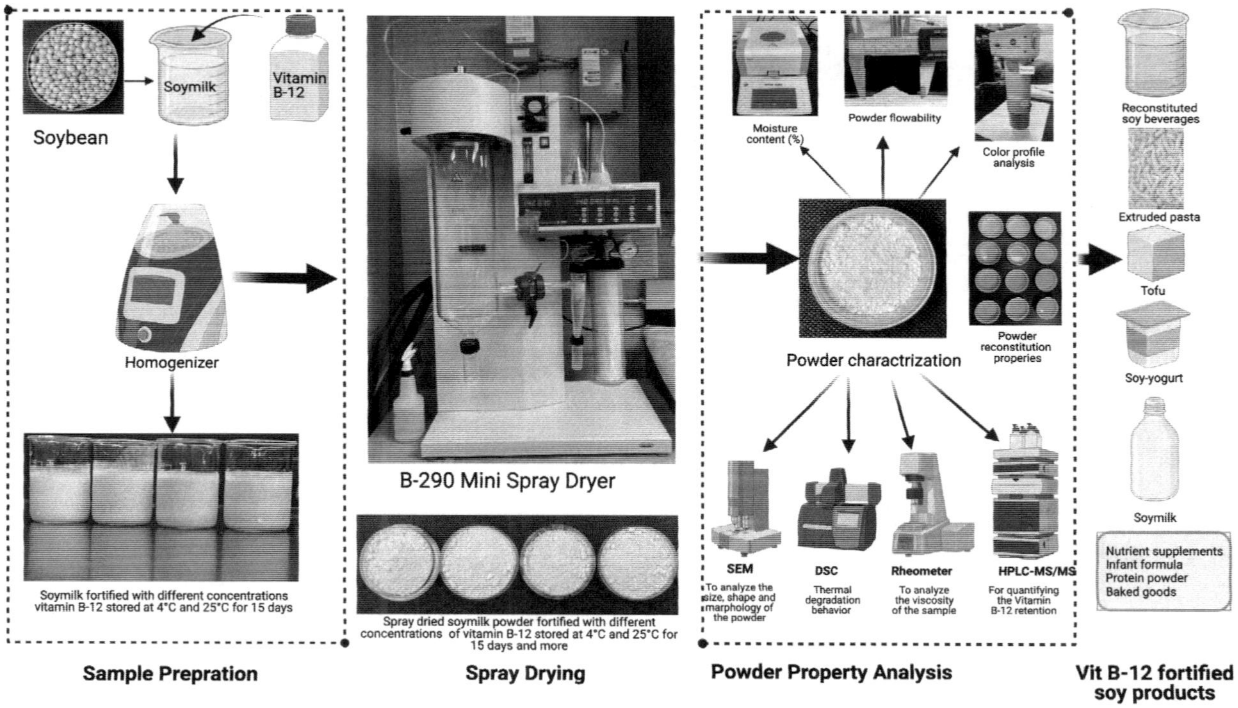

166e Enzyme Engineering of Beta-Glucosidases for Enhanced Thermal Stability and Activity

Emre Erkanli, Edward Chau, Jin Ryoun Kim

Biomass derived from various sources, such as forest and agricultural residues, can be used for production of high-value chemicals and other energy-related products. The biorefinery approach can increase the efficiency of agricultural land usage as well as help to develop more sustainable processes. Lignocellulosic biomass, an abundant biomass resource rich in carbon, is widely used for producing chemicals and biofuels. While hydrolysis of cellulose into its monomeric unit glucose is an important step toward the efficient biorefinery, an enzyme catalyzing this reaction, β-glucosidase, suffers from limitations, such as low activity, low thermal stability and product inhibition. Thus, an approach to engineer β-glucosidase to overcome these limitations would enhance the overall yield and the efficiency of the biorefinery process. Here, we report our approach to engineer β-glucosidases for enhanced thermal stability and activity. To this end, β-glucosidases from *Thermoanaerobacterium aotearoense* (TaBGL) and *Pyrococcus furiosus* (PfBGL) were chosen as parental enzymes. The wild-type TaBGL and PfBGL were recombinantly produced from *Escherichia coli* and purified. The enzymes were then subjected to activity assays using p-nitrophenyl-β-D-glucopyranoside (pNP-Glc) as a substrate. The steady state kinetic parameters of k_{cat}, K_m and K_I were 335 s^{-1}, 0.25 mM and 169 mM, respectively, for TaBGL at 60 °C and pH 6 and 303 s^{-1}, 0.34 mM and 71.2 mM, respectively, for PfBGL at 80 °C and pH 5.5. The half-lives ($t_{1/2}$) for TaBGL and PfBGL were 9 minutes at 60 °C and ~100 days s at 80 °C, respectively. Combinatorial libraries of these enzymes were created either using error prone PCR or guided by a bioinformatic analysis, HotSpot, to enhance thermal stability of TaBGL and increase enzymatic activity of PfBGL at lower temperatures. The agar plate based screening method using esculin hydrate and ferric ammonium citrate has been developed for rapid evaluation of the enzyme mutants for thermal stability and enzymatic activity and used together with 96-well based activity assays with pNP-Glc. Active mutants were identified from the screenings and are being characterized for their activity, thermal stability and glucose inhibition. We will also discuss sequence characteristics important in enzymatic activity, thermal stability and glucose inhibition of β-glucosidases determined from comparative studies based on a collection of literature and our own experimental results.

166f Transient Expression of Varying Gelatin Fragment Lengths in *Nicotiana Benthamiana* to Synthesize Microcarriers for Cultured Meat Production

Justin Wong, Imran Khan, Brandon Pizarro, Liber Mckee, Seongwon Jung, Patrick Negulescu, Somen Nandi, Karen A. McDonald

Microcarriers are widely used to culture adherent mammalian cells in the pharmaceutical industry where the product is usually a high value protein. However, the extension of this technology into cultured meat (CM) production where adherent muscle precursor cells are the product necessitates a microcarrier that does more than just support a high cell density. Ideally, the microcarrier used in CM production would contain no animal derived components, have a low cost of production, and would be food-safe, edible, and contribute to the sensory and organoleptic properties of the final meat product. Gelatin, a denatured form of collagen commonly found in many food products as a gelling agent is a promising biomaterial that fits this profile. However, it is currently produced at large-scale from the skins and bones of pigs and cows and is quite heterogeneous in terms of fragment length and extent of hydroxylation.

In this study we present a process for producing a non-animal derived chicken gelatin by transiently expressing chicken collagen fragments in the plant *Nicotiana benthamiana*. By producing gelatin recombinantly in a plant host, the shortcomings of commercially available gelatin are addressed and the recombinantly produced plant gelatin is comprised of a single fragment length. This allows for systematic screening of various fragment lengths to tune properties such as melting temperature and stiffness with the added benefit of reduced batch-to-batch variability. The unique ability to quickly scale up whole plant production at a relatively low capital and operating cost will also be advantageous in the CM field where price parity with conventionally farmed meat has yet to be achieved.

166g Enhanced Response to Non-V600E BRAF Mutations in Melanoma By Self-Assembled Nanovector Assisted Drug Delivery

Maria Russi, Erik Laurini, Domenico Marson, Suzana Aulic, Sabrina Pricl

BRAF somatic missense mutations are highly frequent in malignant melanomas and, at lower frequency, in a wide range of human cancers. The majority of such mutations locate in the kinase domain, with a single substitution (V600E) accounting for 80% of cases. In our cohort of 1000 melanoma patients, besides the expected wild type and V600E/K carriers, we identified a group of patients harboring non-V600E mutations, including L597S/Q, A598V, T599Dup, and K601E. Our combined computational/experimental results showed remarkable differential in vitro responses of these BRAF mutant isoforms to a range of clinically approved BRAF inhibitors (Dabrafenib, Vemurafenib, Encorafenib) and other investigations drugs (AZ628, PLX8394), with BRAF A598V showing a remarkable positive response to all drug tested.

However, poor drug solubility/bioavailability and general toxicity constitute general challenging hurdles for cancer targeted therapy. The application of nanotechnology for anticancer drug delivery is expected to address these issues and bring new hope for safer and more efficient cancer treatment. In this work, we present the characterization and the results obtained with our original nanomicellar drug delivery system based on an amphiphilic dendron, which could generate supramolecular micelles to effectively encapsulate the different BRAF inhibitors with high drug-loading capacity. The resulting drug-loaded nanomicelles were able to enhance drug potency in different cell models by significantly enhancing cellular uptake. In addition, the drug-loaded nanoparticles decreased significantly the toxicity related to the free drug.

166i Development of an Engineered Co-Culture Consortium for in Situ Depolymerization of Cellulose

Apurv Mhatre, Bethany Kalsheur, Thiagarajan Soundappan, Arul Mozhy Varman

Lignocellulosic biomass is the most abundant untapped renewable resource which can be utilized for the sustainable production of fuels and chemicals. Cellulose is the principal constituent of lignocellulosic biomass and is the most abundant biopolymer on earth. Currently, biorefineries utilize purified β-glucosidase (Bglu) and endoglucanase (Eglc) for breaking down cellulose into its sugars, which is not economical due to the prohibitive cost of purified enzymes. Therefore, efficient and cost-effective methods for hydrolysis of cellulose can play a vital role in making biorefineries economically viable. One pot bioprocess overcomes this drawback by enabling multiple microbial hosts to secrete hydrolases without expensive purification. *Bacillus subtilis* strains are widely used in biotechnology industries for enzyme production and therefore, has been used as a host in this work for developing a one pot process for cellulose. The present work aims to engineer *Bacillus subtilis* for heterologous expression and extracellular export of cellulases to reduce the cost incurred by protein purification.

First, a combination of signal peptides YwmC, SacC, and AmyE with two *β* glucosidases was explored to identify the optimal pair. AmyE- Bglu showed six-fold higher activity than wildtype in the para-Nitrophenyl β-D-glucopyranoside (*p*NPG) assay. Similarly, endoglucanases constructs were developed with Bglc, YwmC, and NprE signal peptide library to demonstrate the complete breakdown of cellulose. Furthermore, we cultured the best performing strains for Bglu secretion and Eglc secretion and isolated the supernatant for further studies. This enzymatic supernatant was further combined to see the synergistic effect of these enzymes on carboxymethylcellulose (CMC). The two enzymes' synergistic effect showed successful cellulose deconstruction into reducing sugars as quantified by dinitro salicylic acid (DNSA) assay. The highest titers obtained were 3.7 g/L of reducing sugar from 10 g/L of CMC. We plan to demonstrate *in situ* depolymerization by using the two screened strains for simultaneous deconstruction, assimilation, and metabolic processing of cellulose into fuels and value-added products. We envision that this approach will play a significant role in providing alternatives for improving the economics of the biorefinery.

166j Metabolic Process Engineering of *Clostridium Tyrobutyricum* Δ*cat*1::*Adh*E2 for Enhanced n-Butanol Production: Effects of Mannitol and Methyl Viologen on Flux Distribution

Jialei Hu, Jun Feng, Shang-Tian Yang

Clostridium tyrobutyricum, a Gram-positive, spore-forming, obligate anaerobic bacterium, is a promising microbial cell factory for butyric acid and *n*-butanol production because of its high metabolic flux toward butyryl-CoA, high butyric acid/butanol tolerance, and excellent ability to use both glucose and xylose present in lignocellulosic hydrolysates. Unlike *Clostridium acetobutylicum* and many other clostridia, *C. tyrobutyricum* uses CoA transferase (encoded by *cat*1), instead of phosphotransbutyrylase (PTA) and butyrate kinase (BK), to convert butyryl-CoA to butyrate, which transfers the CoA to acetate to form acetyl-CoA, thus allowing the reassimilation of acetate. *C. tyrobutyricum* has been engineered to overexpress *adh*E2, encoding bifunctional aldehyde/alcohol dehydrogenase, for *n*-butanol production from various carbon sources. More recently, CRISPR was used to insert *adh*E2 and replace *cat*1 on the genome in *C. tyrobutyricum*. The mutant Ct-Δ*cat*1::*adh*E2 produced *n*-butanol at the highest titer ever reported for any known microorganisms and little butyrate, with acetate and ethanol also produced in higher amounts, resulting in a low butanol yield of ~0.2 g/g glucose due to increased acetate and ethanol production. Compared to ethanol, butanol production requires 2 more NADH. We hypothesized that increasing the intracellular NADH pool would increase butanol production. The effects of using the more reduced substrate mannitol and methyl viologen (MV) as an electron carrier on carbon flux distribution and butanol biosynthesis in CT-Δ*cat*1::*adh*E2 were studied. With mannitol as the substrate, CT-Δ*cat*1::*adh*E2 produced 17.4 g/L butanol at a productivity of 0.36 g/L·h and yield of 0.27 g/g, an increase of 36% compared to glucose as the substrate. With 200 μM MV added in the medium, butanol production from glucose increased to a high yield of 0.29 g/g with significantly decreased acetate and ethanol production. Thus, high-yield n-butanol production can be realized using engineered *C. tyrobutyricum* with mannitol as the substrate or MV as an electron carrier in glucose fermentation.

166k Conversion of CO_2 to Bioplastics By a Sequential Cultivation of *Clostridium Formicoaceticum* and *Pseudomonas Putida*

Opeyemi Bello, Rocky Thapaliya, Forough Doustkhahvajari, Jie Dong

CO_2, a greenhouse gas is currently being emitted in large amounts due to increasing industrialization and population increase, thereby causing global climate change which is a major threat to our planet. Along with climate changes, the world is also facing environmental deterioration due to the accumulation of petroleum-based plastics. In this research, our aim is to sequester atmospheric CO_2 and convert it into polyhydroxyalkanoates (PHAs) through bacteria cultivation of *Clostridium formicoaceticum* ATCC 27076 and *Pseudomonas putida* KT2440. *P. putida* KT2440 can produce PHA but cannot feed on formate as the sole carbon source because it has no pathway to shuttle formate into central metabolism. Hence we employed *C. formicoaceticum* to first covert formate into acetate. *C. formicoaceticum* ATCC 27076 was fermented using different concentrations of pure formate as substrate. Only 5 g/L and 10 g/L produced a significant amount of acetate, 2.4 g/L and 5.3 g/L were produced respectively. These acetates were then used as the sole carbon source by *P. putida* KT2440 to produce PHAs. Afterward, 11 g/L of formate derived from CO_2 was then used and produced 9.6 g/L acetate. The obtained acetate was also used by *P. putida* KT2440 to produce PHAs. This work showed the entire process from CO_2, a deleterious gas in the environment to bioplastics (PHAs). It pointed out a direction of reducing CO_2 level in the atmosphere and in the meantime producing useful biodegradable plastics.

166m Canola Protein Production Using Different Extraction and Concentration Methods: *In Vitro* protein Digestibility, Functional Properties and Process Yields

Cristina Chairez Jimenez, Sergio Román Othón Serna-Saldívar, Cristina Chuck-Hernández

Proteins are important macronutrients in human nutrition and their consumption is relevant to sustain proper growth and maintenance. However, in developing countries protein malnutrition or Kwashiorkor is still one of the chief causes of death among preschool-aged children (Kalu and Etim, 2018). Unfortunately, there are still around 690 to 815 million undernourished people worldwide. The high levels of malnutrition and elevated cost of animal proteins have prompted research into the generation of novel plant-based alternatives that are produced massively. Among the best options, there are raw materials and coproducts generated by the oilseed industry. After oil extraction of most oilseeds, the crushing industries generate defatted protein meals with high protein that are normally channeled to the production of feedstocks.

Canola is the second-largest oilseed crop grown worldwide (USDA, 2020). The canola protein meal (CPM) is the primary byproduct of the oil recovery process containing around 40% protein (Jia et al., 2021). This oilseed was obtained after a long-lasting breeding work or regular rapeseed in Canada with the aim to obtain food-grade oil, namely with low erucic acid (<2%) and glycosylates (<30 µmol/g) contents (Canola Council of Canada, 1990).

Several studies on canola protein reveals the great potential for food application due to the well-balanced essential amino acid profile and the functional properties of the proteins (Gerzhova *et al.*, 2015, Jia *et al.*, 2021, Yang *et al.*, 2021, Witczak, 2021).

Despite a large amount of CPM generated worldwide, its usage as a food ingredient is still limited due to the challenge that exists in separating the protein from the other chemical components (fiber, phenols, glucosinolates, and others) with adequate yields and the generation of desired functional properties for the various sectors of the food industry.

The present investigation aimed to evaluate first different pretreatments to prepare CPM for the further extraction of proteins with/without alcohol-based pretreatment, alkaline extraction and protein separation by isoelectric precipitation and/or, ultrafiltration. Treatments were evaluated in terms of extraction yield, physicochemical and functional properties of extracted proteins and *in vitro* protein digestibility.

To achieve the objectives of this research, four treatments were evaluated: 1) Control: extraction by alkaline process and precipitation with acetic acid (Alk-Ac); 2) pretreatment with ethanol, extraction by alkaline process and precipitation with acetic acid (EtOH-Alk-Ac); 3) alkaline extraction and concentration by ultrafiltration, 5 kDa (Alk-UF); 4) pretreatment with ethanol, extraction by alkaline process and concentration by ultrafiltration, 5 kDa (EtOH-Alk-UF).

The CPM utilized for the different protocols was supplied by AGYDSA Aceites, Grasas y Derivados, S.A. The CPM was ground into a fine meal and chemically characterized by triplicate (moisture, crude protein (Nitrogen *6.25), and crude fat contents according to approved AACC methods 44-15.02, 46-13.01, and 30-25.01, respectively, AACC International, 1999).

The solid extraction yield was calculated by relating total solids recuperated after protein extraction divided by the initial CPM solids. On the other hand, the protein extraction yield was estimated as the total protein obtained from the process in relationship with the initial total protein of the CPM.

All resulting protein pellets or protein concentrates were lyophilized and assayed in terms of moisture and crude protein according to approved AACC methods 44-15.02 and 46-13.01 respectively (AACC International, 1999). The free alpha-amino nitrogen was also determined by the ninhydrin-colorimetric method of Haque and Kito (1989).

Functional properties were also determined. The water absorption and water solubility indexes were assessed according to the centrifugation method of Cheftel *et al.* (1989). The fat absorption capacity (FAC) and emulsion stability were calculated according to Ahn *et al.*, (2005) and Haque and Kito (1989), respectively. Foaming activity, foam stability, and foam density were determined with 3% (w/w) protein water dispersions according to Haque and Kito (1989).

Protein digestibilities of protein extracts were assayed by the enzymatic *in vitro* method suggested by Hsu *et al.* (1977), using protease from *Streptomyces griseus* (Type XIV, >3.5 units/mg solids) instead of discontinued porcine intestinal peptidase.

Statistical analyses were performed using variance procedures and means were compared using Tukey ´s tests ($p < 0.05$) using the Minitab 16 Statistical Software.

Experimental results indicated that the alkaline extraction and ultrafiltration protein concentration generated yield values of protein and solids with 34.95% and 14.38% respectively. Nevertheless, treatments that used acid precipitation comparatively yielded less protein and solids, 18.12-20.08% and 7.18-8.53% respectively.

The functional properties of proteins obtained with the treatment EtOH-Alk-UF showed the best values in terms of water solubility (39.17%) whereas the Alk-Ac had the best water absorption value with 3.86%.

In terms of foaming properties, the proteins obtained from the EtOH-Alk-UF treatment showed the best values, with 505% of foaming activity and 34.37% in foam density. Nevertheless, there were not significant differences among treatments in terms of foam stability.

Regarding protein digestibility *in vitro*, the higher value was obtained by EtOH-Alk-Ac with 95.5% and the lowest value were obtained by treatments that used ultrafiltration for protein concentration (89.9-88.1%).

The extracted canola proteins had acceptable properties for the food industry. The functional properties varied according to the pretreatment, extraction, and concentration processes. The best extraction process was the treatment of ground CPM first with ethanol pretreatment followed by alkaline extraction and ultrafiltration. Treatments precipitated with acetic acid yielded lesser amounts of concentrates and these had comparatively worse functional properties.

REFERENCES

AACC International (1999). Approved methods of analysis. Approved methods of analysis (11th ed.). St. Paul, MN: American Association of Cereal Chemists.

Canola Council of Canada. (1990). Canola oil and meal: standards and regulations. Winnipeg, Manitoba: Canola Council of Canada Publication. 4 p.

Cheftel, J.C.; Cuq, J.L.; Lorient, D. (1989). Propiedades funcionales de las proteínas. In Proteínas

Alimentarias; Editorial Acribia S.A.: Zaragoza, España; Capítulo 4.

Gerzhova, Alina; Mondor, Martin; Benali, Marzouk; Aider, Mohammed (2015). A comparative study between the electro-activation technique and conventional extraction method on the extractability, composition and physicochemical properties of canola protein concentrates and isolates. Food Bioscience, 11(), 56–71.doi:10.1016/j.fbio.2015.04.005

Haque, Z.; Kito, M. Lipophilization of αSl-casein. 2. Conformational and functional effects. (1983)J. Agric. Food Chem. 31, 1231–1237.

Jia, W., Curubeto, N., Rodríguez-Alonso, E., Keppler, J. K., andamp; van der Goot, A. J. (2021). Rapeseed protein concentrate isa potential ingredient for meat analogs. Innovative Food Science andamp; Emerging Technologies, 72, 102758.doi:10.1016/j.ifset.2021.102758

Kalu, R. E., Etim, K. D. (2018). Factors associated with malnutrition among under-five children in developing countries: A review. Global Journal of Pure and Applied Sciences, 24(1), 69–74.

Lin, M. J. Y., Humbert, E. S., andamp; Sosulski, F. W. (1974). Certain functional properties of sunflower meal products. Journal of Food Science, 39, 368–370.

Regenstein, J. M., andamp; Regenstein, C. E. (1984). Food protein chemistry. Orlando, FL: Academic Press Inc.

USDA. (2020). Oilseeds: World markets and trade. In Circular series FOP 06-17. Global oilseed consumption continues to grow despite slowing trade and production (pp. 1–50). https://doi.org/10.1016/S1097-8690(11)70006-3

Wilson W. (2008). Determination of Free Amino Nitrogen in Proteins. Laboratory of Physiological Chemistry. Johns Hopkins Medical School. Baltimore. (En Línea) Disponible en: http://www.jbc.org/cgi/reprint/56/1/191.pdf

Witczak, M., Chmielewska, A., Ziobro, R., Korus, J., andamp; Juszczak, L. (2021). Rapeseed protein as a novel ingredient of gluten-free dough: Rheological and thermal properties. Food Hydrocolloids, 118, 106813.doi:10.1016/j.foodhyd.2021.1068

Yang, Jack; Faber, Iris; Berton-Carabin, Claire C.; Nikiforidis, Constantinos V.; van der Linden, Erik; Sagis, Leonard M.C. (2021). Foams and air-water interfaces are stabilized by mildly purified rapeseed proteins after defatting. Food Hydrocolloids, 112(), 106270–.doi:10.1016/j.foodhyd.2020.106270

166n Effect of the Addition of Hydrolyzed Soybean Protein Isolate, Transglutaminase and $CaCl_2$ on the Production of Panela-Type Cheeses with Partial Substitution of Milk Protein

Ana Maritza Reyes-González, Cristina Chuck-Hernández

The use of soybean protein for partial or total substitution of the milk protein in the production of fresh cheeses such as panela, one of the most consumed in Latin America, is promising to reduce raw material costs and follow current market trends. Nevertheless, it is a challenge from the process point of view to achieve characteristics similar to those made entirely with milk, where casein plays a fundamental role in creating the structure of the product.

Considering vegetable proteins have very different functional characteristics in terms of water absorption, solubility, and interaction with the fat fraction, soy hydrolysates were created to facilitate the incorporation of these proteins through low-molecular-weight peptides in the casein-fat network.

The use of transglutaminase (TG) has also been explored to improve texture and stability, through the formation of isopeptide bonds (Jaros *et al.*, 2006). Additionally, the coagulation process can be greatly influenced by the addition of calcium chloride, as the calcium ions neutralize the charge of the caseins and allow an increased aggregation of the renneted micelles (Dalgeish, 1983). These could favor the incorporation of soy hydrolysates and develop an interpenetrated network of the components during the production of the panela-type cheese structure.

The objective of this research was to evaluate the effect of the hydrolysis of soy isolate, the addition of transglutaminase and calcium chloride on the development of an interpenetrated network during the production of panela-type cheese with partial substitution of 41.6% of milk protein, evaluated on its texture, proximal, and sensory characteristics.

To achieve the objectives of this investigation, the following five treatments were prepared: control without vegetable protein (A); soybean isolate with $CaCl_2$, TG, and hydrolysis (B); soybean isolate with $CaCl_2$ and TG (C); soybean isolate with $CaCl_2$ and hydrolysis (D); soybean isolate with $CaCl_2$ (E).

Milk was pasteurized for 30 min at 63°C and cooled to 32°C. Calcium chloride (0.125% w/v) was added to some treatments and mixed. All treatments with vegetable protein underwent a heat treatment for 15 min at 85°C. The protein hydrolysis was performed with an alkaline protease at pH 8.0 for 3 h at 60°C and was inactivated with heat treatment. Subsequently, the protein solution was added to the milk (corresponding to 1.25% of the total volume and 41.6% of the total protein present in the original raw material) and mixed. Dairy-vegetable protein mixtures were prepared considering a total solids content of 12.5%, using distilled water for the adjustment. According to experimental design, 1.92% TG (of total protein in dairy-vegetable mixtures) was added immediately prior to rennet (0.03% v/v) for the cheesemaking. Content of protein, fat, ash, moisture, sensory analysis, and texture were evaluated.

Experimental results indicated a higher moisture content in treatments with hydrolysis (A with 67% and C with 69%), but lower protein content (39% and 43%, respectively) compared to those with only heat treatment (B with 66% and 44% and D with 66% and 45%, respectively). The missing protein is expected to be present in the whey. All vegetable protein treatments had slightly lower fat content than control, except for C. TG treatments showed slightly lower moisture content and higher yield than their counterparts. A small difference was found regarding the ash content in hydrolyzed treatments.

On the other hand, the sensory analysis showed higher acceptability in A and C (6.0 and 6.3), in comparison with B and D (5.7 and 5.8), but lower with A (7.8). The category *appearance* affected the acceptability the most because the soybean protein in the treatments was visibly distinguishable. Since treatments A and C had low-molecular-weight peptides, the final product had a more homogenized appearance, and therefore a slightly higher general opinion. Additionally, TG seemed to slightly influence the flavor negatively, while improving the appearance.

Texture profile analysis revealed that treatments A and C had the lowest values in all four categories: hardness, cohesiveness, springiness, and chewiness. Samples B and D had similar cohesiveness and springiness values, but higher hardness and chewiness than A. Treatments A and B had higher values in all texture categories than C and D, respectively, due to the presence of transglutaminase.

The findings show the potential of the partial substitution of milk protein with soybean protein implementing processing techniques to create an interpenetrated network during the cheesemaking of panela-type cheeses. Although the protein content and texture profile must be improved in the hydrolysis treatments to achieve better acceptance by consumers, sensory aspects are promising. Transglutaminase along with calcium chloride improved texture (specifically hardness and chewiness), but impaired sensory aspects like flavor, which suggests the concentration might need adjustment. Further research including a microscopic approach is suggested to gain a deeper understanding of the interactions among the factors.

References

Dalgleish, D. G. (1983). Coagulation of renneted bovine casein micelles: dependence on temperature, calcium ion concentration and ionic strength. Journal of Dairy Research, 50(3), 331-340.

Jaros, D., Partschefeld, C., Henle, T., andamp; Rohm, H. (2006). Transglutaminase in dairy products: chemistry, physics, applications. Journal of texture studies, 37(2), 113-155.

499e Engineering the Cyanobacterial Photosynthetic Electron Transport Chain to Improve Photosynthetic Efficiency

Nima Hajinajaf, Aditya Sarnaik, Parker Poole, Danika Kartchner, Amogh Deshpande, Muhammad Faisal, Kira Winsor, Jonathan Nguyen, Abhishek Singharoy, Willem Vermaas, Arul Mozhy Varman

Solar energy harvested by photosynthetic systems including plants, algae, and cyanobacteria is a primary source for energy. Therefore, it is imperative to increase the photosynthetic efficiency of the photoautotrophic microorganisms to secure our food supply for today and the future. To this end, the cyanobacterium *Synechocystis* sp. PCC 6803 (hereafter, *Synechocystis*) serves as a perfect model organism as it is easy to engineer and shares all energy bottlenecks ubiquitous to photosynthesis and respiration in plant cells. In this study, we designed two distinct strategies to engineer the photosynthetic electron transport chain in *Synechocystis* to increase the photosynthetic efficiency. The first approach was directed towards linking the photosynthetic complexes photosystem I (PSI) and cytochrome *bf* (cyt*bf*), to bring them in a closer proximity and thereby, reduce electron loss. We performed computational 3D protein analysis to identify suitable proteins to link cyt *b6f* and PSI. Comprehending the protein orientations and interpeptide distances PetA from cyt *b6f* (C-terminus) and PsaA from PSI (N-terminus) were identified as suitable candidates to be linked. Following these strains expressing the linked proteins were constructed for further experiments.

In the second approach to increase the membrane density of cyt *b6f* complex and thereby reduce the proximity between cyt *b6f* with PSI and PSII, strains were constructed to overexpress the major proteins PetA, PetB, PetC, and PetD. The hypothesis was that overexpressing one of the major proteins would result in increasing the density of cyt *b6f* by leveraging unknown endogenous regulatory systems. Comparative growth studies of the engineered strains with wild type showed that at low light (20 $\mu E\ m^{-2}\ s^{-1}$), there was a 48% increase in growth of the strain expressing PetD compared to WT. Additional studies are underway to get all the strains segregated as well as to test their growth under low, medium, and high light conditions as well as by varying CO_2. In addition, to understand the molecular cause for the increase in photosynthetic yields we will also conduct further studies by using PAM fluorimetry, 77K spectroscopy, and also by tracking the oxygen evolution rates.

250a Engineering a Cell-Free System- Enzyme Fusion Scaffold for Biofuel Production

Matthew Wong, Georges Belfort, Mattheos Koffas

Continuous research effort is being put into alternative energy production, such as biofuels, to reduce our reliance on fossil fuels. One of the stumbling blocks in this research is that production *in vivo* is adversely affected by alcohol toxicity and product inhibition. A solution to this problem has been previously proposed by our lab: remove the enzymes from the cells, immobilize them on resin, and run a cell-free reaction. Our previous research has yielded favorable isobutanol production compared previous *in vivo* systems. An issue we encountered was a bottleneck at the aldehyde conversion step, with much of the aldehyde remaining unconverted. To solve this issue, we proposed a substrate channeling method of engineering dockerin fusions of the pathway enzymes that have targeted binding to a cohesion scaffold. The rationale behind carrying out this fusion is that by binding the enzymes in close proximity to each other, the diffusional length for the aldehyde can be reduced and the local concentration of this intermediate in the presence of the isobutanol producing enzyme will increase. This immobilization is also more effective than the previous one as the enzymes will bind to a specific cohesin on the scaffold, instead of randomly binding to the resin. This system can be interchanged for other enzymes fused to these dockerins, providing a modular system that is also open to optimizations to improve production.

250c Reductive Enzyme Cascades for Valorization of PET Deconstruction Products Guided By the Specificity of Carboxylic Acid Reductases

Madan Gopal, Roman Dickey, Neil Butler, Michael Talley, Mary Watson, Wilfred Chen, Aditya Kunjapur

To better incentivize the chemical recycling of plastic wastes, new chemical transformations must be developed. Polyethylene terephthalate (PET) is a recalcitrant plastic whose deconstruction through chemical and biological means has recently received much attention. However, a limited number of functionalized products have been formed from the PET monomer terephthalic acid (TPA). In this work, we present the design of a biocatalytic cascade to convert TPA, a dicarboxylate, to its corresponding diamine, *para*-xylylenediamine (*p*XYL), by constructing a multienzyme cascade in a cell-free environment. Design of our cascade first begins by probing the substrate specificity of natural carboxylic acid reductases (CARs). CARs catalyze the selective 2-electron reduction of carboxylic acids to aldehydes. In addition to screening many existing and novel CARs towards a range of carboxylate-containing PET deconstruction products, we also compared rates of reduction to the associated aldehydes in a system which included enzymatic NADPH and ATP regeneration. While our highest performing CAR was not able to completely convert 5mM of TPA to terephthalaldehyde (TPAL) at our chosen endpoint of 24-hours, coupling CAR to an ω-transaminase, which converts aldehydes to amines in the presence of an amine donor, drove the conversion of TPA to near completion, and resulted in a high yield and selectivity of the target diamine, *p*XYL. By combining chemical depolymerization with enzymatic functionalization, we show, to our knowledge, the first reports of CAR specificity towards TPA and biocatalytic synthesis of TPAL and *p*XYL. This work lays the foundation for eventual valorization of waste PET to higher value materials, such as nonisocyanate polyurethanes, that can be made from *p*XYL.

250d Computational Design of Phosphotriesterase for Organophosphate Hydrolysis

Jacob Kronenberg, Maria Kulapurathazhe, Jason Chen, Ashwitha Lakshmi, Stanley Chu, Jin Kim Montclare

Organophosphates (OPs) are a class of chemicals which are commonly used as pesticides, but they are highly toxic and pose a threat to human health and the environment. The enzyme phosphotriesterase (PTE) catalyzes the hydrolysis of OPs and is a promising method for detoxification. Fluorination of the enzyme through the incorporation of the non-canonical amino acid *para*-fluorophenylalanine can be used to improve its stability, but decreases the protein's solubility, making expression difficult. Here, we use computational design to design PTE variants with high positive or negative surface charges, allowing improved solubility. We express the designed variants in *Escherichia coli* and purify them using fast protein liquid chromatography. We use dynamic scanning calorimetry and circular dichroism spectroscopy temperature scans to examine the effect of supercharging on the enzyme's thermostability. In order to determine the effect of supercharging on enzyme kinetics and resilience of the variants to high temperatures, each variant is incubated at a range of elevated temperatures, then used to catalyze the breakdown of paraoxon, a model OP. The breakdown rates were analyzed to determine the kinetic parameters and temperature resilience for each variant.

250e The Use of β-Galactosidase to Produce N-Acetyllactosamine As a High Value Molecule Directly from Dairy Whey.

Sandra Kentish, Masih Karimi Alavijeh, Anne Meyer, Sally Gras

Large volumes of whey are generated during the production of cheese, Greek yoghurt and other products from bovine milk. Much of the nutritional value of this whey is still under-utilised, being sent to wastewater or used as animal food. In this work, we focus on the use of this key resource to generate higher value ingredients. In particular, we consider the use of β-galactosidases in transgalactosylation reactions between lactose and N-acetylglucosamine (GlcNAc) to generate N-acetyllactosamine (LacNAc). This functionalised disaccharide is a prebiotic and can be used as a building block for a range of human milk oligosaccharides, which have very high value within infant formula mixtures.

We show that the cations present in dairy whey can have a positive impact on the transgalactosylation reaction. Specifically, concentrations of 100 mM of calcium or magnesium reduce the reaction rate constant for the undesirable hydrolysis reaction and formation of the undesired isomer, Allo-LacNAc at longer timeframes, due to a loss of enzyme activity. Addition of magnesium enhances the selectivity for LacNAc over Allo-LacNAc with no significant reduction in the LacNAc yield.

Further, we show that the β-galactosidase enzyme can be immobilised onto silica particles using a cross-linked layer-by-layer encapsulation method. Immobilization significantly improved the enzymatic LacNAc yield compared to the free enzyme. The immobilized enzyme was successfully reused in eight consecutive reaction cycles with no significant reduction in this yield.

In our laboratory, we have shown that we can generate LacNAc of over 92% purity directly from dairy whey using activated carbon followed by selective crystallisation with ethanol. This LacNAc was used directly in an enzymatic reaction with whey derived sialylated casein glycomacropeptide to form 3'-sialyl-N-acetyllactosamine (3'-SLN). This functionalised trisaccharide can be used to form antiviral drugs that inhibit the binding of sialic-acid-targeting viruses, including several human or avian influenza strains.

Our process simulations compare the economics of a range of different downstream separation designs once LacNAc is produced; Anion-exchange chromatography; activated carbon; GlcNAc crystallization and activated carbon; and selective crystallization. We show that selective crystallisation is the best approach, with a payback period of between 0.6 and 4.4 years for a production scale of 5 tonnes of LacNAc per year.

In conclusion, this body of work provides an exciting opportunity to take a low value dairy process stream and transform it into products of much higher value in an economic and sustainable manner.

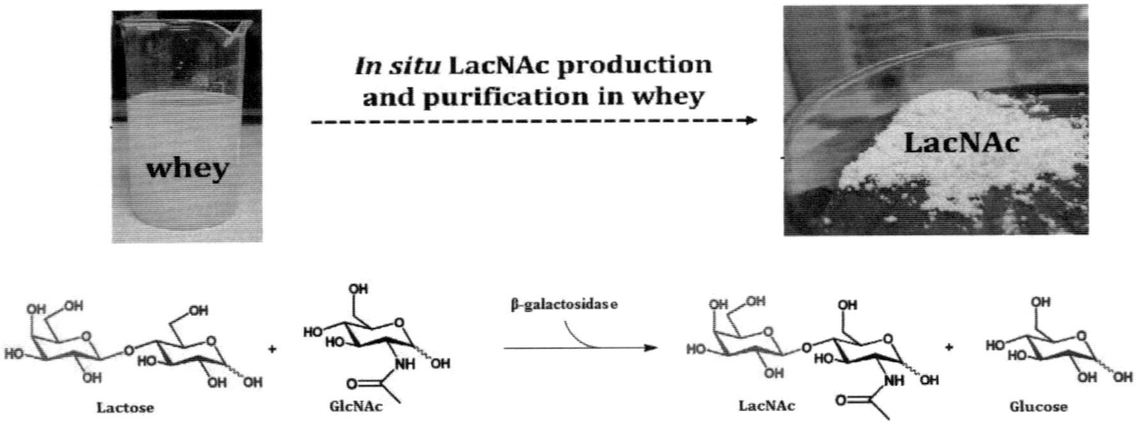

whey
In situ LacNAc production
and purification in whey
LacNAc
β-galactosidase
Lactose
GlcNAc
LacNAc
Glucose

250f Biosynthesis of Triacsin Featuring an *N*-Hydroxytriazene Pharmacophore

Antonio Del Rio Flores, Wenjun Zhang

Triacsins are an intriguing class of specialized metabolites possessing a conserved *N*-hydroxytriazene moiety not found in any
other known natural products. Triacsins are notable as potent acyl-CoA synthetase inhibitors in lipid metabolism, yet their biosynthesis
has remained elusive. Through extensive mutagenesis and biochemical studies, we here report all enzymes required
to construct and install the N-hydroxytriazene pharmacophore of triacsins. Two distinct ATP-dependent enzymes were revealed
to catalyze the two consecutive N–N bond formation reactions, including a glycine-utilizing, hydrazine-forming enzyme (Tri28)
and a nitrite-utilizing, N-nitrosating enzyme (Tri17). This study paves the way for future mechanistic interrogation and biocatalytic
application of enzymes for N–N bond formation.

250g Bioconversion of CO2 to Value-Added Chemicals

Pamela Peralta-Yahya

First and second generation biorefineries have been successful at utilizing microbes to convert sugar and lignocellulosic biomass, respectively, to produce fuels and chemicals. The food vs. fuel dilemma limits the expansion of first generation biorefineries, while in second generation biorefineries, the high cost of lignocellulosic biomass deconstruction stymies the production of low-cost chemicals from this renewable source. In this talk, I will focus on our engineering of biological systems for the production of chemicals from renewables. In particular, I will highlight our recent work towards 3rd generation biorefineries and the direct biological conversion of carbon dioxide to fuels and chemicals.

253a Multi-Scale Modeling of a CHO Production Process Using Integrated Machine Learning Models and Genome-Scale Metabolic Models

Saratram Gopalakrishnan, Jasmine Tat, Miguel Angel Valderrama-Gomez, Fabrice Schlegel, Pablo A. Rolandi, William Johnson, Cleo Kontoravdi, Nathan Lewis

The goal of pharmaceutical bioprocess optimization is the maximization of antibody titer at harvest. Common strategies to achieve this goal include cell line optimization for improved growth and antibody specific productivity, and process optimization for optimal media feeding and timing of hypothermic shift to induce antibody production. Mathematical models that can accurately predict cellular responses and antibody production under changing bioreactor conditions in the bioreactor are a valuable tool that inform interventions and strategies to accelerate cell line and process development. This talk presents a novel dynamic flux balance analysis (dFBA)-based framework to predict concentration profiles of cells, glucose, lactate, amino acids, and therapeutics in a perfusion bioreactor. Because process performance depends on the changing metabolic states over the course of the process, we first characterize the metabolic features of five CHO cell lines engineered to produce three types of antibodies. In four of five cell lines, we find that high producer clones typically grow to higher cell densities than the unselected clones, and that the final titer positively correlates with the final cell density at the end of the growth phase. Interestingly, the final titer negatively correlated with cell density for the fifth cell line, implying that this clone follows a different pathway to achieve optimal antibody titer. Exo-metabolomic characterization of the cell lines revealed two distinct metabolic states corresponding to the growth phase and the production phase, respectively. Upon performing a principal component analysis on the computed fluxes, we found that a shift from growth to production phase was marked by a reduced growth, reduced lactate secretion, and increased antibody production. Hierarchical clustering of gene expression data measured for each cell line revealed that differences in cell state were primarily based on process phase (i.e., growth vs production phase) and that the cell state varied minimally across different cell lines. Upon extracting context-specific models using mCADRE, we found that the growth phase models preferentially activated anabolic pathways such as lipid and nucleotide biosynthesis, whereas amino acid degradation pathways were preferentially activated in the production phase models. Next, we developed a novel framework to simulate the concentration profiles of all metabolites and cells in a perfusion bioreactor using the knowledge gained from the characterization of the five cell lines. The framework expands on the traditional dFBA framework by (i) incorporating nutrient uptake using kinetic rate laws, (ii) constraining fluxes through key metabolic tasks based on process phase, and (iii) limiting scope of the metabolic model based on the process phase. A machine-learning model is trained to predict the process phase based on reactor conditions. The concentration profiles predicted by this multiscale dFBA model are validated using experimental data generated across different bioreactor operating conditions.

253b Mutational Correlation Functions Reveal Patterns of Homologous Recombination across the Bacterial Pangenome

Asher Preska Steinberg, Mingzhi Lin, Edo Kussell

Recombination plays a key role in microbial evolution, and is involved in the spread of antibiotic resistance, antigenic variation, and adaptation to the host niche. However, for many bacteria recombination is so pervasive that it obfuscates clonal relationships, creating challenges for inferring the basic parameters of this critical process from real-world data using traditional phylogenetic methods. We recently developed a non-phylogenetic approach that circumvents these obstacles by analyzing and predicting patterns of correlated mutations in bacterial genomes. By calculating analytical solutions for mutational correlation functions using coalescent theory, and fitting these to measurements from large scale sequencing datasets, we accurately infer the parameters of homologous recombination. Remarkably, the theory allows us to fully characterize the impact of recombination in both an observed sample of sequences and the much larger, unobserved genetic pool with which it recombines. Here, we use this approach to quantify variation in homologous recombination across core and accessory genes (essential genes present in all strains of a given species and niche-adaptive genes present in subsets of strains, respectively). Frequent gene gain and loss confounds phylogenetic analysis of accessory genes, and it has remained unclear if core genes have comparatively higher homologous recombination rates, which could change our fundamental understanding of phylogeny. By analyzing 12 bacterial species using >100,000 genomes, we show that core genes often have higher homologous recombination rates than accessory genes. Moreover, we demonstrate that decreased sequence divergence lowers the recombination barrier for the core genome. Homologous recombination may therefore act concomitantly with purifying selection to optimize genes required for essential cellular function. Our work has far-reaching implications for understanding how the interplay between environmental stress and homologous recombination shapes niche-adaptive genes such as those involved with drug-resistance and antigenic variation.

253c Constraints on Information Flow in Metabolic Networks

Christian Euler, Radhakrishnan Mahadevan

Metabolism, like all open systems, is fundamentally constrained by flows of mass, energy, and information. Mass balance has formed the basis of systems-level metabolic analysis in both FBA and kinetic modelling methodologies, and thermodynamic consistency is now readily considered at scale in metabolic modelling (1, 2). However, informational constraints in metabolic networks remain unexplored, despite the fact that biological systems are awash with information that they have evolved to use to maintain homeostasis, and that understanding such constraints in gene circuits and signal transduction systems has proven useful for their characterization and engineering (3). Uncovering the constraints on information flow within the context of metabolism could enable the design and optimization of next-generation metabolic systems for chemical transformation, including those which make use of small-molecule regulation of enzyme activity in place of/in addition to engineered gene circuits, and those which employ engineered autocatalytic metabolic cycles, as in many novel C1-utilizing systems (e.g. 4). These have the potential to radically transform the space of available feedstocks for bioprocesses, perhaps enabling the valorization of CO_2 and CH_4 at scale in the production of carbon-negative chemicals, for example. However, the regulation and stability of such systemsis not well understood because of the rapid time scale over which small-molecule regulation (SMR) of enzyme activity operates relative to transcriptional regulation. In particular, which metabolites in the network are informational at this level of metabolic control and how that information is naturally transferred (or not, as is relevant to instability) throughout the metabolic and small-molecule regulatory networks (SMRN) are both completely unexplored areas of investigation.

Here, a very simple representative metabolic subsystem - the branchpoint - is modelled under varied constraining scenarios. Sensitivity analysis is used to show that thermodynamic constraints on the consumption of metabolites at branchpoints and saturation kinetics of their consuming enzymes have the combined effect of limiting the amount of information that can be transferred in the metabolic network alone (i.e. without regulation). This is because thermodynamic constraints enforce operation of one downstream branch near equilibrium, causing the others to be pushed far from equilibrium and enforcing their operation near saturation. As a result, fluctuations in upstream fluxes may not be propagated to some pathways downstream of branchpoints in the network, and this represents a loss of information about changing upstream conditions. It is then shown that enzyme inhibition by the metabolites at such branchpoints maximally transfers information, thereby providing a structural mechanism for preventing such information loss. This is because inhibition operates most effectively when the inhibitor is near saturation - the same conditions in which information is lost. To understand if information loss minimization is a 'design principle' of metabolic regulation, a genome-wide analysis of a novel *E. coli* small-molecule regulatory network (SMRN) reconstruction is used to show that inhibitors are indeed significantly more likely to be positioned at branch points with multiple downstream branches and strong thermodynamic constraints. These are points of maximum potential information loss in the network, and the overabundance of inhibitors at such points suggests that indeed minimizing information loss is a broad design goal of metabolic regulation.

Practically-speaking, identification of the structural aspects which render some metabolites informational about upstream fluctuations will no doubt help metabolic engineers to identify metabolites in both native and engineered pathways that they can use to manipulate flux distributions, perhaps dynamically, and also to stabilize engineered metabolic cycles, especially those which are autocatalytic. Understanding the natural constraints on information in metabolic networks is essential to uncovering such native internal sources of metabolic information and the mechanisms by which they correlate fluxes. Both of these must be exploited by engineers to optimize online balancing of metabolite pools for production pathways in which flux correlation is key to stability and productivity maximization. This work will thus begin to enable consideration of the rapid dynamics and related stability criteria of nascent one carbon-assimilation pathways by providing a roadmap to understanding the structures underlying such dynamics and stability.

1. P Salvy, V Hatzimanikatis, The ETFL formulation allows multi-omics integration in thermodynamics-compliant metabolism and expression models. Nat. Commun. 2020 11:1 11,1–17 (2020)
2. CS Henry, LJ Broadbelt, V Hatzimanikatis, Thermodynamics-Based Metabolic Flux Analysis. Biophys. J. 92, 1792 (2007)
3. S Uda, Application of information theory in systems biology. Biophys. Rev. 2020 12:2 12, 377–384 (2020)
4. S Gleizer, R Ben-Nissan, Y M. Bar-On, N Antonovsky, E Noor, Y Zohar, G Jona, E Krieger, M Shamshoum, A Bar-Even, and R Milo. Conversion of Escherichia coli to Generate All Biomass Carbon from CO2. Cell, 179(6):1255–1263.e12, (2019)

Figure caption:

A. In the representative branching system, substrate *S* is produced at a rate *u* and consumed by two downstream enzymes to produce P_1 and P_2. The thermodynamic parameter of the constraining branch, ΔG°_{max} , enforces accumula tion of *S* if this branch is reverse-favoured, $\Delta G^\circ\ max \gg 0$, enforcing operation of the unconstrained, forward-favoured branch in a state far from equilibrium ($P_2/S \rightarrow 0$) and close to saturation. However, if the constraining branch is favourable in the forward direction ($\Delta G^\circ max \ll 0$), accumulation of *S* is not guaranteed. B. Elasticities ($\varepsilon^V{}_S$), or scaled sensitivities, for each branch of the system under varied thermodynamic constraint in the constraining branch. The sensitivity of the constraining branch is maximized for strongly constrained systems, in which this branch is enforced to operate closer to equilibrium. In the same conditions, enforced accumulation of substrate enforces low sensitivity in the unconstrained branch because it is constrained to operate closer to saturation. $K_{eq,2} = 10^4$ here. C. Elasticities (ε^V) for catalysis and inhibition with varying degree of cooperativity as a function of either substrate, *S*, or inhibitor, x_I, concentration, respectively, for the system in (A) solved at steady state for varied *u*. The inhibition elasticity, $|\varepsilon^V{}_{xI}|$, increases to a maximum as the inhibitor concentration increases to saturation, whereas the elasticity for catalysis, $\varepsilon^V{}_S$, decreases to zero as the substrate concentration increases to saturation. η is the degree of cooperativity for inhibition; more cooperativity yields larger maximum elasticities. $K_{eq,1} = K_{eq,2} = 1000$ here. C. Steady-state *S* concentration depends on the constraining thermodynamic parameter. As ΔG°_{max} is increased, $\varepsilon^{VI}{}_S$ grows to a maximum, while $\varepsilon^V{}_S$ is minimized. In highly constrained systems, perturbations to *S* will not be propagated in the unconstrained branch, but inhibition by *S* maximally propagates these perturbations in the same conditions. $K_S = 0.1$, $K_P = 1$, $V_m = 1$, $k = 10$, $K_I = 1$, for representative curves in (B) and (C). E. Lower bound information loss estimated for a given parameter set via lumped elasticities, |a|. For catalysis, |a| decreases asymptotically to 0 as the substrate concentration increases, as indicated by the blue arrow. Information loss via catalysis thus increases rapidly as saturation is approached. Conversely, for inhibition |a| increases to 1 (for $\eta = 1$) as the inhibitor concentration increases, as indicated by the black arrow. Thus, near-perfect information transfer is possible as the inhibitor concentration approaches saturation. Inhibition maximally preserves information at saturation, while catalysis can only do so when substrate concentration is 0. (F) Bivariate density distributions of thermodynamic constraints and out-degrees for reactions producing inhibitors and those producing non-regulatory metabolites. Out-degree is the number of reactions for which a given metabolite is a substrate in our SMRN reconstruction. Reactions producing inhibitors are characterized by large thermodynamic constraints and high out-degrees relative to non-regulatory reactions.

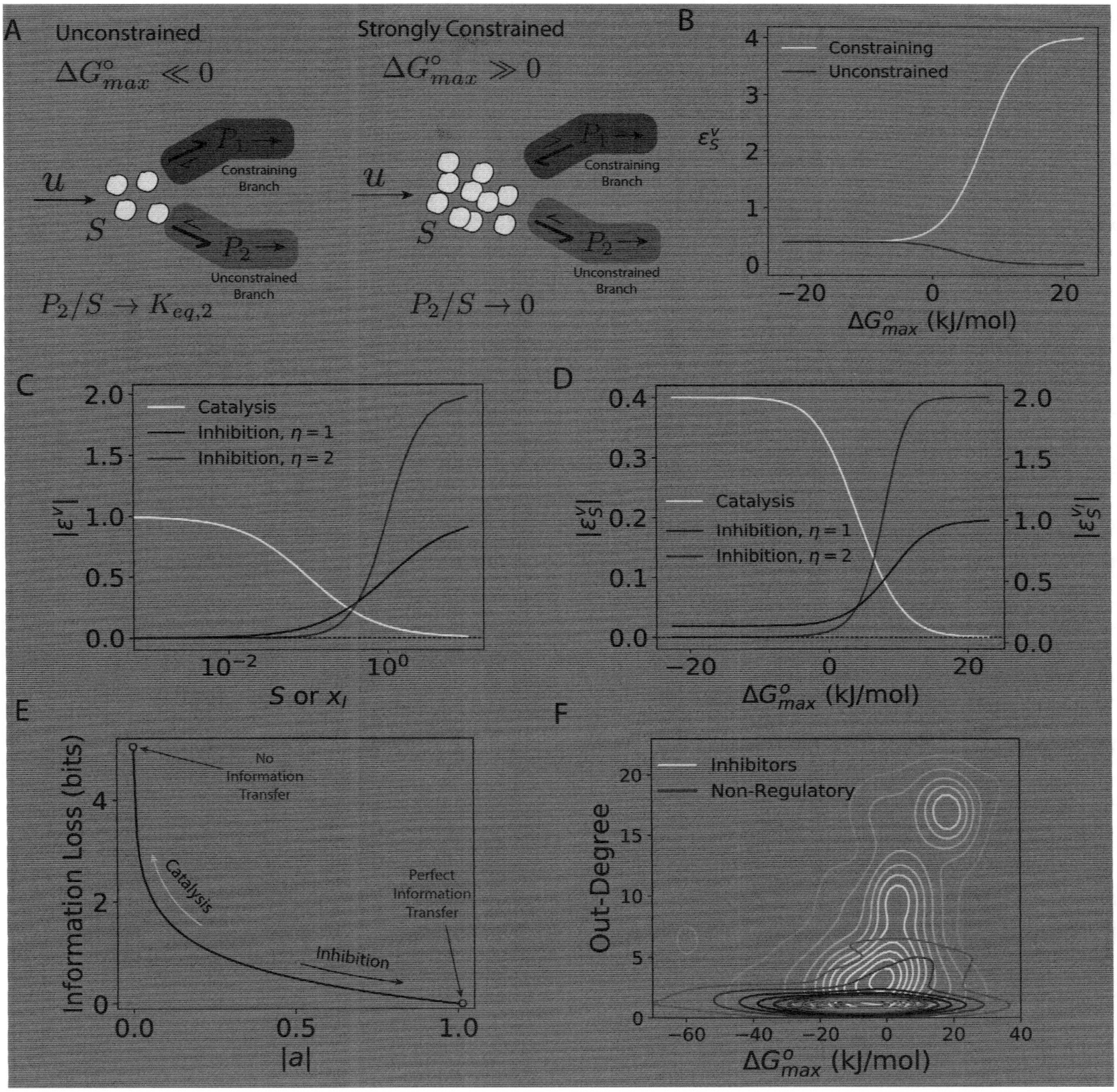
A
Unconstrained
$\Delta G^o_{max} \ll 0$
Strongly Constrained
$\Delta G^o_{max} \gg 0$
Constraining Branch
Unconstrained Branch
$P_2/S \to K_{eq,2}$
$P_2/S \to 0$
B
Constraining
Unconstrained
ε^v_S
ΔG^o_{max} (kJ/mol)
C
Catalysis
Inhibition, $\eta = 1$
Inhibition, $\eta = 2$
$|\varepsilon^v|$
S or x_I
D
Catalysis
Inhibition, $\eta = 1$
Inhibition, $\eta = 2$
$|\varepsilon^v_S|$
$|\varepsilon^{v_I}_S|$
ΔG^o_{max} (kJ/mol)
E
No Information Transfer
Catalysis
Perfect Information Transfer
Inhibition
Information Loss (bits)
$|a|$
F
Inhibitors
Non-Regulatory
Out-Degree
ΔG^o_{max} (kJ/mol)

253d Rapid Discovery of Ribosomal Natural Products Using Synthetic Biology Approaches

Hengqian Ren, Huimin Zhao

Ribosomally-synthesized and post-translationally modified peptides (RiPPs) constitute a major family of natural products that have been found in all three domains of life. To date, RiPPs have been reported with vast structural diversity and hence a broad range of biological functions, including antimicrobial, anticancer, and antiviral activities, making them an ideal source for drug development. The explosion of genomic data and the advancement of genome mining tools have enabled the identification of RiPP biosynthetic gene clusters (BGCs) *in silico*, however, attempts to isolate these compounds often fail due to the negative regulation over these BGCs under standard laboratory conditions, a common problem for traditional natural product discovery. To rapidly explore their biosynthetic potential in nature, here I present the genome mining of RiPPs by leveraging synthetic biology approaches. Candidate RiPP BGCs predicted by bioinformatics are categorized into two groups: the high-fidelity group and the class-defining group. The high-fidelity group comprises BGCs that are annotated as characterized yet highly underexplored RiPP classes. Given the high successful rate of refactored BGCs in heterologous expression, an automation friendly plug-and-play method was developed to refactor the high-fidelity BGCs efficiently. While BGCs in the class-defining group encode unknown and usually complicated biosynthetic machinery, a direct cloning strategy named Cas12a-assisted precise targeted cloning using in *vivo* Cre-lox recombination (CAPTURE) is used to manipulate these BGCs for heterologous expression. Afterwards, the initial product screening is performed by matrix-assisted laser desorption/ionization mass spectrometry (MALDI-MS). Due to the high efficiency and low time consumption of the synthetic biology approaches, the genome mining of RiPP BGCs can be achieved rapidly. The effectiveness of pathway refactoring for high-fidelity BGCs was demonstrated by uncovering a class IV lanthipeptide and four glycocins with interesting structural features and bioactivities, which significantly expanded these two previously underexplored RiPP classes. In addition, two novel classes of RiPPs, named daptides and lipoavitides, were also discovered by the direct cloning approach and characterized with unprecedented biosynthetic machineries. Overall, the synthetic biology based approaches were demonstrated to be highly useful for unveiling novel RiPPs, and have the potential to be developed into a scalable platform for RiPP genome mining.

253f Establishing the Biogenesis Process for a Bacterial Organelle

Carolyn E. Mills, Nolan W. Kennedy, Andre Archer, Svetlana P. Ikonomova, Charlotte Abrahamson, Sasha Shirman, Niall Mangan, Danielle Tullman-Ercek

Identification of novel antibiotic targets is crucial in combatting the growing threat of antibiotic resistance. Ideal targets for antibiotic development are those that specifically exist in pathogenic bacteria, but not healthy biota. Bacterial microcompartments (MCPs) are proteinaceous organelles that specifically aide in the proliferation of enteric pathogens on niche carbon sources, and thus represent excellent candidates for targeted antibiotic development. These MCPs are delimited by a protein shell that encases an enzymatic core, where the core enzymes are responsible for carrying out metabolic processes that aide in proliferation. An abundance of protein-protein interactions (PPIs) are responsible for holding together the MCP shell; however, development of inhibitors against the interactions that make up the MCP has been limited by knowledge of which interactions are essential for proper MCP function.

In this talk, I will describe our recent discovery of the three processes essential to proper formation of the 1,2-propanediol utilization (Pdu) MCP from model pathogen *Salmonella enterica* serovar Typhimurium LT2, and which essential proteins are required for these processes. Further, I will describe how we have used computational modeling tools to understand how different perturbations to the assembly state of the Pdu MCP impact the ability of MCPs to confer benefits to cell metabolism, and how we corroborated these results with experiments. Combined, this information provides a comprehensive picture of the Pdu MCP assembly process, setting the stage for which PPIs we should target for future antibiotic discovery.

253h Dewdrops on the Genome: Regulation of Cell-Identity By Biomolecular Phase Transitions

Krishna Shrinivas

Broader motivation: Human health requires coordination between thousands of cell-types (identities) that all remarkably share an identical genome. A cell's identity i.e. its ability to perform specific yet diverse functions, is dependent on (a) selective ***decoding*** of the genome (gene regulation) (b) spatio-temporal ***organization*** into modules (organelles) and (c) ***interactions*** with other cell-types and environment (signaling) - inherently spanning multiple scales from genes to tissues. Mistakes or dysregulation in this process underlies improper cellular identity and disease, including cancer, auto-immunity, and degenerative disorders.

Key challenge: Recent research, including work I will describe, has uncovered an important role for **phase transitions in cellular function and identity**. However, quantitative models of self-organization, including phase transitions, in living systems remain lacking. This is largely due to the complexity involved - cells contain thousands of interacting, reactive components (DNA, RNA, proteins, lipids) present in crowded heterogeneous environments and modulated by active processes that feedback across scales spanning orders of magnitude.

Talk overview: In this talk, I will outline **an interdisciplinary and collaborative approach** to study **phase transitions in cell function and identity** at the interface of physics and biology, moored by approaches from chemical engineering. My talk will be focused on three aspects of cell-identity outlined above, through the lens of phase transitions and self-organization:

- ***Decoding*** *gene regulation:* I will introduce our discovery of a key role for phase separation, a physicochemical process, in regulation of mammalian gene expression(1). By integrating coarse-grained simulations, polymer chemistry, and non-equilibrium transport with live-cell and biochemical experiments, I will describe a series of collaborative efforts that revealed features of non-coding regulatory elements (2) and identified an important feedback axes between active RNA synthesis and gene expression (3). I will conclude this section by highlighting ongoing research on the role of gene activity in nuclear organization.

- ***Organizing*** *the cell:* Phase separation contributes to organizing the cellular milieu into dozens of co-existing compartments containing hundreds of species. Predicting the phase behavior of such complex fluids has remained a daunting task. In this section, I will outline our recent efforts to address this challenge by integrating statistical thermodynamics, numerical simulation, and approaches from random-matrix theory. I will describe how fluids with dozens of components exhibit surprisingly predictable phase behavior (4), even when the underlying component interactions are drawn randomly from a distribution. I will conclude by highlighting ongoing studies on non-equilibrium processes and linking programmed (or evolved) interactions to emergent self-assembly.

- ***Interacting*** *cell-circuits*: Cells interact with their physiological context to drive signaling that often relies on combinatorial and stochastic cues. In this section, I will describe ongoing collaborations to decipher mechanisms by which cell-types in the sensory system proofread cell-cell connections with high fidelity through self-organization, employing a combination of Monte-Carlo simulations and theory.

Long-term: In the last decade, invention of exciting tools have enabled characterization (sequencing, imaging) and engineering (gene-editing, synthetic biology) of genes and cells - holding great promise for human health and industrial applications. Given the prominent role of self-organization in living systems - I aim to develop quantitative physicochemical models that will provide a rational basis to guide both mechanistic and translational applications of these tools with focus on diseases caused by improper cell-identity. More specifically, in the short-term, I aim to contribute towards the rational design of therapeutics for the expanding list of diseases that are coupled to defects in phase behavior.

References:

1. D. Hnisz=, **K. Shrinivas=**, R. A. Young, A. K. Chakraborty, P. A. Sharp, A Phase Separation Model for Transcriptional Control. *Cell.* **169**, 13–23 (2017).
2. **K. Shrinivas=**, B. R. Sabari=, E. L. Coffey, I. A. Klein, A. Boija, A. V. Zamudio, J. Schuijers, N. M. Hannett, P. A. Sharp, R. A. Young, A. K. Chakraborty, Enhancer features that drive formation of transcriptional condensates. *Mol. Cell.* **75**, 549-561.e7 (2019).
3. J. E. Henninger=, O. Oksuz=, **K. Shrinivas=**, I. Sagi, G. LeRoy, M. M. Zheng, J. O. Andrews, A. V. Zamudio, C. Lazaris, N. M. Hannett, T. I. Lee, P. A. Sharp, I. I. Cissé, A. K. Chakraborty, R. A. Young, RNA-Mediated Feedback Control of Transcriptional Condensates. *Cell.* **184**, 207-225.e24 (2021).
4. **K. Shrinivas[c]**, M. P. Brenner, Phase separation in fluids with many interacting components. *Proc. Natl. Acad. Sci.* **118** (2021), doi:10.1073/pnas.2108551118.

253i Antibody-Lectin Bispecifics for Glyco-Immune Checkpoint Blockade

Jessica C. Stark, Melissa Gray, Simon Wisnovsky, Nicholas Riley, Carolyn Bertozzi

Cancer immunotherapies, including checkpoint blockade antibodies, represent a breakthrough in cancer treatment with long-term remission observed in select patients. Unfortunately, many tumors remain unresponsive to existing immunotherapies, creating an urgent need to therapeutically target additional immune checkpoints that drive cancer progression. Glyco-immune checkpoints – in which cell-surface biopolymers decorated with sugars, or glycans, engage glycan-binding receptors on immune cells – have emerged as immune modulatory pathways that are misregulated in the context of cancer. In particular, recent evidence suggests that upregulation of the sialic acid monosaccharide on cancer cell surfaces allows tumors to resist treatment by engaging inhibitory receptors called Siglecs on immune cells. However, the lack of glycan-binding reagents with high affinity and selectivity has historically prevented targeting of tumor-associated glycans for cancer immunotherapy.

To address this need, we have developed a new class of antibody-lectin bispecifics (AbLecs) targeting tumor-associated glycans for checkpoint blockade. In this approach, glycan-binding domains from immune receptors (e.g., Siglec receptors) are coupled to high-affinity binding domains from FDA-approved antibodies targeting common tumor-associated antigens (e.g., trastuzumab) via knobs-into-holes heterodimerization technology. Western blot and mass spectrometry results indicated correct heterodimeric assembly of eight AbLec candidates, demonstrating the feasibility of the AbLec approach. We found that AbLecs bind cancer cell lines at nanomolar concentrations and can block cognate Siglec receptor binding via flow cytometry. Finally, we showed that AbLecs enhance killing of diverse human tumor cell lines by primary immune cells *in vitro* compared to their FDA-approved parent antibodies. Enhancement of *in vitro* tumor cell killing with AbLecs was dependent on the targeted Siglec and was more potent than the combination of the parent monoclonal antibody and a Siglec-blocking antibody. **These studies provide proof-of-principle for AbLecs as a first-in-class, modular platform technology enabling blockade of glyco-immune checkpoints for cancer immunotherapy.**

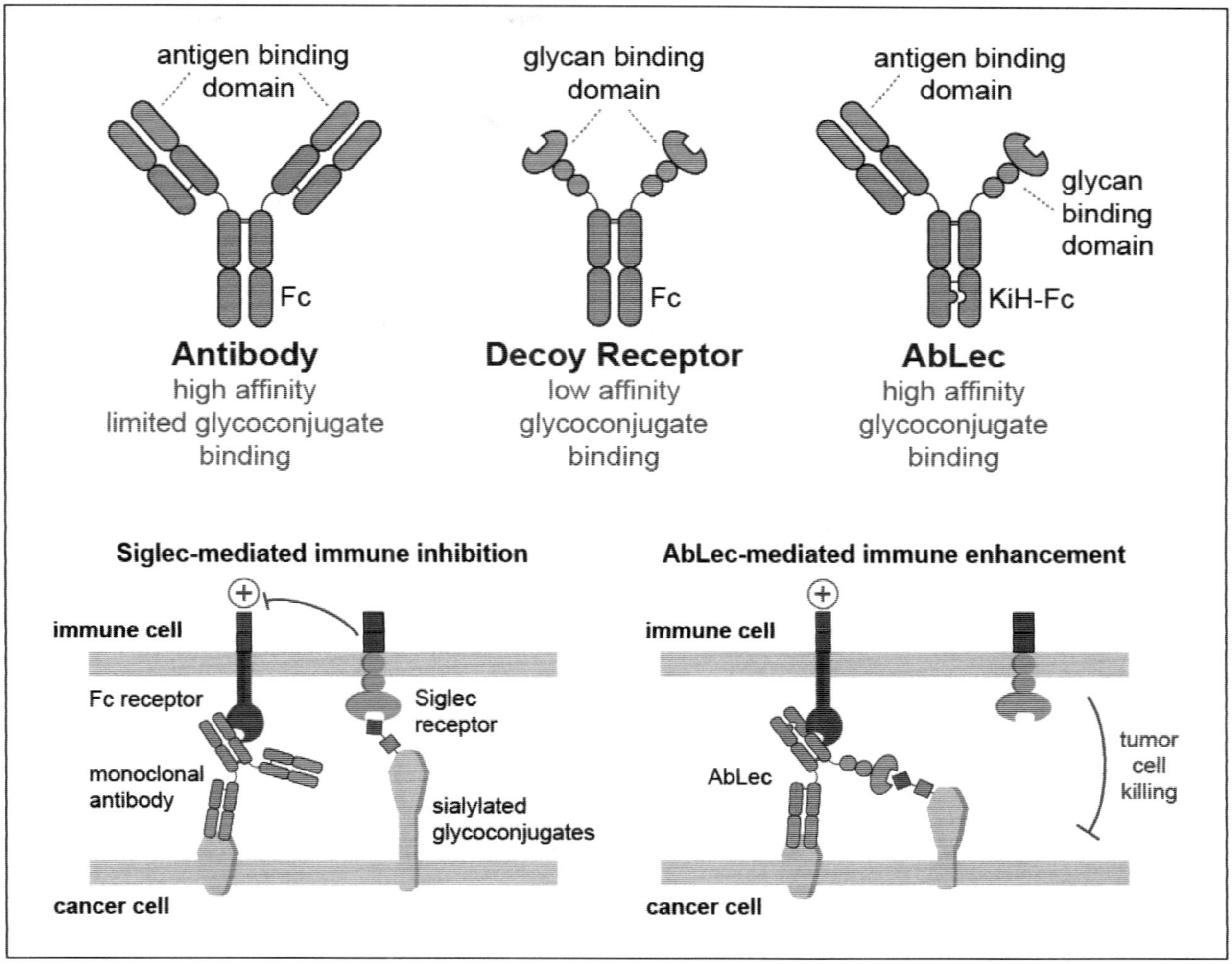
antigen binding domain
Fc
Antibody
high affinity
limited glycoconjugate binding
glycan binding domain
Fc
Decoy Receptor
low affinity
glycoconjugate binding
antigen binding domain
glycan binding domain
KiH-Fc
AbLec
high affinity
glycoconjugate binding
Siglec-mediated immune inhibition
immune cell
Fc receptor
Siglec receptor
monoclonal antibody
sialylated glycoconjugates
cancer cell
AbLec-mediated immune enhancement
immune cell
AbLec
tumor cell killing
cancer cell

596a Image Processing and Immunofluorescent Cell Imaging: Current Challenges and Opportunities

Hawley Helmbrecht, Elizabeth Nance

Background

Immunofluorescent imaging is one of the most common ways to visualize and quantify features of cells of the central nervous system (CNS). While there are many fluorescent imaging methods, including confocal, multiphoton, and light sheet microscopy, and many reviews about optimizing imaging parameters of each microscopy method, there are far less papers analyzing image processing techniques applied after image acquisition. All fluorescent images undergo some type of image processing for presentation in a publication – even a simple brightness adjustment is an image processing technique. However, there is minimal detail about image processing methodologies included in publications using immunofluorescent cell imaging; even when methods are mentioned, it is rare that values for specific variables needed for reproducibility are also included. The field has recognized issues related to reproducibility and repeatability of immunofluorescent images and published multiple perspectives. Often, these perspectives propose methods for improved image acquisition parameter selection or through improved cell thresholding segmentation algorithms, yet, the effect has been nominal on the overall field.

All fields using immunofluorescent imaging will benefit from improved reproducibility and repeatability of image processing methods, but we currently need better methods for standardizing, publishing, and searching for image processing techniques and methodologies. We seek to address immunofluorescent image reproducibility through (1) a high-level look at the prevalence of fluorescent imaging on CNS cells and (2) an in-depth look at publishing practices of image processing methods, with specific emphasis on thresholding and segmentation methods, for quantifying CNS cell morphology. Our high-level look at the prevalence of fluorescent imaging on the major CNS cells - neurons, glia, and vascular cells - with common immunofluorescent imaging techniques shows the widespread application of each imaging method on a every cell of the CNS. The effect of image reproducibility is not cell dependent and addressing this issue would improve image processing methods for all fields studying cell biology. To further identify image processing issues on immunofluorescent imaging of CNS cells, we completed a deeper dive by thoroughly analyzing the papers returned by the earlier search. The deeper dive highlights both the publication practices surrounding image processing methodologies and wide variety of models, software, and quantified features for each cell type. We combine our high-level approach and deeper dive to point out areas for improvement in publication practices behind image processing of immunofluorescent images while also visualizing new connections for researchers to explore new software, models, or feature quantification technique for any cell of the CNS.

Methods

Prevalence of Immunofluorescent Microscopy Methods for Imaging CNS Cells

We identified the major cells of the CNS, neurons, glia, and vascular cells, and then subdivided each category into specific cells or classifications. The neuron group was divided by both functional differences and morphological differences. The glia group was further divided into microglia, astrocytes, and oligodendrocytes. Finally, the vascular cells were divided into endothelial cells, pericytes, and vascular smooth muscle cells. The immunofluorescent microscopy techniques searched in combination with each subgroup of CNS cells are confocal, general fluorescent, fluorescent widefield, multiphoton, light sheet, total internal reflection, and super resolution microscopy. For each CNS cell, we determined the related medical subject heading (MeSH) term, except for the neuronal morphological differences groups which do not have MeSH terms. For the immunofluorescent imaging methods, only confocal microscopy and general fluorescent microscopy have MeSH terms. For all other microscopy techniques, we used the MeSH term "Microscopy, fluorescence" and the type of microscopy as an additional search term. All searches were completed on the same day using PubMed. For the first search, we used the CNS cell word/MeSH term and the microscopy word/MeSH term combined with the join AND. For the second search, we did three searches: a primary search which added the MeSH term "Image Processing, Computer-Assisted" with the word AND, a secondary search looking for either segment OR threshold, and a tertiary search looking for image NOT the MeSH term AND segment OR threshold. Once the search was complete, a CSV file of all paper metadata including title, author, and PubMed ID was saved for deeper analysis.

Deeper Dive Exploration of CNS Cells Imaged with Confocal Microscopy and Tagged with Mesh Term for Image Processing

For all papers returned by our literature search – the search using the MeSH term for confocal microscopy, the CNS cell word or term, and the MeSH term for image processing - we separated all papers published after 2010. For all separated papers, we tagged the CSV metadata of the publication with the following categories: animal or cell model, quantified features, threshold techniques used, segmentation techniques used, and general image processing software used. We then compiled a table of each CNS cell (rows of the table) and under each category (columns of the table) included all tagged features and publication references. After compiling the table, we alphabetized each of the table's cells and applied a two-color heatmap per column for the number of publications that return that specific feature of the category where higher intensity color correlated with higher number of publications.

Results

Confocal microscopy is the most common fluorescent imaging method with multiphoton microscopy as the second most common. Among the cells, sensory receptor cells and endothelial cells have the highest number of overall publications. Additionally, endothelial cells have the most overall use of different microscopy techniques. However, second to endothelial cells, astrocytes also have quite a wide range of fluorescent microscopy techniques used in multiple publications even though they are not the second most published cell. When specifying the search to include papers with the image processing MeSH term, overall publications decrease by a factor of ten for all CNS cells. The decrease in publication prevalence shows that even though authors are regularly using image processing, they are regularly under reporting image processing as a technique within their publications.

After tagging the confocal and image processing publications for specific features, we found that mice are the most common model used and the most common features quantified are count and intensity. It is likely that count and intensity are the most common measured features as they are the most accessible without advanced techniques or software. We also found that the most common segmentation and threshold methods cited in publications were "method unspecified" – the authors claim thresholding and segmentation occurred, and in some cases the software used, but do not include details as to how the processes were carried out. Finally, the most common software used is ImageJ – the open-source image processing software from the NIH – followed by Imaris, with MATLAB being the most frequently used coding language. Prevalence of software use shows the importance of open-source software on cell image processing, but also the wide variety of tools that are available.

Conclusions

Our literature analysis provides an overview of the state of the immunofluorescent cell imaging field and highlights areas where increased methodology publication or publication tagging would make a positive difference. Additionally, our deep dive into specific features shows that image processing methodologies are under-published with both regards to how often they are tagged as image processing and to the variables used for thresholding and segmentation. Our deep dive table can be used by researchers to explore new quantification techniques and methods to apply to their pre-existing work based on similar publications. Immunofluorescent imaging will likely continue to be one of the most popular methods for visualizing and quantifying CNS cells. However, we need to have more rigorous criteria and standards for publication of image processing methods so that we can standardize cell quantification and improve imaging analysis reproducibility.

596c Amplification-Free, Nucleic Acid-Based Detection of Pathogens Using the Lateral Flow Assay Format

Yan Cao, Zhenrong Zheng, Sukantha Chandrasekaran, Jacob Schmidt, Omai Garner, Harold Monbouquette

An amplification-free, nanopore-based nucleic acid detection platform has been demonstrated for rapid, sensitive 16S rRNA sequence-specific detection of pathogenic bacteria. This new technology exhibits an ultralow RNA detection limit of ~100 zM (~10^{-19} M) and has been used to assay *Neisseria gonorrhoeae* over a clinically important range of 10-100 CFU/mL in human urine with no false positives and with sensitivity and specificity of ~98% and ~100%, respectively. The nanopore detector has been integrated into the lateral flow assay (LFA) format for frontend sample processing, and the overall system can detect 10 fM *Escherichia coli* 16S rRNA in <15 minutes. In comparison, current point-of-care (POC) nucleic amplification tests (NAATs) exhibit assay times of 15-30 minutes while studies have shown that a substantial portion of the patient population is unwilling to wait for more than 20 minutes for a test result. In our system, a peptide nucleic acid (*i.e.*, an uncharged polyamide analog to DNA) probe complementary to *N. gonorrhoeae* 16S rRNA, is bound covalently to microbeads. When the charge neutral bead-probe conjugates hybridize with target rRNA, a negatively charged complex is formed that exhibits mobility in an electric field. If the complex is directed to a smaller pore in a glass membrane chip, it will at least partially block it, resulting in a sustained drop in ionic current thereby signaling the presence of the target rRNA and *N. gonorrhoeae*. The single-use component of the system is composed of a lateral flow membrane strip with an integrated, glass chip-based nanopore detector that is preloaded with polystyrene microbeads conjugated with peptide nucleic acid (PNA) probe complementary to the target 16S rRNA. A ~250 µL sample deposited on the test strip quickly flows along the LFA membrane due to capillary action, and target rRNA hybridizes with the complementary PNA probe conjugated to the preloaded microbeads. Subsequent electromechanical detection events measured by conductimetry require low battery power and inexpensive electronics. Further, the strong electric field and an opposing electroosmotic flow at the glass pore mouth together serve as an active mechanism to avert false positives. This simple, rapid, low power and inexpensive amplification-free technology may prove promising for widespread POC use.

596d Enriching the Cell-Free System Based Biosensor Toolbox to Increase Dynamic Range While Retaining Accuracy

Yongchan Kwon, Caroline Copeland, Chloe Heitmeier

Since the cell-free system (CFS) has emerged as an alternative protein synthesis platform to overcome the limitation of the in vivo system, in the synthetic biology era, the CFS receives a spotlight for its inherent versatile characteristics that make the scientists and engineers can repurpose the system to achieve their own research goals outside the cells. With the extraordinary flexibility of the CFS, the system has been expanded to developing biosensors to detect diverse targets from toxic molecules in drinking water to cancer biomarkers in liquid biopsies to viruses. However, these target molecules exist in minuscule amounts in these environmental samples, requiring biosensors to have a low limit of detection, with considerable sacrificing sensing accuracy to reach these low levels. Here, we present ways we have enhanced the output signal of cell-free biosensors while retaining accuracy to achieve low limits of detection. We have improved the output signal by introducing a new fluorescent protein mNeonGreen which shows a significantly brighter output signal compared to commonly used fluorescent proteins such as deGFP, eGFP, and sfGFP. Next, we introduce the use of robust non-native sigma factors, which allow the user more modulation of parts using a cognate promoter compared to a bacteriophage RNAP, and a large reduction in background signal, compared to native *E. coli* sigma factors. Also, we present optimized conditions to enhance the activity of endogenous RNA polymerase to perform analogously to the robust bacteriophage T7 RNAP. Ongoing work in the laboratory seeks to apply this cell-free biosensor toolbox to sense various target molecules. We anticipate that our early findings on the toolbox will accelerate the development of cell-free based biosensors with increasing sensitivity while retaining sensing accuracy.

596f A Novel Inductively Coupled Capacitor Wireless Sensor System for Rapid Detection of Bacterial Growth and Antibiotic Susceptibility

Yikang Xu, Dacheng Ren

Sepsis caused 11 million deaths globally each year. It is paramount that the diagnosis of sepsis is rapid and accurate, study shows that the 96-hour survival rate for antibiotic treatment administered at just 15 hours is only 13.3%. Microbial culturing and Sepsis-Related Organ Failure Assessment are the current leading tools in sepsis diagnosis. Unfortunately, these methods are time consuming and/or lack the capability to identify the causative species and their antibiotic susceptibility. Before the causal pathogens are identified, patients are put on an aggressive empiric antibiotics treatment, which causes adverse side effects and promote the development of antibiotic resistance. The urgent needs for rapid antibiotic susceptibility test (AST) motivated us to develop a new rapid phenotypical assay of antibiotic susceptibility. Here we present a new electrochemical biosensor based on magnetically coupled LC sensors for rapid AST. The engineered LC sensors can be placed in 96 well plates and communicate the reading wirelessly with a receiver coil for signal analysis. The sensors demonstrated rapid detection of growth of *Escherichia coli*, *Staphylococcus aureus*, and *Pseudomonas aeruginosa* in 30 mins, and thus the antibiotic susceptibility to different antibiotics. Drug-resistant strains were used as controls. To our best knowledge, this is by far the quickest AST at phenotype level. This rapid AST has the potential for low-cost point-of-care applications.

630a Transferring CHO Cells from Monolayer to Suspension Culture By a New Design of Experiments - Superlative Box Design

Wanyue Cui, Shijie Liu

A new design of experiments – Superlative Box Design, is adopted to optimize the adaption of an anchorage-dependent, serum-dependent Chinese Hamster Ovary cell line to serum-free suspension culture. This seven-level quadratic response experimental design encompasses the uniformity in the distribution of experiments amongst all the factors of a central composite design while lower the number of experiments required. Six factors are applied in this experimental design with 43 runs plus three more repeating center runs. The cell line adapted to a serum-free medium by serum reduction, chemically defined media ratio increase, and mixing provided by rotation to reach suspension stepwise. Response surface methodology is applied to find the optimum condition.

630c Modeling and Detection of T Cell Exhaustion Markers in a Novel High-Density Centrifugal Bioreactor System with Application in Cancer Immunotherapy Treatments

Brenden Fraser-Hevlin, Kitana Kaiphanliam, Bernard Van Wie, William C. Davis

Cancer is one of the leading causes of death worldwide and furthermore, leads to high medical expenditures each year. As of today, there has yet to be a reliable cure that can eliminate the disease and most traditional treatments such as chemotherapy and radiation have harmful side effects. The need for alternatives has led to the rise of immunotherapy, which involves modifying a patient's immune cells to target cancer cells for destruction. Immunotherapy allows for treatment of cancer with a good long-term prognosis and without the destruction of healthy tissue that occurs in chemotherapy. While immunotherapy is promising, it is not without high costs and significant cell manufacturing obstacles. For most immunotherapy treatments, immune cells must be expanded outside the body after modification using various bioreactor systems. Currently, the processes and systems utilized for cell expansion are costly and inefficient in use of medium. Additionally, in T-cell based treatments, an issue that often arises is the loss of cancer-killing ability during treatment after cell expansion, known as T cell exhaustion. Thus, there is a need for culture processes that are lower-cost, with higher efficiency in use of medium and better maintenance of the cells' cancer-fighting function. Our research group has developed a high-density centrifugal bioreactor (CBR) that maintains suspensions based on a balance of forces in a chamber in which the cells are cultured, shown in Figure 1. Centrifugal force is applied to cells in an outward direction due to the motion of the rotating centrifuge, and fluid forces are applied in the inward direction due to drag and buoyant forces from the continuous flow of medium. The CBR can be used to culture cytotoxic T lymphocytes (CTLs) to high densities, as well as other mammalian cell lines. We developed an updated prototype version of the CBR, and the CTL growth process has been optimized with a kinetic growth model created for the system. More recently, the model is being modified to incorporate the influence of T cell exhaustion on growth kinetics of the effective CTL. Recent work is focused on expanding CTLs in static culture in conditions that have been shown in literature to lead to exhaustion pathways, including limited oxygen and glucose and increased lactate. Here we will report on initial efforts to establish these exhausted cells and confirm the presence of inhibitory receptors and cytokines associated with exhaustion. Initial exhaustion models will also be presented here, along with a comparison to results from large-scale cultures in the physical CBR system. The results of these models will be used to design a process control system and integrated sensors in the CBR that can detect indicators of T cell exhaustion pathways, such as limited oxygen and glucose, and prevent the cells from moving towards an exhaustion pathway. In summary, development of models to optimize the growth of cells in the CBR system, and eventually detect and prevent cell exhaustion from occurring, will allow for major improvements in the efficiency and availability of cancer immunotherapy.

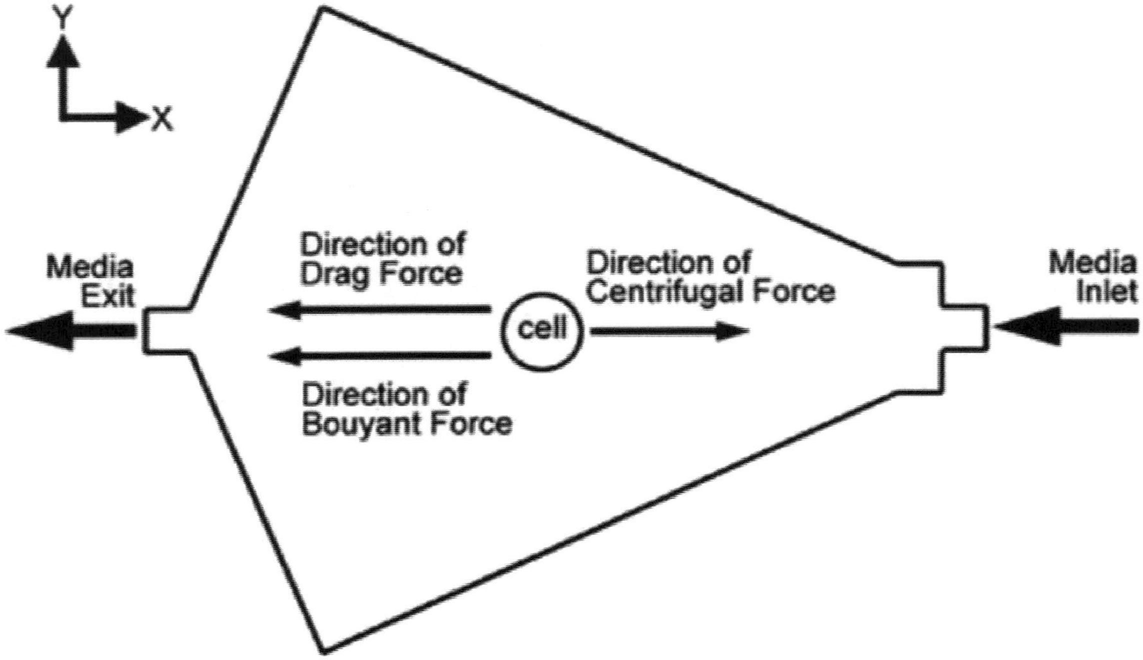
Y
X
Media
Exit
Direction of
Drag Force
cell
Direction of
Centrifugal Force
Media
Inlet
Direction of
Bouyant Force

630d A Deep Examination of Carbon Feeding during the Production of Various Sized Vaccine and Therapeutic Candidates in *Komagataella Phaffi*

Joshua Hinckley, Carmen Elenberger, J. Christopher Love

Komagataella phaffi (*Pichia pastoris*) is a widely used host for heterologous protein expression, owing to its fast growth to high cell density and strong methanol-inducible promoters, such as pAOX1. Strong transgene expression and tight inducibility enabled by pAOX1 have led to its widespread use. However, further optimization of other key process parameters, such as central carbon metabolism, is often tricky. Such factors are highly coupled by a mutual dependence on methanol concentration in *P. pastoris* cultivations. Furthermore, previous studies indicate large protein production requires additional media additives to facilitate proper protein folding, glycosylation, and secretion. In this study, we introduce the use of carbon source metabolism as a pivotal perturbation to investigate such effects *P. pastoris* cultivations have on the production of various sized proteins ranging from low molecular weight nanobodies and insulin glargine to high molecular weight monoclonal antibodies and endogenous human proteins. In particular, these perturbations' effects on protein folding, cellular stress, and post-translational modifications will be addressed. We first explored strategies to tune transgene expression levels and found that specific carbon sources increase transgene expression beyond typical induced methanol expression and increase cellular growth. Next, we identified specific metabolic pathways responsible for cellular development, stress, translation, protein folding, and secretion using high-throughput bulk mRNA and ribosomal sequencing. We used these pathways to determine the optimal combination of carbon sources at the cellular level that reduces cellular stress while increasing protein translation and secretion. Finally, secreted product titers were determined, sequenced using mass spectrometry, and post-translational modifications analyzed using high-performance liquid chromatography. The results from this study indicate the need for optimized carbon source feeding strategies centered around the complexity and the size of the desired protein. This strategy may help accelerate advancements of candidate vaccines and therapeutics from alternative hosts, such as *P. pastoris*, to the clinic.

630e Model-Based Insights on the Effect of Bioreactor pH and Temperature on N-Linked Glycosylation of Mabs Produced By CHO Cells

Jayanth Venkatarama Reddy, Katherine Raudenbush, Ou Yang, Aron Gyorgypal, Antash Chaturvedi, Shishir Chundawat, Marianthi Ierapetritou

Optimization of the monoclonal antibody (mAb) bio-manufacturing process requires efforts in cell line development, media formulation, process monitoring, process control and determining optimal bioreactor operating conditions while meeting critical quality attributes (CQAs) such as those pertaining to N-Linked glycosylation, mAb aggregation and mAb fragmentation [1]. Quality by design (QbD) guidelines set by the FDA emphasizes the need to establish the relationship between process parameters and CQAs when designing a bio-manufacturing process [2]. Choosing the correct pH and optimizing temperature shift strategies is important for bioreactor operation as it can affect N-Linked glycosylation of mAbs, cell growth rate and titer production rate [3, 4]. There is evidence in the literature based on model based investigation of CHO cell-based processes that addition of manganese and galactose has substantial effects on N-Linked glycosylation of the mAbs [5]. However, the literature on model-based efforts to study the effect of bioreactor operating parameters such as temperature and pH on N-Linked glycosylation of mAbs is limited. The effect of such parameters on glycosylation is usually cell line dependent. Hence, more model based efforts can provide insights into developing general trends. Design space evaluation of Temperature and pH is important while designing a bio-manufacturing process. Computational tools such as that presented in this work will provide a platform to do this while utilizing an approach that is in line with the QbD guidelines.

Investigation into process parameters influencing the mAb bio-process was done for a CHO cell line producing a recombinant mAb biosimilar for Trastuzumab by varying pH and Temperature during mid-exponential phase of cell growth to identify metabolic differences that may influence mAb quality. Key metabolites such as glucose, lactate, ammonia as well as extracellular amino acid concentrations were measured along with mAb titer and N-linked glycosylation pattern. A kinetic model was developed to describe the effects of pH and temperature on CHO cell growth, titer production, ammonia profiles and consumption of glucose, aspartic acid and asparagine. An N-linked glycosylation model from the literature [6] was modified to include the effects of pH and temperature and subsequently integrated with the kinetic model for metabolism. Parameter regression was carried out to determine the kinetic and glycosylation model parameters. Sensitivity analysis provides insights into the effect of temperature and pH on the effect of different enzymes in the glycosylation pathway. This model here can be used to improve the understanding of bioreactor operation by capturing the dynamic profile of the protein glycosylation process and further guiding the production of high productivity and high-quality mAb product.

References

1. Bielser, J.-M., et al., Perfusion mammalian cell culture for recombinant protein manufacturing – A critical review. Biotechnology Advances, 2018. 36(4): p. 1328-1340.
2. Rathore, A.S. and H. Winkle, Quality by design for biopharmaceuticals. Nature Biotechnology, 2009. 27(1): p. 26-34.
3. Batra, J. and A.S. Rathore, Glycosylation of monoclonal antibody products: Current status and future prospects. Biotechnology Progress, 2016. 32(5): p. 1091-1102.
4. Yoon, S.K., et al., Effect of culture pH on erythropoietin production by Chinese hamster ovary cells grown in suspension at 32.5 and 37.0°C. Biotechnology and Bioengineering, 2005. 89(3): p. 345-356.
5. Karst, D.J., et al., Modulation and modeling of monoclonal antibody N-linked glycosylation in mammalian cell perfusion reactors. Biotechnology and Bioengineering, 2017. 114(9): p. 1978-1990.
6. Krambeck, F.J., et al., Model-based analysis of N-glycosylation in Chinese hamster ovary cells.

PLOS ONE, 2017. 12(5): p. e0175376.

630f Identification of Heritable Biomarkers That Characterize Resistance to Stress and Improved Productivity in CHO Cell Line Development

Spencer Grissom, Michael Saint-Antoine, Abhyudai Singh, Mark Blenner

Many therapeutic proteins are produced using the Chinese hamster ovary (CHO) cell line due to their natural genetic plasticity, human-like post translational modifications, and superior production of secreted proteins. This genetic plasticity gives way to heterogenous clones that drive cell line development (CLD) where a monoclonal production cell line is identified based off optimized growth, productivity, and product quality. However, this CLD process represents a time and cost barrier to produce these therapeutics and is biased towards clonal populations that perform well in the scaled-down environment that occurs during screening. It fails to identify optimal clones that perform exceptionally well in a larger production environment and associated stress agents. One approach for improving these CLD limitations involves narrowing the clonal pool based on biomarkers, which are genetic states that confer a favorable phenotype. This research describes a workflow for the identification of heritable biomarkers that characterize resistance to stress and improved productivity to enhance the clonal pool during CLD.

To identify suitable biomarkers, a population-based RNA sequencing technique, referred to as MemorySeq, was first used to identify gene expression states whose fluctuations continue for several divisions and were distinct from a noise control. These expression states are considered heritable if their variation significantly exceeded the transcriptome-wide variation in the noise control. This was paired with differential gene expression analysis (DGEA) in the presence of stress characteristic of production conditions. The overlap of heritable expression states from MemorySeq and differentially expressed genes from DGEA with functional analysis may suggest genes that would bias the CLD clonal pool to better performance. The MemorySeq workflow identified nearly 200 heritable expression states and six network communities of co-fluctuating genes, characterized by cellular adhesion, response to chemicals and stimulus, and cell differentiation from GO enrichment analysis. High levels of ammonia, lactate, and osmolality were then introduced in fed-batch format to simulate production cycle media. Day 5 cell samples were used for DGEA and 130 of the heritable genes were differentially expressed in at least one of the stress conditions. Six genes associated with either higher protein secretion, negative regulation of apoptosis, or increased glycosylation were selected from this pool as possible biomarkers for screening. In future work, clones with high expression of one or more of these six genes may be selected and expanded for fed-batch culture to verify the heritability and assess its impact on production performance. If these clones exhibit better performance for extended duration, then this method would significantly reduce the CLD timeline and the adaptability of the clones when grown at production-scale.

630g Synthetic Gene Circuits for Therapeutic Platelet Production

Tara L. Deans

Synthetic biology has transformed how cells can be reprogrammed, providing a means to reliably and predictably control cell behavior with the assembly of genetic parts into more complex synthetic circuits. Using these approaches, we are programming stem cells with novel genetic tools to control genes and pathways that result in changes in stem cell fate decisions, in addition to reprogramming terminally differentiated cells to function as unique therapeutic diagnostic and delivery vehicles.

449a Redox Balance-Based Growth Selection As a Universal Tool for Enzyme Engineering

Han Li

Over 15,000 NAD(P)H-dependent enzymes discovered in Nature represents a rich yet largely untapped resource for catalyst development, with many of these enzymes outperforming any man-made catalysts. To meet the needs of *in vitro*biotransformation and *in vivo* metabolic engineering, these enzymes often need their substrate and cofactor specificity tuned, and their stability and turnover rate improved. However, there's currently no facile ways to accomplish these engineering goals. Traditional methods which typically rely on 96 well-based *in vitro* assays do not provide sufficient throughput to deeply search a large protein sequence space, and do not directly evaluate enzyme's functionality *in vivo*. Here, we report the development of multiple ultrahigh-throughput (~10^9 candidates per round of selection), universal, *in vivo* selection platforms which use cell growth as an easy readout of NAD(P)H-dependent enzyme's activity. These selection platforms are designed based on the general principle of redox balance, where $NAD(P)^+$ to NAD(P)H ratio in *Escherichia coli* cells is drastically perturbed so that cell survival requires a heterologous enzyme that can utilize the desired substrate with sufficiently high activity *in vivo* and concomitantly restore the cofactor redox balance. Using these selection platforms, we have achieved, through just a single round of selection, remodeling of 4-hydroxybenzoate hydroxylase (PobA) to accept a nonnative substrate for natural product biosynthesis, improving the coupling efficiency of cytochrome p450 BM3 for environmental pollutant degradation, and switching the cofactor specificity and enhancing the thermal stability of an industrially important enzyme cyclohexanone monooxygenase (CHMO). Our platforms complete the scope of redox balance-based *in vivo* selection by covering both NADH and NADPH cofactors, and both aerobic and anaerobic conditions. We envision their broad applications in accelerating enzyme engineering.

449b Utilizing Antimicrobial Peptide for Regulation of Global Metabolism and Its Application in Microbial Biosynthesis

Lei Zhuang, Yuxin Liu, Haoran Zhang

Maintaining proper balance between cell growth and target product biosynthesis is a fundamental challenge for metabolic engineering. In this study, we utilize a robust antimicrobial peptide (AMP) to block protein synthesis and thus repress the global metabolism of model bacterium *E. coli* at desired timings, which directs metabolic resources towards the target product biosynthesis. Our results show that, through manipulating the growth conditions (e.g., medium composition), the selected AMP is able to inhibit the cell growth to varied degrees. The induction of AMP expression at different time points during the cell growth also leads to different growth inhibition outcomes, ranging from dramatic to unnoticeable growth freeze. On the other hand, the biosynthesis of a selected target product in the AMP-containing *E. coli* can be engineered to be impacted to a lesser degree by the AMP expression. Thus, the distribution of the metabolic resources between cell growth and target product production can be manipulated through controlled AMP expression. We further explore the use of this strategy for enhancing the biosynthesis of selected products in engineered *E. coli* for facilitating its application in metabolic engineering.

449c Cascaded Dynamic Regulation for Attenuated Metabolic Imbalance and Increased Naringenin Production

Tian Jiang, Chenyi Li, Yusong Zou, Jianli Zhang, Qi Gan, Yajun Yan

The conflicts between cell growth and production often hinder the development of efficient microbial cell factories for the biosynthesis of valuable compounds. Dynamic regulation becomes an efficient approach to relieve the conflicts as it can redirect the carbon fluxes to target production based on cellular status or environmental signals. Naringenin is a valuable natural product with antibacterial, antifungal, and antiviral properties. However, the synthetic pathway of naringenin suffers from the low productivities arising from the imbalance of carbon flux and enzyme expression. Here, we presented a cascaded dynamic regulation system to systematically optimize the naringenin biosynthesis in *Escherichia coli*. The implementation of the cascaded dynamic regulation effectively attenuated the cell burdens and waste of resources caused by constitutive enzyme overexpression, resulting an over 23-fold increase in naringenin titer compared with the original strain. Our results highlighted the great potential of multi-level cascaded dynamic regulation in metabolic engineering.

449d Dynamic Regulation of Metabolic Pathways with Toehold-Gated dCas9 Regulators

Alexander Perl, Manish Rai, Hunter Dalton, Richa Nathan, Mattheos Koffas, Wilfred Chen

In metabolic engineering, once a metabolic pathway is expressed, there is an optimization process that aims to increase the production, titer, rate, and yield of the target molecule with a goal to direct the maximum possible amount of metabolic flux toward the pathway of interest. Unlike static regulatory methods, dynamic regulation of metabolic pathways offers the ability to adapt and respond to intracellular changes, as well as changes in the extracellular environment. Because of this, dynamic approaches have greater potential to regulate metabolic pathways when compared to static ones.

A toehold switch is comprised of 2 components: RNA in a hairpin configuration that blocks a Ribosome Binding Site (RBS) and an endogenous piece of RNA capable of opening the hairpin called a trigger. Our labs were able to take the dCas9 system and incorporate the spacer into a modified toehold switch creating a toehold-gated sgRNA (thgRNA) which acts as an orthogonal, transcriptional regulator when activated by an endogenous trigger strand. This system was used to show transcriptional repression of simple and complex metabolic pathways in *Escherichia Coli* (*E. coli*). We also looked at a variety of factors including: promoter strength, temperature, and induction time on mCherry fluorescence and Violacein production.

More specifically, first we were able to repress a weak promoter driving the expression of Red Fluorescence Protein (RFP) by about 60% when compared to two control groups: No trigger and a non-binding (scramble) trigger. When looking at the strong T7 promoter, only a small change was observed in the experimental group. We also looked at the effect of the T7 thgRNA at repressing the T7 gene in a strain where RFP was chromosomally integrated. Interestingly, despite the increased strength of the chromosomally integrated strain in comparison to the normal T7 promoter, there was significant repression in the experimental group.

Previously, we introduced the violacein pathway, a five-step metabolic pathway, into a pETM6 vector. The expression of the first three genes, VioA, VioB, and VioE are controlled by the weak mutant T7 promoter G6, the VioD gene is controlled by the 4A6 T7 variant, and the VioC gene is controlled by the 3A2 T7 variant. By incorporating these spacers into thgRNAs, we were able to show repression of individual and multiple genes at once with QRT-PCR being used to measure levels of transcriptional repression. All thgRNAs designed in this experiment were expected to perform best at 30°C and induced with .1 mM IPTG. After we expressed the general T7 thgRNA, the thgRNA that indiscriminately targets all five genes, we found that violacein production decreased by ~75%. Interestingly, when the G6 thgRNA was used, the repression dropped to ~67%; unexpected as the G6 thgRNA was found to be more effective at repressing mCherry. When looking at 3A2 repression, we found very little change in the overall levels of violacein or deoxyviolacein despite QRT-PCR showing a 60% decrease in VioC production. In order to look at 4A6 repression, we chose to use the deoxyviolacein pathway, as proviolacein is significantly harder to produce and measure than prodeoxyviolacein. We found a ~30% reduction in deoxyviolacein expression and a ~70% increase in prodeoxyviolacein expression.

The thgRNAs appear to have a greater effect on the weak T7 promoters, as the weakest promoter (G6) had the most repression while the normal T7 promoter showed no statistical repression at all. We were able to show repression of single and multiple genes in the complex Violacein pathway. Our thgRNAs are able to function as a conditional activation of CRISPR-based systems by using highly predictable toehold-mediated strand displacement reactions. Sequence specific unblocking of the spacer allows for both orthogonality and low cross-talk between thgRNAs. Additionally, these devices do not require the screening of large libraries currently needed to create such specific riboregulators as the CRISPR spacers are so sequence specific.

449e MULTI-Sculpt – Multiplex Integration Via Selective, CRISPR-Mediated, Ultra Long Pathway Transformation in Yeast for Natural Product Synthesis

Franklin Gong, Jianing Han, Sijin Li

Yeast has been a versatile model host for complex and valuable natural product biosynthesis via the reconstruction of heterologous biosynthetic pathways. Recent advances in natural product pathway elucidation have uncovered many large and complicated pathways that contain 10-30 genes catalyzing the biosynthesis of structurally complex, valuable natural products in plants and microorganisms. However, the ability to reconstruct ultra-long pathways efficiently in yeast does not match the increasing demand for valuable natural product biomanufacturing. Here, we developed a one-pot, multi-gene pathway integration method in yeast, named MULTI-SCULPT for MULTiplex Integration via Selective, CRISPR-mediated, Ultra Long Pathway Transformation. Leveraging multi-locus genomic disruption via CRISPR/Cas9, newly developed native and synthetic genetic parts, and fine- tuned gene integration and characterization methods, we managed to integrate 21 DNA inserts that contain a 12-gene plant isoflavone biosynthetic pathway into yeast with a 100% success rate in 12 days, with all tested cells yielding the correct, engineered phenotype. This method enables fast and efficient ultra-long biosynthetic pathway integration and can allow for the fast iterative integration of even longer pathways in the future. Ultimately, this method will accelerate the combinatorial optimization of elucidated natural product pathways and accelerate putative natural product pathway characterization heterologously.

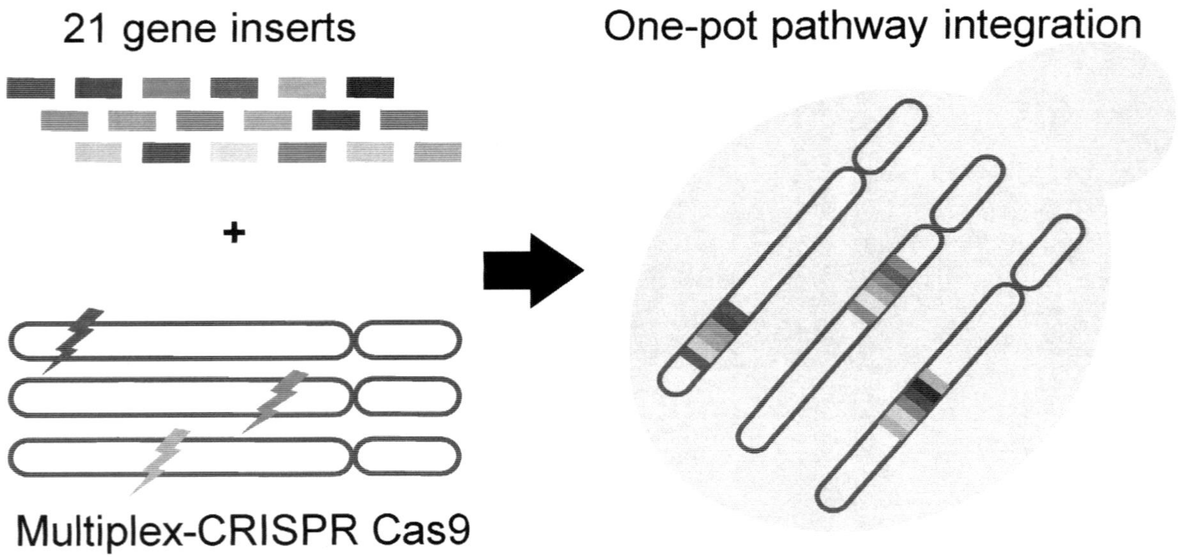

449f Improving Transformation Efficiencies in *Synechococcus* Sp. Via the Novel Combination of DNA Methylation and Plasmid Multimers

Cody Kamoku, Christopher Jones, David Nielsen

Cyanobacteria represent attractive biocatalysts due to their ability to photosynthetically fix CO_2 and convert it into biomass and value-added chemicals. Currently, however, our ability to engineer cyanobacterial metabolism continues to be limited by a lack of versatile and well-characterized genetic tools and associated methodologies. For instance, the delivery genes or pathways of interest is inherently limited by low transformation efficiencies, which slows the overall design-build-test-learn cycle and prohibits applications involving library screening, among others. The availability of self-replicating episomal plasmids can be particularly helpful in this regard, however such parts in cyanobacteria have largely been limited to the RSF1010 broad host range plasmid, which is difficult to work with due to its large size and low copy number, and shuttle vectors derived from native plasmids, whose delivery relies mainly upon conjugation. To help address the need for versatile and easy to use tools for engineering cyanobacteria, we have developed a multi-pronged approach to improve transformation of both self-replicating episomal and chromosomal integration vectors by the synergistic combination of both site-specific methylation and the generation of plasmid multimers. The first approach relies upon incorporation and methylation of the Highly Iterated Palindrome 1 (HIP1) sequence, which is found in virtually all cyanobacteria. HIP1 is an eight base pair sequence (consensus: GCGATCGC) that is highly abundant in cyanobacterial genomes, occurring on average every 1000 bp in *Synechococcus* sp. PCC 7002. Although the exact function(s) of HIP1 remains unknown, it is hypothesized to play a role in DNA recombination, repair, and/or maintenance of chromosome structure. In a previous study, it was shown that HIP1 methylation improved the efficiency of chromosomal integration via natural transformation in *Synechocystis* sp. PCC 6803, which was thought to be due to the bypass of a restriction system targeting the HIP1 sequence[1]. If such were the case, in the traditional paradigm of restriction modification systems, removal or methylation of restriction sites would increase transformation into a host strain. However, on the contrary, here we show that the *addition* and methylation of HIP1 sites to virtually any plasmid is sufficient to increase its natural transformation efficiency in cyanobacteria. To demonstrate this, an *E. coli* strain was constructed to express an integrated copy of the HIP1 methyltransferase gene from *Synechocystis* sp. PCC 6803, and used as the host for amplifying and methylating plasmids for transformation into cyanobacteria. The number of included HIP1 sites on the cyanobacterial vector, their relative spacing, and absolute placement on integrative plasmids was first optimized. From this it was determined that incorporation of HIP1 sites directly flanking the outside of the homology arms resulted in the greatest increase in transformation efficiency. For integration vectors with long homology arms (500 bp), methylation of introduced HIP1 sites increased transformation efficiency ~3-fold, whereas for shorter homology arms (250 bp) a ~100-fold improvement was observed. Meanwhile, we have also demonstrated that replicating plasmids can be transformed into in cyanobacteria via natural transformation using plasmid multimers which can be generated via either in vitro or in vivo methods. Ultimately, two approaches can be synergistically combined to further increase transformation efficiencies in cyanobacteria. To make use of these methods, we have also developed and characterized a series of novel replicating plasmids for use with *Synechcococcus* sp. PCC 7002 and the newly discovered fast growing strain *Synechococcus* sp. PCC 11901. Finally, equipped with these novel tools/protocols we engineer both *Synechococcus* sp. PCC 7002 and PCC 11901 for photosynthetic production of sugars. Given the ubiquitous nature of the HIP1 site, the tools/methods developed here are expected to be readily extended to all other naturally transformable cyanobacteria.

References:

1.Wang B, Yu J, Zhang W, Meldrum DR. Premethylation of foreign DNA improves integrative transformation efficiency in Synechocystis sp. strain PCC 6803. Appl Environ Microbiol. 2015 Dec;81(24):8500-6. doi: 10.1128/AEM.02575-15. Epub 2015 Oct 9. PMID: 26452551; PMCID: PMC4644663.

449g Teaching an Old Dog New Tricks — Elucidating Core Design Principles to Engineer Yeasts As Microbial Factories and Disease Models

Zengyi Shao, Deon Ploessl, Yuxin Zhao

Invited Speaker Placeholder.

21a Engineering a High Affinity Cross-Reactive Antibody to Bind VEGF-a and PlGF-2

Samuel Blackman, Alison Lee, Ahlam Qerqez, Jennifer Maynard

Antibodies that restrict blood vessel growth, such as those binding vascular endothelial growth factor alpha (VEFG-A) have shown promise in treating glioblastoma. When the anti-VEGF-A antibody, Bevacizumab, was given to adult glioblastoma patients, tumor size was temporarily reduced until tumors developed mutational resistance and compensated by overproduction for the loss of VEGF-A effects by overproducing other proliferative cytokines, including placental growth factor (PlGF)[1]. A study on twelve human tumor cell lines identified three (DU4475, Caki-1, SKUT1b) that were growth inhibited by anti-PlGF antibody C9v2[2]. However, subsequent treatment of multiform glioblastoma with the anti-PlGF antibody TB-403 in conjunction with Bevacizumab, which block the PlGF-NRP1 and VEGF-VEFGR1 interactions, respectively, was unsuccessful due to a drug-drug interaction that resulted in blockade of VEGF-A activities only[3].The structural homologies between VEGF and PlGF and the fact that each cytokine is able to bind the VEGFR1 and NRP1 receptors, led us to hypothesize that a single antibody binding these shared epitopes may have therapeutic potential to treat glioblastoma (Figure 1).

Towards this goal, we report the discovery of novel antibodies that block PlGF-NRP1 binding. Two antibodies recognize the human and mouse PlGF alleles with high affinity to support antibody development using mouse models of disease. A third antibody shows cross-reactive binding between human PlGF-1 and PlGF-2 isoforms. These antibodies were discovered using phage display of immune libraries generated from mice immunized with human PlGF-2. The antibodies bind PlGF with high affinities and specificities, as measured by biolayer interferometry, and the epitopes have been initially identified. Their biological activity to block cellular proliferation *in vitro* using DU4475 breast carcinoma cells will be shown and compared to the anti-PlGF antibody C9v2 and Bevacizumab. Future work will include phage display of the immunized human PlGF mouse library for cross reactive clones between VEGF-A and PlGF. In addition, the therapeutic potential of all discovered PlGF antibodies will be tested in a human glioblastoma xenograft mouse model. These data support development of a potent cross-reactive antibody able to simultaneously block PlGF and VEGF activities.

References:

[1] Lee et al *Clin. Cancer Res. Off. J. Am. Assoc. Cancer Res.* **2018**, *24* (19), 4643–4649.

[2] Yao et al *Proc. Natl. Acad. Sci. U. S. A.* **2011**, *108* (28), 11590–11595.

[3] Wang et al *Cancer Chemother. Pharmacol.* **2017**, *79* (4), 661–671.

Figure 1

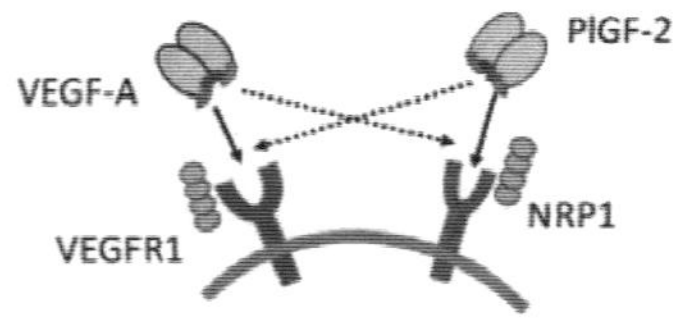

Figure 1. Depiction of observed promiscuous binding of VEGF-A and PlGF-2 to both VEGFR1 and NRP1 cellular receptors.

21b Engineering Pan-Reactive VEGF Antagonists for Neovascular Eye Diseases

Paul Sargunas, Jamie Spangler

Introduction: Neovascularization in the eye drives the pathogenesis of diseases including age-related macular degeneration and diabetic retinopathy, two of the leading causes of blindness in adults. The Vascular endothelial growth factor (VEGF) ligands are major stimulants of neovascularization. Suppression of VEGF ligands binding with their receptors (VEGFRs) have shown clinical benefit in treatment of neovascular eye diseases. Current FDA-approved therapies all primarily target VEGF-A, the most prominent regulator of blood vessel formation, in order to prevent it from binding to VEGFR-2 and subsequently activating signaling. However, VEGF-C is now increasingly implied in pathological angiogenesis, and has been shown to be upregulated following anti-VEGF-A treatment as well as being as potent as VEGF-A in driving retinal neovascularization under hypoxic conditions. No current therapies are bind both VEGF-A and VEGF-C to provide a more complete blockade of neovascularization. We designed an error-prone mutagenic library of the domains 2andamp;3 of VEGFR-2 (VEGFR-2 D23), which binds both VEGF-A and -C, using the yeast surface display platform in order to isolate and characterize high-affinity clones that could bind both VEGF-A and VEGF-C that could act as superior therapeutics for the treatment of neovascular eye diseases.

Materials and Methods: VEGF-A and VEGF-C dimers were produced in a mammalian cell expression system and biotinylated for use as targets in library selections and titrations. We cloned an error-prone mutagenic library of VEGFR-2 into the pCT yeast expression plasmid and transformed the resulting DNA into electrocompetent EBY-100 yeast. We performed five rounds of magnetic- and fluorescence-activated cell sorting (MACS and FACS) against VEGF-A to select for functionally- expressing, high-affinity variants. Variants that bound VEGF-A were then tested for VEGF-C binding. Sequences were extracted from clones that bound both ligands, and the evolved VEGFR-2 D23 mutants were titrated against VEGF-A and VEGF-C in order to determine their affinities for each ligand.

Results and Discussion: VEGFR-2 D23 was found to display on the surface of yeast cells when plasmid DNA encoding the protein sequence with a C-terminal cmyc epitope tag was transformed into competent yeast, but binding was unable to be detected for either VEGF-A or VEGF-C when the biotinylated ligands were incubated with fluorescent streptavidin. An error-prone library with a mutation rate of 1.9 amino acids/variant was cloned and transformed into yeast. We demonstrated the enrichment of the mutagenic library against VEGF-A over 5 rounds of selections (**Figure 1**), and were able to identify several variants with a range of affinities that were able to bind both VEGF-A and VEGF-C with high affinity.

Conclusions: We have identified and characterized high-affinity VEGFR-2 D23 variants that engage VEGF-A and VEGF-C. These clones are being further characterized and will be the basis for a site-directed library based on the crystal structures of VEGFR-2 interacting with VEGF-A and VEGF-C aimed at further increasing the affinity of the antagonist for both ligands.

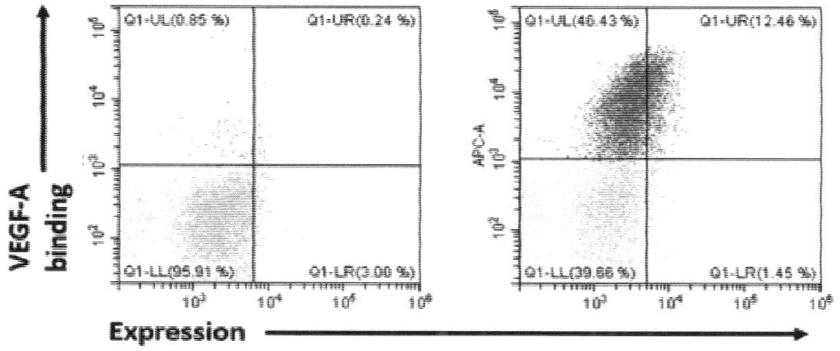

Figure 1: VEGFR-2 D23 mutagenic library enrichment for biotinylated 500 nM VEGF-A monomer binding over 5 rounds of selection, as illustrated by flow cytometry analysis of streptavidin and cmyc tag detection.

21c Discovery and Characterization of Intracellularly Stable hnRNPA2B1 Specific Nanobodies for Live-Cell Imaging and Targeted Protein Degradation

Azady Pirhanov, Yongku Cho

Heterogeneous nuclear ribonuclear proteins, hnRNPs, play a variety of roles in regulating transcriptional, and post-transcriptional gene expression, including RNA splicing, polyadenylation, localization, translation, and turnover. A core member of the hnRNP family, hnRNPA2/B1, is a ubiquitously expressed nuclear protein known to shuttle between the nucleus and cytosol. Moreover, hnRNPA2/B1 proteins are also involved in the formation of membraneless-organelles such as stress granules (SGs) under certain conditions.

Mutations in hnRNPA2/B1 proteins are known to cause multisystem proteinopathy (MSP), a pleiotropic group of inherited disorders that cause neurodegeneration, myopathy, and bone disease. Mutations in hnRNPA2/B1 are also risk factors for the development of amyotrophic lateral sclerosis (ALS) and frontotemporal dementia (FTD). Additionally, hnRNPA2B1 proteins are upregulated in numerous cancers including breast cancer, prostate cancer, and lung cancer. Despite growing evidence implicating hnRNPA2/B1 proteins in a multitude of diseases, our ability to study them in live cells is severely limited due to the lack of intracellularly stable hnRNPA2/B1 specific antibodies.

Nanobodies (sdAbs) are antibody fragments derived from heavy-chain-only antibodies found in camelids and sharks. Owing to their small size (~15 kDa) and unique biochemical properties, nanobodies are emerged as promising alternatives to conventional-antibody derived protein tools, such as single chain-variable fragments (scFvs). Their compact shape and convex antigen binding site (paratope), make nanobodies attractive alternative to probe surfaces that are not accessible to conventional antibodies. Furthermore, nanobodies have superior stability that facilitate their intracellular expression. Full length antibodies and bulkier antibody derived fragments, such as scFvs, are not amenable to intracellular expression, in part due to the reducing cytoplasmic environment that prevents the formation of disulfide bonds needed for their correct folding. The ability to express nanobodies intracellularly opens the door for studying protein-protein interactions, protein aggregations, and protein localizations in live cells under the microscope.

We have developed a high-throughput nanobody discovery platform from synthetic yeast surface display libraries by using antigen coated magnetics beads. Using this platform, we have identified novel hnRNPA2/B1 specific nanobodies and characterized their intracellular stability by expressing them in HEK293 cells. Furthermore, we have developed generalizable and straightforward cellular assays to characterize nanobody specificity in live cells. In addition to using nanobodies for studying hnRNPA2B1 biology in live cells, we have also engineered a panel of modular E3 ubiquitin ligases targeting hnRNPA2/B1 for degradation.

In this presentation, factors affecting successful high-throughput antibody discovery using yeast-surface display will be discussed. We will also talk about the intracellular stability of nanobodies and generalizable methods to characterize nanobody specificity in live cells. Furthermore, we will introduce protein engineering approaches to enhance nanobody affinity. Finally, we will discuss ways of using nanobodies for targeted protein degradation.

21d Directed Evolution of Monoclonal Antibodies Against Tumor Associated Carbohydrate Antigens

Sam Schmidt, Joelle Eaves, Theodore Belecciu, Nathaniel Pascual, Athar Nakissa, Shivangi Chugh, Xuefei Huang, Daniel Woldring

Tumor associated carbohydrate antigens (TACAs) are a promising, yet difficult class of biomolecules to target for theragnostics. Overexpression of TACAs is a unique molecular signature that healthy cells lack. However, most are not recognized as non-self, causing low immunogenicity thereby hindering traditional antibody immunization techniques. This difficulty is increased by the massive, mostly unfunctional sequence space available. We address these challenges using a novel immunization strategy paired with large, rationally designed antibody libraries. Directed evolution of a rationally designed library based on immune-evolved antibodies decreases time and effort spent to achieve a high-affinity, high-specificity binder compared to a naïve library. Moreover, by utilizing a novel immunization technique involving the display of TACAs on multi-valent Qβ nanoparticles, a stronger immunogenic response is elicited from the animal being challenged, causing higher titers of immune-evolved antibodies. Yet even with improved titers, the resulting antibodies commonly show insufficient binding affinity. To overcome this challenge, we utilize dominant antibodies discovered from Qβ nanoparticles as a starting point for our library. In this study, we have generated immune-evolved antibodies which weakly bind NHAcGD2, a ganglioside implicated in the proliferation and invasion of multiple cancers (e.g., neuroblastoma and melanoma). We use this diverse panel of lead antibodies as starting points for cell surface display directed evolution campaigns. To navigate the mutational landscape of these promising monoclonal antibodies more efficiently, our cell surface display libraries are guided by single cell deep sequencing of the immunized animal's B cell repertoire, allowing for native pairing of heavy and light chains. We also leverage in silico experiments to help determine promising candidates. From this, we can identify mutations to increase functionality and drive potent interactions with the TACA target. In this way, we provide a novel platform for discovering antibodies with clinical potential against a notoriously challenging class of biomarkers.

21e Engineering and Design of Tissue Inhibitor of Metalliproteinase-3 for Developing Protein Therapeutics

Linh Do, Maryam Raeeszadeh-Sarmazdeh

Matrix metalloproteinases (MMPs), a disintegrin and metalloproteinases (ADAMs), and ADAM with thrombospondin motifs (ADAMTSs) belong to the zinc-dependent proteases superfamily, responsible for degradation and remodeling of the extracellular matrix (ECM). Overexpression of metalloproteinases (MPs) have been replicated in many fatal and severe health conditions such as cancers, cardiovascular diseases and neurological disorders. Therefore, MPs remain major therapeutic targets to investigate for decades. Many small molecule MP inhibitors have failed to advance to clinical trials because of their off-target side effects and toxicity to patients. Tissue inhibitors of metalloproteinases (TIMP), a protein family of four members, are endogenous inhibitors of MPs. Among all four TIMPs, TIMP-3 is the most unique with the broadest binding spectrum and affinity to various inhibitory substrates. TIMP-3 is the only TIMP able to effectively regulate a large pool of ADAMs, ADAMTSs and MMPs activities, such as ADAM-17. ADAM-17 is also known as tumor necrosis factor α (TNF-α) converting enzyme (TACE) for its role in ectodomains cleavage of TNF-α and almost 100 other membrane proteins including cytokines, growth factors and growth factor receptors. ADAM-17 is considered one of the primary drug targets, especially in regards to inflammatory diseases and cancer. Understanding which mechanism TIMP-3 employs to inhibit certain MPs, via sequence and structure, might provide means to design more efficient, protein-based MP inhibitors.

We used yeast surface display and protein engineering techniques such as directed evolution to identify regions of TIMP-3 interacting with target MPs. We first displayed the library of TIMP-3 mutants on the yeast surface and then analyzed variants' expression level and binding affinity via flow cytometry. Wild type human full-length TIMP-3 displayed on yeast historically displayed low expression levels in our laboratory. We then screened the TIMP-3 mutagenesis library for improved protein expression on yeast surface and MP binding affinity via fluorescent-activated cell sorting (FACS). To characterize our library of mutant's enrichment landscape, we analyzed our DNA sequences, acquired from next generation sequencing after each round of FACS. This study allowed us to explore possible TIMP-3 regions and motifs interacting with desired MPs with high affinity and selectivity.

21f Designing Hyperglycosylated Hemagglutinin Immunogens for Epitope-Focused Influenza Vaccines

Dana Thornlow, Aaron Schmidt

A major goal towards next-generation influenza vaccines with broad, protective coverage is to design immunogens that can focus immune responses toward conserved viral epitopes. On influenza hemagglutinin (HA), there are three such conserved epitopes across subtypes: the stem, receptor binding site (RBS), and the recently identified, "interface" between HA monomers. Selectively eliciting broadly neutralizing antibodies (bnAbs) that recognize these conserved epitopes may lead to a "universal" influenza vaccine.

To immune focus to these conserved epitopes, we concealed or "masked" immunodominant, variable epitopes present on the HA surface by engineering putative N-linked glycosylation sites (PNGs) to create a "glycan shield". This approach was inspired, in part, by the evasion strategy that influenza naturally uses to evade host adaptive immune responses by varying its native glycosylation sites on HA. As a proof-of-principal, we generated hyperglycosylated (hg) immunogens based on the historical H3 Hong Kong 1968 HA (H3HK-68) to conceal highly variable epitopes and selectively expose the conserved RBS. We confirmed PNG placement on the immunogens by assessing binding abrogation with a diagnostic monoclonal antibody panel. We then immunized mice with these hg immunogens, used single B-cell sorting to isolate antibodies, then used downstream biochemical and biophysical assays to characterize their binding. Antibody responses did indeed selectively focus to the HA RBS epitope and could potently neutralize matched virus. However, these antibodies showed limited breadth, binding only to a select few historical H3s.

To increase breadth, we expanded our panel of hg immunogens and to include hg versions of both historical and contemporary, as well as pre-pandemic and circulating influenza HAs. Additionally, we multimerized these hgHAs using ferritin nanoparticles. These nanoparticles can display eight unique HAs on its surface via SpyCatcher-SpyTag ligation to ferritin protomers. Multivalent display of viral antigens on engineered nanoparticles presents them in a geometry that can lead to effective cross-linking of B cell receptors and improve antibody responses. Thus, we anticipate that displaying multiple hgHAs on a single nanoparticle will result in improved breadth of elicited antibody responses. These next-generation, rationally designed hg immunogens are one strategy for eliciting epitope-focused immune responses and a critical step in the pursuit of more broadly protective influenza vaccines.

21g Simplifying Complex Antibody Engineering Using Machine Learning

Peter M. Tessier

Therapeutic antibody development requires selection and engineering of molecules with high affinity and other drug-like biophysical properties. However, maximizing antibody properties such as affinity is often detrimental to other properties such as stability and specificity, which can compromise safety and efficacy. Due to inherent tradeoffs between drug-like biophysical properties, co-optimization of multiple antibody properties remains a difficult and time-consuming process that impedes drug development. This presentation will highlight our recent advances in developing machine learning methods to co-optimize several different antibody properties, including affinity, stability, self-association and non-specific binding. We will also discuss how these methods can be used to greatly expand the exploration of novel antibody sequence space and accelerate the development of highly potent, drug-like antibodies.

498a Engineering Embden-Meyerhof-Parnas (EMP) Glycolysis to Generate Noncanonical Reducing Power

Edward King, Youtian Cui, Derek Aspacio, Frances Nicklen, Linyue Zhang, Sarah Maxel, Ray Luo, Justin Siegel, Erick Aitchison, Han Li

Noncanonical cofactors such as nicotinamide mononucleotide (NMN^+) supplant the electron transfer function of natural cofactors nicotinamide adenine dinucleotide and nicotinamide adenine dinucleotide phosphate ($NAD(P)^+$) at lower cost in cell-free biomanufacturing and enable orthogonal electron delivery in whole-cell metabolic engineering. Here, we redesign the high-flux Embden-Meyerhof-Parnas (EMP) glycolytic pathway to generate NMN^+-based reducing power, by engineering *Streptococcus mutans* glyceraldehyde-3-phosphate dehydrogenase (*Sm* GapN) to utilize this noncanonical cofactor. Through iterative rounds of rational design, we discover the variant GapN Penta with high NMN^+-dependent activity, and GapN Ortho with $\sim 3.4 \times 10^6$-fold switch in cofactor specificity from its native cofactor $NADP^+$ to NMN^+. GapN Ortho is further demonstrated to only function in *Escherichia coli* with the presence of NMN^+, enabling orthogonal control of glucose utilization. Molecular dynamics simulation and residue network connectivity analysis indicate that mutations altering cofactor specificity must be coordinated to maintain the appropriate degree of backbone flexibility to position the catalytic residue. These results provide a strategy to guide future designs of NMN^+-dependent enzymes and the first steps toward an orthogonal EMP pathway with biomanufacturing potential.

498b *De Novo* Biosynthesis of the Non-Standard Amino Acid *Para*-Nitrophenylalanine

Neil Butler, Sabyasachi Sen, Minwei Lin, Aditya Kunjapur

Nitroaromatic functional groups can impart valuable properties to chemicals and biological macromolecules including proteins. In particular, the nitroaromatic amino acid *para*-nitro-L-phenylalanine (pN-Phe) has demonstrated multi-faceted applicability when incorporated into proteins with use cases ranging from immune stimulation to fluorescence quenching. However, the chemical synthesis of pN-Phe does not follow green chemistry principles and impedes provision of pN-Phe to engineered bacterial cells in some contexts. Here, we designed a *de novo* pathway for pN-Phe biosynthesis in *Escherichia coli.*

Initially, we constructed a *de novo* biosynthesis pathway for pN-Phe based upon a literature identified amine oxidizing diiron monooxygenase enzyme, achieving a modest 24h titer of ~150 μM in M9 minimal media. We then performed genome engineering to increase the flux to the pN-Phe pathway, leading to improved 24h titers of ~250 μM in M9 minimal media. However, because the metabolic intermediate pA-Phe was the dominant heterologous product, we further screened a range of putative natural diiron monooxygenases. By expressing one of these diiron monooxygenases within our pathway, we increased pN-Phe titers to 330 μM, which was a 2-fold improvement over our starting titers. By fermenting this strain using a MOPS EZ Rich glucose media, we were able to obtain even higher titers of 820 μM. Given these near millimolar levels of pNPhe obtained from this biosynthetic pathway, we looked to employ strategies to achieve paired biosynthesis and site-specific incorporation of pN-Phe within target proteins. Through combination of these technologies within a strain, it would enable autonomous utilization of pN-Phe within proteins and create an *E. coli* chassis with an expanded 21 amino acid genetic code.

498c Computational Design of CRISPR Guide RNAs to Enable Strain-Specific Control of Microbial Consortia

Austin Rottinghaus, Tae Seok Moon

Microbes naturally coexist in complex, multi-strain communities. However, extracting individual microbes from and specifically manipulating the composition of these consortia remains challenging. The sequence-specific nature of CRISPR guide RNAs can be leveraged to accurately differentiate microorganisms and facilitate the creation of tools that can achieve these tasks. We developed a computational program, ssCRISPR, which designs strain-specific CRISPR guide RNA spacer sequences with user-specified target strains, protected strains, and guide RNA properties. We experimentally verify the accuracy of the strain-specificity predictions in both *Escherichia coli* and *Pseudomonas spp.* and show that up to three nucleotide mismatches are required to ensure perfect specificity. To demonstrate the functionality of ssCRISPR, we apply computationally designed CRISPR-Cas9 guide RNAs to two applications: the purification and engineering of specific microbes through one- and two-plasmid transformation workflows and the targeted removal of specific microbes using DNA-loaded liposomes. ssCRISPR will be of use in diverse microbiota engineering applications. In this talk, we will discuss this promising and impactful technology to engineer diverse microbiota.

Austin Rottinghaus and Tae Seok Moon, US Provisional Patent Application

Austin G. Rottinghaus, Steven Vo, and Tae Seok Moon, Paper under review.

498d A Workflow for the Systems-Level Analysis, Design, and Engineering of Genomically Recoded Organisms

Anush Chiappino-Pepe, Lisa Dratva, Courtney A. Shearer, Huseyin Tas, Jorge A. Marchand, Kamesh Narasimhan, Russel V. Miranda, Alexandra I. Rudolph, George M. Church

Genomically recoded organisms are tightly biocontained and biosolated (*e.g.*, virus resistant) and allow the efficient incorporation of multiple non-standard amino acids, making them attractive platform technologies for biotechnological, industrial, and biomedical applications[1–4]. In the process of recoding, we substitute a set of selected codons by synonymous ones throughout the entire genome[2,5]. Although protein sequences are maintained, we and others have observed considerable fitness reductions upon recoding[2,3,6,7]. To design viable genomically recoded organisms, we need computational methods that can connect genome sequences to cellular phenotypes with enhanced resolution than available methods.

Here, we present a workflow for the rational analysis, design, and engineering of genomically recoded organisms. First, we have developed deep learning models connecting sequence-to-function of regulatory genetic features. Second, we have constructed extended genome-scale models for the integrative analysis of metabolism, expression, and regulation. These models can predict cellular fitness from recoded genome sequences and can integrate large and disparate omics data sets. Finally, we have defined optimization-based methods for the computational analysis of recoding mutations. With this workflow, we can predict at a systems-level the combinatorial effect of recoding mutations on fitness, and we identify sets of mutational bottlenecks constraining growth.

We are applying this workflow to identify fitness-decreasing mutations in genomically recoded strains of *Escherichia coli*. The models developed together with newly obtained RNA-seq data for these strains are helping characterize the metabolic function arising from synthetic genomes. In the future, this workflow will serve to design genomically recoded organisms with minimal growth reduction and will provide a better understanding of genotype-to-phenotype relationships at a cellular level.

References

1. Ostrov, N. *et al.* Synthetic genomes with altered genetic codes. *Curr Opin Sys Biol* **24**, 32–40 (2020).
2. Lajoie, M. J. *et al.* Genomically Recoded Organisms Expand Biological Functions. *Science* **342**, 357–360 (2013).
3. Fredens, J. *et al.* Total synthesis of Escherichia coli with a recoded genome. *Nature* **569**, 514–518 (2019).
4. Robertson Wesley E. *et al.* Sense codon reassignment enables viral resistance and encoded polymer synthesis. *Science* **372**, 1057–1062 (2021).
5. Isaacs, F. J. *et al.* Precise Manipulation of Chromosomes in Vivo Enables Genome-Wide Codon Replacement. *Science* **333**, 348 (2011).
6. Ostrov, N. *et al.* Design, synthesis, and testing toward a 57-codon genome. *Science* **353**, 819–822 (2016).
7. Lajoie, M. J. *et al.* Probing the Limits of Genetic Recoding in Essential Genes. *Science* **342**, 361 (2013).

498e Circuit-Host Interactions: Mechanism and Control

Xiaojun Tian

Failure of modularity remains a significant challenge for assembling synthetic gene circuits with tested modules as they often do not function as expected. Hidden circuit-host interactions, such as growth feedback and resource competition, could significantly impair intended circuit function but are often overlooked. Here, I will present our recent efforts on the quantitative understanding of how the functional perturbation of gene circuits by these hidden interactions depends on network topology, host physiological environment, and resource competition, together with modeling frameworks for predicting circuit behaviors and practical control strategies for engineering robust gene circuits. First, we will discuss how the interference of synthetic gene circuit function by growth feedback depends on circuit network topology and nutrient level. Second, we will demonstrate how resource competition redirected desired successive cell fate transitions following a 'winner-takes-all' rule. Third, we will highlight our control strategies against resource competition, including a division of labor using microbial consortia, negatively competitive regulation (NCR) controllers. Lastly, we will discuss the effects of resource competition on the circuit noise behavior. Overall, our works will help us in the understanding and control of circuit-host interactions toward engineering robust synthetic gene circuits

498f Deregulating Native Bacterial Genetic Circuits As a Potential Anti-Microbial Therapeutic

William Bothfeld, Brian F. Pfleger

The emergence of microbes resistant to multiple antibiotics is an imminent global health threat, generating untreatable "super bugs" that could also be engineered for bioterrorism purposes. Resistance to novel small-molecule antibiotics can quickly evolve and spread to other species through horizontal gene-transfer, mitigating years of research and development, making them unattractive drug candidates. Current development approaches screen large libraries of chemicals against pathogens and are capital and resource intensive. Machine-learning has been utilized to more efficiency discover new antibiotic modes of action, but this approach remains part of the traditional development pipeline. Ideally, we would **1) rapidly identify pathogens through sequencing and 2) synthesize cheap and targeted antidotes on-demand** to combat emerging threats. Towards that end, we anticipate that disrupting gene regulation that controls cellular metabolism and division could be an attractive method to kill pathogenic or bioengineered bacteria. This talk will outline efforts to select strain specific gene targets, tests to deregulate these circuit through internal and external treatments and tests to determine the specificity of treatment, such that only the targeted bacteria would be altered.

694a Mathematical Modeling of Dysregulation of ACE2 Depending on Age-Sex and Blood Pressure and Progression of Fibrosis in COVID19 Survivors

Mohammad Aminul Islam, Ashlee Ford Versypt

Introduction: Severe acute respiratory syndrome coronavirus 2 (SARS-CoV-2) enters the host cells via binding its spike protein with the angiotensin-converting enzyme 2 (ACE2) receptor, which is a crucial component in modulating the balance of the renin-angiotensin system (RAS). RAS is composed of distinct angiotensin peptides and receptors responsible for maintaining the local and systemic inflammation. The dysregulation of ACE2 by the viral infection can shift the balance of RAS towards pro-inflammation and pro-fibrosis. Experimental studies showed downregulation of ACE2 increases collagen deposition in lung tissue, leading to fibrosis [1]. ACE2 receptors maintain the balance between pro-fibrotic ANGII and anti-fibrotic ANG1-7 by converting ANGII to ANG1-7. The downregulation in ACE2 causes higher ANGII resulting in activation and production of anti-inflammatory cytokine TGF-beta in the local tissue. Fibroblasts are recruited by TGF-beta and deposit collagen to repair the damage. But excess deposition of collagen can initiate long-term fibrosis. The fibrotic outcome in COVID19 survivors also depends on age-sex and pre-existing comorbidities.

Methods: We developed a multiscale tissue simulator to investigate the mechanisms of SARS-CoV-2 infection in lung epithelial cells. The overall model contains intracellular viral replication kinetics, ACE2 receptor trafficking, immune response, and tissue damage models [2]. We infect the tissue by placing SARS-CoV-2 viruses in the extracellular space using uniform random and gaussian distribution. Viral particles diffuse through tissue, bind with unoccupied ACE2 receptors on the epithelial cell surface, internalize via endocytosis, replicate through intracellular viral replication kinetics, and export back to the extracellular domain by exocytosis. The viral bound ACE2 is also recycled back to the cell surface after releasing the virus inside the cell. We varied the initial unoccupied ACE2 receptors concentration on the cell surface based on the availability of ACE2 receptors for different age-sex groups [3]. Then, the simulated dynamics of ACE2 receptors for the different age-sex groups are used as input in a separate model of RAS to predict the change in ANGII from homeostasis for normotensive and hypertensive patients. We also consider the effect of feedback signaling of the ANGII-Angiotensin receptor 1 (ANGII-AT1) complex to renin and how the absence of feedback signaling can affect the dynamics of ANGII. The increase of ANGII also increases TGF-beta, which recruits fibroblasts. Fibroblasts are responsible for the excess deposition of collagen. We also consider the effect of negative regulator MMP, produced from macrophages, and can degrade collagen.

Results: Our model reveals that the variation in available ACE2 due to age and sex can significantly change inflammation, tissue damage, and fibrosis. Old males, old females, and mid males are most susceptible to viral infection due to the lower availability of ACE2 receptors, and a higher amount of ANGII and collagen deposition is observed from model simulation. We observed epithelial cell death as the main source of reduction of ACE2 compared to the internalization of ACE2 receptors by viral particles. Since the concentration of ACE2 receptors is high compared to the number of viral bound ACE2 receptors, the downregulation of ACE2 by viral bound ACE2 receptors doesn't significantly affect the RAS. But the change in initial unoccupied ACE2 receptors can affect the viral infectivity of epithelial cells and result in cell death. The death of a large number of epithelial cells can significantly reduce the number of available ACE2 for the RAS network. Our model also identifies that impairment in the feedback signaling of ANGII-AT1 to renin and hypertension can increase the concentration of ANGII and collagen deposition.

Conclusions: Our model quantifies age-sex and blood pressure as major factors for fibrotic outcomes in COVID19 survivors. The predicted results can be used to develop personalized therapeutic strategies for different patients groups and patients with pre-existing comorbidities.

Acknowledgment: This work was supported by the National Institutes of Health grant R35GM133763 and the University at Buffalo.

References

1. Kuba, et al., A crucial role of angiotensin converting enzyme 2 (ACE2) in SARS coronavirus–induced lung injury, Nature medicine8 (2005): 875-879

2. Getz, et al., Iterative community-driven development of a SARS-CoV-2 tissue simulator, BioRxiv (2021)

3. Xudong, et al., Age-and gender-related difference of ACE2 expression in rat lung, Life sciences19 (2006): 2166-2171

694b Cell-Free RNA Liquid Biopsies at Cell Type Resolution in Health and Disease

Sevahn Vorperian, Mira Moufarrej, Stephen Quake

Tissue of origin remains the at-best resolution of liquid biopsy assays; however, pathophysiology occurs at the cellular level. Single cell transcriptomics are enabling dissection of the complex cellular heterogeneity and associated transcriptional programs underlying tissues. Here, we describe two approaches applying single cell transcriptomic data to resolve the cell types of origin of the cell free transcriptome. First, we used Tabula Sapiens, a whole-body transcriptomic cell atlas, to computationally deconvolve the cell-free transcriptome into a sum of cell type specific transcriptomes, to reveal the spectrum of cell types readily detectable in the blood. Second, we used individual tissue transcriptomic cell atlases in combination with the Human Protein Atlas RNA consensus dataset to create cell type signature scores which can be used to infer the implicated cell types from cell-free RNA liquid biopsy for a variety of diseases. Taken together, this work is one of the first direct clinical applications of single cell transcriptomic atlases to date. Our results demonstrate that cfRNA reflects cellular contributions in health and disease from diverse cell types, potentially enabling determination of pathophysiological changes of many cell types from a single blood test.

694c Coupled Quantitative Transcriptomic and High-Throughput Morphological Analysis Predict Biological Effects of Environmental Toxins on Human Lung Cells

Sean Engels, Pratik Kamat, Daniel Haller, Jude Phillip, Lydia Contreras

Exposure to air pollution is linked to diseases of the lungs, brain, and heart and is estimated to contribute to over 7 million deaths annually by the World Health Organization. Although the link between exposure and adverse health effects has been made apparent through epidemiological studies, the role of unique air pollution components in disease pathogenesis needs to be further characterized on the cellular and molecular level. In this study, we conduct quantitative transcriptomics and epitranscriptomics in conjunction with high-throughput biochemical and microscopy analysis to predict effects of environmental toxins on human cells.

Specifically, we examine human bronchial epithelial cells (BEAS-2B cells) as a model system to determine the effects of multiple complex airborne particulate matter (PM) mixtures. We use an interdisciplinary approach involving RNA sequencing and large-scale epitranscriptomics by mass spectrometry, coupled with high-throughput microscopy and a recently developed visually-aided morpho-phenotyping recognition (VAMPIRE) tool, to catalog distinct features of cells exposed to three chemically defined PM mixtures. Using these systems and quantitative biology tools, we have developed a new integrated approach that combines large-scale biological data to understand the interplay between environmentally-induced transcriptional changes and the resulting morphological phenotypes. Collectively, these methods have allowed us to predict outcomes upon exposure to different stresses. We have also used this approach to address the challenges associated with deconvoluting the effects of specific components in PM, which has been a challenge in the field thus far.

We will discuss this novel approach and key findings including (1) distinct morphological traits of single-cell derived clones associated with transcriptomic phenotypes that predict specific cell-fates in the presence of various PM components and how these can be used as prediction factors of cell fates, (2) correlations between pathway activation in the presence of specific pollutants and chemical components, and (3) dynamic changes in RNA modification patterns across a time course of pollutant exposure that suggest an additional complex layer of stress regulation. Together, this systematic approach enables deconvolution of how PM components will uniquely affect lung cell health. We believe this approach can be broadly applicable to investigate subtle changes in cellular responses to different stimuli and pathologies.

694d A Logic-Based Modeling Study of Glucose Mediated Immune Response in Diabetic Kidney Disease

Krutika Patidar, Ashlee Ford Versypt

Introduction: Diabetic kidney disease (DKD) is a complication in 1 out of 3 patients with diabetes. Aberrant glucose metabolism in diabetes leads to immune response and systemic inflammation, which causes structural and functional damage at a cellular scale in the kidney. Complex cellular signaling lie at the core of these metabolic and functional derangements. However, the mechanism underlining the role of inflammation on endothelial cell dysfunction and increased endothelial permeability in DKD is not fully understood. Computational models in systems biology allow the integration of experimental evidence and cellular signaling networks. The present work uses predictive ordinary differential equation (ODE) type logic-based system to model immune response in macrophages and glomerular endothelial cells during diabetic kidney disease progression.

Method: Immune cell crosstalk with glomerular cells in the kidney is studied via a protein signaling network, which is conceptualized using an open-source software package CellNOptR[1]. The network is modeled into a continuous logic-based model by integrating logical rules and normalized-Hill type ODEs in NetFlux[2] software package. The model behavior is governed by two input stimuli, glucose and lipopolysaccharide (LPS) with a value between 0 and 1 based on whether stimulus is inactive, partially active, or fully active. The model equations are governed by reaction and species parameters, where interactions between species are governed by normalized-Hill activation or inhibition functions. The model response is trained and validated against available biochemical data from cell co-culture[3] experiments, and simultaneously model parameters are optimized using a multi-start non-linear least squares optimization technique. The model also takes into account minimization of parameter subspace by careful selection of sensitive model parameters for optimization as well as minimization of prediction error and parameter uncertainty.

Results: The presented biochemical and logic-based network model yields a semi-quantitative time-dependent response to the glucose and LPS stimuli. It provides an insightful link between glucose-dependent fluctuations in glomerular endothelial cell fenestration size, uncoupling of nitric oxide synthase (eNOS/NO) pathway, competitive VEGF receptor activation. Unlike biochemical models, this modeling approach overcomes the unavailability of mechanistic details and rigorous parameter estimation. Model evaluation tests suggest that our dual-cell network is consistent with literature and a subset of model predictions are also validated against data from *in vitro* experiments for short-term exposure to glucose. The immune response model is responsive to intermittent glucose exposure in diabetic mice, and ongoing efforts are made to validate glucose-mediated long-term immune response against *in vivo* data.

Conclusion: This model aids the understanding of cellular damage through pathway analysis and dependencies, even with limited biochemical data. Further, the results suggest that the proposed network model can be used to study the effects of molecular perturbations on cellular scale damage and may be useful in targeting disease progression.

References: [1]Terfve et al. BMC Systems Biology 2012, 6:133. [2] Kraeutler et al. BMC Systems Biology 2010, 4:157. [3]Li et al. Nephron. 2021, 5:145

Acknowledgment: This work was supported by NIH grant R35GM133763, NSF grant 2133411, and the University at Buffalo.

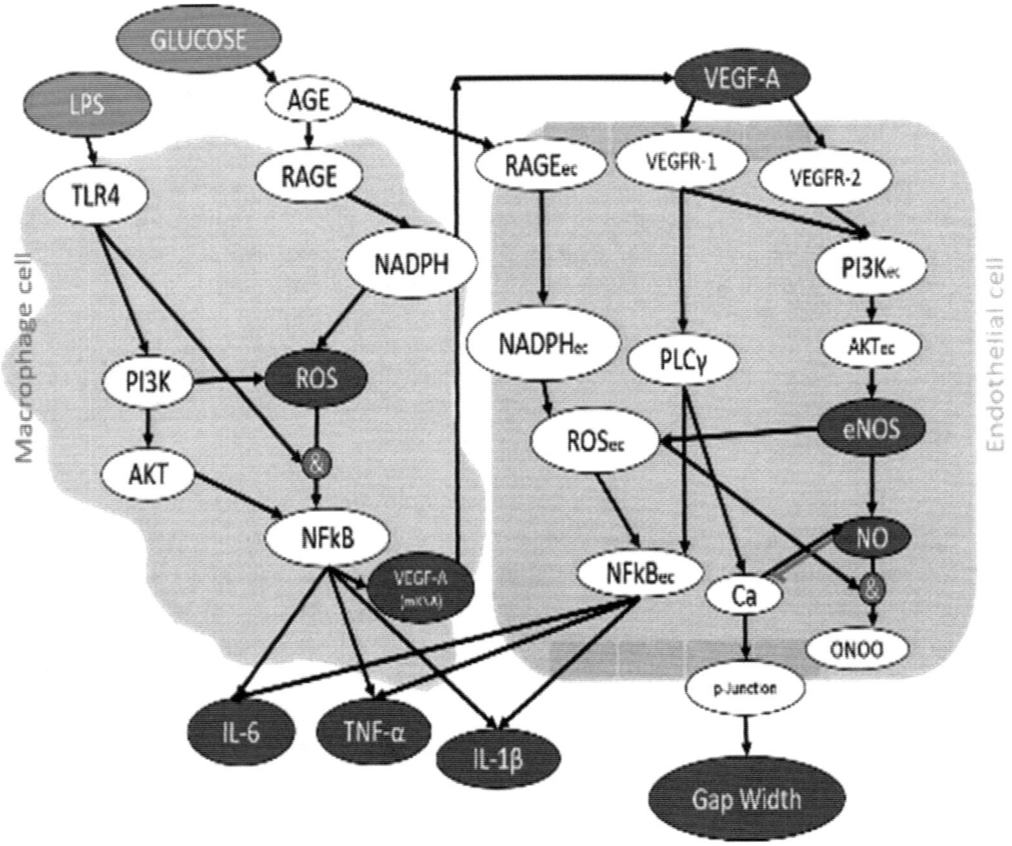

Fig. 1: Logic-based protein interaction network of macrophages and glomerular endothelial cells.

694f Temporal Analysis of Autophagy Rates and Subcellular Morphological Dynamics Using High-Throughput Image-Based Single Cell Profiling

Nitin Sai Beesabathuni, Soyoon Park, Eshan Thilakaratne, Priya Shah

Autophagy is a multi-step intracellular process by which cells recycle their misfolded proteins and damaged organelles into primary building blocks. The cellular materials are sequestered by double-membrane vesicles called autophagosomes, which fuse with lysosomes to form autolysosomes, which then degrade the sequestered material into fatty acids and amino acids. Autophagy is linked to many diseases, including cancer, neurodegenerative diseases, and infectious diseases. The significance of autophagy in various cellular processes is growing, yet comprehensive quantitative characterization of the autophagic process remains a major challenge. Here, we present two new quantitative methods that can lead to a comprehensive characterization of the autophagy pathway.

Established methods to quantify autophagy use steady-state measurements, which provide limited information about the perturbation and the cellular response. We present a theoretical and experimental framework to measure autophagic steps in the form of rates under non-steady state conditions. We use this approach to measure temporal responses to rapamycin and wortmannin treatments, two commonly used autophagy modulators. We quantified changes in autophagy rates in as little as 10 minutes, which can establish direct mechanisms for autophagy perturbation before feedback mechanisms begins. We identified the concentration-dependent effects of rapamycin on the initial and temporal progression of autophagy rates. This new approach enables the quantification of autophagy flux with high sensitivity and temporal resolution and facilitates a comprehensive understanding of this process.

Along with the kinetic rates of the autophagy pathway, cellular morphology is also significantly altered based on the kind of treatment. Profiling cellular morphology can provide additional insights into fundamental mechanisms and would facilitate high-throughput applications like drug screening. We developed an image-based profiling approach to further characterize the autophagy-related phenotypes. We developed a new computational pipeline to segment and track individual cells and extract up to 1000 subcellular morphological features in a time-resolved manner. We identified the key image features that capture the variability and dyanmics of cellular morphology under different autophagy modulators. Moreover, using these image features we can differentiate various kinds of autophagy perturbations. This approach can be used for characterizing phenotypes and understanding the mechanism of action. Furthermore, this work would act as a proof of principle for developing image-based high throughput drug screening technologies. In future work, similar analysis can be expanded for quantifying morphological changes in other organelles such as the mitochondria and lysosomes for characterizing disease state and progression.

In conclusion, these methods would provide a systematic and comprehensive characterization of the autophagy pathway. Additionally, these techniques would assist with high-throughput drug screening and in the development of new therapies to regulate autophagy accurately. As a future direction, we are applying these methods to study various external perturbations such as small molecules drugs and dynamic perturbations such as viruses at the single-cell level.

694g Integrative Approaches to Elucidate Mechanisms of Neurological Disorders

Tamara L. Kinzer-Ursem

Over 30 million people worldwide are living with Alzheimer's disease (AD) and over 45 million people worldwide are estimated to be living with Autism Spectrum Disorders (ASD). The prevalence of both neurological disorders has increased sharply in recent years, having significant negative impact on quality of life and economic burden. Gaps in our understanding of the molecular mechanisms that underlie AD and ASD have exacerbated difficulties in developing effective treatments. In each disorder, dysregulations at the points of neuronal cell connections (synapses) play a critical role. As AD progresses with age, the number and density of synapses in critical brain regions is reduced. At critical developmental timepoints early in life (prior to age 2), synaptic connections are not properly pruned leading to imbalances in excitatory and inhibitory inputs. Our research combines experimental and computational approaches to study the underlying mechanisms of protein translation and signal transduction in synapses. Using multiscale computational approaches we are investigating the role of macromolecular complexes in the timing and spatial dependencies of protein network signaling in synapses. Using novel protein labeling techniques combined with proteomics and high resolution imaging we are investigating the role of protein translation and subcellular localization in synaptic processes. These investigations are generating new insights into the molecular mechanisms underlying the role of tau protein in Alzheimer's disease and the role of scaffolding proteins in Autism Spectrum Disorders.

118a All-Atom Molecular Dynamics Simulations of Poly-Ethylene-Glycol (PEG) and Limp-2 Reveal That Peg Penetrates Deep into the Proposed CD36 Cholesterol-Transport Tunn

Paul Dalhaimer, Kate Blankenship

Poly-ethylene-glycol (PEG) is the most prominent clinically administered synthetic polymer. For example, over 300M people have been administered PEGylated liposome vaccines for SARS-CoV-2. PEG is used in mammals because it has low affinity for most proteins and vice versa. However, this makes it difficult to study the few interactions with proteins that PEG has. On the atomistic level, there are two PEG-protein structures: 1) PEG-LIMP-2 and 2) PEG-aPEG. In the first structure, two monomers of a 1.5 kDA PEG polymer (PEG2) had electron density deep in the postulated cholesterol transport tunnel of LIMP-2, a lysosomal cholesterol transport protein, and member of the CD36 super family of proteins. It is unclear how PEG entered this tunnel. In the second structure, PEG wrapped around a surface-exposed tryptophan on its antibody. Since tryptophan is a rare residue, it is unclear if this PEG-Trp interaction is ubiquitous. To gain deeper mechanistic insight into PEG-protein interactions, we surrounded the LIMP-2 apo structure with 13 PEG chains of 10 monomers each (PEG10), water, and KCl and simulated the system using NAMD. One of the 13 chains penetrated LIMP-2 and came within 3 Å of PEG2. This was possible because of strong hydrogen bonding between multiple oxygens along PEG10 and Arg192, but, most importantly, the clamping of the tertiary structure on PEG10. Clamping stabilized the movements of PEG10 and the leading oxygen of PEG10 was able to penetrate LIMP-2 and head towards to position occupied by PEG2. Phe383 appears to act as a gate for objects to move through this cavity, which continues to the basal/membrane side of LIMP-2. Of all residues, PEG10 molecules had the most sustained interactions with lysine and arginine because of their strong hydrogen-bonding capabilities. These results show that the oxygens of PEG bind residues with high hydrogen bonding capabilities. However, the PEG-protein interaction is likely to be transient unless groups of resides can clamp down on PEG or a cavity that at least part of the PEG chain can enter is in close proximity to lower PEG's entropy.

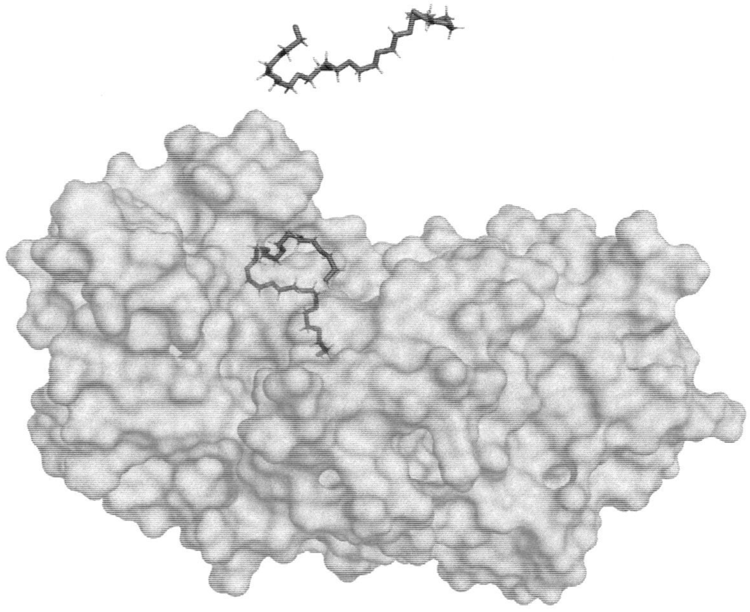

118b Molecular Simulations of Zwitterlation-Induced Conformation and Dynamics Variation of Glucagon-like Peptide-1 and Insulin

Qi Qiao, Lirong Cai, Qing Shao

Zwitterionic materials have shown their ability to improve the circulation time and stability of proteins. Zwitterionic peptides present unique potential because genetic technology can fuse them to any wild-type protein. One critical question is the effect of the fusing zwitterionic peptides on the conformation and dynamics of the original protein domain. To shed light on this question, we investigate the conformation and dynamics of six artificial proteins composed of two small therapeutic polypeptide and protein (glucan-like peptide-1 and insulin) and a zwitterionic (glutamic acid-lysine)$_{10}$ peptide in an explicit solvent using molecular dynamics simulations. The zwitterionic peptide is fused to the N- and C-terminal of the glucan-like peptide-1 and the chain A and B of the insulin. We analyze the conformation and dynamics variation of the polypeptide and protein domain using root mean square deviation, root mean square fluctuation, solvent accessible surface area, and secondary structure distributions. The simulation results show that the zwitterlation induces substantial changes in the conformation of the glucan-like peptide-1 and a moderate change in the conformation of the insulin, while the two polypeptide and protein remain folded. The glucan-like peptide-1 presents a full α-helix conformation when zwitterlated at the C-terminal. The zwitterionic location also plays a role in the conformational change. These zwitterlation-induced conformation variations indicate a comprehensive relationship between zwitterlation and protein stability and activity.

A Comparison of Three Statistical Methods for Transcriptomic Microarray Analysis

D.R. Schultz[1,2]*, *I.S. Frydas* [1,2], *N. Papaioannou* [1,2], *S. Karakitsios*[1,2,5], *D.A. Sarigiannis*[1,2,4,5]

[1] *HERACLES Health and Exposome Research Centre, Centre for Interdisciplinary Research and Innovation, Aristotle University of Thessaloniki, Greece*
[2] *Environmental Engineering Laboratory, Department of Chemical Engineering, Aristotle University of Thessaloniki, Greece*
[4] *Environmental Health Engineering, School for Advanced Study IUSS, Pavia, Italy*
[5] *ENVE.X, Thessaloniki, Greece https://enve-x.com/*

ABSTRACT

Transcriptomics is an effective tool in disease discovery that has undergone large advancements in recent years; available technologies allow for the generation of datasets with tens of thousands of outputs. This requires large computational power and advanced statistical analysis for datasets. Unfortunately, there is debate over the ideal methods to process raw data, analyze and identify differentially expressed genes (DEGs), and generate robust results while limiting Type I and Type II errors. Agilent™ microarray data were obtained from two in-house datasets from multiple chemical exposure experiments (One-Color, SurePrint Zebrafish Gene Expression v3 4 x 44k Microarray, design ID: 026437). Analyses aimed to detect the molecular mechanisms involved in metabolic disorders associated with environmental contaminants in association with the Horizon 2020 OBERON project. 5-day post fertilization (dpf) and 5-week post fertilization (wpf) zebrafish were exposed for 48 hours. In 5 dpf zebrafish, embryos were exposed to two concentrations of the plasticizer bis(2-ethylhexyl) phthalate (DEHP; 25 nM and 10 μM), positive control (amiodarone; 1 μM), and a carrier control [dimethyl sulfoxide (DMSO); 0.1%]. In 5 wpf zebrafish, larvae were exposed to a carrier control (0.1% DMSO), rosiglitazone (RGZ; 10 nM), tributyltin (TBT; 50 nM), and triclosan (TCS; 0.5, 1.0, 3.0 nM). Comparisons were completed between three statistical method pipelines to analyse a transcriptomic data set for DEG discovery and functionality comparisons; Linear Models for Microarray Data (LIMMA; R software), moderated t-test (Agilent GeneSpring™ software), and Significant Analysis for Microarray (SAM; R software). R is an open-source software that allows extensive control of statistical pipelines and parameters and the ability to handle complex statistical designs, although it requires basic knowledge of programming languages. GeneSpring™ is a commercially available, user-friendly analysis software that is efficient for basic statistical analyses with the benefit of freely available technical support. Results found that there was inconsistent accordance between the three analysis methods and varied highly between the comparisons. In accordance with a previous report (Chrominski and Tkacz 2015), this outcome is expected in real data sets (not artificially generated).

KEYWORDS: Systems biology, statistical analysis, computational biology, data science

1. INTRODUCTION

Omics is the overarching term given to various different branches of biology that characterize and quantify different groups of biological molecules that translate into the structure, function, and dynamics of an organism or organisms. While this includes various fields of research, it is constantly expanding as the field itself also expands. However, 'omics' most frequently refers to the analysis of genes/DNA (genomics), RNA (transcriptomics), proteins (proteomics), lipid (lipidomics), sand metabolites (metabolomics). In combination, this molecular network can influence an organism's phenotype and has provided major contributions to understanding diseases (Tan et al 2016). Although these technologies have vastly catapulted our understanding of the biology of organisms, omics techniques generate huge quantities of data that require advanced statistical software and methods for analysis.

Transcriptomics involves the analysis of the transcriptome, which is the complete set of RNA that is produced by the genome of an organism under specific conditions. This includes messenger RNA (mRNA), which are the templates for protein synthesis and therefore reflects the genes that are actively expressed. Microarrays are a frequently used technology in analysing the mRNA in an organism and can assess tens of thousands of genes in a single application. In order to do so, fluorescently labelled mRNA is hybridized to an array with a set of complementary short nucleotide oligomers ("probes"). The resulting fluorescence intensity emitted from each probe location is indicative of the transcript abundance for that probe sequence. The genes associated with each probe can then be determined and information such as differentially expressed genes (DEGs) amongst conditions can be assessed. However, standard analysis methods for transcriptomics data have yet to be established.

There is no lack of methodologies existing to analyze these complex data sets, however there are a couple pipelines that are more widely accepted in the field. R software is regularly used to handle transcriptomics data. R is an open source statistical analysis and graphical software that is highly utilized because it is free and allows far greater manipulation capabilities versus commercially licensed softwares. Nevertheless, the R interface requires basic computer coding capabilities to edit generic pipeline scripts to a script that is tailored specifically to ones' needs. Also, trouble-shooting is generally done by using online support forums. Another common software is Agilent GeneSpring™, which is a commercially available statistical analysis software which is quite user friendly and has tech support through the company. Unfortunately, GeneSpring™ is quite rigid in its data manipulation capabilities, by default assumes parametric data so does not always offer non-parametric options and can be quite costly.

In R, data were processed using Linear Models for Microarray Data [LIMMA, [4]] and Significant Analysis for Microarray [SAM; [6]]. LIMMA utilizes linear model fitting in order to control for study design and applies empirical Bayesian functions to calculate the gene-wise test statistic (moderated t-test statistic). This method provides more statistical power as it borrows information from inter-gene comparisons. LIMMA, although it aims to overcome shortcomings based on parametricity assumptions, is still theoretically a parametric test. Conversely, SAM employs a non-parametric adjustment on the t-test that administers an ad hoc modification before analysis. For each gene, SAM produces a test statistic value based upon the observed value's deviation from the expected value. SAM determines significance based on the deviation of the observed data from the expected value using numerous permutations. SAM is a method that was generated specifically to process transcriptomic gene expression between groups. In certain regards, SAM is like a t-test, although SAM utilizes non-parametric statistics mainly owing to the fact that microarray data are not normally distributed in the vast majority of cases (Tusher, Tibshirani, and Chu 2001).

2. METHODOLOGY

During experiments, 3-day post-fertilization (dpf) zebrafish (*Danio rerio*) were exposed to a carrier control [dimethyl sulfoxide (DMSO); 0.1%], positive control (amiodarone; 1 μM), and bis(2-ethylhexyl) phthalate (DEHP; 25 nM and 10 μM) for 48 hours. Alternately, 5-week post-fertilization (wpf) larvae were exposed to a carrier control [dimethyl sulfoxide (DMSO); 0.1%), rosiglitazone (RGZ; 10 nM), tributyltin (TBT; 50 nM), and triclosan (TCS; 0.5, 1.0, 3.0 nM) for 48 hours. Experiments were comprised of 4 replicates per treatment with approximately 17 larvae at sampling (5 dpf) or 1 larvae (5 wpf). Replicate larvae were snap frozen until RNA extraction was completed.

RNA was extracted using a modified Invitrogen TRIzol method and stored at -20°C until microarray analysis. Subsequently, samples were processed following the One-Color Microarray-Based Gene Expression Analysis (Low Input Quick Amp Labeling) Protocol version 6.9.1 supplied by Agilent Technologies. Samples were hybridized using Agilent's Gene Expression Hybridization Kit (Agilent 5188-5242) to Agilent SurePrint G3 Human Gene Expression v3 8 x 60k Microarray Kit, design ID: 072363 (Agilent Technologies Inc., CA). Microarrays were read using Agilent SureScan Microarray Scanner™ (G2600, Agilent Technologies, Inc., CA) and analyzed using GE 1200 one-color protocol in the Agilent Feature Extraction™ software. After feature extraction, the raw data was exported and analyzed with GeneSpring [moderated t-test (MTT)] and R (SAM, LIMMA) based on results found in the literature (Chrominski and Tkacz 2015). Figure 2 outlines the general data analysis pipeline followed in the analysis.

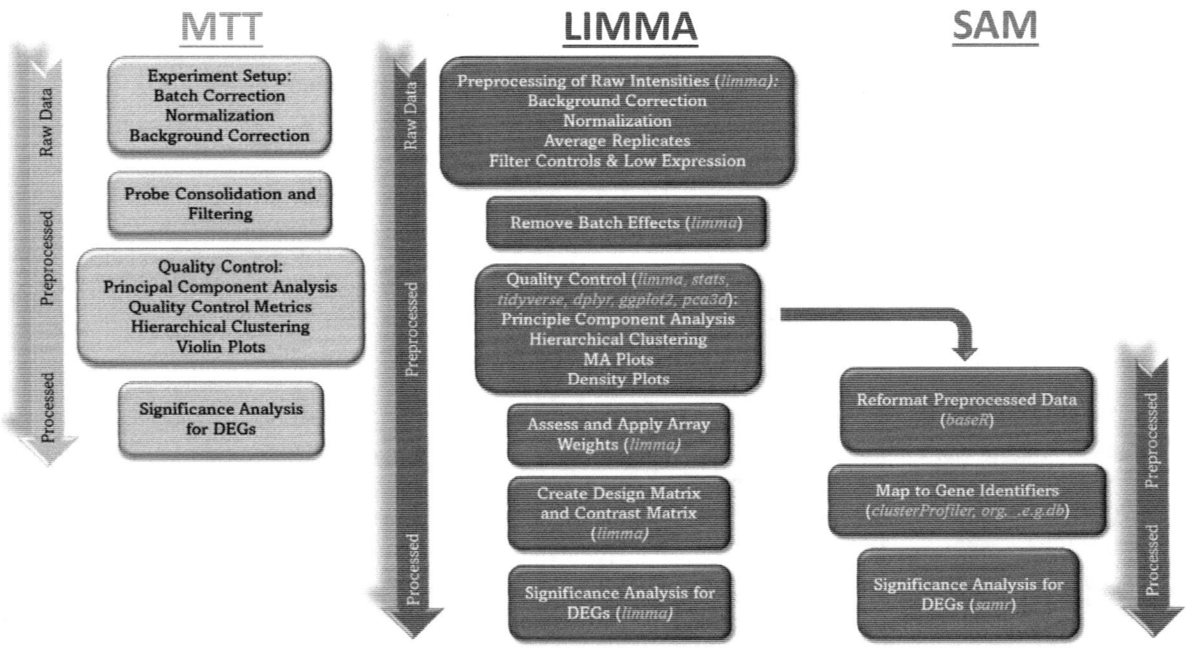

Figure 2: Statistical analysis pipeline for moderated t-test (MTT), linear model for microarray data (LIMMA), and significance analysis for microarrays (SAM). Brackets indicate the R packages used for the analysis.

A moderated t-test (MTT) was completed in GeneSpring™, which is a test that is modelled after LIMMA's moderated t-test. However, this analysis is not model-based and therefore does not account for study design. LIMMA and SAM methods were utilized in R. Across all methods, preprocessing (background correction, normalization, probe filtering, outlier analysis, and batch effect analysis) was undertaken where possible.

Data for the GeneSpring™ MTT was first batch corrected using a parametric *ComBat* algorithm with quantile normalization in the R Script Editor extension. General overlap of the red and black lines of the density plots generated from the algorithm indicates that downstream analyses can be completed as parametric (Figure 3) as the difference between a parametric and non-parametric correction will have a low level of deviation (<5%). Following batch correction, background correction was applied to the median of all samples. Normality was then assessed using the Shapiro-Wilk test although there is no test to analyze equal variance – in order to determine variance, a violin plot was generated to visually compare the sample datasets. Quality control (outlier analysis) was completed using principal component analysis (PCA), sample correlation analysis, and flags on parameters generated by GeneSpring™. Outliers were removed from the dataset before continuing. After preprocessing, a gene-wise analysis was set up to combine multiple probes coding for the same gene and so replicate genes are not presented individually. Non-detected and compromised probes, as well as control probes, were removed. Control probes are utilized for grid alignment and quality control and are not informative past this point. Unannotated and predicted genes were removed in R as GeneSpring™ does not have this capability.

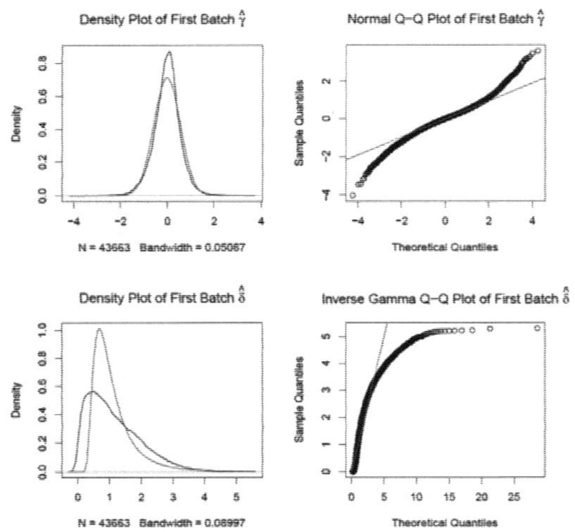

Figure 3: Graphical output from ComBat *algorithm in GeneSpring™ R Script Editor. General accordance of red and black lines indicate that data can be analysed with parametric batch correction assumptions.*

A fold change cutoff of 1.5 and adjusted p-value cutoff of < 0.05 were applied [7,8]. To determine DEGs, a moderated t-test was implemented using a Benjamini-Hochberg false discovery rate (BH) as well as a Westfall Young Permutative (WYP) multiple testing correction. The BH is the mot frequently used multiple testing correction used in omics analysis, but the WYP provides a good balance between Type I and Type II error [2] and was used here in order to balance out other pitfalls of the GeneSpring™ pipeline in terms of lack of manipulability.

In R, data were analyzed using the limma package with built-in analyses specific for One-Colour Agilent microarray data. Raw intensitites were inspected and background correction was undertaken ("normexp", offset = 50) as well as quantile normalization ("quantile"). Batch effects and outliers were visualized using multiple packages (stats, tidyverse, ggplot2, pca3d). Raw and preprocessed data were then compared using boxplots (graphics), log-intensity ratio to log-intensity average (MA plot; limma), and expression densities between arrays (limma) to visualize the efficacy of the preprocessing methods. In the current pipeline, array quality weights were applied in order to correct for larger residual values (limma). Before analysis, the batch effect were removed using the built-in limma correction. Consequently, the analysis was set up for the design matrix and contrast matrix for all relevant comparisons. Weights were applied to the samples and then a t-statistics were completed. After generation of the DEGs, lists were tidied by filtering unannotated genes and mapped to known genes (org.Dr.eg.db, clusterProfiler).

For SAM analysis, data was required to be preprocessed and reformatted. Analysis was therefore done using limma and GeneSpring™ preprocessed data as described above. After reformatting, genes were mapped (org.Dr.eg.db, clusterProfiler) and then processed for individual treatment comparisons (samr; response type = "Two class unpaired", number of permutations = 1000). Selection of the delta value [the different values of delta for which the numbers of genes and the estimated false discovery rates (FDRs) are computed] requires manual assessment and application for each comparison. Following these steps, the algorithm was run to obtain the list of DEGs.

In order to compare the results of each analysis, we used three test combinations (Table 1). Further, we were interested in total detected DEGs by each method, total overlapping DEGs across the analyses, and mapped DEGs to be used in future biomarker discovery,

Table 1: Test parametres used in the analyses for moderated t-test (MTT), linear models for microarrays (LIMMA), and significance analysis for microarrays (SAM).

MTT	LIMMA	SAM
Benjamin-Hochberg	Benjamin-Hochberg	GeneSpring™ preprocessed data
Benjamin-Hochberg	Benjamin-Hochberg	LIMMA preprocessed data
Westfall-Young Permutative	Benjamin-Hochberg	LIMMA preprocessed data

3. RESULTS & DISCUSSION

Table 2 outlines the total number of DEGs found by the different analysis methods as well as those that could be mapped (mapped to an EntrezID using GeneSpring™ or org.Dr.eg.db in R). Compared to control treatments, the MTT detected the most DEGS almost all tests, except for the DEHP comparisons using SAM with GeneSpring™ preprocessed data. Table 3 and 4 suggest that preprocessing plays a large role on the number of DEGs found.

Table 2: Total detected DEGs and mapped genes (EntrezID) between moderated t-test (MTT), linear models for microarrays (LIMMA), and significant analysis of microarrays (SAM) across the three analyses. MTT was mapped using GeneSpring (*) and LIMMA/SAM were mapped using R (°; org.Dr.eg.db annotation package). Values that are underlined represent the greatest number of DEGs detected.

Condition	Comparison	GeneSpring: MTT		R: LIMMA		R: SAM
		Total DEGs	Mapped*	Total DEGs	Mapped°	
BH; GeneSpring preprocessed data	Amiodarone 1 µM	8951	<u>7324</u>	3024	1715	6971
	DEHP 25 nM	8878	8092	2813	1606	<u>9846</u>
	DEHP 10 µM	8961	8154	2848	1578	<u>9866</u>
BH; LIMMA preprocessed	Amiodarone 1 µM	8951	<u>7324</u>	3024	1715	1327
	DEHP 25 nM	8878	<u>8092</u>	2813	1606	1155
	DEHP 10 µM	8961	<u>8154</u>	2848	1578	1187
WYP (MTT); BH (LIMMA); LIMMA preprocessed data	Amiodarone 1 µM	6918	<u>5585</u>	3024	1715	1327
	DEHP 25 nM	2383	<u>1818</u>	2813	1606	1155
	DEHP 10 µM	2862	<u>2218</u>	2848	1578	1187

Table 3: Total overlapping mapped DEGs between MTT and SAM (using GEneSpring™ preprocessed data) and percent of overlapping DEGs.

Comparison	Overlapping DEGs	% MTT	% SAM
Amiodarone 1 µM	5313	72.5	5313
DEHP 25 nM	6376	78.8	6376
DEHP 10 µM	6454	79.2	6454

Table 4: Total overlapping mapped DEGs between LIMMA, and SAM (using LIMMA preprocessed data) and percent of overlapping DEGs.

Comparison	Overlapping DEGs	% LIMMA	% SAM
Amiodarone 1 µM	1037	76.2	98.5
DEHP 25 nM	1099	68.4	95.2
DEHP 10 µM	1107	70.2	93.3

Table 5 outlines the total number of overlapping DEGs across methods per comparison as well as percent of common mapped DEGs as a percentage per comparison. The second comparison, using BH for both MTT and LIMMA, as well as using LIMMA preprocessed data for SAM resulted in the most common DEGs as well as the most overall percentage overlap across tests. If we assume that overlapping genes across all three tests are more likely to be true positives, the parameters used in the second test comparison appear optimal.

Table 5: Common mapped DEGs (EntrezID) as well as percentage of common DEGs from moderated t-test (MTT), linear models for microarrays (LIMMA), and significant analysis of microarrays (SAM) across the three analyses conditions. Underlined values indicate the greatest number of DEGs detected.

Condition	Comparison	Common DEGs	% MTT	% LIMMA	% SAM
BH; GeneSpring preprocessed data	Amiodarone 1 μM	1097	15.0	61.6	15.7
	DEHP 25 nM	<u>1045</u>	12.9	62.7	10.6
	DEHP 10 μM	1037	12.7	63.4	10.5
BH; LIMMA preprocessed	Amiodarone 1 μM	<u>1173</u>	16.0	65.8	85.4
	DEHP 25 nM	1031	12.7	61.9	86.8
	DEHP 10 μM	<u>1077</u>	13.2	65.9	87.6
WYP (MTT); BH (LIMMA); LIMMA preprocessed data	Amiodarone 1 μM	1136	20.3	63.7	82.7
	DEHP 25 nM	640	35.2	38.4	53.9
	DEHP 10 μM	794	35.8	48.6	64.6

Results from dataset 1 found that the MTT detected many more DEGs than LIMMA, likely owing to its preprocessing parametres, which resulted in SAM having high numbers when using GeneSpring™ preprocessed data. However, SAM appeared to perform better using LIMMA preprocessed data, as the total number of common DEGs were higher even though the first comparison found significantly more total DEGs for MTT and SAM. The average percent overlap was also higher. Finally, the BH multiple testing correction appeared to be preferable when comparing across methods in this analysis as slightly more DEGs were found when the BH was applied versus the WYP.

In dataset 2, results were not consistent with findings from dataset 1. The BH multiple testing correction applied to the MTT resulted in almost no detected DEGs (Table 6). LIMMA consistently found the most DEGs across comparisons and LIMMA preprocessed SAM also resulted in more DEGs. Using the WYP multiple testing correction detected more DEGs than the BH. Understandably, both MTT analyses using BH resulted in zero overlapping DEGs across comparisons. Comparisons using the WYP multiple testing correction resulted in a single exposure group with overlapping DEGs, which was the analysis that resulted in the highest total DEGs in general (Table 7).

Table 6: Total detected DEGs and mapped genes (EntrezID) between moderated t-test (MTT), linear models for microarrays (LIMMA), and significant analysis of microarrays (SAM) across the three analyses. MTT was mapped using GeneSpring™ (*) and LIMMA/SAM were mapped using R (°; org.Dr.eg.db annotation package). Underlined values represent the greatest number of DEGs detected.

Condition	Comparison	GeneSpring: MTT		R: LIMMA		R: SAM
		Total DEGs	Mapped*	Total DEGs	Mapped°	
BH; GeneSpring preprocessed data	RGZ 10 μM	0	0	3	<u>3</u>	2
	TBT 50 μM	4	3	112	<u>55</u>	0
	TCS 0.5 nM	0	0	143	<u>63</u>	1
	TCS 1 nM	0	0	137	<u>63</u>	3
	TCS 3 nM	1	1	202	<u>114</u>	1
BH; LIMMA preprocessed	RGZ 10 μM	0	0	3	3	<u>54</u>
	TBT 50 μM	4	3	112	<u>55</u>	5
	TCS 0.5 nM	0	0	143	<u>63</u>	62
	TCS 1 nM	0	0	137	<u>63</u>	<u>63</u>
	TCS 3 nM	1	1	202	<u>114</u>	96
WYP (MTT); BH (LIMMA); LIMMA preprocessed data	RGZ 10 μM	25	19	3	3	<u>54</u>
	TBT 50 μM	30	19	112	<u>55</u>	5
	TCS 0.5 nM	248	<u>176</u>	143	63	62

	TCS 1 nM	10	8	137	63	63
	TCS 3 nM	6	4	202	114	96

Table 7: Total overlapping mapped DEGs between MTT (WYP), LIMMA, and SAM (LIMMA preprocessed data) and percent of overlapping DEGs.

Comparison	Overlapping DEGs	% MTT	% LIMMA	% SAM
RGZ 10 μM	0	0.0	0.0	0.0
TBT 50 μM	0	0.0	0.0	0.0
TCS 0.5 nM	7	4.0	11.1	11.3
TCS 1 nM	0	0.0	0.0	0.0
TCS 3 nM	0	0.0	0.0	0.0

When LIMMA and SAM were compared (Table 8), there again was high overlap in DEGs. As expected, the MTT using BH had low overlap with SAM (Table 9) while using the WYP resulted in a small number more overlapping DEGs (Table 10).

Table 8: Overlapping, mapped DEGs between LIMMA and SAM (LIMMA preprocessed). Greatest values are depicted as underlined.

Comparison	Overlapping DEGs	% LIMMA	% SAM
RGZ 10 μM	2	66.7	3.7
TBT 50 μM	4	7.3	80.0
TCS 0.5 nM	55	87.3	88.7
TCS 1 nM	57	90.5	90.5
TCS 3 nM	90	78.9	93.8

Table 9: Overlapping, mapped DEGs between MTT (BH) and SAM (GeneSpring™ preprocessed).

Comparison	Overlapping DEGs	% MTT	% SAM
RGZ 10 μM	0	0.0	0.0
TBT 50 μM	0	0.0	0.0
TCS 0.5 nM	0	0.0	0.0
TCS 1 nM	0	0.0	0.0
TCS 3 nM	1	100	100

Table 10: Overlapping, mapped DEGs between MTT (WYP) and SAM (GeneSpring™ Preprocessed).

Comparison	Total	MTT	SAM
RGZ 10 μM	0	0	0.0
TBT 50 μM	1	5.3	20.0
TCS 0.5 nM	7	4.0	11.3
TCS 1 nM	0	0.0	0.0
TCS 3 nM	0	0.0	0.0

Main findings from dataset 2 suggest that the WYP multiple testing correction may be more stable and consistent when using GeneSpring™ software to preprocess and analyse data. This was surprising, considering the WYP multiple testing correction, which controls the family-wise error rate, is considered more conservative than the BH, which controls for false discovery rate. However, potentially due to the resampling-based step-down procedure, it may be more powerful in this context. Moreover, as seen in dataset 1, preprocessing methods appeared to have a large influence on DEG detection. Considering normalization methods were constant (quantile), the discrepancies must be owing to batch corrections (ComBat versus limma) and/or background correction methods.

Results suggest that tailored analyses are likely required, where each researcher must decide and balance Type I and Type II error. Chrominski and Tkacz (2015) found that real arrays had a low level of accordance across analysis using SAM, LIMMA, and a Student's T-Test (among others);

however, they found that LIMMA and SAM continuously performed better than the Student's T-Test on artificial datasets. Therefore, this could also be the case for the MTT as well.

Some important considerations when planning an analysis are that GeneSpring™ MTT is very rigid in its data manipulation capabilities, particularly when analysing non-parametric data, although it is more user friendly. On the contrary, R allows for greater manipulation of data and more robust pre-processing options and therefore, lends much better insight into the nuances of the data. Unfortunately, because it requires at least basic knowledge of coding and a reasonable amount of input effort and time to manually tailor generic pipelines, there is an obvious barrier to entry. LIMMA is a very frequently used method with numerous help forums and generic pipelines available. LIMMA can also account for co-variates and experimental design through modeling. Alternately, SAM requires preprocessing of the raw data with additional packages/software such as LIMMA or GeneSpring™ but is generally able to handle small samples sizes.

Unfortunately, there are many factors that can influence these types of analyses. For example, it is important to note that LIMMA models the data between the genes and so probe filtering based on variance is not recommended because this will still lend information to the model and removal could limit the extrapolation between genes. On the contrary, in GeneSpring™ MTT, probe filtering is important because the more comparisons you have, the more information you will lose due to your multiple testing corrections as the algorithm is unable to use the intergene information as efficiently as LIMMA. Also, the methodology of the batch effect corrections is an important distinction as well – while the *ComBat* function is available in R, it is generally not recommended for use. With regards to SAM, because it is based on permutation testing, it is difficult to obtain a completely replicable outcome i.e. each analysis, even when set to a high number of permutations, will provide a slightly different outcome each time the analysis is run albeit in this data set, the difference was generally less very small. Therefore, it is difficult to fully compare these procedures and selection of an appropriate method is researcher- and lab-specific.

REFERENCES

[1] Bandyopadhyay, S., Mallik, S., & Mukhopadhyay, A. (2014). A Survey and Comparative Study of Statistical Tests for Identifying Differential Expression from Microarray Data. IEEE/ACM Transactions on Computational Biology and Bioinformatics, 11(1), 20. https://doi.org/10.1109/TCBB.2013.147

[2] Dudoit, S., Shaffer, J. P., & Boldrick, J. C. (2003). Multiple Hypothesis Testing in Microarray Experiments. Https://Doi.Org/10.1214/Ss/1056397487, 18(1), 71–103. https://doi.org/10.1214/SS/1056397487

[3] Pandey A, Priyanka H. Tripathi, Ankita H. Tripathi, Satish Chandra Pandey, Saurabh Gangola (2019) Chapter 2 - Omics technology to study bioremediation and respective enzymes. Academic Press, Pages 23-43, https://doi.org/10.1016/B978-0-12-818307-6.00002-0.

[4] Smyth GK (2004) "Linear Models and Empirical Bayes Methods for Assessing Differential Expression in Microarray Experiments," Statistical Applications in Genetics and Molecular Biology: Vol. 3: Iss. 1, Article 3.

[5] Tan, S., Begley, P., Mullard, G. et al. (2016) Introduction to metabolomics and its applications in ophthalmology. Eye 30, 773–783. https://doi.org/10.1038/eye.2016.37.

[6] V. Tusher, R. Tibshirani, and G. Chu (2001) Significance analysis of microarrays applied to transcriptional responses to ionizing radiation. Proc. Natl. Acad. Sci. USA., 98:5116–5121.

[7] Raouf,A. et al. (2008) Transcriptome analysis of the normal human mammary cell commitment and differentiation process. Cell Stem Cell, 3, 109–118.

[8] Peart,M.J. et al. (2005) Identification and functional significance of genes regulated by structurally different histone deacetylase inhibitors. Proc. Natl Acad. Sci. USA, 102, 3697–3702

118d Extending the Bioavailability of Hydrophilic Antioxidants for Metal Ion Detoxification Via Crystallization with Polysaccharide Dopamine

Ryan Miller, Youngsam Kim, Chang Gyun Park, Young Jun Kim, Hyunjoon Kong

Although metal ions, such as silver and gold, have been shown to have strong antimicrobial properties, their potential toxic effects on human and environmental health have gained interest with improved understanding of their mechanisms to promote oxidative stress. More specifically, silver ions released from nanoparticle formulations can depolarize the mitochondrial membrane, increase intracellular reactive oxygen species (ROS) production, and neutralize endogenous antioxidant activity. The prolonged survival of ROS can denature lipids and proteins and lead to abnormal cell behavior. As such, a drug delivery system that can maintain ROS-homeostasis through the scavenging of overproduced ROS would nullify the toxic effects of metal ion accumulation.

Polymer-directed assembly of small molecule antioxidants has attracted attention as a way to control drug release and efficacy. These systems usually rely on a polymeric excipient that can help form amorphic solid dispersions that decrease the crystallinity of hydrophobic antioxidants resulting in improved bioavailability. However, for hydrophilic drugs such as N-acetylcysteine (NAC), both a direct scavenger of ROS and a prodrug for glutathione, an increase in crystallinity is desired for prolonged release and redox control. In our system, to further increase NAC release we designed a polymer excipient with thermodynamically favorable interactions with NAC, that can form stabilized colloidal antioxidant crystals. Herein we report a unique antioxidant crystal that dissolves sustainably through polymer-stabilization using sodium hyaluronate conjugated with dopamine (HA-dopa).

In our approach, we explore the role of dopamine incorporation into crystal stabilizing polymers and quantify the balanced trade-off between drug–polymer interactions and competing polymer–polymer interactions to yield the most stable colloidal form. The addition of HA-dopa in the crystallization of NAC resulted in a colloidal crystal assembly with a 5-fold extended release profile. We hypothesize that the extended release of NAC from the colloidal crystals would directly control the intracellular oxidative stress levels and lead to maintained cellular function. We explored this hypothesis with two model systems exposed to $AgNO_3$; cardiomyocytes to explore metal ion effects on human health and *Daphnia magna* to explore metal ion effects on ecological health. In both systems we illustrate that the colloidal crystals maintain intracellular ROS levels equal to healthy populations but no treatment or treatment with non-stabilized crystals have elevated levels. Furthermore, we show that deviations from ROS-homeostasis lead to deviations in metabolic function, ATP levels, and abnormal contracting/beating behavior. We propose that the results from these studies would be impactful for improving the therapeutic efficacy of a broad array of drug compounds administered in oxidative environments.

118e Lentiviral-Based Fluorescent Reporters for Assessing Human Lung Fibroblast Activation in Response to Microenvironmental Stimuli in Multidimensional Culture Systems

Samantha Cassel, April Kloxin

Introduction:

Idiopathic pulmonary fibrosis (IPF) is a progressive disease with no cure, characterized by the uncontrolled accumulation of scar tissue in the lower regions of the lung, leading to irreversible stiffening of tissue, decreased organ function and eventual respiratory failure. One significant driver of this progression is the persistent activation of fibroblasts. Fibroblast activation is a critical process for healthy healing after tissue injury, facilitating extracellular matrix (ECM) regeneration and wound contraction. However, when this phenotype becomes persistent, it leads to excess secretion of ECM proteins and crosslinking of collagen, as well as misregulated crosstalk with other critical cell types, further driving disease progression.[1] While current model systems for IPF have provided critical insights, there is still much unknown about the mechanisms that drive initiation and progression of human fibrotic disease. Consequently, therapeutic options for IPF are limited with only a few drugs that slow progression. Robust and relevant human culture models are needed to better understand these mechanisms and to identify potential therapeutic targets and strategies for halting disease progression.

A key challenge for human model system design is the need for tools for real-time assessment of cellular responses to microenvironmental stimuli. While the dynamics of cell-microenvironment interactions are thought to be instrumental in disease progression, cell behavior assessment tools are often limited to common endpoint, destructive methods incapable of assessing these dynamics. Increasing the number of experimental timepoints for more temporal resolution often requires prohibitively high sample counts for adequate assessment. Further, fibroblast populations tend to be quite heterogeneous, and the population averages from these techniques can mask critical differences occurring within cell sub-populations. There exists a need for innovative approaches to assess temporal changes in fibroblast activation behavior, from a range of donor types, in physiologically relevant, multidimensional culture systems.

Approach and Experimental Methods:

To address this need, we have established an approach to monitor the responses of human lung fibroblasts using a lentiviral-based dual-fluorescence reporter system that allows for quantification on a single-cell basis. In this system, cells constitutively express red fluorescence and dynamically express green fluorescence conditional on expression of alpha smooth muscle actin (αSMA), a common indicator of fibroblast activation.[2] We have implemented three versions of this reporter system with varying green fluorescence half-lives, allowing us to elucidate multiple response timescales.

Stable reporter cell lines were produced from human lung fibroblasts derived from healthy and IPF, male and female patients for facile investigation of intrinsic differences driven by sex and disease state. Utility of the reporter system was assessed by culturing these reporter cell lines in microenvironments with well-characterized activating biochemical and mechanical stimuli. Reporter fibroblasts were cultured on top of rat tail collagen I (ColI) coated polyacrylamide hydrogels (Matrigen, E~2kPa) or ColI-coated glass-bottomed well plates (Sigma-Aldrich, CellVis) to track changes in activation over time. During culture on both substrates, fibroblasts were treated with TGFβ1, a well-characterized activating cytokine, and fluorescence expression over time was assessed on a single-cell and population basis using live-cell imaging confocal microscopy, as well as endpoint flow cytometry and immunostaining.

Results and Conclusions:

In initial assessments, reporter cell lines demonstrated marked differences in the baseline expression of αSMA between healthy and IPF tissue-derived cells, demonstrating the reporter's ability to assess variation due to intrinsic factors. Further, when reporter cells were treated with activating stimuli (hyperphysiological stiffness, TGFβ1), they expressed increased green fluorescence intensity that correlated with αSMA expression and an activated phenotype. With this work, we have demonstrated that we can assess dynamic responses of lung fibroblast cell lines to biochemical and mechanical changes within the cell microenvironment. Future studies will allow for cell behavior investigations of both cell lines and primary patient-derived cell populations in more complex synthetic microenvironments that capture additional structural elements of diseased tissue, towards establishing a robust, representative human fibrosis culture model for understanding disease mechanisms and identifying new potential therapeutic targets.

References:

1. C. Hewlett, J.A. Kropski, and T.S. Blackwell. *Matrix Biology*. **71**, 112-127 (2018)

2. Alimperti, P. Lei, J. Tian, and S.T. Andreadis. *Gene Therapy*. **19**, 1123-1132 (2012)

118f Invited Talk: Leveraging Self-Assembly of Recombinant Fusion Proteins to Design Biomimetic Protocells

Yeongseon Jang

The living cell could have evolved from a simpler precursor, protocell – a self-assembled particulate where chemical and biological processes take place around or within. We develop a systematic engineering approach to construct a protein-powered synthetic protocell based on the self-assembly of recombinant globular fusion proteins into cellular compartment-like structures. Amphiphilic recombinant fusion proteins made of globular protein, leucine zippers, and elastin-like polypeptide (ELP) self-assemble into coacervates or vesicles depending on environmental conditions. In this talk, I will discuss our understanding of the structural properties of protein vesicles, such as stability upon external stimuli triggers, membrane structures, and permeability, which are critical to integrating and engineering their biological functionalities. Our ongoing work to make giant protein vesicles towards synthetic cells with de novo protein synthesis through in vitro transcription and translation and sensing of biochemicals will also be discussed. Creating a synthetic protocell would provide an understanding of how complex biological cells come from non-living molecules and a powerful practical tool to create sophisticated cellular functions in synthetic materials.

317a Lactate Inhibits the Growth of Human Pluripotent Stem Cells without Affecting Pluripotency

Demetrios Stoukides, Emmanuel Tzanakakis

Introduction: Advances in human pluripotent stem cell (hPSC) biology have intensified efforts for the development of cellular therapies against a gamut of currently incurable diseases. Bioprocess for the generation of stem cell-derived therapeutics entails two stages: hPSC expansion with preservation of their competence for downstream differentiation and directed specification to the cell or tissue type of interest [1]. The metabolism of hPSCs under propagative mode is glycolytic resulting in the production of excess lactate. Lactate is known to inhibit the growth of mammalian cells commonly employed in biopharmaceutical processes [2]. This metabolic product also induces signaling activity in highly proliferative tumor cells [3]. Nonetheless, the effects of lactate on the growth and pluripotent state of hPSCs, particularly in the context of biomanufacturing of cell therapeutics, remain understudied. To this end, results are presented from recent work combining planar and stirred suspension hPSC cultivation with biochemical and molecular analyses.

Materials andamp; Methods: Lactate feed experiments were performed to establish if hPSC culture performance is affected by elevated lactate over 6-day passages in line with the lactate buildup observed in bioreactor cultures. The H9 human embryonic stem cells (hESCs) and iPSC(IMR90)-clone 4 (IMR90) human induced pluripotent stem cells (hiPSCs) were used in these studies. Experimental endpoints were subjected to qPCR, immunostaining, flow cytometry, and western blotting for relative expression of pluripotency genes (e.g., NANOG, OCT4) as well as genes relevant to lactate transport and hPSC metabolic activity (e.g., monocarboxylate transporters (MCTs), uncoupling protein 2 (UCP2)). Pluripotency was also assessed by directed differentiation of hPSCs to the three germ layers.

Results/Discussion: H9 hESCs and IMR90 hiPSCs in 5 g/L of exogenous lactate exhibited an 11.5- and 11.7-fold increase, respectively, compared to 21.4- and 22.4-fold for the control groups with no additional lactate over 6 days while viability was maintained over 85% for all conditions. Moreover, hPSC aggregates cultured at elevated lactate displayed cyst-like protrusions and membrane blebbing, an attribute we have observed to be a negative prognosticator of successful directed differentiation. Lactate exchange between cells and their environment is mediated primarily by MCTs 1 – 4, which are associated with the transfer of ketone bodies and lactic acid in and out of the cells[4]. MCT-1 and -2 are responsible for shuttling lactate inside the cells while MCT3 and MCT4 facilitate its removal. The expression of MCT1 and MCT3 was lower in both cell lines cultured with added lactate. Immunostaining and western blotting did not reveal appreciable changes in MCT3 among the control and 5 g/L lactate groups. Results for the expression of other MCTs were hPSC-line specific. For example, MCT2 and MCT4 increased in H9 hESCs, but the opposite was true in IMR90 hiPSCs. Moreover, we observed the behavior of the uncoupling protein 2 (UCP2), which is known to remove four-carbon substrates from the mitochondria, thereby reducing the tricarboxylic acid (TCA) cycle activity and the production of ATP. Literature reports UCP2 regression to coincide with the metabolic shift from glycolysis to oxidative phosphorylation when hPSCs are differentiated Additionally, pluripotency genes such as NANOG and OCT4 remained unchanged and there was even upregulation of NANOG in the presence of lactate. The maintenance of the pluripotent state was further assessed by subjecting cultured hPSCs to directed differentiation toward endoderm, ectoderm and mesoderm. Acknowledging that these results pertain to the short-term effects that lactate may have on hPSCs, ongoing experiments are aiming to investigate global transcriptomic changes inflicted over a series of passages via RNA sequencing.

Conclusion: Our study shows that lactate, a common metabolic product in mammalian cell bioprocessing, negatively impacts the growth of hPSCs in culture. Lactate also induces changes in the transport and metabolic networks of hPSCs. Elucidating the cascade of molecular events that lactate accumulation imposes on cells, will inform strategies for lactate control in hPSC cultures, efficient expansion, and robust directed differentiation to therapeutically useful cell types.

References

1. Jacobson, E.F., et al., *Non-xenogeneic expansion and definitive endoderm differentiation of human pluripotent stem cells in an automated bioreactor.* Biotechnol Bioeng, 2021. **118**(2): p. 979-991.
2. Hassell, T., S. Gleave, and M. Butler, *Growth inhibition in animal cell culture. The effect of lactate and ammonia.* Appl Biochem Biotechnol, 1991. **30**(1): p. 29-41.
3. Kumagai, S., et al., *Lactic acid promotes PD-1 expression in regulatory T cells in highly glycolytic tumor microenvironments.* Cancer Cell, 2022. **40**(2): p. 201-218 e9.
4. Halestrap, A.P., *The monocarboxylate transporter family—Structure and functional characterization.* IUBMB Life, 2012. **64**(1): p. 1-9.
5. Zhang, J., et al., *UCP2 regulates energy metabolism and differentiation potential of human pluripotent stem cells.* EMBO J, 2011. **30**(24): p. 4860-73.

317b Skin Tissue Derived Neural Crest Stem Cells and Metabolic Requirements for Multipotency

Pihu Mehrotra, Izuagie Ikhapoh, Pedro Lei, Georgios Tseropoulos, Yali Zhang, Jianmin Wang, Song Liu, Stelios Andreadis

Neural crest (NC) cells are a multipotent stem cell population that give rise to a diverse array of cells in the body, including peripheral neurons, Schwann cells (SC), smooth muscle cells, and melanocytes. Neural crest (NC) induction during embryogenesis takes place in response to two key signaling events: the Wnt pathway that is active at the neural plate border; and BMP signaling from the developing mesoderm. In our lab, we have isolated NC-like stem cells (KC-NCs) from adult skin tissue, and have shown that these resemble the NCs that originate during development, including their differentiation potential. Hence, these cells serve as a safe and efficient patient-specific source for cell therapy. However, KC-NCs are very transient, and lose their stem cell potential over time in culture limiting their clinical use. Methods to find ways to preserve the multipotency for these cells is thus crucial to exploit their regenerative potential.

In this regard, we mimicked the developmental microenvironment during NC induction and treated the cells with BMP2 and the Wnt agonist CHIR99021 to preserve multipotency of KC-NCs. Combinatorial Wnt/BMP2 signaling sustained the expression of transient NC genes including Sox10, Pax7, AP2A, HNK1 and TrkC in culture. Subsequently, this treatment enhanced the differentiation of KC-NCs to melanocytes, Schwann cells and smooth muscle cells, as indicated by increased expression of MITF, KROX20 and αSMA respectively. We discovered that Wnt/BMP signaling rewired the metabolic landscape of NCSCs to confer multipotency. This was achieved by activating anaerobic glycolysis in KC-NCs, which increased lactate production, glucose consumption and sensitivity to insulin stimulation. Energy demands of multipotent KC-NCs were evaluated by employing the Seahorse ATP Rate assay. Indeed, cells treated with CHIR/BMP2 derived most of their ATP from glycolysis, and the mitochondrial activity was diminished. Taken, together, this study elucidates how developmental events control the carbon metabolism and bioenergetics state of NCs, which is important in understanding metabolic disorders arising from dysregulation of NC induction and differentiation during embryogenesis. We also describe a simple method using a cocktail of chemicals and growth factors for maintaining KC-NC multipotency for treatment of demyelinating and other diseases.

317c Impact of Oncostatin M and Tumor Necrosis Factor-α on Chondrogenic and Immunosuppressive Capacity of Human Mesenchymal Stem Cells in Pellet Culture and 3D Hydrogel Constructs

Carly Battistoni, Qinghua Xu, Rithika Athreya, Alyssa Panitch, Julie C. Liu

Statement of Purpose: Mimicking the disease conditions of osteoarthritis (OA) in a laboratory setting is important to better understand biological processes and for efficacy of potential treatments. There are various ways to simulate OA *in vitro*. One common approach is to add inflammatory cytokines found in OA to cell culture media. This method is employed often in ex plant models and has gained traction for studying 2D cell culture and 3D hydrogels. The combination of oncostatin M (OSM) and tumor necrosis factor-α (TNF-α) at 10 ng/mL and 20 ng/mL, respectively, has been employed in ex plant models and validated *in vivo*. However, the impact of the combination of these cytokines on pellet culture and hydrogel systems has not been studied. Mesenchymal stem cells (MSCs) can be differentiated down cartilage lineage and are known to be immunosuppressive. Additionally, as the environment strongly impacts cell response (e.g., to mechanical stimuli), comparing MSC response in both pellet and hydrogel constructs will provide insight into contributions from more than just cytokine stimuli. In this work, we investigate MSC differentiation and response in the presence of OSM and TNF-α in both pellet culture and blended collagen I/II hydrogels, a system we previously developed and showed promoted chondrogenesis, without and with encapsulated hyaluronic acid (HA).

Results: Human MSCs were maintained in pellets and hydrogels for 2 weeks. For pellet culture, five groups were studied: 1) negative control (base media), 2) positive control (chondrogenic media with 10 ng/mL of TGF-β3), 3) OSM and TNF-α ("OT", positive control with 10 ng/mL OSM and 20 ng/mL TNF-α), 4) TNF-α ("T", positive control with 20 ng/mL TNF-α), and 5) OSM ("O", positive control with 10 ng/mL OSM). Collagen I/II hydrogels (col I/II) and col I/II gels with HA (1.5 MDa) at 1 mg/mL ("HMW1") were maintained in both positive media and OSM and TNF-α media.

After 2 weeks, pellets and constructs were harvested and analyzed for matrix production (glycosaminoglycan, GAG, content) and for cell proliferation (DNA quantification). As shown in Figure 1 below, the presence of OSM and TNF-α did not limit the amount of GAG produced (1a and 1c). The increase in GAG production for HMW1 compared to col I/II was anticipated as HA provides further cues for cartilage production. However, the similar GAG production from HMW1 with cytokines was unexpected. The most surprising result was the DNA for the hydrogel groups (Figure 1d). The DNA for gels exposed to cytokines was statistically higher and on the order of ~3x more for both col I/II and HMW1 with cytokines compared to their non-cytokine counterparts. Although not statistically significant, the DNA for positive and cytokine groups for pellet culture were also higher as compared to the negative control pellets (Figure 1b). The increase in DNA suggests that the cells are proliferating more in the cytokine-stimulated groups as compared to non-stimulated groups. Additionally, the similar GAG levels for the pellet groups in the presence of cytokines suggest that overall the MSCs were still secreting cartilage extracellular matrix components (Figure 1a).

Conclusions: *In vivo*, cells respond to many biological and mechanical cues simultaneously. Simplifying the system *in vitro* to parse out individual contributions can aid in understanding how specific factors impact cell responses. Here, we found that the addition of cytokines to pellet or hydrogel systems resulted in similar amounts of overall GAG production. However, the increase in cell proliferation suggests that the cells are more active in response to the added cytokines and were likely compensating to maintain the same amount of matrix leading to less GAG per cell – especially for the hydrogel groups. Moving forward, we plan to quantify various genes related to cartilage differentiation, inflammation, and immunosuppression to elucidate what else is happening in the system.

Figure 1. Matrix quantification and cell proliferation for pellets and hydrogels. (a) GAG content for pellets, (b) DNA content for pellets, (c) GAG content in hydrogels, and (d) DNA content for hydrogels. For hydrogels: "no" indicates no added cytokines and "cyto" indicates added OSM and TNF-α. Different letters represent statistical significance ($p < 0.05$).

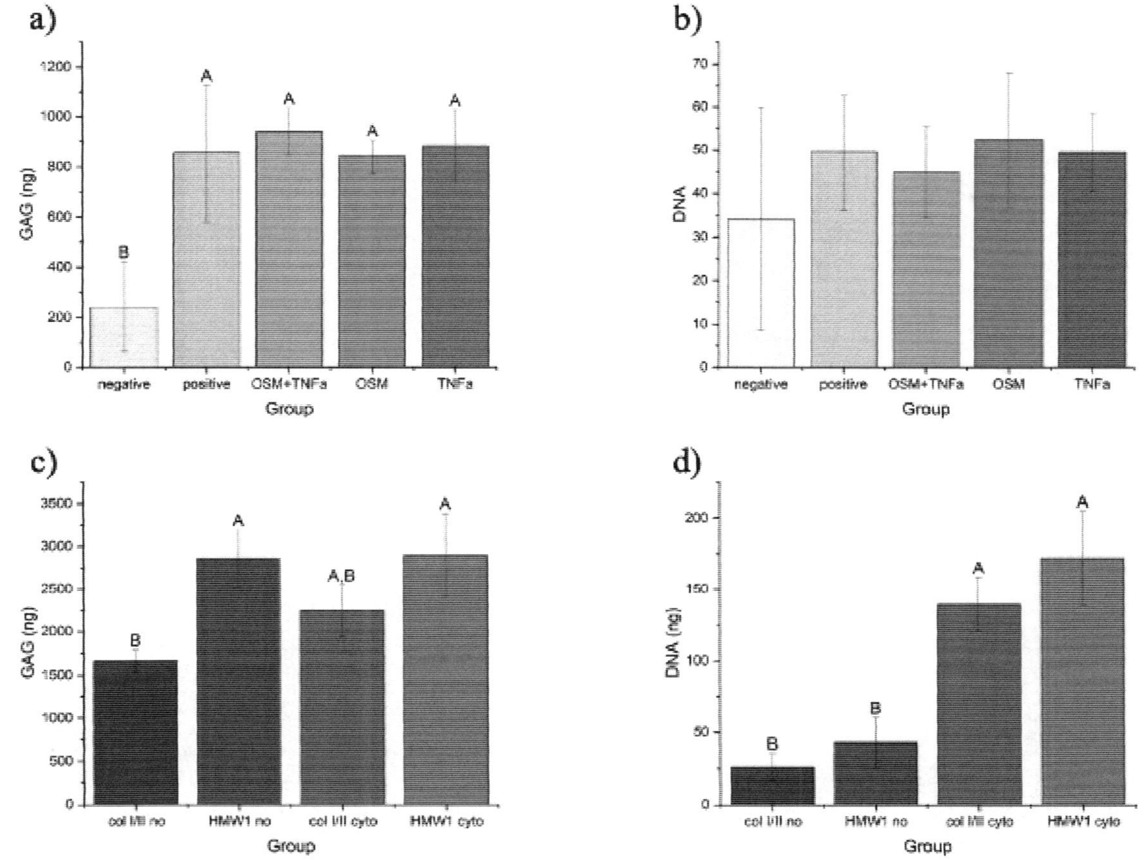

317d Surface Engineering of Auxetic Scaffolds for Neural and Vascular Differentiation from Human Stem Cells

Xingchi Chen, Chang Liu, Xiaolin Wang, Tristan Driscoll, Changchun Zeng, Yan Li

The complexity and dynamic microenvironments of the extracellular matrix (ECM) can change the interactions with stem cells which can regulate stem cell fate decisions. Stem cells can sense the biophysical and biochemical properties of their microenvironment (such as biomechanical properties andgrowth factors) and transduce the extracellular signaling to the intracellular nuclei to induce gene transcription. Negative Poisson's Ratio (NPR) (i.e., auxetic) materials have attracted high attention in biomedical applications due to their unique biophysical properties. Poisson's ratio describes the degree of a material that contracts (or expands) transversally when axially strained, and most materials have Poison's ratio of 0.3-0.5. The NPR materials also have high porosity, intendent resistance, and shear resistance, which are attractive properties in neurovascular tissue engineering. Human induced pluripotent stem cells (hiPSCs) have the ability to differentiate into all types of cells. The potential applications of hiPSCs include the investigation of neural and vascular tissue degeneration to treat blood-brain barrier dysfunction. In this study, six different polyurethane auxetic foams with different modulus (700-2700 Pa) and Poisson's ratio (-0.3 to -0.5) were fabricated and characterized to study the hiPSC lineage-specific differentiations. The biocompatibility of the foams was regulated by modifying the surfaces with chitosan and heparin. Chitosan modified foams were shown to promote cell viability. Then, the vascular and neural differentiation of hiPSCs on different foams with distinct elastic modulus and Poisson's ratio were investigated. Heparin modification can prompte endothelial and neural differentiation. In the foam with modulus of 365 ± 50 Pa and Poisson's ratio of –0.196, the neural specific markers were mostly expressed when cultured with the chitosan-modified scaffolds. In addition, endothelial cell-specific markers CD31 and ZO-1 were mostly expressed in the foam with modulus of 1400 Pa and Poisson's ratio of -0.297. This study has significance in understanding biophysical regulation of human pluripotent stem cell fate decisions and in establishing 3-D neurovascular tissue models for drug screening and neurological disease modeling.

317e A Streamlined Protocol for Highly-Efficient Reprogramming to Induced Motor Neurons

Nathan Wang, Brittany Lende, Kate E Galloway

Cellular reprogramming converts common, accessible cells into rarer, inaccessible cell types, such as neurons. We recently developed a chemical-genetic cocktail that increases reprogramming of mouse embryonic fibroblasts (MEFs) to induced motor neurons (iMN) at 100-fold higher rates over the base delivery of six reprogramming transcription factors. This specific cocktail increases reprogramming by increasing the population of hyperproliferating, hypertranscribing cells (HHCs), which reprogram at near-deterministic rates. However, we observed significant variance in the reprogramming efficiency across batches of cells, which may be due to variability in media conditions as well as the differential delivery of six different viruses which each encode a transcription factor. To improve this process, we aimed to simplify the reprogramming process by developing a streamlined protocol for highly efficient reprogramming. Specifically, we sought to reduce the number of viruses required for co-delivery of reprogramming factors and the high-efficiency genetic cocktail. Additionally, we explored eliminating costly and potentially variable protein supplements during reprogramming.

To achieve this goal, we identified the minimal set of transcription factors sufficient to induce robust MEF-to-iMN reprogramming: Ngn2, Isl1, and Lhx3 (NIL). Interestingly, we find that by reducing the number of transcriptions factors used from the original six to just three (NIL), we massively increase the population of HHCs and reprogramming yield, increasing the yield of iMNs another 20-fold. We explored optimizing the transcription factor stoichiometry by rearranging the order of the transcription factor cassette. We found that varying stoichiometries reprogramming factors (NIL) have minimal effects on the yield from our highly efficient reprogramming protocol. Our results suggest that total levels rather than stoichiometric expression of reprogramming factors dictate iMN reprogramming yields. In addition, to simplify our protocol and remove extra sources of variability, we explored reprogramming in the absence of neurotrophic factors. We find that neurotrophic factors are not required for robust iMN reprogramming in our high-efficiency system, minimizing both cost and complexity of the process.

Taken together, this work outlines a highly efficient and robust iMN reprogramming protocol that is consistent with our previous finding that expanding the population of highly-plastic cells in reprogramming increases reprogramming yields. With this massive increase in reprogramming yield, we can readily obtain sufficient number of cells to perform biochemical assays, allowing us to precisely map the molecular mechanisms of cell-fate transitions during high-efficiency reprogramming.

317f Light-Controlled Enhancement of Glucose-Stimulated Insulin Secretion of Optogenetically Engineered Human Pancreatic Beta-Cells

Zijing Chen, Emmanuel Tzanakakis

Diabetes is a group of disorders linked to the lack of insulin-producing pancreatic β-cells or insulin resistance exhibited by glucose-absorbing tissues. The transplantation of β-cells or islets is a promising avenue for restoring normal blood glucose homeostasis in diabetes patients. Yet, the number of cells required to reconstitute glucose-stimulated insulin secretion (GSIS) – whether by directly transplanting islet cells or employing a bioartificial pancreas – poses significant challenges on maintaining optimal cell viability and function. To mitigate this issue, optogenetic tools can be applied to lower the oxygen and nutrient demand as fewer cells are required with insulin production upregulated by 2- to 3-fold in rodent β-cells, as we reported [1][2]. Optogenetics allow modulation of cellular processes in a non-invasive and reversible manner with high specificity and fewer side effects in comparison to pertinent pharmacological treatments. The secondary messenger cyclic adenosine monophosphate (cAMP) plays a central role in the regulation of GSIS in pancreatic β-cells. The increased cAMP activates the protein kinase A (PKA), which phosphorylates and enhances the function of voltage-dependent calcium channels, allowing for greater Ca^{2+}influx and therefore stimulation of membrane depolarization for subsequent release of insulin granules. We engineered human β-cells to express a blue light photoactivatable adenylyl cyclase (bPAC) for photoinduced augmentation of insulin secretion at various extracellular glucose levels. Alginate hydrogels were adopted to encapsulate and facilitate the delivery of the optogenetically modified β-cells.

We have demonstrated that murine insulinoma cells can form PIs in stirred-suspension spinner flasks and in low-adhesion plates under orbital shaking [3]. To this end, human EndoC-βH3 cells were seeded in 50-100 mL culture medium in spinner flasks and cultured at 37 °C and 5% CO_2 with agitation speed set to 70 rpm. For up to nine days, cells formed aggregates and were maintained with half-volume media changes every 3 days. The average aggregate size increased from 67.8 ± 7.8 mm on day 3 to 94. 8 ± 19.9 mm on day 9 as determined from the quantification of PI images acquired by bright field microscopy. A significant increase in size was noted on days 7 and 9 of the cultures by 7% and 40% respectively. The PIs were infected with an adenoviral vector, AdbPAC, carrying the bPAC-cMyc-IRES-mCherry cassette (cMyc tag for immunodetection of bPAC; mCherry: fluorescent reporter coexpressed with bPAC; IRES: internal ribosomal entry site) under the CMV promoter. The expression of mCherry was detected by fluorescence microscopy in about 50% of cells in monolayers and PIs. Given the similarities in the molecular circuitries governing GSIS in rodent and human β-cells, we hypothesized that expression of the PAC to human β-cells will provide a similar handle for controlling insulin function. Insulin secretion and intracellular cAMP were measured under different glucose conditions. Insulin release was upregulated by 2.5-fold stimulated by the addition of glucose at 20 mM compared to 0.5 mM. A 2-fold and 1.5-fold increase of secreted insulin was observed by raising [cAMP]i with forskolin at 20 mM glucose, and 0.5 mM glucose level, respectively.

For the delivery of the optogenetically engineered β-cells, we set out to develop a scaffold featuring wireless circuitry and an LED for optostimulation in a closed-loop system. Alginate, which exhibits excellent biocompatibility, minimal cytotoxicity and benign gelation conditions, has been used successfully for encapsulating islet cells as reported by numerous groups. The scaffold is designed to harbor PIs and an LED with a magnetic reed for remote control during *in vitro* testing. Alginate polymers can be crosslinked with calcium ions. Using $CaCl_2$, the gelation rate of alginate is rapid, but the polymer gels are opaque rendering them undesirable for optogenetic applications, which rely on light transmissibility. To control the gelation rate of 1.5% (w/v) alginate and obtain homogeneous hydrogels, we used $CaCO_3$ and 50-200 mM glucono-δ-lactone (GDL) (1:2) as the crosslinking agents, which resulted in a slower, more controlled gelation process leading to structural uniformity and notably higher transparency. An increase in hydrogel opacity was observed along with a faster gelation time with higher Ca^{2+} concentration. Gelation with $CaCO_3$ and GDL improved scaffold transparency by 80% compared to $CaCl_2$-crosslinked alginate gels at higher Ca^{2+} concentration. The Young modulus of the hydrogels was determined by a dynamic mechanical analyzer (RSA3, TA Instruments, New Castle, DE) with a 35-N loading cell at room temperature. Scaffolds formed using $CaCl_2$ showed concentration-dependent mechanical properties. Young's modulus ranged between 4 kPa – 25 kPa for 25 mM – 100 mM $CaCl_2$. For reference, the Young modulus of murine pancreas is 4.2 kPa and of human normal pancreatic tissue is 1 kPa, whereas higher modulus values are reported for pancreas under pathological conditions [4]. Our premise is that hydrogels with mechanical properties resembling those of the native pancreatic tissue will provide an environment that is more akin to the physiological milieu potentially contributing to enhanced functionality of the optogenetically engineered cells. Moreover, a blue microLED light that can be activated remotely with a radio-frequency identification (RFID) reader was also encapsulated in the alginate disk scaffold. Viability of encapsulated EndoC-βH3 and incubated with Calcein AM dye was 80% when determined by confocal microscopy an hour after the gelation. Ongoing experiments aim to determine the glucose stimulation and release rate of insulin from bPAC-expressing cells in hydrogels.

Next-generation technologies for diabetic management are expected to incorporate glucose-sensing, insulin-releasing cells. The transplantation of β-cells is shown to restore normal blood glucose level autonomously unlike treatments entailing exogenous insulin injections. Through the expression of optogenetic switches like bPAC, GSIS can be enhanced without the use of pharmacological agents with potentially adverse side effects. Molecular customization of human β-cells with optogenetic moieties, scalable cultivation of the engineered islet cells as PIs, and encapsulation within an LED-harboring hydrogel are key elements in the design and development of bioartificial pancreas devices. Such technologies will be a significant step toward translating optogenetic control-based solutions to therapeutic modalities for diabetes patients.

References:

[1] F. Zhang, E. S. Tzanakakis, 'Amelioration of diabetes in a murine model upon transplantation of pancreatic β-cells with optogenetic control of cyclic adenosine monophosphate.' *ACS Synth. Biol.* 8(10): 2248-2255, 2019.

[2] F. Zhang, E. S. Tzanakakis, 'Optogenetic regulation of insulin secretion in pancreatic β-cells.' *Sci. Rep.* 7(1): 1-10, 2017.

[3] L.T. Lock, S. G. Laychock, E.S. Tzanakakis, "Pseudoislets in stirred-suspension culture exhibit enhanced cell survival propagation and insulin secretion", *J. Biotechnol.* 151(3): 278-86, 2011.

[4] A. Nabavizadeh et al. "Noninvasive Young's modules visualization of fibrosis progression and delineation of pancreatic ductal adenocarcinoma (PDAC) tumors using harmonic motion elastography in vivo", *Theranostics*10(10):4614-26, 2020.

317g Modeling liver organogenesis from human stem cells in vitro and in vivo; new approaches in liver tissue engineering

Natesh Parashurama

During liver organogenesis (LION), E9.5 collectively migrating hepatoblasts (MHs) arise from the E9.0 liver diverticulum (LD) and penetrate the surrounding mesoderm (MES) tissue. The role of the LD in LION has been largely ignored in current human pluripotent stem cell (hPSC) studies. We hypothesize that the LD and surrounding MES form a functional tissue complex (LD-MESC) that triggers LION by simultaneously coordinating exponential growth, collective migration, and differentiation. This can be incorporated into protocols to improve the induction of migrating hepatoblasts.

We first employed systems biology of scRNA-seq data to analyze in vivo LION and support our hypothesis that many biological events are coordinated at the LD-MESC stage. Our analysis of early LION demonstrates that hypoxia, early gene regulatory network (GRN) activation, ALB transcription, exponential growth, collective migration, and metabolic programming are linked. Further, we observe a wide-range of upregulated pathways in signaling (e.g., Hippo), metabolic changes (e.g., oxidative phosphorylation), and migration pathways. To model the LD-MESC concept in vitro, we designed a novel in vivo transplant system, such that the LD could be induced in vivo from definitive endoderm (DE), in the vicinity of mesenchymal cells (MES). Cells were transplanted subcutaneously to model hypoxic growth. For our in vitro studies, we developed a 3-stage protocol with a unique medium formulation. Our system is unique because it employs a single medium throughout, is under completely under hypoxic conditions, and contains no formal instructive factors, thus relying on spontaneous differentiation of gut tube endoderm cells. We performed microscopy and molecular analysis.

The in vivo LD-MESC model demonstrated that subcutaneous transplantation resulted in cord formation similar to hepatic cord morphogenesis, exponential growth, liver gene expression, with no evidence of blood vessels. This is the first reported hPSC system that mimics these important characteristics of early LION. Our 14 day monolayer protocol demonstrated a uniform population of hPSC-derived Hepatoblasts (HBs). Next, we demonstrated successful compaction, and when placed in matrigel droplet culture, we demonstrated collective outward, 3D collective hepatoblast migration for the first time. These cells produced higher ALB, PROX1, and demonstrate Albumin + hepatic cords, and a ~5 fold increase in overall growth. To determine the mechanism by which migration occurs, we evaluated numerous signaling pathways in a chemical screen of growth/migration. W found VT (Hippo) and SU5416 (VEGFR2) were found to inhibit migration, agreeing with the transcriptomic data for murine MH. Interestingly, Hippo pathway is in agreement with our murine MH transcriptomic data, mediates migration and growth of hPSC-MH in vitro. These data substantiates the LD-MESC model, establishes a novel protocol which generates early HBs, and demonstrates a new in vivo model based upon LION principles.

631a A Novel Microfluidic Device to Decipher the Interplay of Cellular Contractility and Electric Field during Galvanotaxis

Anindya Sen, Soontorn Tuntithavornwat, Bishwa Ranjan Si, Se Jong Lee, Kaustav Bera, Konstantinos Konstantopoulos

Introduction: The movement of cells under an electric field, a process known as galvanotaxis, is responsible for a myriad of (patho)physiological processes ranging from wound healing [1], angiogenesis [2] to cancer metastasis [3]. A lot has been studied about the galvanotactic response of different cells on flat 2D substrates. However, little is known about the signaling pathways induced by an electric field when cells are confined in the microchannels (capillaries) in the body. This is especially important since the cells are known to utilize distinct signaling pathways in confined spaces and on unconfined substrates [4]. In this study, we have developed a novel microfluidic device wherein, we simultaneously observe the migration and reversal of directionality of HT1080 fibrosarcoma cells under EF and investigate the signaling pathways that are associated with these processes. We believe that our work will help us in developing unique strategies for combatting cancer, in particular, and in gaining insights into different physiological processes, in general.

Materials and Methods: A polydimethylsiloxane (PDMS)-based microfluidic device was designed and fabricated using photolithography techniques and replica molding to create microchannels mimicking the narrow capillaries that the cancer cells travel through during metastasis. This was achieved by seeding cells in a central chamber of 3 μm width and 10 μm height (**Fig. A**), whereby the cells moved freely away from the reservoir into the microchannels. Ag/AgCl electrodes embedded in agarose were used at both ends of the device to apply DC electric fields across the microfluidic device using a programmable potentiostat. HT1080 fibrosarcoma cells were used in this study. Cell migration was visualized and recorded with time-lapse phase-contrast microscopy. The activity of proteins of interest was observed by using confocal microscopy along with fluorescently tagged proteins, such as FLIM-FRET biosensors or LifeAct tagged with mCherry. Pharmacological inhibitors were used to decipher migration mechanisms.

Results and Discussion: In the absence of an electric field, HT1080 fibrosarcoma cells migrated freely away from the central reservoir into the microchannels **(Fig. B, upper)**. In the presence of electric field, we observe that the HT1080 cells, which were originally migrating towards the anode reversed their direction of migration after the EF was switched on, whereas the cells migrating towards the cathode maintained their direction of motion (**Fig. B, lower**). HT1080 cells subjected to EF undergo a phenotypic transition from a predominantly protrusive phenotype to a non-protrusive phenotype characterized by blunt, dense bundle of actin and blebs at the poles **(Fig. C)** in both the maintaining and reversing side of the device.

Since such phenotypic change is influenced by RhoA-ROCK-actomyosin-based contractility, we quantified RhoA activity in cells migrating in the maintaining and reversing side using the Förster resonance energy transfer (FRET)–based RhoA activity biosensor combined with confocal fluorescence lifetime imaging (FLIM). We found that the cells undergoing repolarization under EF have more RhoA activity unlike the cells on the maintaining side. This signified a difference in the role of RhoA-based contractility during the maintenance and reversal of migration direction under EF **(Fig. D)**. On the maintaining side of the device, we found that EF facilitates the ROCK-inhibited migration of fibrosarcoma cells **(Fig. E),** in other words, EF bypasses the role of actomyosin-based contractility while maintaining the direction of cell migration under EF. However, on the reversing side, the inhibition of RhoA facilitates cell reversal, that is, contractility is acting as a hurdle for the EF to repolarize the cell **(Fig. F)**.

Conclusion: We have designed and fabricated a novel microfluidic device that enables us to study the migration and reversal of directionality of cells in the presence of a direct-current electric field under physiologically relevant confinement. Using this device, we have uncovered the differences in the role that actomyosin-based contractility plays in maintaining and reversing the directionality of cells under EF. Our research will give more perspective into the different biophysical pathways that cells employ during galvanotaxis.

References:

[1] Huttenlocher, A., andamp; Horwitz, A. R. (2007). Wound healing with electric potential. *New England Journal of Medicine*, *356*(3), 303-304.

[2] Zhao, M. (2009, August). Electrical fields in wound healing—an overriding signal that directs cell migration. In *Seminars in cell andamp; developmental biology* (Vol. 20, No. 6, pp. 674-682). Academic Press.

[3] Zhu, K., Hum, N. R., Reid, B., Sun, Q., Loots, G. G., andamp; Zhao, M. (2020). Electric fields at breast cancer and cancer cell collective galvanotaxis. *Scientific reports*, *10*(1), 1-11.

[4] Paul, C. D., Mistriotis, P., andamp; Konstantopoulos, K. (2017). Cancer cell motility: lessons from migration in confined spaces. *Nature Reviews Cancer*, *17*(2), 131-140.

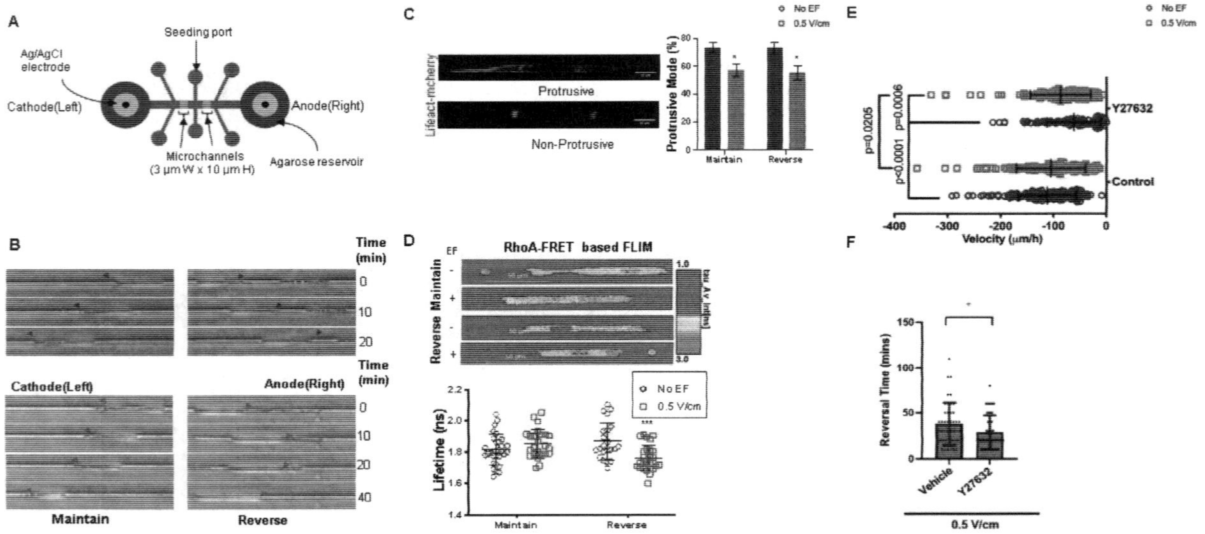

(A) Schematic of the microfluidic device. HT1080 cells migrate away from the central seeding area into the microchannels **(B, upper)** in the absence of EF and **(B, lower)** preferentially move towards the cathode under EF. **(C)** EF induces a phenotypic change from protrusive to non-protrusive in HT1080 cells. **(D)** Quantifying RhoA activity using RhoA-FRET-based FLIM for cells maintaining and reversing their direction of motion under EF. Lesser lifetimes indicate greater RhoA activity. **(E)** EF largely rescues contractility-inhibited migration of HT1080 cells. **(F)** Inhibition of contractility causes cells to reverse faster under EF. Y27632 (10 μM) is the pharmacological inhibitor of ROCK.

631c Development of Heterogeneous Tumor Tissue-Mimicking Glioblastoma Organoids

Seungjo (Joe) Park, Alexandra D. Avera, Yonghyun (John) Kim

Glioblastoma multiforme (GBM) is the most aggressive type of brain tumor that originates from glioblastoma stem cells (GSCs). In the brain, GSCs are supported by a tumor microenvironment (TME) residing in the perivascular niche and the hypoxic niche. The GBM TME is highly heterogenous and exhibits complex cell-to-cell interactions. However, most *in vitro* models do not adequately capture such heterogeneity. Here, we describe the development of a simple, matrix-free, and tissue-like GBM organoids (GBOs) using patient-derived xenograft GBM lines in small-scale bioreactors. Shear stress was optimized to produce highly reproducible millimeter-scale GBOs within 4-5 weeks. Our GBOs exhibited high stemness and strong cell-to-cell interactions compared to conventional tumorsphere cultures. They displayed spatial gradients of HIF-1α positive hypoxic cores where CD133-positive GSCs resided, as well as spatially heterogeneous expression of Notch signaling. We also observed a self-established, hierarchically organized, and heterogeneous TME with GBM transdifferentiation into endothelial cells, pericytes, and astrocytes. Collectively, we demonstrate the ability to biomanufacture uniformly sized GBOs that recapitulate *in vivo* GBM TME features.

631d Comparison between Two Different 3D Printed Tumor-on-Chip Devices As Anti-Cancer Testing Platforms

Salvador Gallegos Martínez, Itzel Lara, David Choy Buentello, Grissel Trujillo de Santiago, Mario Moisés Álvarez

The recapitulation of the tumor niche in tumor-on-chip systems will greatly contribute to accelerate cancer research in the fronts on fundamental research, pharmacological testing of new anti-cancer compounds, and personalized cancer medicine.

Here we describe the development and comparison of two micro/mini fluidic platforms aimed to contribute to the advance of tumor-on-chip research and based on 3D-printed devices.

First, we describe a simple and robust method for the fabrication, maturation, and extended culture of large breast cancer (MCF7 and MCF7/fibroblasts) spheroids (~800 μm in diameter) in a 3D-printed mini continuous stirred tank reactor (mini-CSTR) (Figure 1 A-B). In brief, MCF7 cell suspensions (5×10^4 cells) were incubated in batch culture for three days in ultra-low-adherence (ULA) plates to form discoid cell aggregates (700–800 μm in diameter). These microtissues were then transferred into the mini-CSTR and continuously fed with culture media for an extended time (~30 days). The spheroids progressively increased in size during the first 20 days of perfusion culture to reach a steady diameter. At low residence times (~22 h) and glucose concentrations of 1.1 g L^{-1}, the spheroids reached a diameter of 700-800 μm. We characterized the spheroid morphology and architecture (i.e., shape, size, and cell viability at different radial zones within the spheroid), and the evolution of expression of relevant tumor-related genes (i.e., ER, VEGF, Ki67, Bcl2, LDHA, and HIF-1α) in spheroids cultured for 30 days.

This mini-CSTR culture strategy enables the simple and reproducible fabrication of relatively large spheroids and offers great potential for studying the effects of diverse effectors on tumor progression. Moreover, we used the mini-CSTR system to evaluate the effect of the continuous perfusion of three anti-cancer drugs (i.e., docetaxel, paclitaxel, and doxorubicin) on large MCF7-derived spheroids. Our results show that the use of this agitated system as a tumor-on-chip platform may be a useful and practical tool for testing the efficacy and safety of novel anti-cancer drugs and may enable personalized medicine applications.

In addition, we introduce 3D-printed microfluidic system (Figure 1 c-D) that can be generically used to culture tumor-tissues under well-controlled environments. The system is composed by three compartments. The left and right compartments have one inlet and one outlet each, providing means to continuously feed two liquid streams to the system. The central compartment is designed to host a cell-laden hydrogel where a microtissue can be confined and cultured. A transparent lid and bottom can be adapted to enable visual inspection under a microscope.

We conducted fluorescent and saline tracer experiments to characterize the hydraulic performance of the system. In addition, we cultured MCF7-spheroids embedded in a cell-laden hydrogel construct (placed in the central chamber), to illustrate the use of this system to sustain long term micro-tissue culture experiments. We also present experimental results that illustrate the flexibility and robustness of use of this 3D-printed device for tumor-on-chip experiments that including pharmacological testing of individual and combined anti-cancer compounds (i.e., docetaxel and docetaxel/paclitaxel).

These "open-source" organ-on-chip systems are intended to be a general-purpose resource to facilitate/democratize the development of tumor-on-chip applications.

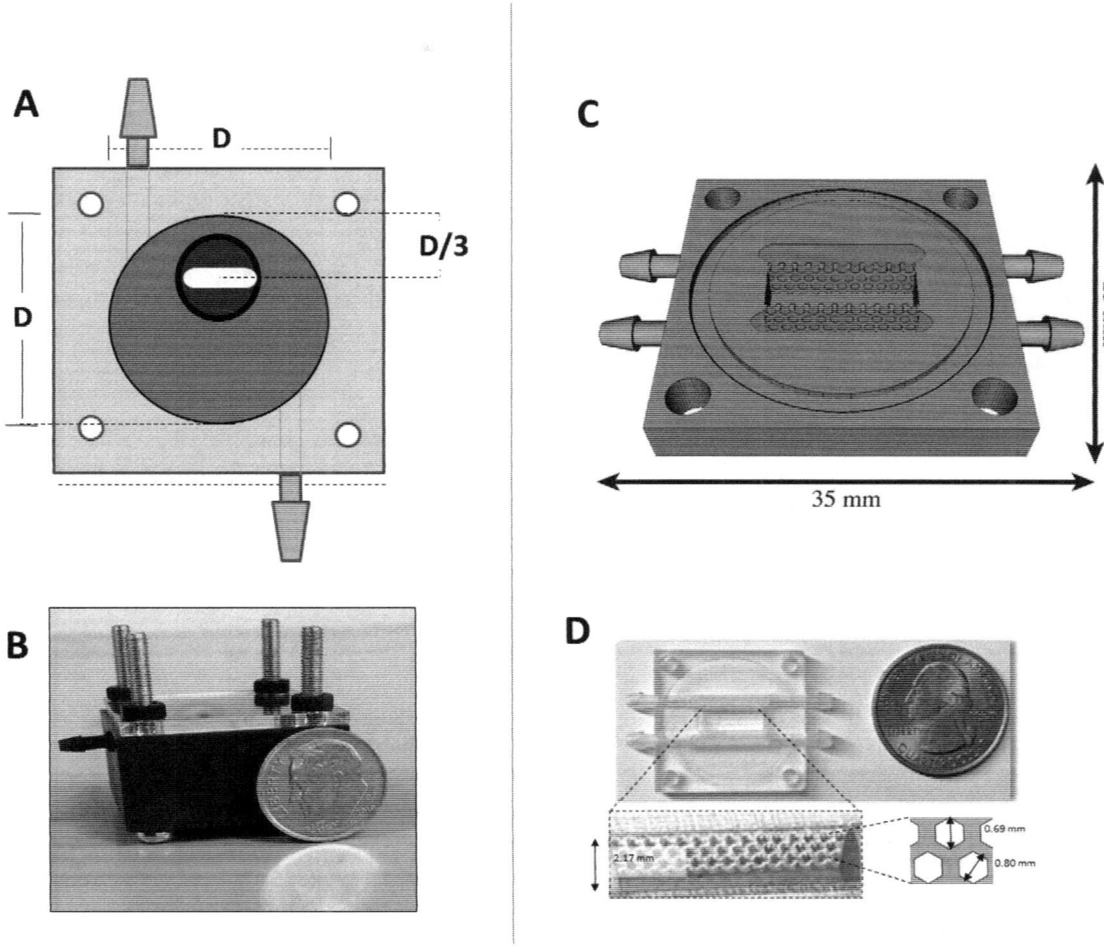

Figure 1. Comparing of two tumor-on-chip systems: (A-B) A miniaturized continuous stirred tank system, and a (C-D) 3D-printed device that simulates arterial and venous circulation

631e Islet-on-Chip Model for Type 2 Diabetes

Connor Wiegand, Ravi Krishnamurthy, Ipsita Banerjee

Diabetes is an increasingly prominent disease that develops from the dysfunction of the insulin production within pancreatic islets. This dysfunction stems from autoimmune elimination of the insulin producing beta cells (Type 1) or a toxic environment that hinders insulin production or effectiveness (Type 2). Testing of possible treatments for preventing or mitigating the complications of diabetes is hindered by expensive and inadequate animal models. The human-on-chip project seeks to replace these models and aid in the transition from initial testing to clinical trials by mimicking the biochemical and biophysical environment of organs on a fluidic device. This project aims to develop a type 2 diabetic islet-on-a-chip model utilizing both primary islets and human induced pluripotent stem cells (hiPSCs).

The islet-on-chip platform is based on the commercially available Organ-on-Chip fluidic device from Micronit that utilizes three glass layers to form a 2-chamber system partitioned by a polyester membrane. The two chambers can be individually accessed through a dedicated perfusion channel that allows for customizable flow configurations through the system. The polyester membrane was designed for culturing adherent cells with the options of cell seeding either through perfusion flow or prior to chip assembly. This aspect of the base chip design is not conducive to culturing pancreatic islets, since retention of spheroidal islet architecture is critical for continued islet function. We introduced a novel strategy for culturing 3D spheroidal islets in the microfluidic device, by supporting the islets with a hydrogel substrate for its structural stability and micropatterning the hydrogel on the membrane. A COMSOL-based in silico flow field model informed this system design for optimizing flow rate and micropattern design to retain adequate oxygen level in the islet vicinity, while minimizing shear stress. Our islet-on-chip system retained high viability and function of glucose stimulated insulin secretion (GSIS) with primary human islets over an extended period of 28 days of perfusion culture under normal 'fasting' condition. We expect this extended culture of healthy islets will be instrumental in modeling disease conditions and tracking disease progression/ reversal over time. The progression of type 2 diabetes was modeled within the formed chip, by simulating the diseased states of glucotoxicity, lipotoxicity, and glucolipotoxicity through long term exposure to pathological glucose levels, elevated free fatty acids, and combinations of the two. With these methods, the resulting islets showed lower insulin expressing cells and loss in functional secretion of insulin in response to glucose changes, which is indicative of a diabetic state. In parallel to testing with primary islets, hiPSC-derived islet organoids were developed to replace the primary human islets as a regenerative cell source. As with the primary islets, the fluidic culture system maintained good survival and functional glucose-stimulated insulin responsiveness in the hiPSC-derived islets over a 14-day span. The end goal of this project is to test of type 2 diabetes treatments, such as Pioglitazone, to determine the reversibility of the toxic states and how well it simulates treatment of the disease in patients.

631f A Multi-Cellular Brain-on-Chip with Integrated Vasculature

Rebecca L. Pinals, Alice E. Stanton, Adele Bubnys, Dong Shin Park, Nhat Truong, Emre Abgas, Oyku Cerit, Benjamin James, Robert Langer, Li-Huei Tsai

Our understanding of brain function under healthy and diseased states is dependent upon, and often limited by, the models and tools with which we study them. Previously, our group has developed an *in vitro* blood-brain barrier model derived from human induced pluripotent stem cells (iPSCs). We have expanded this system to a multi-cellular integrated brain (miBrain) composed of all major brain cell types derived separately from iPSCs (neurons, endothelial cells, pericytes, astrocytes, oligodendrocytes, and microglia) encapsulated together in an engineered hydrogel. These three-dimensional brain-tissue models are uniquely suited to study complex cellular interactions and emergent tissue behaviors.

Herein, we present our efforts toward engineering a fully vascularized miBrain platform in which iPSC-derived endothelial cells self-assemble to form brain microvasculature. We compare the miBrain co-culture system to brain microvascular endothelial cell (BMEC) monocultures in terms of morphological features and expression levels of key junction markers. We further investigate vascularization of the miBrain when adapted to a different form factor, microfluidic devices, that will ultimately enable rapid functional readouts and nutrient delivery via perfusable flow to support longer term culture periods. Finally, we explore the ability of this system to model nanoparticle-based delivery in the human brain microenvironment.

631g Invited talk: Biodegradable Microcapsule Designer Using Silk Fibroin Technology

Muchun Liu

Open use of polymeric microcapsules for agricultural applications and household applications makes it inevitable to release microplastic waste into the environment. Conventional microplastics are known to last for a long time; they travel in various ecosystems and accumulate in living organisms along food chains. However, stringent performance criteria and the large availability of new materials that can retrofit existing manufacturing processes hinder the replacement of commonly used cargo with biodegradable alternatives. Here, we demonstrate that microencapsulation in commonly available silk biopolymers can be achieved by modulating the protein protonation and chain relaxation at the point of material assembly.

315a Endoplasmic Reticulum (ER) Stress Decreases the Efficacy of Delivered siRNA Via Cholesterol Dysregulation

Ryan Splichal, Christina Chan, S. Patrick Walton

RNA-based therapeutics are increasingly important tools for the treatment and prevention of diseases. Because therapeutics are routinely developed and tested under idealized conditions, it is important to understand how clinical comorbidities can impact the effectiveness of delivered RNA. Obesity, diabetes, altered immune function, bacterial infections, and nutrient deprivation are examples of comorbidities that could be present in recipients of RNA-based therapeutics and reduce their efficacy. For example, the COVID-19 mRNA vaccines developed by Pfizer and Moderna showed lower antibody titers in obese vaccine recipients. The physiological impacts of these comorbidities vary, but all result in stress on the endoplasmic reticulum (ER). ER stress develops when there is a buildup of unfolded or misfolded proteins in the ER. In human cells, there are three proteins that sense ER stress: ATF6, IRE1, and PERK. Upon activation, these proteins begin cascades that lead to either prosurvival or apoptotic changes in cell function.

In response to ER stress, ATF6 is translocated to the Golgi, where it is activated. Membrane-bound transcription site-2 protease (S2P), the same protein that activates ATF6, activates sterol regulatory element binding proteins (SREBPs), which are the global regulators of cholesterol biosynthesis. When IRE1 is activated, it splices X-box binding protein 1 (XBP1) mRNA. The protein coded by this spliced mRNA (XBP1s) is translated into a transcription factor that controls transcription of genes that alleviate ER stress by increasing cholesterol biosynthesis. Overall, ER stress causes an increase in the amount of cellular cholesterol. Increased cholesterol subsequently reduces the fluidity of the cell and ER membranes. This results in increased rigidity in vesicles trafficking among the cell membrane, ER, and Golgi. Cholesterol rich vesicles are less permeable and quickly convert into lysosomes for degradation of the contents, preventing further ER stress but also potentially sequestering/destroying delivered RNAs. In this work, we demonstrate that sequestering of RNA inside membrane bound vesicles due to increased cholesterol levels from the ER stress response results in decreased effectiveness of delivered siRNAs.

HeLa cells constitutively expressing EGFP were treated with increasing levels of Tunicamycin, a common ER stress inducer, followed by transfection with Lipofectamine 2000/siRNA complexes targeting EGFP. Fillipin III staining for cholesterol, flow cytometry, and confocal laser scanning microscopy were used to show that cells under ER stress have increased membrane cholesterol and reduced EGFP silencing. In addition, transfected siRNAs are colocalized with cholesterol. Addition of AEBSF, an inhibitor of cholesterol biosynthesis, restored silencing.

Our results demonstrate that when designing RNA based therapeutics, ER stress must be considered. Therapies that must precisely control the level of protein produced in a cell will have altered efficacy in patients with high levels of ER stress.

315b Anionic Bioconjugation Enables Intracellular Protein Delivery with Ionizable Lipid Nanoparticles

Azmain Alamgir, Matthew DeLisa, Christopher Alabi

Protein-based therapeutics comprise a rapidly growing subset of pharmaceuticals, accounting for 20% of all drugs on the market today. However, due to their large, hydrophilic structures, delivery of proteins across cell membranes has been a longstanding issue in the field. As such, protein-based drugs are limited primarily to extracellular applications. The potential of proteins as intracellular therapeutic agents has been highlighted in the use of novel protein scaffolds that inhibit "undruggable" biological targets and more recently by gene-editing machinery for treating human disease. Translating many of these promising technologies for clinically relevant applications requires the development of delivery methods capable of translocating proteins into the cytosol of a cell for functional activity. Several protein delivery approaches have been investigated over the years, yet none have been approved for clinical *in vivo* use.

To overcome these delivery challenges for ultimately translating protein therapeutics for intracellular applications, we were interested in adopting lipid nanoparticle formulations that have found great success in the delivery of nucleic acids in many clinical models. It has long been demonstrated that mixing nucleic acids in the presence of cationic lipids results in the formation of hydrophobic nanoparticle structures within which nucleic acid cargos are packed. The formed lipid nanoparticles (LNPs) can then mediate the entry of nucleic acid cargos into cells for their intended therapeutic effect. To utilize LNPs for protein delivery, we explored the possibility of modifying protein surface charge to mimic the anionic nature of nucleic acids, which would enable electrostatic complexation with cationic lipids to form LNPs.

To this end, we sought to utilize a bioconjugation-based approach for attaching anionic chemical moieties onto surface-exposed residues of protein cargos. We designed a panel of lysine-reactive chemical linkers containing anionic sulfonate groups and a self-immolative, cleavable disulfide bond. The designed compounds would hypothetically provide an anionic "cloak" to protein cargos to enable LNP formation with cationic lipids. Once protein cargos are delivered into cells, the reductive environment of the cytosol will cause cleavage of the disulfide bond, which can further self-immolate to release the protein cargo back in its native form. To test our hypothesis, we turned to super-folder green fluorescent protein (sfGFP) as a model protein to study chemical modification and delivery. We find that anionic modification of sfGFP results in successful sfGFP internalization within cells using Lipofectamine, an off-the-shelf cationic lipid transfection reagent widely used for plasmid transfection. We further expand upon these studies by exploring protein delivery with LNP formulations utilized for *in vivo* delivery of siRNA and mRNA into cells. We find that LNPs formulated with anionically-modified sfGFP and an ionizable lipid called MC3 leads to robust increase in intracellular fluorescence. Furthermore, proteins delivered with LNPs exhibit negligible cytotoxicity, with most cells retaining viabilities above 90%.

Taken together, we demonstrate that anionic bioconjugation of proteins can enable intracellular protein delivery with LNPs. Our exciting proof-of-concept results lay the groundwork for a generalizable, bioconjugation-based platform that can be employed for the delivery of a wide range of protein cargos. Future work will involve applying this technique to deliver functional therapeutic proteins, with the ultimate goal of translating protein therapeutics for intracellular applications.

315c Wireless Iontophoresis Lens Mediated Ocular Drug Delivery

Andrea Zuccaro, Sabrina Coffman, Naomi Addai Asante, Nicolas Hamouche, Donald S. Sakaguchi, Surya Mallapragada, Metin Uz

Age related macular degeneration (AMD) is the most common cause of vision loss in the elderly population (65 years or older).[1,2] The number of AMD cases is anticipated to reach over 288 million worldwide.[1,2] Neovascular AMD, leading to severe visual loss due to choroidal neovascularization, comprises 10% of all AMD cases.[3,4] The current clinical treatment for neovascular AMD treatment relies on inhibiting the vascular endothelial growth factor (VEGF) pathway by using intravitreal injection of anti-VEGF drugs (i.e., Bevacizumab: 1.25 mg/injection or Aflibercept: 2.0 mg/injection) every 4-6 weeks.[3,4] However, intravitreal injection-based approach ***(1)*** *is invasive,* ***(2)*** *is not practical for self-administration,* ***(3)*** *requires frequent office visits for injections,* ***(4)*** *suffers from short drug half-life, and* ***(5)*** *causes complications due to repeated injections (i.e.,* endophthalmitis, retinal detachment, retinal hemorrhage, globe perforation cataract formation, and secondary glaucoma*).*[5–7] *Therefore, there is a critical need for alternative, safe and non-invasive approaches enabling efficient, precise, and accurate drug delivery into the posterior segment of the eye for AMD treatment.*

In this study, we develop a non-invasive self-administered wireless iontophoresis device in the form of a single-use contact lens for efficient delivery of anti-VEGF drugs, Bevacizumab and Aflibercept, into the posterior segment of the eye for AMD. For this purpose, the site-specific and local drug reservoirs are integrated with wireless antenna/coil design in a single lens platform. The wireless iontophoresis lens is fabricated using our previously patented, simple polymer casting and microfluidics-based flexible electronics fabrication method.[8–10] Briefly, predetermined locations of drug reservoirs and microchannels for wireless antenna coils are created on molds using computerized numerical control. The microchannels on the mold are filled with conductive graphene nanoplatelet solution (prepared by ethanol thermal reaction) using syringe pump controlled microfluidic approach. Following graphene circuit formation, the drug suspension at initial dose (Bevacizumab: 1.25 mg or Aflibercept: 2.0 mg, based on applied doses in clinic) is used to fill the site-specific drug reservoirs. After drying, PMMA solution (10 wt% in chloroform) is cast on the mold with wireless graphene antenna and drug. Upon drying and film formation, the film is peeled off to transfer the drug and graphene antenna.[8–10]

The integrity of wireless antennas and drug reservoirs are validated using scanning electron microscopy (SEM). The conductivity/sheet resistance of the graphene antennas are evaluated via 4-point probe measurement. The functionality of the antenna is tested using a light-emitting diode electrically connected to the wireless array as a visual indicator in PBS (pH 7.4 at 37 °C) mimicking cell culture using a waveform/function generator, oscilloscope, and network analyzer. The frequency characteristics of the wireless antenna are assessed via network analyzer. The drug loading efficiency and release profiles are evaluated by collecting release samples at predetermined time points under different wireless electrical stimulation (WES) parameters in PBS (pH 7.4 at 37 °C) using drug specific ELISA kits. Visual release tests on agarose gels (2 wt% in PBS, pH 7.4), mimicking the eye tissue, are also conducted using fluorescence dye attached drug loaded lens.

The *in vitro* biocompatibility therapeutic efficacy of the drug loaded lens is assessed against VEGF expressing *in vitro* AMD model using retinal pigment epithelium cells (ARPE-19 from ATCC) by detecting the changes in VEGF expression using ELISA and PCR tests. We also tested the efficient penetration of the drugs to the posterior segments using fluorescence dye attached drug loaded lens on *ex vivo* pig eyes using fluorescence imaging of the pig eye tissue sections.

Our results demonstrated the successful fabrication of the wireless iontophoresis lens with high resolution and small feature size using the patented method.[8–10] The initial prototypes for iontophoresis application demonstrated controlled release of model drugs under different electrical stimulation conditions without changing the temperature and pH of the environment. The same platform showed efficient drug penetration into the agarose gel, mimicking the eye tissue. We also demonstrated that retinal pigment epithelium cells (RPE) can attach, grow, and proliferate on these platforms without any biocompatibility issues. Our *ex vivo* tests on pig eyes clearly showed that the drugs can be pushed to the retina without causing any visual damage to the cornea and the lens.

In conclusion, this wireless iontophoresis lens technology enables non-invasive WES directed on-demand ocular drug delivery, self-administered by the patient, preventing frequent office visits, and reducing treatment burden for AMD. The local drug reservoir technology also provides therapeutic stability enabling prolonged shelf-life and precise dose control, minimizing dose related side effects. The successful development of this technology will change the paradigm in traditional iontophoresis concept and clinical practice by introducing a wireless and self-administered contact lens platform not only for AMD but also for the potential treatment of other ocular conditions addressing the economic and societal needs of patients.

References:

1. Wong, W. L. *et al.* Global prevalence of age-related macular degeneration and disease burden projection for 2020 and 2040: a systematic review and meta-analysis. *Lancet Glob Health* **2**, e106-116 (2014).
2. Friedman, D. S. *et al.* Prevalence of age-related macular degeneration in the United States. *Arch Ophthalmol* **122**, 564–572 (2004).
3. Arepalli, S. andamp; Kaiser, P. K. Pipeline therapies for neovascular age related macular degeneration. *International Journal of Retina and Vitreous* **7**, 55 (2021).
4. Al-Khersan, H., Hussain, R. M., Ciulla, T. A. andamp; Dugel, P. U. Innovative therapies for neovascular age-related macular degeneration. *Expert Opinion on Pharmacotherapy* **20**, 1879–1891 (2019).
5. Gaudana, R., Ananthula, H. K., Parenky, A. andamp; Mitra, A. K. Ocular Drug Delivery. *AAPS J* **12**, 348–360 (2010).
6. Gote, V., Sikder, S., Sicotte, J. andamp; Pal, D. Ocular Drug Delivery: Present Innovations and Future Challenges. *J Pharmacol Exp Ther* **370**, 602–624 (2019).
7. Patel, A., Cholkar, K., Agrahari, V. andamp; Mitra, A. K. Ocular drug delivery systems: An overview. *World J Pharmacol* **2**, 47–64 (2013).
8. Uz, M. andamp; Mallapragada, S. Methods, apparatus, and systems for fabricating solution-based conductive 2D and 3D electronic circuits. (2021).
9. Uz, M. *et al.* Fabrication of Two-Dimensional and Three-Dimensional High-Resolution Binder-Free Graphene Circuits Using a Microfluidic Approach for Sensor Applications. *ACS Appl. Mater. Interfaces* **12**, 13529–13539 (2020).
10. Uz, M. *et al.* Fabrication of High-resolution Graphene-based Flexible Electronics via Polymer Casting. *Sci Rep* **9**, 10595 (2019).

315d Formulation and Scale-up of Fast-Dissolving Lumefantrine Nanoparticles for Oral Malaria Therapy

Madeleine Armstrong, Leon Z. Wang, Kurt Ristroph, Chang Tian, Jiankai Yang, Lirong Ma, Shuhong Jiang, Santipharp Panmai, Donglu Zhang, Karthik Nagapudi, Robert K. Prud'homme

Lumefantrine (LMN) is one of the first-line drugs in the treatment of malaria due to its long circulation half-life, which results in enhanced effectiveness against drug-resistant strains of malaria. However, LMN's therapeutic efficacy is diminished due to its low bioavailability in the fasted state when dosed as a crystalline solid. LMN is therefore dosed with food for achieving adequate adsorption making it sub-optimal for pediatric administrations. The goal of this work was to produce low-cost, highly orally bioavailable, stable LMN powders applicable for global health applications which can be dosed either in the fed or fasted state.

We report the development of a LMN nanoparticle formulation and the translation of that formulation from laboratory to industrial scale. We applied Flash NanoPrecipitation (FNP) to develop nanoparticles with 90% LMN loading with sizes of 200-250 nm. The final powders are readily redispersible and are stable over accelerated aging conditions (50°C, 75% RH, open vial) for 4 weeks giving equivalent and fast drug release kinetics in both simulated fed and fasted state intestinal fluids making them suitable for pediatric administration. The nanoparticle-based formulations increase the bioavailability of lumefantrine 4X *in vivo* when compared to the control crystalline formulation. The three continuous processes (FNP, tangential flow ultrafiltration, and spray drying) were optimized at small and pilot scale to produce the nanoparticles. We describe the translation of the laboratory-scale process at Princeton University to the clinical manufacturing scale at WuXi AppTech (Shanghai, China).

315e Oral Delivery of High Isoelectric Point Therapeutic Proteins Using pH-Responsive Nanoscale Complexation Hydrogels

Fabiola Chapa Villarreal, Heidi Oldenkamp, Avha Mohanty, Alex Chiu, Nicholas Peppas

Protein therapeutics represent a rapidly growing pharmacological sector, with over 200 FDA-approved products[1]. However, due to poor stability and large molecular weight, protein administration is almost exclusively limited to injection. These frequent injections can be very painful and are often associated with fear and major side effects at the injection site. Achieving oral delivery is ideal and would almost certainly improve patient compliance but is fraught with challenges. Efforts to develop delivery platforms have been limited by the conditions in the gastrointestinal tract, which greatly reduce the activity of the proteins, and thus, their bioavailability is negligible if delivered in an unprotected state[2]. Environmentally responsive polymer carriers, such as those that swell and collapse in response to changes in pH, have been explored for the development of an oral delivery platform for protein drugs. These pH-responsive carriers rely on charge interactions for the complexation behavior that facilitates protein loading and release[3]. This characteristic presents a difficulty with proteins that exhibit a high isoelectric point, as the protein is likely to remain bound to the anionic hydrogel carrier at the pH of the small intestine instead of being repelled. This interaction must be overcome to allow protein release and thus, therapeutic efficacy. The aims of our work were to explore strategies for increasing release of high isoelectric point therapeutic proteins from anionic hydrogel nanoparticles under neutral conditions.

Copolymeric nanoparticle systems containing acrylamide, itaconic acid, and crosslinked with N,N'-methylenebisacrylamide were synthesized via inverse emulsion polymerization, a protocol adapted from Zhong et al[4]. Analysis with FTIR and potentiometric titration confirmed the expected chemical composition. Nanogel swelling was characterized with dynamic light scattering and zeta potential was measured with electrophoretic light scattering. With dynamic light scattering, it was confirmed that the synthesized nanogels exhibit the expected swelling behavior with increasing pH values, swelling from a hydrodynamic diameter of approximately 85 nm at pH 3 to 105 nm at pH 9 in 1X PBS. Electrophoretic light scattering demonstrated that both formulations exhibit a negative zeta potential, with the value becoming more negative with increasing pH values. Cytochrome C was used as a model high isolectric point protein and the nanoparticles exhibited loading efficiencies of this protein up to 95%. Minimal protein release at pH 4 was observed while more than 50% of the protein was released within 5.5 hours at pH 7.4. *In vitro* studies using Caco-2 cells were performed to analyze the cytocompatibility of the nanoparticles at concentrations up to 5mg/mL. Further testing on protein transport studies across the epithelium barrier and implementation of strategies to enhance protein uptake after delivery to the intestinal lumen must be explored. The synthesized nanogels show promise for oral delivery applications of high isoelectric point proteins.

This work was supported by NIH grant number R01 EB022025 and the Cockrell Family Regents Chair in Engineering (UT Austin). F.A.C.V. was supported by the Conacyt/ConTex Fellowship, and the Cockrell School of Engineering Fellowship (UT Austin). H.F.O. was supported by the National Science Foundation Graduate Research Fellowship and the Archie W. Straiton Endowed Graduate Fellowship in Engineering #2 (UT Austin).

References

1. Fosgerau, K. andamp; Hoffmann, T. *Drug Discovery Today* 20, 122–128 (2015).

2. Gupta, S. *et al. Drug Delivery* 20, 237–246 (2013).
3. Brannon-Peppas, L. andamp; Peppas, N.A *Journal of Controlled Release* 16, 319–329 (1991).
4. Zhong, J.X., et al.*J Biomed Mater Res Part A* 106A, 1677– 1686 (2018).

315f Evaluation of Vegetarian Softgels Using SeaGel® Technology for Pharmaceutical Applications

Jin Zhao, Becca Putans, Benjamin Roscoe, Michael Evangelista, Thomas Watson, Teegan Penkala, Ian Gillespie, Dean Lee, Koudi Zhu, Michael Baumann

Softgels continuously serve as a preferred oral drug and dietary supplement dosage forms. As softgels can easily encapsulate a variety of liquid and semi-solid formulations, they have been used to deliver poorly soluble active ingredients and provide enhanced bioavailability. Although gelatin softgels currently dominate the pharmaceutical industry, there are numerous shortcomings with the material. Gelatin has the potential to crosslink with active ingredients leading to reduced solubility of the capsule shell. Gelatin is sensitive to high temperature and humidity conditions, presenting challenges with storage and shipment of gelatin softgels. There is also a growing consumer preference for non-animal products to suit different dietary needs (vegan, halal, kosher, etc). In this study, a poorly soluble active pharmaceutical ingredient (API), Ibuprofen, was encapsulated using either gelatin as the shell forming material or a plant-based material based on IFF's SeaGel® technology (noted as SeaGel® softgels).

This study used the USP I method to study the dissolution behavior of Ibuprofen from gelatin and SeaGel® capsules. Additionally, evaluation of stability of both gelatin softgels and SeaGel® softgels was performed at accelerated storage conditions (40 ºC, 75% RH) as well as long term storage conditions (30 ºC, 65% RH) for more than 2 months.

For freshly produced SeaGel and gelatin softgels, it was observed that the plant based SeaGel® softgels have slightly slower delivery at dissolution time less than 80 minutes. However, the f_2 similarity factor was greater than 52 at all time points which indicates an average difference of less than 10% between SeaGel softgels and gelatin softgels. Additionally, when stored at accelerated or long term storage condition for more than 1 month, SeaGel® softgels showed equivalent dissolution profile as the profile of freshly produced SeaGel softgels, while gelatin softgels showed significantly slower dissolution behavior than its profile of freshly produced gelatin softgels. Furthermore, the mechanical properties of gelatin softgels indicated possible cross-linking and migration of fill components and varies more significantly than those of SeaGel® softgels as a function of aging time and aging conditions. This is consistent with similar findings of other scientists. [1]

In summary, dissolution of an active in the dosage form is critical for absorption and bioavailability, and for product quality and consistency assessments during drug product life cycle. SeaGel® softgels showed approximately equivalent API dissolution behavior as that of gelatin capsules at the time of capsules were produced but better stability than that of gelatin softgels.

References

1. Chen, T. Humidity Influence to the Thermal and

Mechanical Properties of Gelatin Capsules. https://www.tainstruments.com/pdf/literature/TA409.pdf.

315g Design and Function of Peptide-Based Materials

Handan Acar, Seren Hamsici, Gokhan Gunay

Fibrillar self-assembling peptide aggregates are abundant in nature and are used in various synthetic technologies in biomedical research and material science. The high level organization of the peptide aggregates emerges from small changes in the amino acid sequence that provides non-covalent interactions, such as electrostatic forces, hydrogen bonds, hydrophobic effects, and aromatic stacking. Discovery of peptide sequences that assemble into nanostructures with desired properties and functions in the proper context is a significant challenge for the field. Studies based on editing the peptide sequences of natural designs typically have an initial focus on a particular type of interaction for the property of interest, which may restrict the achievable material properties. As a result, there is a need for a strategy that can optimize the exploratory scope with the benefits of editing known sequences to map and search new peptides for new functionalities.

We addressed this need with a new molecular discovery tool. Rather than attempt a brute-force approach, we instead identify a new strategy for discovering and studying intermolecular interactions; "co-assembly of oppositely charged peptide' (CoOP), a framework that `encourages' peptide assembly by mixing two oppositely charged hexapeptides.1This results in a reasonable expansive design space of fibrillar assembly's within that space leading to substantially increased efficiency in the exploration. The one-dimensional assemblies that form through this unique CoOP approach are well suited for study via a set of reliable quantitative tools. We showed that the peptides that are designed with CoOP strategy have wide application areas. The peptides that self-assemble into the fibrillar structures created hydrogels with different mechanical properties, based on their aggregation kinetics and affinity. Additionally, we showed the use of these peptides as vaccine adjuvants.2 We showed that the CoOP approach represents a unique, simple, and elegant framework that can be used to identify the structure-property relations of self-assembling peptide-based materials.

[1] Hamsici, S., White, A. D. andamp; Acar, H. Science Advances (2022)

[2] Gunay, G.Hamsici, S., andamp; Acar, H. BioRxiv (2022)

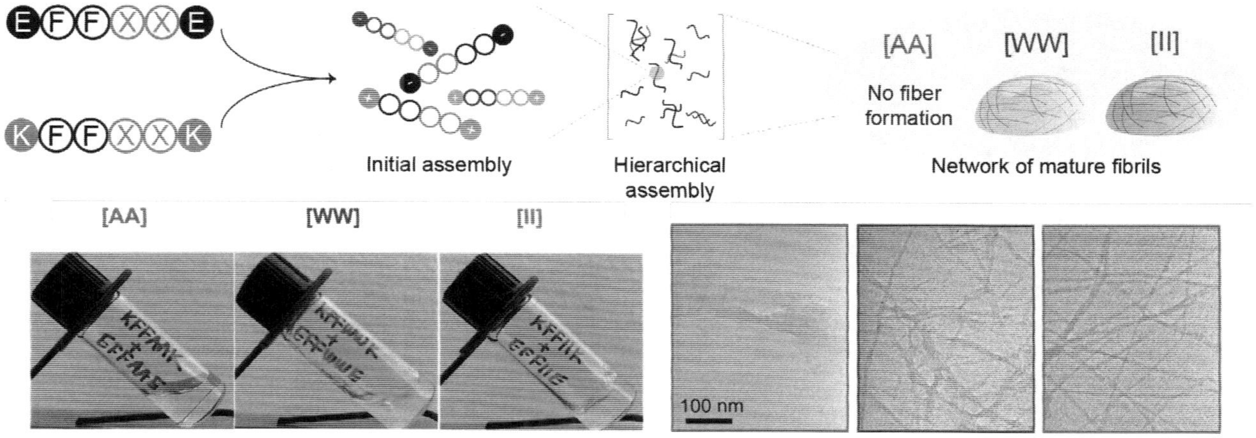

555a Polysalicylic Acid Polymer Microparticle Decoys Therapeutically Treat Acute Respiratory Distress Syndrome

Emma Brannon, William J. Kelley, Michael W. Newstead, Alison Banka, Kathryn E. Uhrich, Colleen E. O'Connor, Theodore J. Standiford, Omolola Eniola-Adefeso

Acute lung injury (ALI) and acute respiratory distress syndrome (ARDS) remain problematic due to high mortality rates and lack of effective treatments. Neutrophilic injury contributes to mortality in ALI/ARDS. Here, technology for rapid ARDS intervention is developed and evaluated, where intravenous salicylic acid-based polymer microparticles, i.e., Poly-Aspirin (Poly-A), interfere with neutrophils in blood, reducing lung neutrophil infiltration and injury in vivo in mouse models of ALI/ARDS. Importantly, Poly-A particles reduce multiple inflammatory cytokines in the airway and bacterial load in the bloodstream in a live bacteria lung infection model of ARDS, drastically improving survival. It is observed that phagocytosis of the Poly-A microparticles, with salicylic acid in the polymer backbone, alters the neutrophil surface expression of adhesion molecules, potentially contributing to their added therapeutic benefits. Given the proven safety profile of the microparticle degradation products—salicylic acid and adipic acid—it is anticipated that the Poly-A particles represent a therapeutic strategy in ARDS with a rare opportunity for rapid clinical translation.

555c Fabrication and Characterization of Composite Particles for Controlled Release Drug Delivery Applications

Michael L. Felder, Daniel Kupor, Hanieh Safari, Omolola Eniola-Adefeso

Many orally and intravenously delivered drugs suffer from issues such as high clearance rate, low bioavailability, and systemic side effects. As such, there has been a push to develop targeted and controlled release drug delivery systems. These systems typically center on a polymeric or liposomal carrier, into which the drug of choice is loaded. However, these systems are primarily composed of the polymer or lipid material and are difficult to implement in the clinic due to their poor stability and negative side effects of the carrier. Recently, a means to fabricate metal-templated composite microparticles out of two key bile salts, cholate and deoxycholate, was discovered, whereby the particle itself was composed of greater than 95% (w/w) the active drug material. Our work aims to understand the mechanism underlying this process and to expand on it, where we utilize a modified double emulsion solvent evaporation process. We hypothesize the charge distribution on the ester group at carbon 17 of the steroid backbone interacts with the metal or acid within the inner water phase to form a particle nucleus, thereby initiating a growth mechanism to create particles on the micron scale. Thus, we anticipate metals or acids that allow charge transfer, when incorporated into the inner water phase, can facilitate this templating process. Ethyl acetate is used as the oil phase and the drug of choice is used in the outer water phase. Drugs in the outer water phase consist of water-soluble bile salts. The metal or acid, via the oil/water interface, then initiates the nucleation/growth mechanism. Particles have been successfully fabricated using a variety of bile salts. These particles, seen in **Figure 1**, have exhibited controlled release properties in a DI water degradation study. Furthermore, particles encompass the nucleating material, such as gold or salicylic acid, within the particle matrix. Bioactivity assays completed via ELISAs have shown that particles release active drug molecules. This work is significant as it provides alternative strategies for particle drug delivery systems that do not involve a polymeric or liposomal carrier. Future work will center on optimizing fabrication parameters to better control particle size, shape, and drug release.

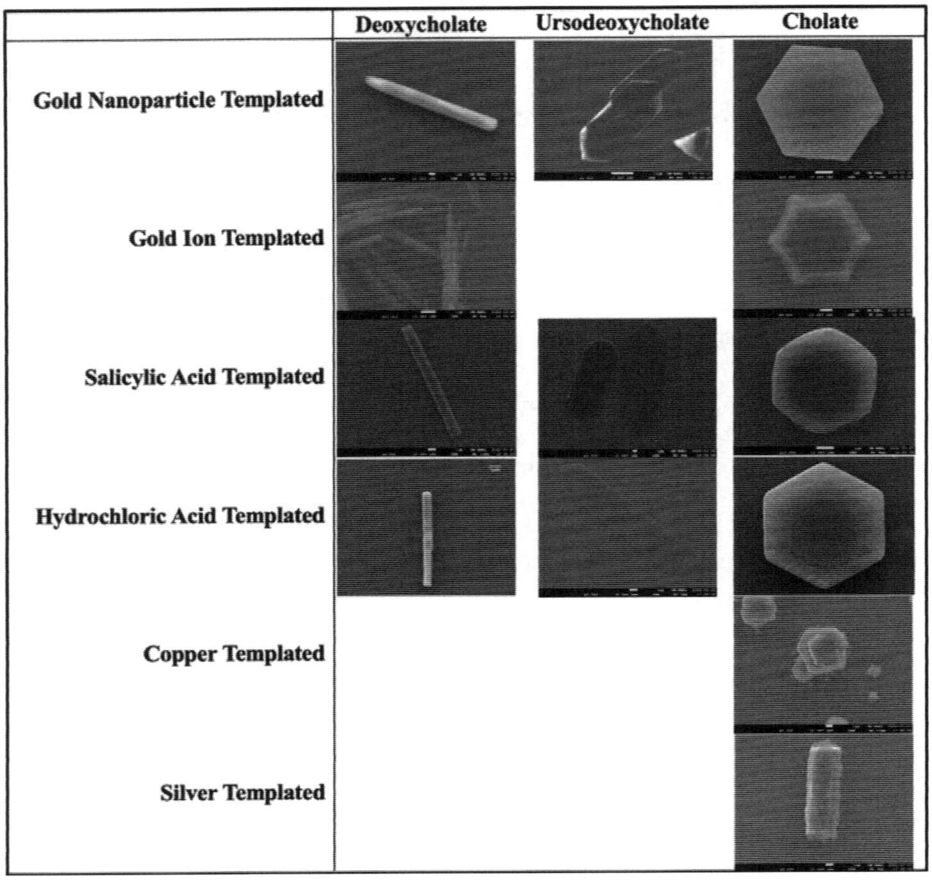

Figure 1: Scanning electron microscopy (SEM) images of particles fabricated via the modified double emulsion solvent evaporation process are shown.

555d Silk Fibroin Nano- and Micro- Particles As a Potential System for a Novel Hemoglobin-Based Oxygen Carrier (HBOC)

Marisa Pacheco, Jostin Armada, Marina P Fernandez-Campa, Jack W. McNamara, Michael O'Brien, Nickolas A. Davies, Bruce D. Spiess, Whitney Stoppel

Traumatic injury, diabetes, anemia, stroke, and tumor growth can lead to local or systemic lack of oxygen[1,2]. Blood transfusions are the gold-standard for delivery of O_2 when enriched respiratory delivery is not sufficient; however, blood donations and human-derived blood products are in limited supply, require extensive pathogen screening, and require cold-chain storage[3]. Because of these limitations, many options for therapeutic O_2 carriers (OCs) are under investigation. These include hemoglobin-based oxygen carriers (HBOCs) that make use of recombinantly produced hemoglobin or hemoglobin from other species, as well as synthetic perfluorocarbons[4-6]. Many of these current systemically delivered OCs struggle with issues such as long-term organ toxicity, dose inefficiency, and rapid clearance rates (<12 hours)[7]. Thus, despite years of efforts, alternative O_2 carrying therapeutics are needed that prolong O_2 delivery without leading to systemic toxicity.

For the development of an all-natural HBOC, we investigate a novel source of hemoglobin. The hemoglobin from teleost fish (salmon, tuna, etc.) is unique due to the Root effect. The Root effect is a phenomenon in which an acidic pH decreases the affinity of hemoglobin for O_2, leading to efficient delivery of hemoglobin-bound O_2 to the hypoxic microenvironments[8]. Teleost fish leverage the Root effect to regulate their buoyancy, improve their vision, and maintain proper oxygenation of their tissues[9]. We aim to leverage this evolved advantage of salmon hemoglobin to locally and efficiently deliver O_2 to hypoxic tissues within humans. To improve shelf-life and dose efficiency, as well as avoid toxicity, the hemoglobin proteins are encapsulated in a matrix of an all-natural protein carrier. Silk fibroin is a protein isolated from *Bombyx mori* cocoons that is well established as a cytocompatible, inert natural polymer[10,11]. Silk fibroin can be processed into multiple biomaterial formats including nano- and micro- particles, and subsequently degraded into simple amino acids[12].

In this work, we analyze the separation and purification of salmon hemoglobin from whole blood, establish a design space to control silk particle size, analyze hemoglobin encapsulation, assess the stability of the silk-hemoglobin particles as a function of their size and storage condition, and measure oxygen loading and release *in vitro* using a custom plug-flow reactor system equipped with dissolved oxygen sensors.

Methods: Hemoglobin is isolated and purified from whole salmon blood using immobilized metal affinity chromatography (IMAC). Purity is assessed using SDS-PAGE, while UV-vis is used to assess levels of methemoglobin. Silk particles are formed via phase separation with polyvinyl alcohol (PVA) as described previously[13,14]. The phase separation occurs as a function of solvent evaporation. To test for shifts in particle size, we varied the sonication amplitude (0, 8, 12, 16, 25, 40%), silk concentration (2, 20, 50, 70 mg/mL), silk degumming time (30, 60, 90 min, inversely proportional to silk molecular weight) and hemoglobin concentration. Samples were assessed for particle size, stability, and morphology using dynamic light scattering (DLS) and scanning electron microscopy (SEM). To visualize hemoglobin encapsulation, hemoglobin was combined into the soluble silk solution prior to particle formation. Immunofluorescent imaging was then performed to confirm colocalization of the hemoglobin into the silk particles. To test the stability of the silk particles containing hemoglobin, we incubate both liquid and lyophilized conditions at temperatures ranging from -80°C to 40°C. Weekly, samples are assessed for changes in hemoglobin conformation using UV-vis and changes in particle size, stability and morphology using DLS and SEM.

Results: We found that hemoglobin was isolated with less than 25% of total isolated protein being methemoglobin. When attempting to control particle size, we found that we can generate particles ranging from 200 nm–10 um. We also observed the following trends: (i) increases in sonication amplitude and/or time lead to smaller particle sizes (ii) silk concentration is negatively correlated to average particle size, and (iii) silk degumming time is positively correlated to average size. Moderate levels of polydispersity (range of 0.2-0.4) were observed across all samples, prompting investigation into strategies to narrow polydispersity. While altering variables involved in particle formation like the sonication probe diameter and probe to sample volume ratios can impact polydispersity, we are also investigating post-processing strategies, like filtration. Hemoglobin was observed to be incorporated in particles by fluorescent microscopy and no significant changes in particle size were observed as a function of the hemoglobin inclusion.

These results show the control of particle size and level of hemoglobin incorporation, indicating the potential of this system to serve as a novel HBOC. These results are also being used in parallel efforts that seek to simulate O_2 release from the hemoglobin rich regions of the particles using a two tier COMSOL® reaction-diffusion model. Data from this model can drive experimental design for *in vitro* characterization assays and release experiments. Release experiments aim to evaluate O_2 delivery profiles in unique microenvironments using an *in vitro* perfusion flow system equipped with Presens® O_2 sensors. Preliminary results suggest that acidic microenvironments characteristic of injury or tumor sites prompt burst release of oxygen, loss of hemoglobin cooperativity, and particle degradation in both nano- and micro- particles, while in healthy physiological pH ranges the particles persist in solution and are capable of sustained release. Current work in the field still aims to understand how to optimize particle size for different targets and circulation times[15], so we continue to model and test both silk micro- and nano-particles in preliminary modeling and *in vitro* work. Future work aims to assess safety of the key candidate particle formulations *in vivo* via intravenous (IV) injections and subsequent clearance by the kidneys, liver, and spleen along with the systemic immune response to injection.

References

1 Catrina, S.-B. andamp; Zheng, X. Hypoxia and hypoxia-inducible factors in diabetes and its complications. *Diabetologia* **64**, 709-716, doi:10.1007/s00125-021-05380-z (2021).

2 White, N. J., Ward, K. R., Pati, S., Strandenes, G. andamp; Cap, A. P. Hemorrhagic blood failure: Oxygen debt, coagulopathy, and endothelial damage. *J Trauma Acute Care Surg* **82**, S41-S49, doi:10.1097/TA.0000000000001436 (2017).

3 Roberts, D. J., Field, S., Delaney, M. andamp; Bates, I. Problems and Approaches for Blood Transfusion in the Developing Countries. *Hematol Oncol Clin North Am* **30**, 477-495, doi:10.1016/j.hoc.2015.11.011 (2016).

4 Natarajan, C., Signore, A. V., Kumar, V. andamp; Storz, J. F. Synthesis of Recombinant Human Hemoglobin With NH(2) -Terminal Acetylation in Escherichia coli. *Curr Protoc Protein Sci* **101**, e112, doi:10.1002/cpps.112 (2020).

5 Jani, V. P. *et al.* Polyethylene Glycol Camouflaged Earthworm Hemoglobin. *PLoS One* **12**, e0170041, doi:10.1371/journal.pone.0170041 (2017).

6 Jägers, J., Wrobeln, A. andamp; Ferenz, K. B. Perfluorocarbon-based oxygen carriers: from physics to physiology. *Pflugers Arch* **473**, 139-150, doi:10.1007/s00424-020-02482-2 (2021).

7 Buehler, P. W., D'Agnillo, F. andamp; Schaer, D. J. Hemoglobin-based oxygen carriers: From mechanisms of toxicity and clearance to rational drug design. *Trends Mol Med* **16**, 447-457, doi:10.1016/j.molmed.2010.07.006 (2010).

8 Rummer, J. L. andamp; Brauner, C. J. Root Effect Haemoglobins in Fish May Greatly Enhance General Oxygen Delivery Relative to Other Vertebrates. *PLoS One* **10**, e0139477-e0139477, doi:10.1371/journal.pone.0139477 (2015).

9 Verde, C., Vergara, A., Giordano, D., Mazzarella, L. andamp; Di Prisco, G. The Root effect – a structural and evolutionary perspective. *Antarctic Science* **19**, 271-278, doi:10.1017/S095410200700034X (2007).

10 Nguyen, T. P. *et al.* Silk Fibroin-Based Biomaterials for Biomedical Applications: A Review. *Polymers* **11**, 1933, doi:10.3390/polym11121933 (2019).

11 Pham, D. T. andamp; Tiyaboonchai, W. Fibroin nanoparticles: a promising drug delivery system. *Drug Deliv* **27**, 431-448, doi:10.1080/10717544.2020.1736208 (2020).

12 Wongpinyochit, T., Johnston, B. F. andamp; Seib, F. P. Degradation Behavior of Silk Nanoparticles-Enzyme Responsiveness. *ACS Biomater Sci Eng* **4**, 942-951, doi:10.1021/acsbiomaterials.7b01021 (2018).

13 Rockwood, D. N. *et al.* Materials fabrication from Bombyx mori silk fibroin. *Nature Protocols* **6**, 1612-1631, doi:10.1038/nprot.2011.379 (2011).

14 Vargas Montoya, N., Peterson, R., Ornell, K. J., Albrecht, D. R. andamp; Coburn, J. M. Silk Particle Production Based on Silk/PVA Phase Separation Using a Microfabricated Co-flow Device. *Molecules* **25**, 890 (2020).

15 Di, J. *et al.* Size, shape, charge and "stealthy" surface: Carrier properties affect the drug circulation time in vivo. *Asian Journal of Pharmaceutical Sciences* **16**, 444-458, doi:https://doi.org/10.1016/j.ajps.2020.07.005 (2021).

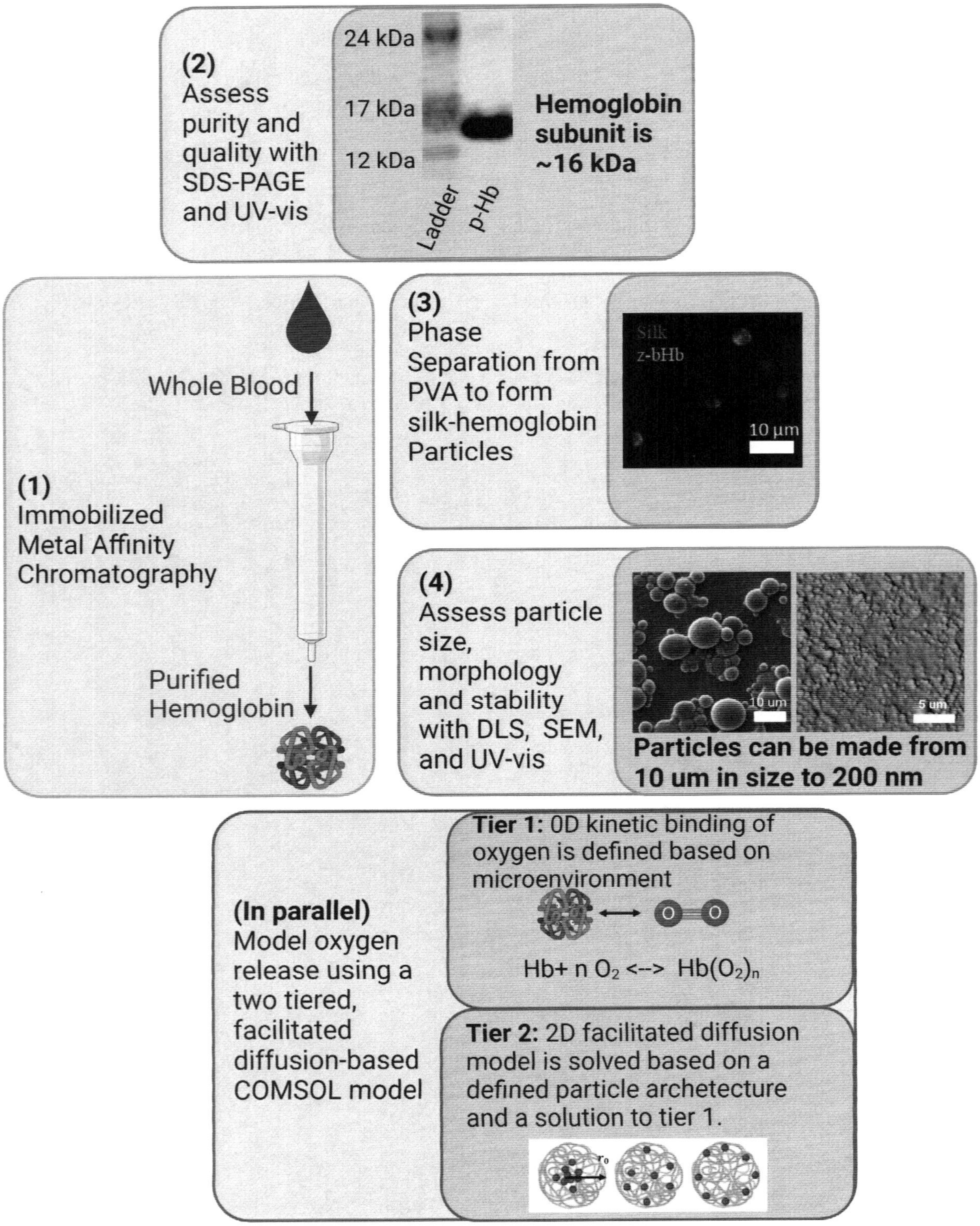

Figure 1: Graphical Abstract

555e Hemoglobin Encapsulated Metal Organic Framework Nanoparticles As an Oxygen Therapeutic with Ultrahigh Encapsulation Efficiency

Xiangming Gu, Andre Palmer

Motivation: Hemoglobin (Hb)-based oxygen carriers (HBOCs) constitute one major class of artificial red blood cell (RBC) substitute, which has gained considerable attention in treating hemorrhagic shock, ameliorating traumatic brain injury and suppressing tumor growth via oxygenation of hypoxic tissues. Unfortunately, prior generations of HBOCs yielded disappointing clinical outcomes due to significant safety concerns. Various strategies have been developed to reduce/prevent extravasation of cell-free Hb including Hb polymerization, surface conjugation of Hb, and liposome encapsulation of H. All of these strategies increase the molecular radius of Hb, so that it is unable to extravasate through the blood vessel wall. It was observed that chemical modification of Hb via polymerization and surface conjugation approaches drastically decrease the flexibility of Hb, which affects cooperative oxygen (O_2) binding and release. Although liposome encapsulation had a negligible impact on cooperative O_2 binding/release, low Hb encapsulation efficiency remains a significant issue that impeded scale-up and potential commercialization. Those underlying flaws associated with liposome encapsulated Hb nanoparticles underscore the need to develop next generation Hb nanoparticles with enhanced uniformity, high Hb content, and high encapsulation efficiency without jeopardizing cooperative O_2 binding and release. Metal organic frameworks (MOFs) are crystalline materials synthesized by the coordination of metal ions and organic linkers. Zeolitic imidazolate frameworks (ZIFs) are a subclass of MOF, which exhibits outstanding solvothermal stability due to their unique structure compared to other MOFs. ZIF-8, the prototypical ZIF, preserves its crystallinity and zeolite-like porosity when exposed to boiling water, organic solvents, and several biological buffers. Very recently, ZIF-8 has emerged as a promising material in biological sensing, drug delivery, and HBOC fabrication. In this study, bovine Hb (bHb) was successfully encapsulated using ZIF-8 precursors into ZIF-8P-Hb nanoparticles as a potential O_2 therapeutics.

Methods: Bovine Hb (bHb) was purified via tangential flow filtration (TFF). Briefly, Hb was purified via a two-stage TFF system with hollow fiber cartridges with MWCOs of 500 and 50 kDa (Repligen Corporation, Rancho Dominguez, CA). The purified bHb was concentrated to > 200 mg/mL and stored at −80 °C for future use. To synthesize bHb encapsulated ZIF-8 precursors-based (ZIF-8P-Hb) nanoparticle. Initially, 50 mL of deionized (DI) water was used to fully dissolve 100 mg of $Zn(NO_3)_2 \cdot 6H_2O$, chased by a 1 mL bolus addition of bHb solution (250 mg/mL) with continuous stirring (500 rpm) for 5 min. 827.9 mg of Hmim powder was then added to the above mixture. The reaction proceeded for 1 h at 25 °C followed by overnight stabilization at 4 °C. Then, the formed ZIF-8P-Hb nanoparticles were purified by washing with TFF for 6~10 diacycles using a hollow fiber cartridge with 500 kDa pore size first with DI water and buffered exchanged into phosphate buffered saline (PBS, 0.1 M) for further use. Bare ZIF-8 nanoparticles were synthesized following a similar protocol without the addition of bHb and were suspended in DI water.

Results: The size of ZIF-8P-Hb nanoparticles was primarily controlled by Hmim:Zn molar ratio, flow rate during TFF processing, concentration of EDTA, and concentration of zinc nitrate. It was found that adding Hmim directly into the reaction vessel regulated particle size, which also affected the crystalline structure of the particle. Furthermore, we demonstrated that the high molar ratio of Hmim:Zn could be used to better control nucleation of ZIF-8P-Hb nanoparticles. The monodisperse size distribution was a result of the rapid nucleation rate facilitated by the relatively high molar ratio of Hmim:zinc, and TFF operated at relatively low flow rate did not exert a strong impact on the size distribution.

Takeaways: In this study, we established a scalable purification platform to manufacture ZIF-8 and ZIF-8P-Hb nanoparticles via TFF. The optimized synthesis protocol yielded relatively low batch-to-batch variance with respect to most biophysical properties including hydrodynamic diameter, zeta potential, oxygen equilibria, and oxygen offloading rate constant. The higher bHb loading and bHb encapsulation efficiency of ZIF-8P-Hb nanoparticles compared to prior attempts in the literature is critical in order to potentially use these materials to treat hemorrhagic shock.

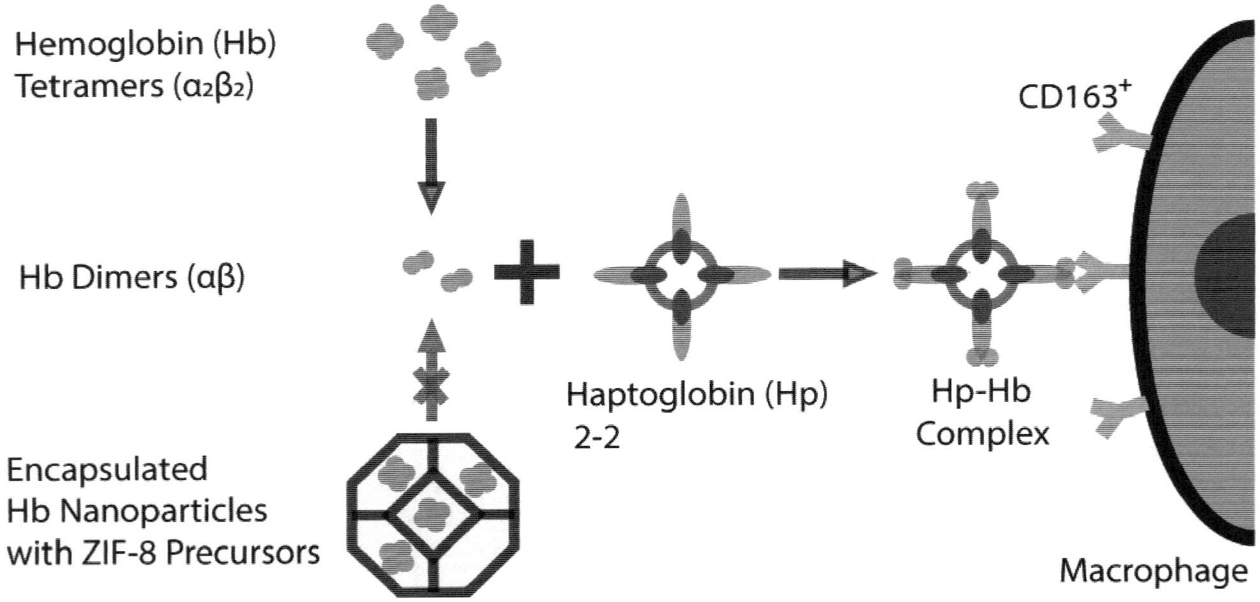

555f Peptide Amphiphile Micelles As a Biomaterials Platform for Universal Influenza Vaccination

Megan Schulte, Ava Steele, Mariam Morafa, Paul Anderson, Rachel Olson, Bret Ulery

Despite 48% of adults receiving the flu vaccine this past year, there were still at least 4 million cases and 40,000 hospitalizations due to influenza during the 2021-2022 flu season. The yearly influenza vaccine protects against three or four specific influenza viruses, using forecasting to predict which strains will have the most significant impact in the upcoming year. Unfortunately, the predictions are not always accurate, sometimes reducing the efficacy of the flu vaccine for a given season. A universal vaccine could address this problem with broader protection against many, or even all, strains of influenza. A possible way to generate a universal vaccine is to focus on viral genomic regions that have considerable homology across various influenza strains. From those, subunit vaccines, contrary to the current live attenuated influenza vaccines, can be generated that consist of just the antigenic peptide epitope(s) associated with the virus. While these do not possess the complications associated with the replication and infection potential of the virus itself, peptides are known to be poorly immunostimulatory on their own, requiring a drug delivery system to enhance their efficacy.

A variety of biomaterials systems have been studied for subunit vaccine delivery, though a specific biomolecular material in peptide amphiphile micelles (PAMs) possess a number of exciting features for use in this application. PAMs are made of amphiphilic lipopeptides that self-assemble into nanoparticles in aqueous environments. Our previous research has shown that PAMs possess the capacity to self-adjuvant incorporated peptides to improve their overall vaccine potential. Building on this foundational work, we have focused on using our platform to deliver the $M2_{1-24}$ antigen, a potential universal influenza peptide vaccine candidate consisting of the first 24 residues of the M2 protein which are highly conserved and known to contain a B cell epitope. To further enhance the immune response to the incorporated antigen, adjuvants can also be packaged with the micelles, either associated with the micellar corona or core depending on adjuvant properties.

To fabricate various PAMs for this research, peptide amphiphiles (PAs) containing the $M2_{1-24}$ antigen were synthesized using solid-phase peptide synthesis. These PAs were hydrated with and without the TLR-2 agonist Pam_2CSK_4 adjuvant for which their micellization was characterized by standard techniques including the critical micelle concentration assay and transmission electron microscopy. Mice were immunized with formulations consisting of peptide or PAMs with or without adjuvant in PBS at days 0 and 21. Blood was collected 14 days after each vaccination and enzyme-linked immunosorbent assays were conducted on the samples to quantify antigen-specific antibody production induced by each vaccine formulation. PAMs supplemented with adjuvant showed similar levels of M2-specific IgG production as the positive control (*i.e.*, M2 peptide plus Pam_2CSK_4). Even PAMs without adjuvant showed high levels of IgG production; however, antibodies were specific to an antigen on the peptide amphiphile that was not restricted to the M2 antigen, which could reduce protection against a live viral challenge. The inherent immunogenicity of the PAMs, even without adjuvant, is a positive, though ongoing work is underway to better understand how to restrict antigenic drift and shift within peptide amphiphiles and help to improve their overall performance as a universal influenza vaccine.

555g Invited talk: Lipid-based systems: From therapeutics to in vitro models

Christina Bailey-Hytholt

Lipid constructs are ubiquitous and advancing healthcare technologies through advancements in therapeutics and in vitro model systems. Lipid nanoparticles (LNPs) and liposomes serve as delivery vehicles for nucleic acids and small molecules by improving transfection efficiency, targeting, and stability of their loaded cargo. Additionally, lipid structures, including lipid vesicles and lipid bilayers, are useful tools to understand molecular interactions at biological interfaces. This talk will discuss strategies for using lipids for both therapeutic and in vitro model applications. Methods to manufacture LNPs and liposomes, characterization, stability, and in vitro studies will be discussed for nucleic acid (e.g. messenger RNA (mRNA)) loaded LNPs and small molecule (e.g. antifungal) loaded liposomes. This talk will also discuss the formation and use of multi-lipid bilayers representative of placental trophoblast lipid bilayers for pharmaceutical and environmental toxicant interaction studies.

665a Differential Gene Expression Analysis of Primary Human Hepatocytes and iPSC-Hepatocyte like Cells By RNA Seq

Neeti Gandhi, Kyle Akers, Lauren Wills, T. M. Murali, Padmavathy Rajagopalan

Introduction

Stem cell research has expanded in the last two decades after the discovery of induced pluripotent stem cells (iPSCs). Advantages to iPSCs include retrieval by non-invasive procedures, maintenance of host genotype, and differentiation into various cell types such as iPSC-hepatocyte-like cells (iHLCs). However, disadvantages to using iPSC-derived cells include teratoma formation, tumorigenic properties, and genetic mutations. Specifically, iHLCs have been reported to perform similar functions as primary human hepatocytes (PHHs) such as glycogen storage, albumin secretion, urea production, hepatocyte morphology, and cytochrome P450 (CYP450) expression. However, these liver functions are lower in iHLCs in comparison to PHHs. Although, PHHs are widely used in in vitro liver models, they are difficult to obtain, require invasive retrieval, and have significant donor variability. For these reasons, iHLCs exhibit significant potential in being used instead of PHHs. Due to the reported differences between iHLCs and PHHs, our goal is to understand the transcriptome-wide gene expression differences between these cell types using an RNA Seq analysis. Our focus was to assess the specific pathways and key genes that are upregulated in PHHs compared to iHLCs, specifically looking at pathways related to hepatic functions and drug metabolism.

Methods

RNA extraction occurred immediately upon thawing iHLCs and PHHs and then 48h after a type 1 collagen sandwich (CS) culture was assembled for both cell types. iHLCs were cultured as monolayers for 7 days to enable maturation. CS cultures were assembled 24h after PHH seeding and 7 days after iHLC seeding. RNA extraction was carried out using the RNeasy Plus Micro Kit (Qiagen). Sequencing was conducted at the Biocomplexity Institute of Virginia Tech. RNA Seq samples from the four groups were compared by using DE-Seq2. The sequenced library was aligned to the Gene Ontology (GO) database and Kyoto Encyclopedia of Genes and Genomes (KEGG) database. Genes with false discovery rate (FDR) adjusted p-values < 0.05 were selected as sets of differentially expressed genes and were selected as over- represented or under-represented in each comparison.

Results

We focus our results primarily on PHHs and iHLCs at 48h after CS assembly to investigate differences in pathways associated with hepatocyte function and drug metabolism aligned with the KEGG database. The analysis of RNASeq data showed an upregulation of 44 out of a total of 80 genes involved in bile secretion. Specific genes in this pathway exhibited up to a 455-fold change increase in PHHs compared to iHLCs. Upon immunostaining, bile canaliculi were shown to be present in PHHs. The TCA cycle in PHHs was upregulated compared to iHLCs with 19 out of a total of 29 genes expressing significantly higher fold-changes. Our experimental analysis shows a 2-2.5-fold increase in normalized urea production in PHHs compared to iHLCs. Additionally, the RNA Seq analysis resulted in three drug metabolism pathways that were upregulated in PHHs compared to iHLCs. Metabolism of xenobiotics by CYP450 enzymes displayed an upregulation of 48 out of a total of 66 genes. Significantly higher fold-changes of 43 out of a total of 61 genes were expressed in the metabolism of drugs by CYP450 enzymes pathway. Lastly, drug metabolism by other enzymes showed upregulation of 46 out of a total of 76 genes. Of note, CYP2E1 in PHHs was significantly higher expressed by 21810-fold. Experimental results show PHHs in CS cultures exhibit a 3.7-fold increase in CYP2E1 activity compared to iHLCs.

Conclusions and Future Work

The RNA Seq data and aligned experimental analysis shows down-regulated enzyme expression in iHLCs compared to PHHs. We have identified specific pathways involved in hepatocyte function that are upregulated in PHHs in CS culture models, compared to iHLCs in similar culture models. Future work will include using RNA Seq to understand iHLC maturation cultured in multi-cellular organoids.

665b Transcription of Proximal RNAs Can Regulate Gene Expression By Modulating Transcriptional Condensate Dynamics

Pradeep Natarajan, Krishna Shrinivas, Halima Schede, Jonathan E. Henninger, Ozgur Oksuz, Arup K. Chakraborty, Mehran Kardar, Richard A. Young, Phillip A. Sharp

Bio-molecular factories called transcriptional condensates facilitate the transcription of mRNAs in eukaryotic cells. These condensates are dense assemblies of proteins including transcription factors and transcriptional coactivators. They form through a phase separation mechanism at genomic regions having a high density of protein-binding DNA called super-enhancers (SE). Transcriptional condensates are important from a therapeutic standpoint as they control several genes associated with cell identity and cancer.

Many transcriptional proteins have positively charged disordered domains. Through these domains, they can interact with the negatively charged phosphate backbone of RNAs via screened electrostatic interactions and undergo complex coacervation. At low RNA concentrations, the RNA and protein condense together into a dense phase due to attractive interactions between the oppositely charged polymers. At high RNA concentrations, the dense phase dissolves due to a buildup of negatively charged RNA, which repel each other. The result is a re-entrant phase diagram at equilibrium where protein partitioning into the dense phase initially increases and then decreases upon titrating the RNA concentration.

However, living cells operate far away from equilibrium. Enhancer RNAs (eRNAs) are actively produced from SE regions and can interact with transcriptional proteins. Phase field equations based on a free energy can describe the transport of proteins due its interactions with other molecules. The rate of eRNA transcription depends on the local concentration of transcriptional proteins. We used phase field equations for proteins coupled with reaction diffusion equations for actively produced RNA species to model the spatiotemporal dynamics of concentrations of eRNAs and proteins. The transcription of mRNA happens over a slower time scale and the total amount of mRNA produced (gene expression) depends on the dynamics of protein concentration at the SE regions.

Transcriptional condensates reside in an environment where several other RNA species including long non-coding RNAs are also being actively transcribed. Prior studies show that active transcription of proximal RNAs can affect mRNA transcription from neighboring genes. However, the mechanism is not well understood. We hypothesize that localization and active transcription of proximal RNAs can regulate the dynamics of protein concentrations at the SE regions, which in turn affects mRNA transcription.

To investigate the consequences of this mechanism and make testable predictions, we modified the phase field + reaction diffusion equations described before to account for multiple RNA species. We investigated how the distance to the proximal RNA locus and its rate of transcription affects the dynamics of protein concentration at the SE region.

In the absence of active RNA transcription, we predict that proximal RNAs near SE regions accelerate recruitment of transcriptional proteins by electrostatic attraction when the local protein concentrations are small. Once enough protein accumulates to form a dense phase, some proximal RNA can also jump from its locus to the SE region due to the same electrostatic attraction and further speeds up protein recruitment. This jumping effect however does not happen if the RNA locus is too far away.

When there is active transcription, we predict that actively transcribing proximal RNAs can "accelerate" protein recruitment to the SE regions during early times in the dynamics due to favorable electrostatic interactions. If the transcription rates of the proximal RNAs are large, there is an accumulation of this RNA at later stages in the dynamics, which leads to unfavorable electrostatic repulsion between the RNA species and "shuts down" protein recruitment. The tussle between these two effects determines the dynamics of protein concentration at the SE region and the total amount of mRNA transcribed. Our model predicts that when transcription rate of the proximal RNAs is low, the "acceleration" effect dominates and they promote transcription from neighboring genes. On the other hand, quickly transcribing proximal RNAs repress transcription of neighboring genes.

In summary, we hypothesize that proximal RNAs can fine-tune mRNA transcription through its interactions with molecules in transcriptional condensates. This could be a novel mechanism by which proximal RNAs regulate expression of neighboring genes in cis, in a manner that depends only on their total charge and not their specific sequence or secondary structure. The consequences of this mechanism could explain several puzzles related to gene regulation by non-coding RNAs, such as (1) why non-coding RNAs promote gene expression in certain cases and repress gene expression in others (2) why the sequence of certain non-coding RNAs don't affect their impact on gene expression and (3) why certain non-coding RNA loci are present at conserved genomic positions with respect to target genes.

665c Dissecting the Role of a Tumor Suppressor Protein in Enabling Highly-Efficient Reprogramming

Adam Beitz, Brittany Lende, Kate Galloway

Reprogramming is the transformation of a cell from one lineage to another by the forced overexpression of lineage-specifying factors. Regardless of the initial or final cell type, the tumor suppressor protein p53 serves as a barrier to reprograming. p53 maintains genomic integrity in response to DNA damage, earning it the moniker of "guardian of the genome." Active p53 induces cell cycle arrest and apoptosis, which limit reprogramming yields. A wide variety of methods to inhibit p53 have been employed to enhance reprogramming efficiency, including RNA interference, gene knockout, chemical inhibition, and "dominant negative" mutants of p53. We find that not all methods produce equal results. We find that inclusion of mutant p53 in the reprograming cocktail produces the greatest yield increase, inducing up to a 100-fold increase in induced motor neurons from fibroblasts. Curiously, this p53 mutant enables rapidly proliferating cells to accumulate wild-type p53. Combining this mutant with RNA interference to knockdown expression of wild-type p53 reduces this accumulation and reduces reprograming efficiency, indicating an indispensable role for wild-type p53 in highly-efficient reprogramming that our mutant p53 cannot fulfill. How then does wild-type p53 both limit and promote reprogramming? By combining mathematical modeling with targeted mutations of the p53 protein, we are defining the specific protein-protein and protein-DNA interactions that allow p53 to promote high efficiency reprogramming. Together, our data suggest a dual role for p53 in cell-fate transitions that functions as a band-pass filter to restrict transitions. Understanding how p53 promotes cellular plasticity may help to identify processes that promote reprogramming and that can be enhanced without oncogenic mutants. Our observations also implicate a role for p53 in oncogenic transitions and may help to identify specific interactions that could be interrupted to limit cancerous transformations.

665d Short Chain Fatty Acid Influence on Normal and Estrogen Deficient Bone Remodeling

Carley V. Cook, Brenda J. Smith, Ashlee Ford Versypt

Introduction: Osteoporosis is a costly and debilitating disease in which bone resorption chronically exceeds bone formation. Estrogen deficiency, a major contributor to osteoporosis, influences the cytokines involved in bone health through T cell populations such as anti-inflammatory T regulatory cells (Tregs) and inflammatory T helper (Th)17 cells [1]. A promising area of research to treat osteoporosis involves the use of prebiotics or probiotics to increase the amount of short chain fatty acids (SCFAs) produced by gut metabolites. These SCFAs have been shown to positively influence bone health indirectly, by increasing Tregs. The Tregs then modulate osteoblast production (bone forming cells) via Wnt signaling [2]. However, in a previous model developed in our lab we determined that Wnt signaling alone could not reproduce the result of SCFAs on bone health pointing to other influences [3, 4]. We hypothesize that dietary induced increases of SCFAs could prevent or reverse bone loss caused by estrogen deficiency.

Methods: In this work we will explore other SCFA influences, model estrogen deficiency via T cell populations and cytokines, and apply the SCFA action to an estrogen deficient case. We will utilize newly acquired data to inform and expand our multicompartment ordinary differential equations (ODEs) model of SCFA influence on bone. Our initial model includes compartments for the gut, blood, and bone with ODEs corresponding to SCFA production, naïve CD4+ T cells, and Treg cells. TGF-β and Wnt signaling are also considered in the bone compartment along with equations corresponding to bone cell populations such as Osteoblasts, Osteoclasts, and Osteocytes.

Because our initial model is unable to replicate the total effect of increase SCFAs with Wnt signaling alone, we will consider the possibility of SCFAs having a direct action on bone health as well as other indirect actions caused by the increase in Tregs. We previously explored possible direct actions of SCFA on bone health and determined that they could influence osteoblasts proliferation, osteoclasts differentiation, and osteoclasts death. Here we will apply a new data set to the model to better understand the possible direct actions of SCFAs. Our initial model also only considers the indirect impact of Tregs via Wnt signaling. The new data will allow us to determine other cytokine changes in the bone compartment that could be related to the change in Tregs. These cytokine changes would add new terms to the bone cell population ODEs that are proportional to the Treg population.

Knowing that estrogen influence T cell populations, we will then alter the bone compartment of the model to include and ODE equation for Th17 cells. Here, we will focus on capturing the proportional relationship between Tregs and Th17 cells as well as the inflammatory action of Th17 cells. For the inflammatory action of Th17 cells we will fit new parameters to the existing equations that allow Th17 cells to influence osteoclast production and survival. Using the newly altered model we will simulate an estrogen deficient case by altering parameters related to estrogen in bone health such as increasing osteocyte death, reducing the ratio of Treg to Th 17 cells, and increasing inflammatory cytokines. Lastly, we will artificially dose this case with SCFAs to explore the action of SCFAs on estrogen deficient bone loss.

Results: Using the model, we will identify other potential actions of SCFAs on bone health. We will also be able to see the effect of estrogen deficiency on the ratio of Treg and Th17 cells within the bone compartment. Lastly, we will be able to predict if and at what amounts increased SCFAs could counteract estrogen deficient bone loss. These insights on bone health could lead to improved treatment and prevention protocols for those at risk for osteoporosis.

Acknowledgment: This work was supported by the National Institutes of Health grants R35GM133763 and R15AT010725 and the University at Buffalo.

References:

1. Pacifici, et al., Role of T cells in ovariectomy induced bone loss—revisited, *J Bone Miner Res* (2012).

2. Tyagi, et al., The microbial metabolite butyrate stimulates bone formation via T regulatory cell-mediated regulation of WNT10B expression. *Immunity* (2018).

3. Islam, et al., Mathematical modeling of the gut-bone axis and implications of butyrate treatment on osteoimmunology, *Ind Eng Chem Res* (2021).

4. Cook, et al., Mathematical modeling of the effects of Wnt-10b on bone metabolism, *bioRxiv* (2022).

665e An Optimal Regulatory Grammar for Dynamic Gene Control in Space and Time

Sahla Syed, Bomyi Lim

Proper gene control across space and time is crucial for the seamless execution of various cellular functions, as too much or too little gene expression in the wrong place or time leads to developmental defects and disease phenotypes. While the importance of gene regulation is well acknowledged, systematic characterization of dynamic gene control is lacking. What is the "normal" range of gene expression, and how does transcriptional kinetics play a role? We employ quantitative live imaging and mathematical modeling to examine the governing equation of dynamic gene control for normal development by perturbing regulatory DNAs. Through systematically modulating 3 Dorsal and 1 Twist transcription factor binding sites in the *snail* enhancer, we found that the mutations in these binding sites caused a drastic reduction in the maximum intensity, resulting in a reduction in total mRNA production. Using an equilibrium binding model, we characterized the synergistic capabilities of each binding site by quantifying the contribution of each binding site to the total mRNA production. Moreover, we implemented the Hidden Markov Model and determined the time each cell spent in the on and off states of transcription. Through this, we identified distinct mechanisms by which transcription factors regulate proper gene expression during development.

665f Modeling Supercoiling-Dependent Feedback As a Transcriptional Coordinator to Understand and Engineer Biological Circuits.

Christopher Johnstone, Kate Galloway

Cells coordinate complex behaviors through precise spatiotemporal control of gene expression and a myriad of interwoven control circuits. Gene- and cell-based therapies require synthetic circuits to similarly coordinate dynamic patterns of gene expression across a large population of cells. However, stable, robust expression of synthetic circuits in mammalian cells is challenging due to significant extrinsic and intrinsic expression noise. In particular, the stochastic nature of transcription makes coordinating expression across multiple genetic elements challenging. Spatial variation in the nucleus and biochemical dynamics in transcriptional condensates may contribute to bursting but provide limited parameters for tuning transcriptional noise.

Mechanical sources of gene regulation, such as the waves of positive and negative supercoiling induced by the motion of RNA polymerases, can be a powerful force driving expression changes at the kilobase scale. In yeast and human cells, transcriptionally-induced supercoiling demarks bounds of gene activity. In particular, transcriptional activity dictates the strength of contact domains such as gene-to-gene interactions, indicating a role for transcription in forming and maintaining interactions at the kilobase scale.

By modeling the underlying biophysical interactions that drive RNA polymerase activity--including the motion and stalling of polymerases acting under the influence of supercoiling and the supercoiling-dependent initiation process--a context-dependent understanding of biological circuit behavior emerges. In addition to being sensitive to the surrounding boundary conditions, the relative placement, orientation, and promoter strength of genes affect the simulated behavior. Applying the resulting stochastic-differential equation model to the behavior of simple two-gene genetic circuits, we identify that supercoiling-mediated feedback drives differences in gene expression via changes in polymerase stalling and transcriptional burst frequency. By further simulating the behavior of mutually-inhibitory toggle switches, we find that supercoiling-dependent feedback provides an additional, tunable source of control that can be layered on traditional synthetic regulatory schemes.

Applying our model of supercoiling-dependent feedback to native systems, we find that accumulated supercoiling well explains the observed behavior of the *her1-her7* segmentation gene network in zebrafish, with supercoiling-dependent polymerase initiation predicted to play a key role in establishing robust, periodic oscillations.
In sum, our model both predicts design rules for creating synthetic regulatory circuits and and provides explanatory power into the context-dependent mechanical regulation of native circuits. With these tools, we may be able to design robust sensors and actuators of cell state.

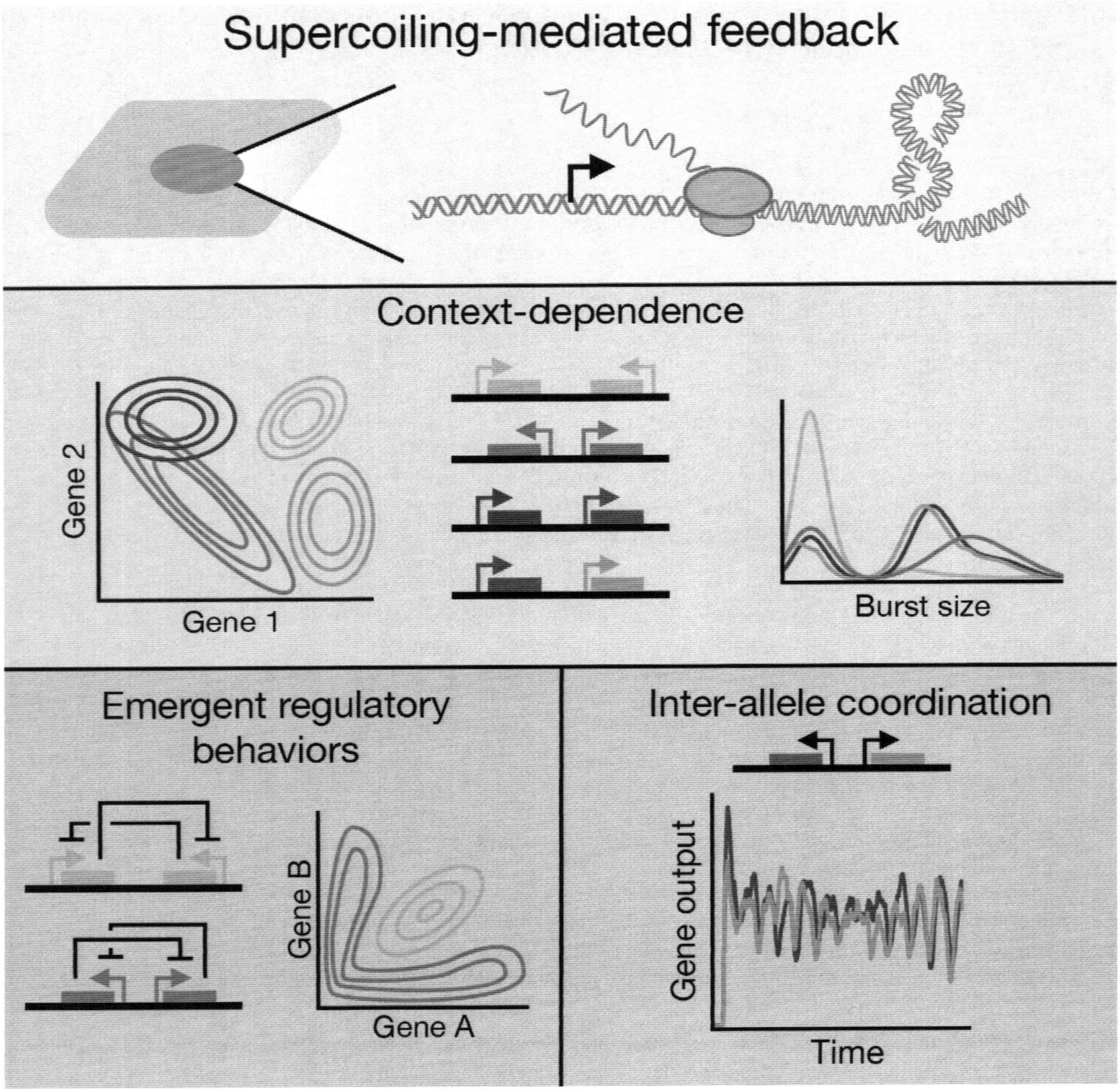
Supercoiling-mediated feedback
Context-dependence
Gene 2
Gene 1
Burst size
Emergent regulatory behaviors
Gene B
Gene A
Inter-allele coordination
Gene output
Time

665g Invited Talk: Biomechanics of Epithelial Tissue Homeostasis, Collapse, and Eversion

Vani Narayanan, Purboja Purkayastha, Bo Yu, Richard Dickinson, Daniel Conway, Tanmay Lele

Organization of cells into functional tissues is essential to development and wound healing. Irregular or uncontrolled growth and assembly of cells leads to pathologies such as tumor formation and cancer. In recent years, model tissues grown in culture into functional cell clusters called organoids have been widely used in drug development and regenerative medicine, as well as in vitro studies of morphogenesis, host-pathogen interactions, and tissue repair. When cultured within 3D extracellular matrices, epithelial cells spontaneously grow into a spherical organoid called an acinus, which consists of a contiguous monolayer of polarized cells surrounding a fluid-filled lumen. The morphology and stability of the acinar structure depends on the tensile forces between cells in the monolayer and lumen pressure created by the osmotic gradient across the monolayer. In this work, we describe our studies on the mechanical stability of MDCK epithelial acinar structures. Upregulation of Rho GTPase activity in already assembled acini with lumens results in a collapse of the luminal space, while inhibition of Rho kinase results in a luminal expansion. These changes can be induced directly by modulating Rho or indirectly by several different types of perturbations. Remarkably, after collapse, the acinus everts itself and achieves apical-out polarity, as opposed to the normal apical-in polarity. Myosin activity is high on the basal surface and low on the apical surface pre-eversion. Likewise, apical surfaces are more curved than basal surfaces. We interpret these findings with a vertex-based theoretical model of the acinus. Our experimental and theoretical results suggest that acinar eversion is mechanically driven by the surface tension difference between the apical and basal surfaces of the cells in the monolayer.

60b Porous Polytetrafluoroethylene Supported Silicalite Nanosheet Laminated Membrane Volatile Fatty Acid Separation from Anaerobic Bioreaction Broth

Zishu Cao, Landysh Iskhakova, Xinhui Sun, Maobing Tu, Junhang Dong

Volatile fatty acids (VFAs) with C_2–C_6 short chains such as acetic, propionic, butyric, valeric, and caproic acids are products from anaerobic fermentation of a variety of organic wastes. These VFAs are important feed stocks for synthesizing many high-value chemicals and materials. However, the economic viability of such VFA production from wastes has been limited by their generally low yield and difficulty in separation and recovery from complex solutions containing dissolved salts and solid contents. This presentation discusses the fabrication of a highly hydrophobic, molecular sieve silicalite nanosheet membrane laminated on macroporous polytetrafluoroethylene (PTFE) film and experimental investigation on its performance in pervaporation (PV) separation of VFAs from the saline solutions. The silicalite nanosheet laminated membrane (SNLM) has been proven highly efficient and robust for PV desalination of high-concentration brines. In this work, we focus on the SNLM formation on the chemically, structurally, and thermally stable porous PTFE and demonstration of its extraordinary hydrophobicity and membrane stability for enhanced VFAs separation from the aggressive solutions. We will also investigate the effects of operating parameters on the PV membrane separation of VFAs, including the feed pH, operating temperature, and chemical compositions.

60c *Enhancing 'Dark' CO2 Fixation in Succinate Fermentations Via Hollow Fiber Membrane Carbonation and Strain Engineering*

Amanda Godar, Timothy Chase, Dalton Conway, Yen-Jung Lai, Bruce Rittmann, Xuan Wang, David Nielsen

Although photoautotrophic microbes can naturally fix CO_2 to biomass and useful chemicals via the Calvin-Benson-Basham Cycle, overall rates are limited by slow carboxylation kinetics by RuBisCO, low cell densities and growth rates, and poor light penetration. In contrast, 'dark' fermentations offer the potential to fix CO2 in both a RuBisCO and light-independent manner, by employing alternative electron donors and carboxylases. Heterotrophic fermentations represent a promising option here, due to the availability of many optimized strains that are capable of growing quickly and achieving high product yields. Due to low energy costs and desirable kinetic parameters, the highest potential for heterotrophic fixation of inorganic carbon (C_i) occurs during anaerobic fermentations via the reductive TCA (rTCA) cycle. In the rTCA cycle, C_i is combined with phosphoenolpyruvate (PEP) by PEP carboxylase (*ppc*) or PEP carboxykinase (*pck*) to form oxaloacetate (OAA); the key precursor to succinate, along with malate, fumarate, aspartate and other bioproducts of interest. That said, to provide the needed C_i, industrial succinate fermentations currently require the media be supplemented with significant levels (up to 20 – 50 g/L) of bicarbonate (HCO_3^-) or carbonate (CO_3^{2-}) salts; levels far above what is required by the cells, with the excess going unused or even released as CO_2. Moreover, production of HCO_3^- and CO_3^{2-} salts from CO_2 or natural ores is thermodynamically unfavorable and thus energetically costly, which reduces the overall sustainability of the fermentation process. To address this, we have recently been investigating how Hollow Fiber Membranes (HFM) carbonation can instead be used to achieve efficient, bubble-free, and on-demand delivery of CO_2 in support of succinate fermentation by *Escherichia coli* and other microbes. Other benefits of the HFM delivery system include increased energy efficiency, high specific surface area to volume ratios, little off gassing, reduced shear stress (previously from bubbles passing the cells), prevention of foam formation, and reduced stripping of water and other volatile media components.

Here, using composite HFMs comprised of a nonporous polyurethane core and a microporous polyethylene outer layer we first demonstrate how net gas delivery rates can be controlled by modulating internal gas pressure within the membrane lumen and/or membrane surface area. Then, based on this, novel HFM fermenters were designed and constructed using optimized membrane surface areas to ensure that biofixation rates are adequately met. Further optimization of the HFM fermenter design and the development of a dynamic pressurization schedule was investigated to further enhance titer, rate, and yield (TRY) metrics. In fermentations with engineered *E. coli,* succinate titers and yields remained on par with published results using HCO_3^- or CO_3^{2-} salts, whereas experimental CO_2 fixation rates reached up to 125 mg/L/h: a 1.5 - 12.5 fold increase relative to typical, photosynthetic processes. Importantly, meanwhile, HFMs showed little sign of biofouling, suggesting that they should be well-suited for continued reuse in long-term applications. Lastly, multiple strategies were also investigated to enhance strain performance through both transporter and enzyme engineering. Overall, HFM-based CO_2 delivery represents a more sustainable alternative to the use of HCO_3^- or CO_3^{2-} salts in succinate fermentations, and likely also other 'dark' fermentations.

60f Science of Cutting Multi-Layer Systems: An Inspiration from Our Kitchens

Udita Ringania, Sunny Kumar, Tuhin Chakrabortty, M. Saad Bhamla

Cutting, slicing, and dicing are terms most of us are familiar with from our adventures in kitchen. In this talk, we will explore the topic of cutting, specifically in kitchen, with an aim to leverage the physics to optimize the force exerted in the process. While the existing literature presents work on physics of cutting, these are often on homogenous materials. However, most of the materials in kitchen are complex which comprises of multiple layers with different elastic modulus, fracture toughness and relative thickness like skin, pulp and seeds in fruits and vegetables. Hence, in this work we experimentally capture the force profiles while cutting and slicing of natural foods with a multilayer system for various fruits and vegetables. We also compare our experimental data with the existing force information available in the literature to estimate the optimum force by our experiments and compare to the existing data. It must be noted here that most of the existing data do not account for the multilayer system. Apart from that there is a huge variability in the nomenclature and the respective forces and the method of measuring these forces in literature, which makes the comparison a daunting task. Finally, we mathematically model this system by modification to the Atkin's force model for cutting of homogeneous system, incorporating the complexities for a multilayer system and compare with our experimental results.

60g [Keynote] Recent Progresses in Particular Membrane Separation Techniques for Bioprocessing and Food

Kamalesh Sirkar

This talk will focus on a number of particular types of membrane separation techniques and technologies relevant for bioprocessing and food processing applications. Specifically, I will focus on the following techniques/technologies: membrane-based nondispersive solvent extraction of bioproducts from their dilute solutions; ultrafiltration separation of mixtures of proteins having molecular weight ratios between as low as 1.03 to 3+; pervaporation separation for dilute biofuel/biochemical systems; continuous hollow fiber membrane-based crystallizations of active pharmaceutical ingredients; organic solvent reverse osmosis. I will illustrate examples of separations by each of these membrane processes where significant progress has been achieved.

554b Engineering of Cyanobacterial Rubisco for Sustainable Production of Essential Amino Acids

Muhammad Faisal, Aditya Sarnaik, Apurv Mhatre, Muhammad Javaid Asad, Ryan Davis, Arul Mozhy Varman

Global demand for high quality digestible protein is expected to significantly exceed population growth over the coming decades. To address this alarming issue we engineered RuBisCO, a key abundant soluble protein found in all photoautotrophs to act as a sink for the accumulation of two essential amino acids methionine and lysine. Specifically, the large sub-unit of RuBisCO in the model cyanobacterium *Synechocystis* sp. PCC 6803 was engineered for this purpose.

First, by utilizing Grantham's distance matrix, leucine and isoleucine in the RuBisCO amino acids sequence were replaced with methionine, while arginine and histidine were replaced with lysine. Towards this, four RuBisCO variants, cbbL1, cbbL2, cbbL3, and cbbL4 were created *in silico*. Further, *in silico* studies revealed that more than 80 % sequence similarity and 0.98 structure similarity were maintained by all protein variants. The *rbcL* variants encoding the engineered RuBisCO were integrated into the *rbcL* locus of *Synechocystis* 6803. Comparative growth studies of the engineered strains showed that there was no significant difference in the growth rate compared to the wild type. Estimation of chlorophyll a and carotenoid pigments also confirmed that there was no extra cellular stress on the engineered strains.

Secondly, a quantitative absorbance-based assay and a novel RNA-sensor based fluorescence assay for RuBisCO were developed. Both the assays confirmed that the function and properties of the RuBisCO variants were unaffected, indicating successful protein engineering. In parallel, the engineered *rbcL* genes were expressed in *Escherichia coli* cells to conduct amino acids analysis of the protein variants. GC MS analysis of the partly purified insoluble RuBisCO protein from the *E. coli* cell lysate exhibited ~75% increase in methionine and ~65% increase in lysine content as compared to that of the wild type. In conclusion, we have successfully engineered and demonstrated the increased accumulation of essential amino acids in cyanobacteria by utilizing a protein as an amino acid sink.

554c Engineering *Aureobasidium Pullulans* to Overexpress Malate Dehydrogenase in the Reductive Tricarboxylic Acid Pathway for Poly (L-malic acid) Production from Glucose

Zhen Qin, Jun Feng, Shang-Tian Yang

Poly (L-malic acid) (PMA) can be produced by the yeast-like fungus *Aureobaisidum pullulans*. The high water-solubility, non-toxicity, and biocompatibility of PMA make it an ideal drug carrier for the pharmaceutical industry. The monomer of PMA, L-malic acid (MA), is an important C4 platform chemical and food additive. *A. pullulans* belongs to the order Dothideales of the family Dothideaceae, which has been applied in biotechnology for decades. However, PMA fermentation by *A. pullulans* requires further improvement to realize industrial production. Fermentation yield from the substrate, PMA titer, and productivity can be enhanced through metabolic engineering and process optimization. Among the metabolic pathways for PMA biosynthesis, the reductive tricarboxylic acid (rTCA) pathway that converts pyruvate to oxaloacetate and transfers to malate by a malate dehydrogenase (MDH) has the highest theoretical yield of 2 mol MA/mol glucose, which is the target for metabolic engineering of *A. pullulans* to enhance PMA production from glucose.

A PMA producing strain *A. pullulans* ZX-10 previously isolated in our lab. was used in this study. A mutant strain *A. pullulan-mdh* overexpressing a cytosolic malate dehydrogenase (*mdh*) in *A. pullulans* ZX-10 was obtained. The effects of *mdh* overexpression on PMA production was evaluated in shake-flask fermentation, which showed that *A. pullulan-mdh* gave a 16% higher PMA yield from glucose or 0.50 g/g compared to 0.43 g/g for the wild-type strain. The fermentation process was scaled up in a 1.5-L bioreactor and further optimized by fed-batch fermentation. Techono-economic analysis showed that MA can be produced from glucose at $1.83/kg, which is significantly lower than the current market price of >$2/kg for MA produced from current petrochemical-based processes.

554d Acetic Acid Production from Glucose and Formate By the Co-Culture of *Clostridium Formicoaceticum* and *Lactobacillus Lactis*

Rocky Thapaliya, Opeyemi Bello, Jie Dong

Clostridium formicoaceticum, is a mixotrophic bacteria that produce acetic acid from different carbon sources such as glycerol, lactic acid, formic acid, fructose, and carbon dioxide. *Lactobacillus lactis* is a facultatively anaerobic, gram-positive bacteria that convert glucose to lactic acid. At initial, mixotrophic fermentation with *C. formicoaceticum* was assessed to identify the best possible lactate and format concentrations and their molar ratios to maximize the carbon conversion and acetate titer. The media with different concentrations of format and lactate and with different molar ratios of them screened in a serum bottle to identify the growth of bacteria and production of acetate. The target was to achieve the production of acetate at a high yield of >0.9 g/g substrate (lactate and Format) under anaerobic conditions at 37°C. The outcome of 0.91 g/g acetate is obtained from the culture with 2 g/L of formic acid and 5 g/L of lactic acid in combination. This was used to guide in the design and optimization of cocultured fermentation of *C. formicoaceticum* and *L. lactis.*

Further, the coculture fermentation of *L. lactis* and *C. formicoaceticum* was done to evaluate the interaction between them under various culture conditions such as pH, substrate concentration, nutrition level, etc. initially with wild type strain and then with the engineered *L. lactis* that has increased lactate production. Substrate (glucose and formate) consumption, production (acetate) production, and cell density of each strain were monitored throughout the culture period. Additionally, a reporter gene, mCherry, was loaded into *L. lactis* to enable the quantification of cell population ratio. The interactions between *L. lactis* and *C. formicoaceticum* and their effects on community (commensal and symbiotic) growth under optimized conditions were studied and found to enhance acetate production from glucose and formate at a high yield of >0.9 g/g substrate. This result has also guided us in the design and optimization of the consortia fermentation for butanol production.

554e From Screening to a Scalable System: Investigating the Impact of a Peptide Additive on Struvite Formation and Morphology

Jacob Hostert, Olivia Kamlet, Quincy Spitzer, Paola Giammattei, Zihang Su, Naomi Kane, Julie N. Renner

Precipitation of struvite ($MgNH_4PO_4 \cdot 6H_2O$), a slow-release fertilizer, provides a means of recycling phosphate from wastewater streams. In this work, a high-throughput struvite precipitation method is developed to investigate the effects of a peptide additive. The reactions occurred in small volumes (300 μL or less) in a 96-well plate for 45 minutes. The formation of struvite was monitored by fitting absorbance at 600 nm over time to a first order model with induction time, with the addition of peptide inducing significant changes to the yield parameter and formation constant in that model. The impact of struvite seed dosing was also investigated, highlighting the importance of optimization when peptide is present. The composition of the precipitate was confirmed through Fourier-transform infrared spectroscopy, while morphology and crystal size were analyzed through optical microscopy. Crystals had a higher aspect ratio when precipitated with the peptide. Finally, the utility of the high-throughput platform was demonstrated with a 2^5 full factorial design to capture the effects and interactions of: magnesium dose, mixing time, seed dose, pH, and temperature. The peptide was then expressed on a membrane protein in *E. coli* to provide a sustainable and cost-efficient means of using this system in the future. Overall, this study quantifies novel effects of a sequence-defined peptide on struvite formation and morphology *via* a newly developed high throughput platform, and offers a means for scalable implementation.

554f Recent Advances in Utilization of Fermented Plant Proteins and Their Resulting Hydrolysate and Peptide Fractions for Their *In Vivo* and *in Vitro* Antidiabetic Effects.

Bababode Kehinde, Poorva Sharma

Considerable attention has been given to the probiotic fermentation of plant foods and its proteolytic effect on their proteins in recent years. This is attributable to the neoteric demand for functional plant-based diets as alternatives to their animal counterparts based on socio-religious, environmental, and philosophical concerns around the globe. On another note, there has been an increasing trend in cases related to metabolic syndrome diseases, especially those associated with diabetes and its related health issues. However, the recognition of fermented plant foods as natural and biofunctional materials for the alleviation of health disorders has heightened their dietetic and therapeutic demands in recent years. Plant food materials have been observed to bear a significant load of numerous phytochemicals that are fit for employment as nutraceuticals. Accordingly, fermentation of plant proteins have been experimentally identified to be pharmacologically beneficial both in their isolated and hydrolyzed forms. Further purification into peptides has also been studied to enhance their bioactivites. In this study, recent investigations are discoursed regarding the recent progresses in the adoption of fermented plant proteins, their hydrolysate fractions, and sequence peptides. The novel adoption of singular or mixed microbial cultures for proteolysis operation with subsequent chromatographic fractionation methodologies was studied. In addition, efficient sequencing techniques such as edman degradation, single-mass fluorosequencing, and tandem mass spectroscopy prior to their in vivo and in vitro biofunctional characterizations are also examined. Fermented plant food proteins and their products were found to be usable for the therapeutic management of diabetes through mechanisms such as inhibition of digestive enzymes and the enhancement of insulin secretion with an overall decrease in serum glucose level.

554g [Keynote] Strain and Bioprocess Engineering to Enhance and Investigate Carbon Efficient Biochemical Production

David Nielsen

Using metabolic engineering and synthetic biology tools/strategies, our group focusses on developing novel microbes capable of producing value-added chemicals with bulk, specialty, and even nutritional applications. Meanwhile, to drive production metrics and minimize wasted carbon, a key, current focus is on the use or reuse of CO_2 as (co-)substrate. This objective, however, is met by several technical and analytical challenges resulting from the low and pH-dependent aqueous solubility of CO_2. Accordingly, we have also been exploring novel bioprocess designs aimed at enhancing CO_2 delivery or promoting its in situ recycling. As one approach, using various model and emerging cyanobacterial strains, we have engineered photosynthetic biocatalysts capable of producing amino acids and bioplastic monomers directly from CO_2 and light. Alternatively, as demonstrated using engineered *E. coli* strains with optimized flux through the reductive TCA cycle, 'dark' fermentation processes can also be developed for high-rate fixation of CO_2 as an essential co-substrate. Lastly, in a departure from the conventional focus on using single strains, we have also been investigating the engineering of synthetic co-cultures for developing carbon efficient bioprocesses that further benefit from metabolic 'division of labor' and facile tunability. Specifically, by pairing *E. coli* strains with net CO_2-evolving vs. CO_2-fixing metabolisms, we demonstrate how specific co-product mixtures can be produced from biomass-derived sugars with greater carbon efficiency and less waste.

663a A New Class of Xylose-Based Green Solvents Produced from Renewable Biomass

Anastasia Komarova, Zezhong Li, Jeremy Luterbacher

Organic solvents represent more than half the mass of all raw materials required for the production of chemicals. Concurrently, solvent use in industry is facing increasing regulatory and commercial pressure due to tightening legislation and an evolving awareness of health and safety issues. In parallel, growing environmental concerns over the source and life cycle of common solvents encourage the development of bio-based alternatives that can compete with petroleum-derived counterparts.

Lignocellulosic biomass is a promising feedstock for the production of renewable solvents since it is available in large quantities and does not compete with food crops. Several biomass-derived solvents have been developed and commercialized so far. However, their use in industry is still considered problematic due to various reasons such as high production cost (GVL, Dimethyl isosorbide), high flammability (2-Me-THF), instability in harsh conditions (Cyrene, cyclic carbonates). Newly emerging candidates that aim to replace hazardous analogs are also facing challenges. In this regard, acetal-stabilized xylose derivatives that can be inexpensively produced directly from biomass at over 90% yield (on a xylan basis) by aldehyde-assisted fractionation (AAF) technology developed in our lab could be interesting candidates for use as solvents[1].

In this work, we explored diformylxylose (DFX) and its derivatives isolated during AAF of biomass as potential solvents by assessing their performance in model reactions, as well as measuring their physical and solvatochromic parameters. DFX was found to have similar solvation properties to conventional polar aprotic solvents such as NMP, DMF, DMAc, while DFX's derivatives, for example, dipropylxylose (DPX) and dibutylxylose (DBX) were closer to medium-polarity solvents such as acetone and THF, according to a comparison of Kamlet-Abboud-Taft parameters. The high polarity and high hydrogen-bond accepting ability of DFX also make it a promising component of deep eutectic solvents, several of which were prepared in this work as demonstrations. In addition, determined Hansen Solubility Parameters indicate that DFX is the most similar to Dimethyl Isosorbide while being potentially much less expensive. At the same time, DPX and DBX can complement the list of conventional ethers and ketones like di-n-butyl ether, methyl isobutyl ketone (MIBK), but bringing new physical properties and, possibly, a safer profile.

Notably, we demonstrated that DFX's performance in alkylation, cross-coupling, and hydrogenation reactions was comparable with conventional problematic analogs and in some cases outperforming existing green alternatives[2]. DBX and DPX also possessed the right solvation properties to promote some of the model reactions. DFX was also found to have potential in many other applications including preparation of electrodes, synthesis of metal-organic frameworks, polymer synthesis, solid-phase peptide synthesis, biomass processing, liquid exfoliations, microfiltration membrane preparation, etc.

A preliminary toxicological assessment (bacterial Ames test) performed on DFX showed that this molecule is unable to cause mutations both directly and indirectly and cannot be considered a mutagenic compound. This shows that DFX is a good candidate for follow-up in-depth toxicology assessment (e.g. in vivo) needed for the approval of the compound for public use.

Finally, we demonstrated that DFX and its derivatives can be produced sustainably from biomass, specifically corn cobs in a kg-scale. This shows an excellent example of how waste biomass can be valorized into valuable chemicals with relatively low cost and high efficacy. A life cycle assessment has been performed, which demonstrated high competitiveness of this new class of solvents against conventional analogs.

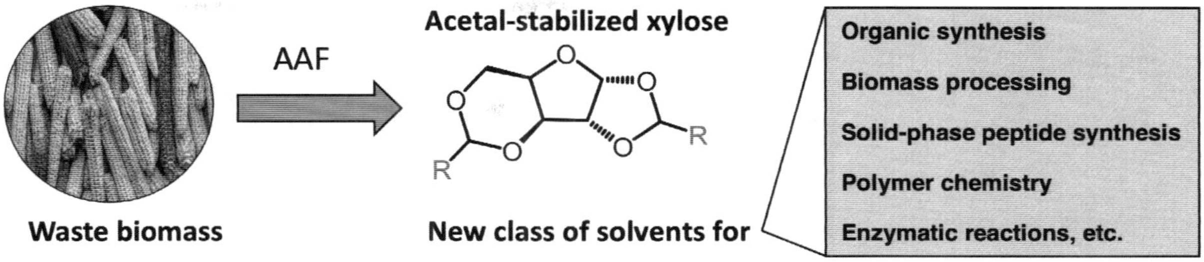

Figure 1. Production of xylose-based solvents from waste biomass

663b Recombinant Production and Purification of Green Fluorescent Protein (GFP)-Fused Metal Binding Protein for Palladium Nanoparticle Synthesis

Shadrach Ibinola, Hazim Aljewari, Imann Mosleh, PhD, Robert Beitle, Rita Tejada Vaprio

In lieu of chemical and physical methods, biologically guided synthesis is increasingly used as a cost-effective method for the fabrication of nanoparticles (NPs). Recently, a palladium metal binding sequence Pd4 (TSNAVHPTLRHL) has been demonstrated to be instrumental in controlling the production of palladium (Pd) nanoparticles. This work aims to provide an analysis of the catalytic turnover frequency of NPs derived from $(Pd4)_3$-GFPuv and correlate these values to the level of recombinant purity. An *E. coli* expression system was employed for expression of three copies of palladium binding sequence fused with green fluorescence protein (GFP). This protein was purified with chromatography from cell lysate. Transmission electron microscopy (TEM) was utilized to characterize the synthesized NPs from clarified lysate and pure protein sample, and the catalytic property of fabricated NPs from the purified protein was analyzed by a model Suzuki-Miyaura coupling reactions. It was observed that the purified metal binding protein had are three times higher turnover frequency compared to the crude lysate previously published.

663c Renewable Bio-Graphene Synthesis and Its Application As an Anti-Corrosive Coating

Suriya Narayanan Ramasubramanian, Hema Ramsurn, Gabriel LeBlanc

Graphene, the thinnest 2D material ever known have been widely used in many applications including sensors, displays, solar, panels, etc. State-of the-art for synthesis of graphene is chemical vapor deposition (CVD) which uses high vacuum and hydrogen for removing the oxide layers and etching the surface for graphene formation. High vacuum makes the process expensive and hydrogen use require careful handling due to its flammability. In this study, graphene is grown over metals at atmospheric pressure in a novel method where hydrogen and vacuum are not required. Unlike CVD which uses a gaseous carbon source, a solid renewable carbon source (biochar) is the graphene precursor. Biochar is prepared from biomass model compound (cellulose) by hydrothermal carbonization (HTC) at 300^0 C and 8.5 MPa in subcritical water. The solid carbon source is then placed on the metal (Fe, Co) foils and heated to 1000^0 C using a quartz tube furnace. On cooling, it was observed that graphene can form on both sides of the metal foil through the carbon dissolution and precipitation mechanism. Usually, at the high temperatures, the carbon atoms from biochar dissolve into the metals and upon cooling, carbon precipitates to form graphene. From the cobalt-carbon phase diagram, cobalt forms a cobalt carbide which is unstable at room temperature and dissociates into graphene. Due to the absence of carbides over the surface, graphene coated cobalt can theoretically provide a better corrosion resistance. XRD analysis showed a sharp peak corresponding to graphene. Interestingly, Raman analysis revealed that there were no defects on the graphene coated cobalt as no significant D band was observed. Based on the 2D/G ratio, both single and multilayer graphene were found on Co. On the other hand, from the phase diagram of iron-carbon, the solubility of carbon in iron is 1.4wt% at 1000^0 C where a mixture of γ-Fe, Fe_3C and graphitic carbon are present. On cooling to the eutectic temperature (727^0 C), the carbon atoms either precipitate to form graphene or form a stable iron carbide. From the XRD analysis, both graphene and iron carbide are detected over the iron foil. From Raman analysis, the 2D/G ratio was less than 1, corresponding to multilayer graphene. The anti- corrosive property of graphene was then tested in 3.5 wt% NaCl. Open circuit potential (OCP), Electrochemical Impedance Spectroscopy (EIS) and Linear Polarizing Resistance (LPR) technique were employed to analyze the corrosion behavior of metals and graphene coating in the presence of an electrolyte. Linear Polarizing Resistance (LPR) technique was used to generate Tafel plots by scanning the measured OCP from -300mV to +300mV vs OCP. The auto fit method was then applied to extrapolate the Tafel slopes to determine the corrosion current from which the corrosion rates of metals with and without graphene can be calculated. Initial corrosion measurements show that iron and cobalt metal without graphene showed a corrosion rate of 0.195 and 0.06 mmpy (millimeter per year). Graphene coated iron and cobalt showed a corrosion resistance of 0.20 and 0.61 mmpy. In the case of iron, non-homogenous coverage of graphene and the presence of the carbide phase gave rise to micro-galvanic corrosion, hence showing no improvement in corrosion resistance. Usually, multilayer graphene provides more corrosion resistance than monolayer graphene. Since both monolayer and multilayer graphene were present on Co, the monolayer did not give provide resistance to the electrolyte and hence the corrosion rate was similar to that of pure Co. Operational parameters will now be fine-tuned so that better multilayer coverage can be obtained on the iron foils for better corrosion resistance.

663d Thermoplastic Degradation By Yellow Mealworm Gut Microbial Communities and Isolates

Lummy Monteiro, Jyoti Singh, Ross Klauer, Kevin Solomon, Mark Blenner

To address the significant thermoplastic waste problem, we sought a biological strategy for plastics depolymerization from the gut microbiomes of the yellow mealworm. While bacterial community members have been identified, the specific pathways responsible for biodegradation remain to be elucidated and the potential contributions of fungal members are unexamined. Additionally, emerging evidence suggests that nutrient supplementation enhances plastic metabolism up to 70% and gives rise to a gut community structure distinct from that without additional nutrients. However, it is unclear if nutrient supplementation induces microbes to participate in the plastic degradation or if it supports a more optimal community composition.

In our study, we measured consumption rates of PS, LDPE, and HDPE of 20.4, 12, and 1.1, mg per (100 larvae) per day, respectively. However, nutrient supplementation enhanced plastics consumption by 158.8, 60, and 232.1 %, respectively. Antibiotic and antifungal selection studies suggest that fungal communities in the mealworm gut microbiome are likely to play an important role in the plastic consumption and that the inter-kingdom relationship between bacteria and fungi may be antagonistic. Fourier transform infrared spectroscopy (FTIR) analysis of frass (excrement) from mealworms fed PS and LDPE revealed incorporation of oxygen not found in untreated controls. Thermogravimetric analyses (TGA) of frass also confirmed the changes in physical properties. Finally, gel permeation chromatography (GPC) of the frass confirmed decreased apparent polymer molecular weight. Taken together, these results demonstrate that the plastics being ingested by the larvae are being oxidized and depolymerized. Microbiome community analysis via 16s and ITS sequencing revealed a rich consortium of bacteria and fungi. Mealworm diet led to unique community structures adapted to degradation of the fed plastic substrate, where nutrient supplementation frequently selected for taxa that were not observed in plastics only or nutrient only controls, suggesting currently unrecognized interactions.

Finally, characterization of the cultivable microbial community and isolates revealed a rich landscape of plastic depolymerization activities from microbes known to degrade plastics as well as several new genera that have never before been reported for their plastic degrading capabilities. We will present evidence strongly supporting the depolymerization of PS and LDPE, and synergistic activities of combinations of isolates for enhanced degradation.

663e Genome Mining Unveils Class of Ribosomal Peptides with Two Amino Termini

Hengqian Ren, Shravan R. Dommaraju, Chunshuai Huang, Haiyang Cui, Yuwei Pan, Marko Nesic, Lingyang Zhu, David Sarlah, Douglas A. Mitchell, Huimin Zhao

Ribosomally synthesized and post-translationally modified peptides (RiPPs) are a family of natural products growing rapidly in the genomics era. However, genome mining of first-in-class RiPPs remains an unsolved challenge. The few successful examples of bioinformatically discovered, first-in-class RiPPs exclusively focused on profiling homologs of known RiPP biosynthetic enzymes. Here we report a new RiPP class defined by an unusual (2S)-2-(dimethylamino)propylamine (Dap)-modified C-terminus, which we term the daptides. Biosynthetic gene clusters (BGCs) associated with daptide production were bioinformatically identified through a class-independent approach using analysis of the locally encoded RiPP precursor peptide with a common substrate-binding domain called the RiPP Recognition Element (RRE). The validity of our predicted BGCs were confirmed using synthetic biology techniques for heterologous expression of a representative BGC from Microbacterium paraoxydans DSM 15019. We then employed mass spectrometry, NMR, and synthesis of authentic standards to determine the structural modifications to the peptide C-terminus. Derived from a C-terminal threonine residue, the class-defining Dap residue is installed over three steps by an oxidative decarboxylase, aminotransferase, and methyltransferase. The genes encoding these enzymes, along with two genes encoding RRE domains, are found in nearly 500 daptide BGCs. In-depth characterization of a domain of unknown function-RRE fusion protein that is essential in the oxidative decarboxylation step reveals its function as a scaffold protein that directs the reaction in a leader peptide-dependent manner. This discovery of a class-defining post-translational modification in RiPPs may serve as a prototype for unveiling additional first-in-class RiPPs through genome mining.

663f Identifying the Genetic Mechanisms of Diatom Silicification for the Production of Novel Biomaterials

Hugh Purdy, Michelle O'Malley

Biological systems present attractive alternatives for traditional industrial processes in the search for new and more environmentally friendly technologies. The production of silica-based materials is one such area for which biological processes exist that may be able to supplement, replace, or improve current manufacturing approaches. Specifically, the unicellular, eukaryotic algae know as diatoms are recognized for their ability to synthesize cell walls made almost entirely of hydrated silica at ambient temperatures and pressures. Additionally, these cell wall structures typically contain a high degree of intricate micro- and nano-scale patterning. As such, there is increasing interest in the biosynthetic potential of diatoms for producing advanced silica-based materials, with applications ranging from catalyst support to drug delivery to micro-optics.

However, the engineering of diatom-derived materials is in its infancy, in part due to an incomplete understanding of the mechanisms underpinning siliceous cell wall synthesis. To help address this lack of knowledge, we are beginning investigations into the genetics of diatom silicification, with the ultimate goal of more precisely controlling and engineering this process. Transcriptomic and comparative genomic approaches in both model and non-model diatoms will enable us to identify novel genes influencing the intracellular silica-deposition process that leads to cell wall formation. For example, analysis of the transcriptional response of diatoms to conditions of silica-deprivation, as well as other nutrient stresses, is guiding our identification of genes regulating morphological stress responses, such as loss of spines or pores in the cell wall. Likewise, transcriptional analysis of diatom resting spore formation, in which the silica morphology drastically alters, will help to reveal the genes governing this change. The combination of these 'omics approaches with morphometric and materials-characterization techniques (e.g. sold-state NMR) will additionally allow us to identify relationships between diatom genes and the material properties of the cell-wall. Lastly, a small but growing set of genetic tools for use in diatoms enables us to directly probe candidate silica-associated genes to assess their role in cell wall formation. Through these and other approaches we are working to reveal genetic targets and control mechanisms to engineer diatom silicification for the production of advanced biomaterials.

663g Dose-Controllable Long-Term Drug Delivery Implant

Yoonjee Park, Xingyu He, Zheng Yuan

Repetitive intravitreal injections of drug or intravitreal injections of sustained-release corticosteroid implants (Ozurdex, Retisert, and Iluvien) are the current standard of care for the chronic diseases. However, these methods involve complications due to the high initial dose at the time of injection, known as burst release.

In this study, we have developed a size-exclusive nanoporous biodegradable polymer capsule for dosage-controllable drug delivery implants to avoid the burst release and control the dose only when needed. Light-activated liposomal drug (methotrexate) was enclosed in the polymer capsule. Methotrexate (MTX) was released through the capsule membrane when the implant was activated by laser, leaving the liposomal drug in the capsule. We have optimized the porosity and the pore size for MTX release kinetics, and tested the stability and the safety for 6 months in vivo rabbit eyes. We also demonstrated MTX release in vivo by irradiating near infrared laser (NIR, 1064 nm) through the lens of rabbit eyes. Lastly, we also demonstrated biodegradable sustained-release monoclonal antibody implant, which releases monoclonal antibodies slowly for 8 months to date. The technology allows us to avoid monthly invasive injections, which is the current standard of care for most chronic diseases that require antibodies as therapeutics. The drug delivery system that we developed is a platform technology which can be applied to different and/or multiple drugs. More importantly, the dose-controllability can avoid any side effects associated with high dose or uncontrollable dose.

314a Beginning with the End in Mind: Integrating Cell-Free Protein Synthesis with Coarse-Grained Simulation for Site-Specific Optimization of Protein Therapeutic Pegylation.

Brad Bundy, Emily Zhao, Mehran Soltani, Addison K. Smith, J Porter Hunt, Joshua W. Wilkerson, Thomas Knotts IV

Problem in Perspective: PEGylation is the only FDA-approved polymer to attach to therapeutic proteins for enhanced pharmacokinetics. However, PEGylation often comes at a high price where most of the protein therapeutic activity is lost (e.g. only 7% of the activity is retained when Interferon alpha is PEGylated[1]). This is due to an inability to perform site-specific PEGylation optimization due to the repeating nature of amino acids in proteins. For example, the commonly targeted primary amine of lysine is typically exposed on multiple locations of a protein and thus PEG is attached at different locations at different efficiencies depending on accessibility. Even with a significant setback in activity, PEGylated protein therapeutics are often more effective due to the longer retention times. However, there is significant potential to obtain enhanced pharmacokinetic retention with PEGylation and retain or even enhance protein therapeutics activity with location-optimized PEGylation. Due to the many variables at play and the inaccuracy of design heuristics, an integrated rapid simulation and high-throughput experimental approach is required.

Methods: Coarse-grained protein simulation is engineered to include PEG to allow rapid assessment of protein function and stability when PEG is theoretically attached at any and all locations on the target protein. Cell-free Protein Synthesis screening using noncanonical amino acids for site-specific covalent PEGylation of protein therapeutics enables high-throughput screening and validation of simulation results.

Results: Here we present the results of engineering coarse-grain protein simulation to include covalently attached PEG molecules and experimentally screening the impact of site-specific PEGylation using cell-free protein synthesis and noncanonical amino acids. These methods are applied to multiple protein targets with different sizes of PEG molecules. Remarkably, the coarse grain simulation is highly accurate in predicting the optimal location for PEGylation and the cell-free experimental method is able to rapidly verify the prediction. In this presentation we report our most recent results combined with recently published result from our labs[2-5]. Overall, this method has the potential to transform PEGylation. Rather than add PEG as an afterthought, in the future researchers will be able to optimize PEGylation at the beginning of protein therapeutic discovery and development. This would enable the development of more effective protein therapeutics with fewer administrations and side effects in the future.

References:

1) Bailon P, Palleroni A, Schaffer CA, Spence CL, Fung WJ, Porter JE, Ehrlich GK, Pan W, Xu ZX, Modi MW, Farid A. Rational design of a potent, long-lasting form of interferon: a 40 kDa branched polyethylene glycol-conjugated interferon α-2a for the treatment of hepatitis C. Bioconjugate chemistry. 2001 Mar 21;12(2):195-202.

2) Zhao EL, Soltani M, Smith AK, Hunt JP, Knotts IV TA, Bundy BC. Assessing Site-specific PEGylation of TEM-1 β-lactamase with Cell-free Protein Synthesis and Coarse-grained Simulation. Journal of Biotechnology. 2022 Jan 5.

3) Smith AK, Soltani M, Wilkerson JW, Timmerman BD, Zhao EL, Bundy BC, Knotts IV TA. Coarse-grained simulation of PEGylated and tethered protein devices at all experimentally accessible surface residues on β-lactamase for stability analysis and comparison. The Journal of Chemical Physics. 2021 Feb 21;154(7):075102.

4) Wilkerson JW, Smith AK, Wilding KM, Bundy BC, Knotts IV TA. The Effects of p-Azidophenylalanine Incorporation on Protein Structure and Stability. Journal of Chemical Information and Modeling. 2020 Sep 23;60(10):5117-25.

5) Wilding KM, Smith AK, Wilkerson JW, Bush DB, Knotts IV TA, Bundy BC. The locational impact of site-specific PEGylation: streamlined screening with cell-free protein expression and coarse-grain simulation. ACS synthetic biology. 2018 Feb 16;7(2):510-21.

314b Structural Characterization of an Effector-Biasing Interleukin-2 Immunocytokine

Joseph Gould, Elissa Leonard, Jamie Spangler

Background: Interleukin-2 (IL-2) has been used as a cancer therapeutic for roughly three decades. However, severe side-effects such as vascular leak syndrome combined with the vanishingly short serum half-life of the cytokine have limited the effectiveness of IL-2 therapy. Adding to this complexity, IL-2 signals through activation of either a high-affinity heterotrimeric receptor consisting of the IL-2 receptor-α (IL-2Rα, also CD25), IL-2Rβ, and common gamma (γ_C) chain or an intermediate affinity heterodimeric receptor that lacks IL-2Rα. IL-2Rβ and γ_C mediate signaling, whereas IL-2Rα regulates affinity of the receptor complex. Since T_{Reg}s robustly express IL-2Rα, whereas naïve effector cells (Effs) do not, T_{Reg}s have dramatically higher sensitivity to the IL-2 cytokine.

To address shortcomings of IL-2 therapy, previous studies have explored linking cytokines such as IL-2 to antibodies to improve both pharmacokinetics and targeting, in molecules called immunocytokines (ICs). Taking these ideas a step further, we have developed ICs which fuse antibodies against IL-2 to the cytokine itself. One of these molecules, denoted F10 IC, biases the IL-2 response toward immunostimulatory activity. Encouragingly, F10 IC has proven to be safe and effective as a cancer therapeutic in mouse models of melanoma. However, numerous molecular and mechanistic features of F10 IC have yet to be elucidated. Improved understanding of IL-2 ICs promises to advance their therapeutic design and development. Here, we present a structure-based characterization of the F10 IC molecule, and propose a model in which interactions between the antibody and cytokine result in a functionally relevant conformational change to IL-2.

Methods: High-resolution characterization of the interaction between the F10 antibody fragment and IL-2 was carried out using x-ray crystallography. Structural features of the full-length ICs were elucidated using cryogenic electron microscopy (cryo-EM). Single particle reconstruction methods were employed to generate intermediate-resolution maps into which IC models were built. Structural studies were supplemented with immune cell subset expansion profiling in mice following administration of our immunostimulatory IC versus IL-2 alone.

Results: Structural analysis revealed F10 (present as an scFv for crystallographic studies) as a direct steric inhibitor of the alpha subunit of the receptor complex. Further analysis demonstrated that F10 scFv binding resulted in a conformational change to IL-2, resulting in a conformation more similar to the receptor bound state. This feature was also observed in an immunostimulatory anti-IL-2 antibody that recognizes the mouse cytokine. Cellular subset and animal studies demonstrated the superiority of F10 IC over IL-2 alone in the selective expansion of CD8+ T cells and NK cells.

Implications: Our study demonstrates the mechanism by which the F10 IC biases IL-2 signaling in an immunostimulatory manner. The structural and functional insights our studies revealed will provide actionable data for further rational design of cytokine therapies. Future studies will focus on structure-informed improvements to the ICs, and therapeutic evaluation in animal models of cancer. Continued development of IL-2-containing ICs represents a fundamental advance for cytokine therapy, with improved pharmacokinetic properties and signaling specificity compared to unconjugated IL-2.

Acknowledgments: This research was funded by NIH R01EB029455, DoD Concept Award W81XWH-18-1-0735, and a Melanoma Research Alliance Young Investigator Award to Dr. Jamie Spangler.

314c Novel Experimental and Computational Approaches for Conformational Antibody Engineering

Matthew Smith, Alec Desai, Jennifer Zupancic, Peter Tessier

A common feature of many neurodegenerative disorders is the misfolding of disease-specific peptides and proteins into related types of pre-fibrillar oligomers and amyloid fibrils. Antibodies that recognize protein aggregates with strict conformational specificity are important for mechanistic, diagnostic and therapeutic applications. Despite progress in using *in vitro* antibody display methods such as yeast surface display for isolating conformational antibodies, a key limitation is that the insoluble nature of many protein aggregates (e.g., amyloid fibrils) prevents their use as soluble antigens and requires their immobilization on magnetic beads for magnetic-activated cell sorting (MACS) of yeast-displayed libraries. The use of conventional MACS methods to sort antibody libraries is undesirable because the binding selections are strongly influenced by the antibody expression level on the yeast surface and avidity effects in addition to the intrinsic antibody affinity, which leads to isolation of sub-optimal antibodies with modest affinity and/or conformational specificity.

Here we report two novel methods for addressing the limitations of conventional MACS methods to isolate high-quality conformational antibodies. First, we have developed a quantitative fluorescence-activated cell sorting (FACS) method for directly selecting high-quality conformational antibodies against multiple types of amyloid fibrils. Our approach uses quantum dots (QDs) functionalized with antibodies to capture amyloid fibrils, and the resulting QD-amyloid conjugates are used in a similar manner as conventional soluble fluorescently-labeled antigens for multiparameter FACS selections. Notably, we find that this approach is robust for isolating high-quality conformational antibodies against tau and a-synuclein fibrils with combinations of high affinity, high conformational specificity and, in some cases, low off-target binding that rival or exceed those of tau (zagotanemab) and a-synuclein (cinpanemab) clinical-stage antibodies. The generality of this approach is expected to simplify the generation and affinity maturation of high-quality conformational antibodies against diverse types of protein aggregates and other insoluble antigens (e.g., membrane proteins) that are compatible with presentation on antibody-functionalized QDs.

Second, we have developed a novel MACS methodology for generating high-quality conformational antibodies against Alzheimer's Abeta fibrils in the native IgG format. Our directed evolution approach uses targeted mutagenesis, yeast surface display, cell sorting and deep sequencing to identify antibody candidates with optimized binding properties. Notably, we find that this approach results in robust isolation of IgGs with higher affinity, higher conformational specificity and lower off-target binding than multiple clinical-stage Abeta antibodies, including aducanumab and crenezumab. This antibody engineering platform can be readily applied to generate high-quality conformational antibodies against diverse types of peptide and protein aggregates linked to neurodegenerative diseases.

314d Antibody Affinity and Specificity Co-Optimization Via Machine Learning

Patrick Kinnunen, Emily K. Makowski, Lina Wu, Jie Huang, Jennifer J. Linderman, Peter M. Tessier

Therapeutic monoclonal antibodies (mAbs) are used to treat a variety of diseases and show promise for treating many others. However, current antibody design methods often fail to reliably and rapidly generate therapeutic mAbs with favorable biophysical properties, including high affinity and low nonspecific binding. These failures result from difficulties exploring the large design space without well-understood sequence-property relationships for biophysical properties that impact clinical success. Furthermore, biophysical properties often exhibit strong tradeoffs, meaning that enhancing one property diminishes others. In particular, antibody affinity and specificity (non-specific binding) are often inversely correlated, as they are both mediated in part by the physicochemical properties of the complementarity determining regions (CDRs) of antibodies. Therefore, it is critical to develop methods for co-optimizing multiple antibody biophysical properties to improve generation of therapeutic antibodies with drug-like properties.

To better understand antibody sequence-property relationships, high-throughput screening of antibody libraries is commonly combined with next-generation sequencing (NGS). In applications of this technique, large repertoires of antibody mutants are screened based on their ability to bind to a specific reagent, labeling them as either positive or negative. Each bin of mutants is subsequently sequenced, providing the amino acid string of each antibody in each bin. Though this data represents a monumental step towards revealing sequence-property relationships, its use for rational therapeutic antibody design remains substantially limited. This is because the binary deep sequencing data obscures intra-class variation and does not provide information regarding the continuous nature of binding. Continuous measurements are necessary to tune properties that exhibit strict tradeoffs such as affinity and specificity.

In this work, we introduce a method that enables the co-optimization of antibody affinity and specificity by applying protein sequence feature-extraction methods and supervised dimensionality reduction methods for the extraction of continuous predictions of binding from binary deep sequencing data. We focused on the optimization of emibetuzumab, a stage 2 clinical anti-cancer antibody with high on-target (affinity) and off-target binding. We combined NGS and high-throughput screens to generate datasets consisting of thousands of antibody sequences with CDR mutations labeled for both affinity and specificity. We next applied a variety of machine learning methods to extract features from each of our labeled sequences. Using these feature sets, we were able to use linear discriminant analysis (LDA) or more complex neural networks (NNs) to further characterize the data. We found that using both LDA and NN classification methods enabled extraction of continuum projections of our antibody features. These projections correlated strongly with low-throughput, continuous measurements from selected clones in our library. Thus, by applying learned projections to each of the sequences in our library, we were able to estimate the continuous affinity and specificity of each clone and characterize the tradeoff between these two properties. Our analysis also facilitates multi-objective (Pareto) optimization of affinity and specificity.

We next identified maximally optimized sequences along the Pareto frontier, which comprises the set of antibodies that cannot improve specificity or affinity without sacrificing the other property. We predicted new clones containing in- and out-of-library mutations that lie at or beyond the Pareto frontier in order to improve both affinity and specificity at the same time. Using this method, over half of the antibodies we generated exhibited improved affinity and specificity compared to wild-type emibetuzumab, some of which went beyond the Pareto frontier in the original library.

Our combination of feature extraction and LDA- or NN-based dimensionality reduction provides two key advantages in antibody design. First, by characterizing tradeoffs between biophysical properties at scale, we can rationally select the combination of properties most likely to exhibit clinical efficacy, as opposed to optimizing one property at the expense of others. Secondly, some of our feature sets enable extrapolation to novel mutational space. We expect this methodology will facilitate optimization of therapeutic antibody biophysical properties and accelerate rational therapeutic antibody development.

314e Combining Protein Sequence and Structure Pretraining

Hugh Yeh, Kevin Yang

Proteins efficiently and precisely perform complex tasks under a wide variety of conditions. This combination of versatility and selectivity makes them not only critical to life, but also to a myriad of human-designed applications. Engineered proteins play increasingly essential roles in industries and applications spanning pharmaceuticals, agriculture, specialty chemicals, and fuel. The ability of a protein to perform a desired function is determined by its amino acid sequence, often mediated through folding to a three-dimensional structure. Machine-learning methods that predict fitness will enable the engineering of new protein functions.

Large pretrained protein language models have advanced the ability of machine-learning methods to predict protein structure and function from sequence, especially when labeled training data is sparse. Most modern self-supervised protein sequence pretraining combines a uses a neural network model trained with either an autoregressive likelihood or with the masked language modeling (MLM) task introduced for natural language by BERT (bidirectional encoder representations from transformers). For example, ESM-1b is a 650M-parameter transformer model, and CARP-640M is a 640M-parameter convolutional neural network (CNN), both trained on the MLM task using sequences from UniRef. While large pretrained models have had remarkable success in predicting the effects of mutations on protein fitness, predicting protein structure from sequence, and assigning functional annotations, pretraining often provides very little benefit on the types of datasets tasks often encountered in protein engineering. Furthermore, pretraining on only sequences ignores other sources of information about proteins, including structure.

Previous work shows that a graph neural network (GNN) that encodes structural information in addition to sequence context can also be used for masked language modeling of proteins, achieving much better performance on the reconstruction task than sequence-only models with a fraction of the data and model parameters (Figure 1). In this work, we show that a GNN MLM pretrained on 19 thousand sequences and structures outperforms sequence-only pretraining on protein engineering tasks where a structure is available, including zero-shot mutant fitness prediction and tasks in the FLIP (Fitness Landscape Inference for Proteins) benchmark. We then show that using the output logits from a fixed CARP-640M, pretrained on 42M sequences, as input to the GNN further improves performance on both the pretraining MLM task and on downstream tasks.

Historically, most pretraining methods for proteins have treated the amino-acid sequence as text and borrowed methods from natural language processing. However, proteins are not sentences, and protein sequence databases contain additional data that can be useful for pretraining, including structure, annotations, ligand, substrate, or cofactor information, and free text. Integrating this information into pretrained models will be essential to leveraging all available information for protein engineering.

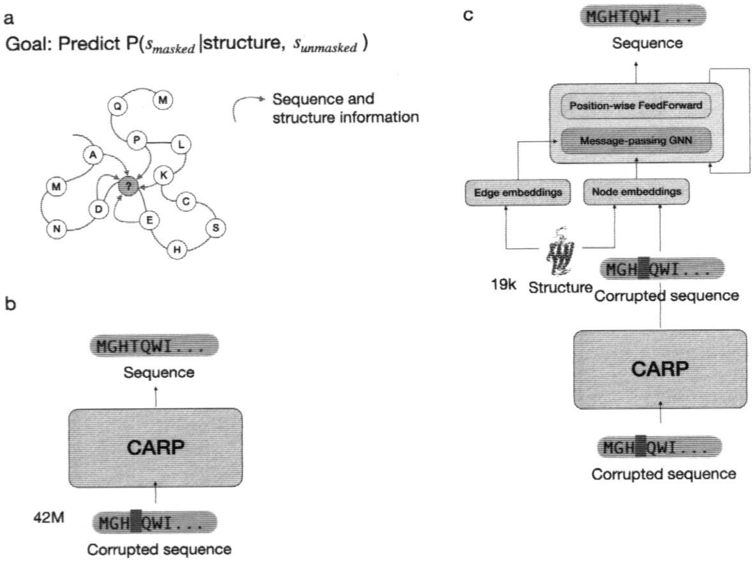

d

Model	Perplexity
GNN	4.96
Sequence only	7.06
Pretrained logits -> GNN	4.08

Figure 1. (a) A structure-conditioned masked language model predicts masked amino acids from sequence and structural context. (b) CARP is pretrained on 42 million sequences from UniRef50. (c) Using the outputs from CARP as the amino acid embedding in a GNN structure-conditioned masked language model. (d) Performance on the pretraining task.

314f Computational Modeling of Sars-Cov-2 Spike RBD Binding to Human ACE2 Receptor Using Molecular Simulations and Machine Learning

Veda Sheersh Boorla, Chen Chen, Ratul Chowdhury, Deepro Banerjee, Victoria S Cavener, Ruth H Nissly, Abhinay Gontu, Nina R Boyle, Shubhada K Chothe, Lindsey LaBella, Padmaja Jakka, Santhamani Ramasamy, Kurt J Vandegrift, Meera Surendran Nair, Suresh V Kuchipudi, Costas D. Maranas

SARS-CoV-2 is a highly virulent pathogen which gains entry to human cells by binding with the cell surface receptor – angiotensin converting enzyme (ACE2). We used biophysics based molecular modeling to computationally contrast the binding interactions between human ACE2 and coronavirus spike protein receptor binding domain (RBD) of the SARS-CoV-1, SARS-CoV-2, and bat coronavirus RaTG13 using the Rosetta energy function. We found that the RBD of the spike protein of SARS-CoV-2 is highly optimized to achieve the strongest binding of the loft against human ACE2 (hACE2) which is consistent with its enhanced infectivity[1]. On top of its high natural infectivity, SARS-CoV-2 has continued to evolve further into several novel variants with increased fitness and potential for immune evasion. Most of these variants had mutations which lead to amino acid changes in the RBD with significant effects on the receptor binding strength and dynamics. Reliably predicting the effect of amino acid changes on the ability of the RBD to interact more strongly with the hACE2 can help assess the implications for public health and the potential for spillover and adaptation into other animals. We introduced a two-step framework that first relies on recording molecular dynamics (MD) trajectories of RBD–hACE2 variants to collate binding energy terms decomposed into Coulombic, covalent, van der Waals, lipophilic, generalized Born solvation, hydrogen bonding, π–π packing, and self-contact correction terms. In the second step, we implemented a neural network to classify and quantitatively predict binding affinity changes using the decomposed energy terms as descriptors. The neural network model achieved a validation accuracy of 82.8% for classifying single–amino acid substitution variants of the RBD as worsening or improving binding affinity for hACE2 and a correlation coefficient of 0.73 between predicted and experimentally calculated changes in binding affinities[2]. However, the need for computationally expensive MD simulations limited the use of this model for a large-scale variant surveillance. Hence, we developed a convolutional neural network (CNN) model trained on protein sequence and structural features to predict experimental RBD-hACE2 binding affinities of 8,440 variants upon single and multiple amino acid substitutions in the RBD or ACE2. This deep learning model achieved a classification accuracy of 83.28% and a Pearson correlation coefficient of 0.85 between predicted and experimentally calculated binding affinities in five-fold cross-validation studies and predicted improved binding affinity for most circulating variants. We used the trained CNN model to exhaustively screen for novel RBD variants with combinations of up to four single amino acid substitutions and suggested candidates with the highest improvements in RBD-ACE2 binding affinity for human and animal ACE2 receptors. We found that the binding affinity of RBD variants against animal ACE2s follows similar trends as those against human ACE2. White-tailed deer ACE2 binds to RBD almost as tightly as human ACE2 while cattle, pig, and chicken ACE2s bind weakly. The model allows testing whether adaptation of the virus for increased binding with other animals would cause concomitant increases in binding with hACE2 or decreased fitness due to adaptation to other hosts.[3]

References:

1. Chowdhury, Ratul, et al. 'Computational biophysical characterization of the SARS-CoV-2 spike protein binding with the ACE2 receptor and implications for infectivity.' Computational and

structural biotechnology journal 18 (2020): 2573-2582.
2. Chen, Chen, et al. 'Computational prediction of the effect of amino acid changes on the binding affinity between SARS-CoV-2 spike RBD and human ACE2.' Proceedings of the National Academy of Sciences 118.42 (2021).
3. Chen, Chen et al. "A CNN model for predicting binding affinity changes between SARS-CoV-2 spike RBD variants and ACE2 homologues." bioRxiv 2022.03.22.485413.

61a *In Vitro* Continuous Evolution of Proteins Via Bioautomata

Tianhao Yu, Haiyang Cui, Nilmani Singh, Huimin Zhao

Protein engineering aims to identify variants with properties useful for biological, industrial, or medical purposes.[1] As a powerful protein engineering tool, continuous evolution enables gene diversification and screening in an iterative manner without human intervention. However, *in vivo* continuous evolution can rarely be applied to proteins whose phenotypes are unrelated to cell growth. It is also challenging to adapt existing continuous evolution methods for new target proteins.[2] To address these limitations, we developed a flexible and versatile *in vitro* continuous evolution workflow by integrating synthetic biology, robotics automation, and machine learning. While automation can expedite the directed evolution workflow and generate data at a capacity beyond manual work, it alone is not enough to perform directed evolution continuously. Thus, we use a machine learning model to predict and infer beneficial mutations and guide the design of continuous evolution. As a proof-of-concept, we used phytase as a case study to develop the iterative Design-Build-Test-Learn (DBTL) cycle in a biofoundry named BioAutomata.[3] The cycle was initiated with an informed small library designed by EVmutation[4] with the size of 96. We automated the entire workflow of constructing the variant library using site-directed mutagenesis (SDM) using two-step PCR. The SDM accuracy reached as high as 90%, with automated primer optimization. The subsequent experimental steps including transformation, colony picking, protein expression and assay measurement were all fully automated using biofoundry without the need of human intervention. The collected data was then used to train a machine learning model under the low-N framework.[5] The beneficial mutants predicted by the machine learning model were constructed for subsequent cycles. Such DBTL cycle took 4 days and can be carried out iteratively, with new data being generated and fed to the machine learning model in each cycle and leading to increased prediction accuracy. This *in vitro* continuous evolution workflow should be generally applicable to any target proteins even without prior knowledge or high-throughput screening methods.

(1) Yang, K. K.; Wu, Z.; Arnold, F. H. Machine-Learning-Guided Directed Evolution for Protein Engineering. *Nat. Methods* **2019**, *16* (8), 687–694.

(2) Wang, Y.; Xue, P.; Cao, M.; Yu, T.; Lane, S. T.; Zhao, H. Directed Evolution: Methodologies and Applications. *Chem. Rev.* **2021**, *121* (20), 12384–12444.

(3) HamediRad, M.; Chao, R.; Weisberg, S.; Lian, J.; Sinha, S.; Zhao, H. Towards a Fully Automated Algorithm Driven Platform for Biosystems Design. *Nat. Commun.* **2019**, *10* (1), 5150.

(4) Hopf, T. A.; Ingraham, J. B.; Poelwijk, F. J.; Schärfe, C. P. I.; Springer, M.; Sander, C.; Marks, D. S. Mutation Effects Predicted from Sequence Co-Variation. *Nat. Biotechnol.* **2017**, *35* (2), 128–135.

(5) Hsu, C.; Nisonoff, H.; Fannjiang, C.; Listgarten, J. Learning Protein Fitness Models from Evolutionary and Assay-Labeled Data. *Nat. Biotechnol.* **2022**.

61b Endoplasmic Reticulum Sequestration Empowers Phosphorylation Profiling on the Yeast Surface

Jose Ezagui, Brittney Russell, Yave Mairena, Lawrence A. Stern

Cells sense cues in their surroundings using a variety of receptor proteins, which turn an extracellular recognition event into an intracellular response. In many cases, the currency of intracellular response is transient post-translational modification. Chief among these modifications is phosphorylation, the addition of a phosphate to a serine, threonine, or tyrosine residue. There is great interest in understanding phosphorylation/dephosphorylation events that yield responses in cell signaling networks, but to date, methods to characterize the phosphoproteome and interactome rely on challenging low-throughput methods. In this talk, we will discuss our recent progress in developing and applying a yeast surface display system for high-throughput screening of kinase/substrate reactions using an endoplasmic reticulum co-localization strategy. We will demonstrate the ability to 1) display full receptor intracellular domains on the yeast surface, 2) robustly enrich phosphorylated intracellular domains from dilute pools using magnetic selections, 3) profile substrate specificities for a tyrosine kinase, and 4) demonstrate the ability to co-localize multiple units, leading to cascade-dependent phosphorylation of an intracellular substrate. Further progress in generalizing this system will also be discussed.

61c Redesigning Known Proteins to Detect Insulin and IL-6 across a Four Order of Magnitude Concentration Range

Alan Richard, Robert Pantazes

Certain features of biosensors represent an uncommon challenge to rational protein engineering methods. Most such methods solely focus on making the best possible protein: the one with the strongest affinity or the highest catalytic efficiency. However, a well-designed biosensor must be able to detect its analyte of interest across a range of concentrations. If the sensor only uses very high affinity proteins, it will become saturated at low analyte concentrations and unable to provide meaningful information at higher concentrations.

As part of an NSF EPSCoR project, we are working with a multijurisdictional team to design in-line, continuous, insulin and interleukin-6 (IL-6) detecting biosensors. The sensors must be able to detect the analytes across concentrations ranging from 10^{-9} to 10^{-5} molar. To design the protein recognition elements of the sensors, we began with a literature review that identified a total of 13 high affinity proteins and peptides that bind either insulin or IL-6. The structure of each was predicted using RosettaFold, followed by the prediction of each complex using Haddock. The results were compared to literature information about the protein structures and complexes when that information was available to validate the workflow. Next, molecular mechanics calculations using CHARMM and Rosetta and molecular dynamics calculations using VMD were performed to determine the hotspot residues in each complex that were most important to binding. Finally, mutations to non-hotspot residues were identified which are predicted to slightly destabilize the hotspot interactions. It is our hypothesis that binding variants with a range of affinities by identifying mutations around the critical residues.

This presentation will discuss the findings from each step of our workflow, as well as provide a detailed overview of the final designs selected for experimental validation.

61d Correlating Protein Motor Structure with Function Via Novel Biophysical Approaches

Pushkar Lele

The revolution in single-particle cryo-EM techniques has provided unparalleled resolution in visualizing membrane proteins without crystallization. In bacteria, these methods have been successfully extended to study a variety of molecular motors. In particular, several high-profile publications[1,2,3] have revealed astonishing details that appear to suggest conserved symmetries in the structures of motors likely involved in cell division and outer membrane maintenance (TolQ-TolR), those that mediate transport of nutrients (ExbB-ExbD), and those involved in flagellar motility (MotA-MotB) and gliding motility (GldL5M2). On the basis of structural similarities, the prevailing view is that these motors are likely rotary in nature. This hypothesis hinges on MotA-MotB: a MotA pentamer assembles around a MotB dimer forming ion channels. The flux of protons through these channels is believed to power the rotation of the pentamer around the dimer to enable motility. I will discuss innovations in biophysical approaches that are enabling this hypothesis to be tested. I will elaborate on how bacterial proprioception[4] is likely to be coupled with MotA-MotB function. I will conclude with the broader implications of successfully explaining MotA-MotB operation for our understanding of bacterial physiology.

[1] Santiveri et al, Cell, 2020; [2] Deme et al, Nature Microbiology, 2020; [3] Chang et al, Nature Structural andamp; Molecular Biology, 2020; [4] Antani et al, Nature Communications, 2021.

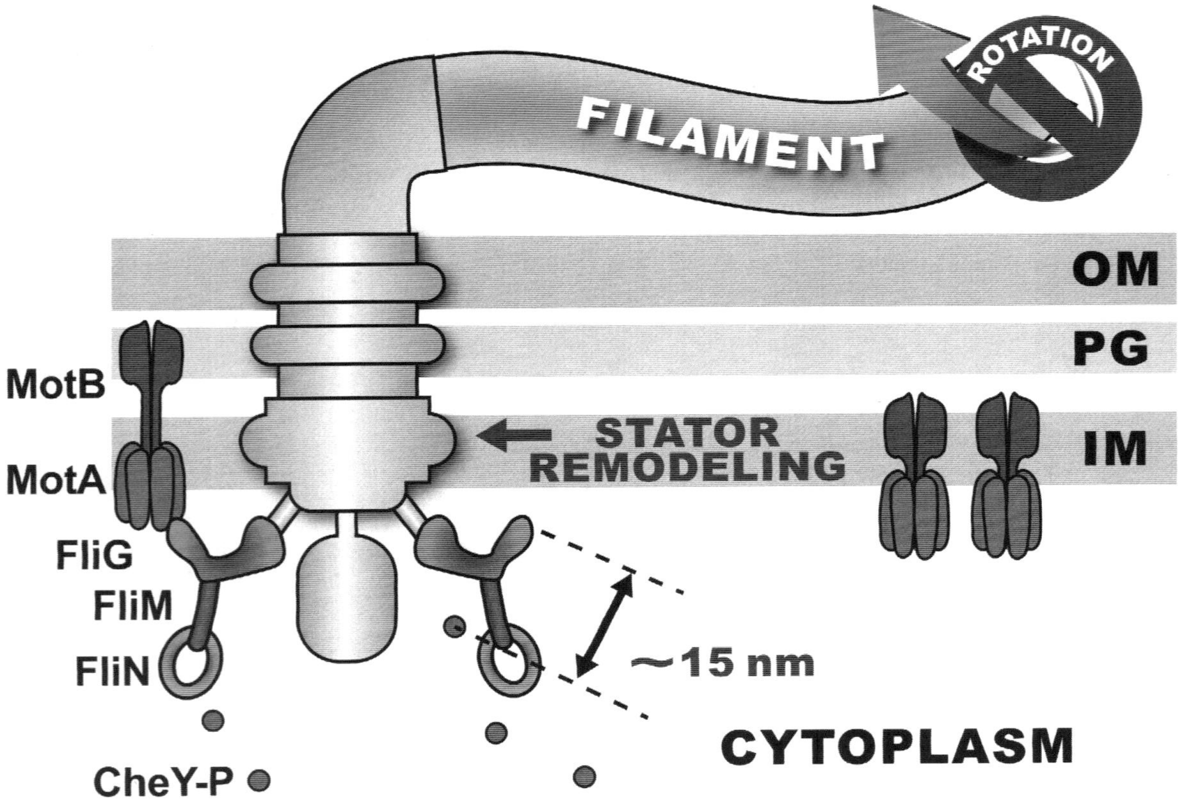

61e Engineering Dual Noncanonical Amino Acid Incorporation in Yeast

Priyanka Lahiri, James Van Deventer

The biosynthesis of proteins containing noncanonical amino acids (ncAAs) provides opportunities for dissecting basic biological processes and engineering proteins with "druglike" features using chemistries not found in the 20 canonical amino acids. The production of proteins with more than one ncAA provides opportunities for simultaneously introducing multiple new features into a protein, such as new molecular recognition and crosslinkability. Here we describe our progress in introducing multiple, distinct ncAAs into proteins prepared in the yeast *S. cerevisiae*. To complement the use of the UAG (amber) stop codon in yeast, we evaluated suppression of the UGA (opal) stop codon using three distinct orthogonal translation systems (based on *E. coli* tyrosyl-tRNA synthetase/tRNA, *E. coli* leucyl-tRNA synthetase/tRNA, and *M. alvus* pyrrolysyl-tRNA synthetase/tRNA pairs). Following mutation of the tRNAs of the orthogonal translation systems and complementary conversion of amber codons to opal codons in ncAA incorporation reporter systems, we observed stop codon readthrough for all three UGA suppression systems without detectable crossreactivity from native translation components. In general, stop codon readthrough efficiencies at UGA codons decreased by a factor of two or more when compared with readthrough efficiencies at corresponding UAG codons. We also observed that readthrough efficiencies at UGA exhibited position-dependent changes in efficiencies similar to those observed at UAG codons. Furthermore, previously identified yeast single-gene knockout strains known to increase readthrough efficiencies at UAG codons supported similar enhancements in readthrough efficiencies at UGA codons (by approximately 2-fold). We then used combinations of two orthogonal translation systems to investigate the feasibility of dual ncAA incorporation at UAG and UGA codons within the same mRNA transcript. Two different system combinations supported the production of full-length proteins based on fluorescent protein reporters with estimated overall efficiencies in the range of 1 to 5 percent of wild-type protein production levels (*E. coli* tyrosyl-tRNA synthetase/tRNA and *E. coli* leucyl-tRNA synthetase/tRNA pairs or *E. coli* tyrosyl-tRNA synthetase/tRNA and *M. alvus* pyrrolysyl-tRNA synthetase/tRNA pairs). We also investigated the application of the dual incorporation system in yeast display format by preparing a synthetic antibody containing two ncAAs. The full-length antibody containing two distinct ncAAs was detected and retained its binding function (though with a decrease in target binding affinity of approximately one order of magnitude). Ongoing work focuses on applications of the dual ncAA incorporation system as well as improvements in the efficiency of ncAA incorporation. Routine access to proteins containing multiple chemical features, especially in yeast display format, will further expand opportunities for the discovery and application of "druglike" proteins.

61f Engineering Genetically-Encodable Oxygen-Independent Fluorescent Reporters Using Lov Proteins

Nolan Anderson, Arnab Mukherjee

The Green Fluorescent Protein (GFP), DsRed, and their derivative fluorescent proteins have revolutionized our biochemical understanding of cellular processes over the last several decades. These fluorophores are used as platforms to build genetically-encodable sensors for a wide range of phenomena and analytes, including pH, ATP, NADH, protease activity, and gene expression. However, these fluorescent proteins require an oxygen-independent post-translational modification to develop fluorescence, rendering this family of probes ineffective in anaerobic organisms. To illuminate this anaerobic "shadow," we have engineered light-, oxygen-, and voltage-sensing (LOV) proteins to act as anaerobically compatible fluorescent reporters. LOV proteins bind the ubiquitously available flavin as a chromophore, eliminating the need for exogenous cofactors, including oxygen. However, LOV proteins suffer poor brightness and spectral diversity, and do not have a reliable design paradigm for facile sensor construction.

Toward making LOV proteins highly useful tools for synthetic biology, we have greatly reduced the size of iLOV, eliminating 10 residues making it one of the smallest fluorescent proteins available, while retaining 86.5 ± 2.2% of wildtype iLOV fluorescence intensity. We demonstrate its ability to track localization of proteins through several cellular compartments. Further, we have created a circularly permuted version of iLOV to facilitate construction of molecular sensors.

We also present the development of a LOV-based sensor for protease activity, as well as one sensing adenosine triphosphate (ATP), demonstrating its utility to construct highly adaptable sensors. These engineering efforts are strongly facilitated by adapting sensors based on GFP, indicating a "plug-and-play" paradigm of sensor engineering.

61g G-Protein Coupled Receptors: Protein Engineering Approaches to Improve Expression and Determine Protein-Protein Interactions

Anne Robinson

The adenosine receptor subfamily of G-protein coupled receptors is an important family of membrane receptors that modulate blood pressure, and more recently have been implicated in cancer, neurodegenerative diseases, and diabetes, making them a significant fraction of drug discovery efforts. As multipass membrane proteins, G protein coupled receptors remain challenging but important targets of drug discovery, and our efforts have enabled the high-level expression of the adenosine family of receptors and in-depth biophysical characterization of their binding to G-proteins and lipids, which I will highlight in this talk. Adenosine A_{2A} receptor ($A_{2A}R$), one of four subfamily members, has a longer C-terminus than the other adenosine receptor subtypes, which may contribute to its exceptional trafficking to the plasma membrane, but its flexibility has been attributed to protein-protein interactions. Chimeric A_1 and A_3 receptors with $A_{2A}R$ C-termini showed improved localization to the plasma membrane and are functionally active, with greater than three-fold higher yields than previously reported from other heterologous expression systems. Interestingly, these chimeras are able to bind to their native G proteins, suggesting the specificity of interaction does not rely on the C-terminus. To investigate this further using purified $A_{2A}R$ and Gαs, we determined the affinity and binding rates via surface plasmon resonance (SPR). Together with *in vivo* signaling data, our results suggest that loss of downstream signaling for the truncation may result from a reduced GPCR-G protein complex interaction, which may be important for interpretation of structure-based drug discovery screening. We have also investigated the interactions with lipids and cholesterol, in light of the cholesterol binding motif that has been identified in Class A GPCRs, and which has been implicated in a role in allosteric binding. Using site-directed mutagenesis and both in vivo and in vitro binding studies, we determined that cholesterol has a specific effect on signaling, as well as a bulk effect on $A_{2A}R$ function.

117a Elucidating the Complex Interactions between Stability, Oligomericity, and Activity of α-Amino Ester Hydrolase

Colton Lagerman, Martha Grover, Ronald Rousseau, Andreas Bommarius

α-Amino ester hydrolases (AEHs, EC 3.1.1.43) comprise a small class of enzymes capable of enantioselective production of semi-synthetic β-lactam antibiotics. AEH activity was first described in 1972 by Takahashi et al. [1], but AEHs from *A. turbidans* [2], *X. citri* [3], and *X. campestris pv. campestris* [4], have only recently been isolated, expressed in E. coli, and characterized in terms of substrate specificity, solvent effects, and mutability. Despite their rapid kinetics and high selectivity, AEH use is complicated by rapid deactivation and low stability [5]. Improvements in AEH stability has so far been limited, and further insight into the nature of AEH in solution is necessary.

Several studies have shown that *X. citri* and *A. turbidans* AEH monomers associate in a tetrameric structure in crystalline form. *X. citri* AEH monomers have also been shown to associate as a tetramer in solution, and dissociation of the tetramer into AEH monomers eliminates activity [6]. The tetrameric association has been assumed to be present and necessary for activity in all AEHs; however analytical ultracentrifugation (AUC) studies of *X. campestris* AEH show that only 15% of AEH in solution appears as a tetramer under conditions where the enzyme is also active. Furthermore, mutant AEHs designed for increased monomer stability demonstrate higher tetrameric association (up to 30%) as well as up to 70% higher activity toward cephalexin hydrolysis under the same conditions. This suggests that unlike previous studies, the AEH monomer is active, but association exists between monomer stability, tetrameric association, and kinetic activity.

This study provides further insight into the relationship between the stability, oligomericity, and activity of *X. campestris* AEH by comparing the deactivation kinetics of the wildtype AEH to mutant AEHs of varying stability and oligomericity. In addition, AEH tetramer was isolated from AEH monomer, and the activity of both forms were analyzed. Several mutants were analyzed using differential scanning fluorimetry (DSF) to determine melting temperatures of the monomers (T_m), ampicillin hydrolysis assays to determine activity levels and deactivation kinetics, and AUC to determine oligomeric states. Finally, AUC was used to determine oligomeric fractions of *A. turbidans* and *X. citri* AEHs in solution to compare along with literature data of kinetics and stability to *X. campestris* AEH. The relationships between stability, oligomericity, and activity as well as comparisons among the class of AEHs is discussed herein.

1. Takahashi, T., et al., *Enzymatic synthesis of cephalosporins.* J Am Chem Soc, 1972. **94**(11): p. 4035-7.
2. Polderman-Tijmes, J.J., et al., *Cloning, sequence analysis, and expression in Escherichia coli of the gene encoding an alpha-amino acid ester hydrolase from Acetobacter turbidans.* Appl Environ Microbiol, 2002. **68**(1): p. 211-8.
3. Barends, T.R.M., et al., *The Sequence and Crystal Structure of the α-Amino Acid Ester Hydrolase fromXanthomonas citriDefine a New Family of β-Lactam Antibiotic Acylases.* Journal of Biological Chemistry, 2003. **278**(25): p. 23076-23084.
4. Blum, J.K. and A.S. Bommarius, *Amino ester hydrolase from Xanthomonas campestris pv. campestris, ATCC 33913 for enzymatic synthesis of ampicillin.* J Mol Catal B Enzym, 2010. **67**(1-2): p. 21-28.
5. Blum, J.K., M.D. Ricketts, and A.S. Bommarius, *Improved thermostability of AEH by combining B-FIT analysis and structure-guided consensus method.* J Biotechnol, 2012. **160**(3-4): p. 214-21.

6. Kato, K. and A. Kakinuma, *Dissociation and Reassociation of Xanthomonas Alpha-Amino-Acid Ester Hydrolase.* Agricultural and Biological Chemistry, 1980. **44**(7): p. 1663-1664.

117b Orthogonal Glycolytic Pathway Enables Directed Evolution of Noncanonical Cofactor Oxidase

Edward King, Sarah Maxel, Yulai Zhang, Karissa Kenney, Gregory A. Weiss, Ray Luo, Han Li

Noncanonical cofactor biomimetics (NCBs) such as nicotinamide mononucleotide (NMN^+) provide enhanced scalability for biomanufacturing. However, engineering enzymes to accept NCBs is difficult. We established a self-sufficient growth selection to evolve enzymes that can utilize NMN^+-based reducing power. This is based on an orthogonal, NMN^+-dependent glycolytic pathway in *Escherichia coli* which can be coupled to any reciprocal enzyme to recycle the ensuing reduced NMN^+ (NMNH). With a throughput of $>10^6$ variants per iteration, the growth selection discovered a NADH oxidase variant with ~10-fold increase in NMNH catalytic efficiency and enhanced activity for other NCBs. Importantly, molecular modeling and experimental validation suggest that the mutations, instead of directly contacting NCBs, optimize the enzyme's global conformational dynamics to resemble that with the native cofactor bound. Thus, restoring the enzyme's conformation states, through deep navigation of protein sequence space by high-throughput evolution, provides a universal route to engineer NCB-dependent enzymes.

117c Rank-Ordering of Known Enzymes As Starting Points for Re-Engineering Novel Substrate Activity Using a Convolutional Neural Network

Vikas Upadhyay, Veda Sheersh Boorla, Costas D. Maranas

Retrosynthetic approaches have made significant advances in predicting synthesis routes of target biofuel, bio-renewable or bio-active molecules. The use of only cataloged enzymatic activities limits the discovery of new production routes. Recent retrosynthetic algorithms increasingly use novel conversions that require altering the substrate or cofactor specificity of an existing enzyme to complete the pathway. Although suggesting such novel conversions is only limited by the computational expanse of the search algorithm, the identification and re-engineering of the enzymes needed to complete the pathway is currently the bottleneck in the implementation of these designs. Here we present EnzRank, a convolutional neural network (CNN) approach, to rank-order existing enzymes in terms of their suitability to undergo successful protein engineering through directed evolution or *de novo* design towards the desired specific activity. EnzRank was inspired by the recent rapid progress of machine learning-based approaches in predicting enzyme classes, properties, and functions with tools such as alphafold2, DeepEC, and SDN2GO. EnzRank requires encoding both the similarity of the native vs. novel substrate(s) and the plasticity of the targeted enzyme to assign a probability score that a given enzyme has activity on a given substrate. The calculated probability score by EnzRank of the enzyme-substrate pairs can then be used to rank-order all candidate enzymes as to their potential to exhibit activity (even residual) for the novel substrate. Therefore, one must account for both substrate and enzyme information simultaneously to create a rank-order list to select a starting enzyme for any *de novo* reaction step in a pathway design, which is our primary goal. It is important to note that even though numerous algorithms predict EC classifications given an enzyme sequence, this level of detail is not sufficient for our goal as all candidate enzymes would presumably be classified with the same EC number. The varying length of the protein sequences makes it challenging to describe the protein features for the machine learning models. Moreover, only certain parts of the protein, such as specific residues, are involved in the enzyme-substrate activity rather than the complete protein structure. Hence, the physicochemical properties of the entire protein sequence do not seem to be the appropriate feature for predicting activity due to the noise information from the portions of the sequence that are not involved in the enzyme-substrate activity. Thus, the extraction of the local residue patterns involved in actual enzyme-substrate interaction is necessary to make an accurate prediction. We obtained known active enzyme-substrate pairs from the BRENDA database. These data also generated scrambled negative pairs by using the Tanimoto-based chemical similarity index to pair enzymes substrates that are entirely dissimilar to their native substrate. We created a total of 11,080 enzyme-substrate pairs and 11,076 generated negative pairs. Next, we used the holdout method to train and cross-validate the CNN method (using an 80:10:10 split for training and validation and testing data) on ten random data splits. EnzRank achieves an average recovery rate of 81.6% and 74.9% for positive and negative pairs and performs similarly on test data split with 81.5% positive and 73.8% negative recovery. We demonstrate a graphical user interface (which will be made publicly available at https://github.com/maranasgroup/EnzRank) that predicts enzyme-substrate activity from customized user input of novel substrates as SMILES strings and enzyme sequence information.

117d A High-Throughput Activity Screen for Reprogramming Proteases (HARP)

Samantha Martinusen, Ethan Slaton, Cassidy Simas, Benjamin Stone, Carl Denard

Protease dysregulation is a hallmark of many human diseases[1–3]. Due to this, controlling proteolytic activity is the goal of many therapeutics. Protease-targeting therapeutics are primarily small molecules discovered via binding-based high-throughput screens, often resulting in molecules with an active site bias[4]. Targeting protease active sites with small molecules remains challenging and often lacks enzyme selectivity due to the highly similar active topologies of related proteases[2]. Additionally, active site binding primarily results in enzyme inhibition. This approach limits the scope of current therapeutic approaches and can negatively affect proteolytic involvement in homeostatic pathways. Efforts have been made to discover distal regulators that modulate protease activity beyond inhibition, but these sites are challenging to find, and the screens require extensive structural analysis[5]. A robust functional screen to isolate protease modulators as a direct result of functional outcomes is needed to address these challenges.

We have developed the High-Throughput (HT) Activity screen for Reprogramming Proteases (HARP) system, a functional HT screening platform that links protease activity and modulation to observable fluorescent phenotypes on the yeast surface. HARP allows one to (1) modulate the activity of an exogenous protease using protein-based binders, (2) select protease modulators from large ligand libraries in a high-throughput fashion, (3) relate modulation phenotypes directly to the physicochemical properties of the ligand, and (4) map the binding epitopes of the modulator to infer their mechanism of action. We have used HARP to (1) optimize the activity of human and viral proteases in yeast and (2) isolate protease inhibitory nanobodies (Nbs) from large synthetic libraries.

HARP includes three gene cassettes, one containing an active protease sequence, one containing the protease's target substrate(s), and one containing an Nb modulator sequence **(Figure 1A)**. Using the system to screen human proteases active in yeast is achieved by selecting a high anti-FLAG, low anti-HA phenotype using a two-color stain, as the HA tag is lost when the substrate is cleaved. If one wants to screen for an inhibitory effect of a protease: Nb interaction, selection of the fluorescent phenotype high anti-FLAG and anti-HA signals is used. This methodology can be used to observe and induce modulation beyond inhibition, for instance, by reprogramming substrate selectivity and activating proteases on non-naïve substrates. We have observed significant activity of PLpro and 3Cl, the two main proteases involved in SARS-CoV-2 disease pathology, and MMP8, a protease associated with cancer tumor proliferation. Using previously published nanobodies, we have also used HARP to inhibit PLpro[6] and MMP8[7] **(Figure 1B)**.

This system can be expanded to discover modulators by introducing an Nb modulator library and sorting cell populations for a desired modulatory phenotype using fluorescence-activated cell sorting (FACS)**.** In a proof-of-concept, we show that we can recover an MMP8 specific nanobody inhibitor from a large population of non-binders and non-inhibitors. We aim to show early results of isolating Nbs from a large synthetic Nb library that are highly selective against disease-related proteases (i.e., BACE1, IDE, MMP8). Additionally, we are further expanding HARP to discover substrate-selective modulators by incorporating multi-display systems.

The use of HARP to screen and discover protease modulators that reprogram activity with increased specificity and selectivity opens many doors for advancement in protease targeting, therapeutic development, and distal site regulation. Using a functional screen as the basis for modulator discovery allows us to select modulatory outcomes such as activation and enhancement that are not discoverable in current systems. As we explore Nb-based modulation and isolation from large Nb libraries, we aim to discover modulators with favorable kinetics and potencies that rival current modulators. With each new application of this system, we look forward to tuning the HARP and finding new avenues for controlling proteases.

References: (1) Hampel, H., et. al., *Biol. Psychiatry*, (2021). (2) Coussens, L.M., et.al., *Science*, (2002). (3) López-Otín, C., et. al., *J. Biol. Chem.*, (2008). (4) Schmidt, M.F., et. al., *Trends Biotechnol.*, (2009). (5) Maianti, J.P., et. al., *Nat. Chem. Biol.*, (2019). (6) Armstrong, L.A., *PLOS ONE*, (2021). (7) Demeestere, D., et. al., *Mol. Ther.*, (2016).

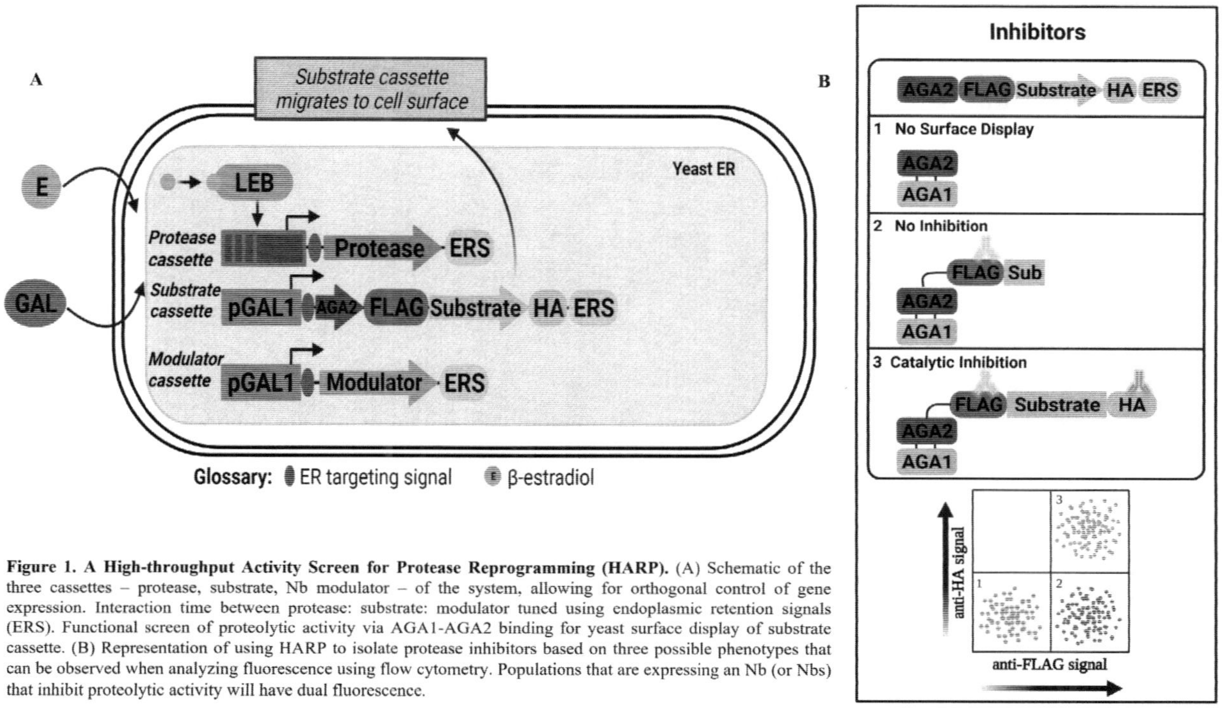

Figure 1. A High-throughput Activity Screen for Protease Reprogramming (HARP). (A) Schematic of the three cassettes – protease, substrate, Nb modulator – of the system, allowing for orthogonal control of gene expression. Interaction time between protease: substrate: modulator tuned using endoplasmic retention signals (ERS). Functional screen of proteolytic activity via AGA1-AGA2 binding for yeast surface display of substrate cassette. (B) Representation of using HARP to isolate protease inhibitors based on three possible phenotypes that can be observed when analyzing fluorescence using flow cytometry. Populations that are expressing an Nb (or Nbs) that inhibit proteolytic activity will have dual fluorescence.

117e Effects of Sequence Features on Machine-Learned Enzyme Classification Fidelity and Implications on *De Novo* enzyme Design

Md Sakib Ferdous, Nigel Reuel

Assigning enzyme commission (EC) numbers using sequence information alone is the focus of classification algorithms where statistics, homology and machine-learning based methods are used. This work benchmarks performance of four recent algorithms and specifically maps fidelity as a function of sequence features such as chain length and amino acid composition (AAC). This enables determination of optimal classification windows for *de novo* sequence generation and enzyme design. In this work we developed a parallelization workflow which efficiently processes >500,000 annotated sequences through each candidate algorithm and a visualization workflow to observe the performance of the classifier over changing enzyme length, main EC class and AAC. We applied these workflows to the entire SwissProt database to date (n = 565254) using two, locally installable classifiers, ECpred and DeepEC, and collecting results from two other webserver-based tools, Deepre and BENZ-ws. It is observed that all the classifiers exhibit peak performance in the range of 300 to 500 amino acids in length. In terms of main EC class, classifiers were most accurate at predicting translocases (EC-6) and were least accurate in determining hydrolases (EC-3) and oxidoreductases (EC-1). We also identified AAC ranges that are most common in the annotated enzymes and found that all classifiers work best in this common range. Among the four classifiers, ECpred showed the best consistency in changing feature space. These workflows can be used to benchmark new algorithms as they are developed and find optimum design spaces for the generation of new, synthetic enzymes. Preliminary work is presented on generating candidate sequences for therapeutic enzyme design with a classifier used in its optimal prediction space.

117f Engineered Cynr Regulon for Enzymatic Biocatalysis Enables Milk Oligosaccharides Biosynthesis Using Azido Sugars

Chandra Kanth Bandi, Mohit Kumar, Shishir Chundawat

Autonomous detection of azide-tagged biomolecules like azido sugars can enable the fields of chemical biology and glycoengineering. Here, we report an engineered synthetic promoter system called cyn regulon, and complementary *Escherichia coli* engineered strains, to selectively detect azide ions and autonomously induce downstream expression of reporter genes for various glycoengineering applications (1). The engineered cyn operon allowed highly tunable reporter green fluorescent protein (GFP) expression over three orders of inducer azide ion concentrations (0.01–5 mM) and rapidly induced GFP expression by over 600-fold compared to the uninduced control. We showcase the superior performance of this engineered cyn-operon over the classical lac-operon for recombinant protein production. Finally, we highlight how the cyn-operon synthetic biology toolkit can enable glycoengineering-based applications by facilitating *in vivo* activity screening of mutant carbohydrate-active enzymes (CAZymes), called glycosynthases, using azido sugars as donor sugar substrates and various acceptor sugars (e.g., lactose) to synthesize targeted prebiotics like human milk oligosaccharides (HMOs).

1. Bandi, C. K., Skalenko, K. S., Agrawal, A., Sivaneri, N., Thiry, M. andamp; Chundawat, S. P. S., 'Engineered Regulon to Enable Autonomous Azide Ion Biosensing, Recombinant Protein Production, and in Vivo Glycoengineering,' *ACS Synthetic. Biology*. 10, 682–689 (2021). DOI: 10.1021/acssynbio.0c00449. Weblink: https://pubs.acs.org/doi/10.1021/acssynbio.0c00449

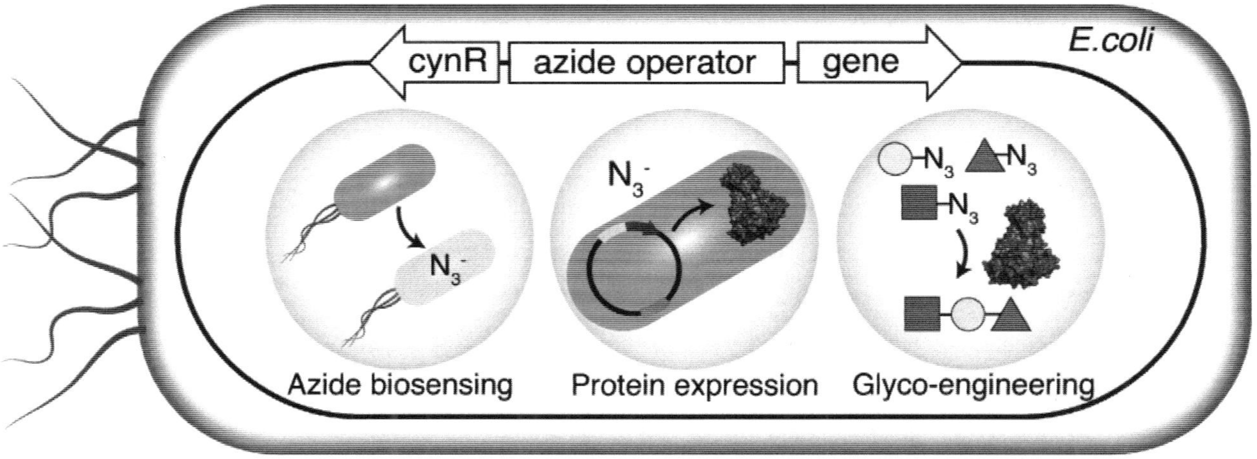

117g Data Driven Biomolecular Engineering Using High Throughput Measurements of Protein Function

Timothy A. Whitehead

Massively parallel, high throughput measurements of protein function have revolutionized protein science and engineering. Such measurements enable machine learning and other data-driven approaches for the efficient navigation of functional sequence space. In this talk I will describe the technology my lab has built for determining quantitative sequence-function relationships in enzymes for tens of thousands of variants using growth-based selections and screens. Then, I will show data-driven insights for improvement of enzymes involved in biomass catabolic pathways, secondary metabolic pathways, and fungal bioluminescence. I'll end the talk by describing a complementary bottoms-up strategy to regulate enzymes using small molecules, with the engineering enabled by high throughput screening.

713a A High-Throughput Screen for Bacterial Microcompartment Closure

Carolyn E. Mills, Nolan W. Kennedy, Curt Waltmann, Monica Olvera De La Cruz, Danielle Tullman-Ercek

Realizing the potential of microbes as cellular factories requires the development of orthogonal strategies for tuning pathway performance. In nature, spatial organization of biological functions is common, even in organisms as simple as bacteria. The capacity to engineer such spatial organization is thus an important goal in synthetic biology. Bacterial microcompartments (MCPs) are proteinaceous organelles that specifically aid in the proliferation of bacteria on niche carbon sources. These MCPs are delimited by a protein shell that encases an enzymatic core, where encapsulation of enzymes is hypothesized to confer several benefits to metabolic pathway performance, including increased pathway flux, sequestration of toxic intermediates, and access to a private cofactor pool. The genetically-encoded nature of both the core and shell of these organelles makes MCPs attractive protein engineering targets; however, the low throughput of existing MCP characterization techniques has limited the scope of many engineering efforts to date. Here, we describe our discovery that closure of the 1,2-propanediol utilization (Pdu) MCP from model pathogen *Salmonella enterica* serovar Typhimurium LT2 is mediated by vertex protein PduN. We find that, in the absence of PduN, extended Pdu microtube (MT) structures form, leading to an elongated cell phenotype that can be rapidly screened by both flow cytometry and high-throughput microscopy. We leverage this phenotype-assembly link to characterize how different point mutant libraries of PduN impact MCP closure, presenting the first high-throughput platform for screening MCP assembly. Further, we use molecular modeling on the protein-protein interface responsible for incorporating PduN into the Pdu MCP to understand the specific protein-protein interactions responsible for its incorporation, contextualizing the results of our point mutant libraries

713b Surface Functionalization of Recombinant Barley Stripe Mosaic Virus-like Particles for Biomedical Applications and Nanomaterial Synthesis

Akash J. Vaidya, Julia Donlevie, Oluwatoni Akin-Adenekan, Kevin Solomon

Rod-shaped plant viruses are RNA-templated nanoparticles with densely packed surfaces that can be engineered for various nanotechnological and medical applications. While some rod-shaped plant viruses, such as tobacco mosaic virus (TMV), receive a great deal of research focus, barley stripe mosaic virus (BSMV) remains relatively unexplored. We recently used a bacterial host to prepare BSMV virus-like particles (VLPs) that demonstrated superior palladium-deposition capacity relative to TMV. The recombinantly expressed BSMV coat protein spontaneously self-assembled into disk multimers *in vivo,* which further formed helical nanorods upon expression of an RNA template containing the viral origin of assembly sequence for TMV. This heterologous expression platform circumvents the need for host plant infectivity, thus greatly expanding the achievable range genetically engineered variants. However, bacterial hosts cannot accommodate protein modifications and processes of interest, and their products suffer from toxic lipopolysaccharide contaminants that hinder medical applicability. We addressed these limitations by producing BSMV VLPs in yeast for the first time. We also leveraged the engineering flexibility granted by each recombinant expression platform to perform rapid design-test-build-learn cycles on various mutant VLPs. This accelerated engineering workflow guided the fusion of single residues, peptides, and proteins at the particle surface. I will discuss how these various modifications can be used to tune physicochemical and biological properties of BSMV VLPs and support particle stability, purification, immunogenicity, and cellular interactions.

713c Incorporation and Assembly of a Light-Emitting Enzymatic Reaction into Model Protein Condensates

Muyang Guan, Mikael V. Garabedian, Marcel Leutenegger, Benjamin S. Schuster, Matthew C. Good, Daniel A. Hammer

Eukaryotic cells achieve specialized functions through subcellular compartmentalization, such as the sequestration and concentration of enzymes. Biomolecular condensates provide a strategy for compartmentalization through liquid-liquid phase separation (LLPS) of macromolecules such as proteins and nucleic acids, resulting the formation of a condensed phase isolated from the surrounding cellular environment. Emerging evidence suggests a role for LLPS and biomolecular condensates in human pathologies like neurodegenerative diseases, cancer, and viral replication. To study the enzymatic activities within biomolecular condensates, we utilized a model protein condensate system built from the RGG domain of the *C. elegans* P granule protein LAF-1. Herein, we demonstrate the incorporation of an ultrabright luciferase (NanoLuc) into model condensates that enables studies by bioluminescence microscopy and luminometry. We show that incorporation of NanoLuc significantly increased the overall reaction rate of the two-phase system compared to NanoLuc dispersion in the homogeneous phase. NanoLuc was estimated to be concentrated 10-fold inside these condensates. The resulting enzymatic reaction showed a diffusion-limited behavior, and the condensed phase behavior and light output was affected by condensate viscosity. As our model condensates have low viscosities, we expect the reaction to be diffusion-limited in most condensates and propose such a condensate-NanoLuc system for high-throughput screening (HTS) to identify drug molecules that alter the condensate viscosity. Further, we show that recruitment of complementary NanoLuc fragments into the condensates promoted their reconstitution into a functional full-length enzyme, indicating a liquid environment favoring protein-protein interactions (PPI). Lastly, we demonstrate cargo recruitment at the condensate-water interface through an engineered surfactant protein with coiled-coil interaction domain and achieve localized NanoLuc enzymatic activity therein, as well as the localization of other cargos. Overall, in this work, we demonstrate spatiotemporally regulated enzymatic activities through LLPS and protein self-assembly.

713d Structure Modeling and Design of Coiled-Coil Protein Origami

Ratnakshi Mandal, Mostafa Bedewy, Won Min Park

Nanotechnology requires methodologies for high-throughput and scalable nanofabrication. A promising "bottom-up" approach is biomolecular self-assembly because of the programmability of biomolecules in their length, conformation, and intermolecular interactions. For example, DNA molecules have been used to fabricate strikingly complex and well-controlled nanostructures, using the method of DNA origami. The use of proteins as building blocks for origami, instead of DNA, can offer several advantages that include low cost and scalability for manufacturing as well as programmable biological functionality. Coiled-coils protein motifs, which have been widely used as versatile toolkits to create protein assemblies in various systems, can serve as DNA-like building blocks for origami. They form oligomers through the protein-protein interaction with controllable binding affinity, specificity, orientation, and oligomeric states. The sequence-to-structure relationships of coiled-coil motifs are well established, which have enabled the de novo design of synthetic coiled-coil sequences that orthogonally interact. Indeed, orthogonally interacting coiled-coil motifs have been modularly arranged to assemble well-defined protein nanostructures that consist of connected coiled-coil helices. This method is termed coiled-coil protein origami (CCPO), which is based on the modular recombination of coiled-coil motifs into fusion proteins. Mediated by specific and orthogonal intra- or inter-strand dimerization of the coiled-coil modules, the coiled-coil fusion proteins are expressed, folded, and form a nano-assembly. In CCPO, modeling has been used as a powerful tool for the characterization of origami structures in atomistic resolution. However, there are no reliable tools to assess the model quality, which has limited structure prediction without experimental data. In this study, we demonstrate a method of comparative modeling that we use to build CCPO structure models, predict the radius of gyration, and evaluate the models. The comparative CCPO models were built by templating with coiled-coil crystal structures or models and refining through MD optimization with simulated annealing. We investigated the effective MD optimization length to improve the modeling quality and derived a new method for model valuation. The modeling results were validated by small-angle X-ray scattering data, and our model evaluation method showed significantly improved predictive power than the methods implemented in comparative structure modeling tools. The modeling method demonstrated in this study is fast and does not require a significant amount of computational resources. Thus, it can enable high-throughput computational screening of a library of CCPO designs with enhanced structure characterization. We extended our approaches to various CCPO designs, which showed the potential as a useful design tool for computational CCPO design.

713e Assembly of a Modular and Tunable Worm-like Protein Nanostructure Using a Bottom-up Approach

Antonio Goncalves, Wilfred Chen, Millicent O. Sullivan

Protein nanostructures have garnered interest as carriers in drug and intracellular protein delivery for their enhanced biocompatibility and stability compared to most synthetic particle systems. Studies into the design of an effective drug carrier suggest that modulating the shape of the particle is a key parameter in nanoparticle design. Where there is a balance in tuning high-aspect ratio nanoparticle's maximal lengths between lengths small enough for effective extravasation and internalization by endocytosis and, long enough for an effective circulation half-life through avoiding renal clearance and minimizing macrophage uptake. However, most high-aspect ratio protein nanostructures, like nucleoprotein-based nanorods, either form through self-assembly onto native RNA templates which limits the capacity for tuning length scales or, through a further *in vitro* disassembly and re-assembly step onto an engineered RNA template. The latter top-down assembly strategy instills the capacity to modify particle lengths but has been shown to regularly result in a heterogenous mixture. Furthermore, these nanostructures commonly lack the capacity for extensive and patterned surface decoration, usually limited to incorporating bioconjugation chemistries at solvent-exposed termini or, randomly reacting solvent-exposed lysine or cysteine side chains. Therefore, we have designed a multi-pot stepwise bottom-up approach for generating a worm-like protein nanostructure with homogenous and tunable control of structure length and, control of the position and valency of multiple protein cargos and a Click-compatible non-standard amino acid (nsAA), using orthogonal Catcher-Tag isopeptide-ligating protein-peptide pairs in tandem with elastin-like polypeptides and the controlled incorporation of 4-azido-L-phenylalanine. With this platform we demonstrate that we can modularly control the valency of multiple protein cargos as well as their position relative to each other within the structure. And that this assembly strategy allows for the incorporation of multiple different alkyne-functionalized ligands with control of their local density and relative positions to each other by virtue of the multi-pot nature of this assembly approach. We anticipate that by using this assembly strategy, high-aspect ratio worm-like protein nanostructures can be assembled for any desired application due to the inherent "plug-and-play" nature of the assembly strategy with respect to protein cargos, Click-compatible nsAA positioning and particle length.

713f Microbially Synthesized Polymeric Amyloid Fiber Promotes β-Nanocrystal Formation and Displays Gigapascal Tensile Strength

Jingyao Li, Fuzhong Zhang

Dragline spider silk stands out as one of the strongest and toughest natural materials and has motivated decades of research to produce recombinant silk fibers at macroscales. Despite these efforts, achieving gigapascal tensile strength with higher than 150 MJ*m^{-3} toughness in recombinant fibers remained challenging. On the other hand, the ability of amyloid peptides to self-assemble into stable β-sheet nanofibrils had made them potential candidates for material innovation in nanotechnology. However, such nanoscale feature was rarely translated into attractive macroscopic properties. Inspired by the structure of natural spidroins, we propose a strategy that combines the mechanical superiorities of spider silk and amyloid nanofibrils by fusing amyloid peptides with spidroin-derived flexible linkers. These microbially synthesized polymeric amyloid proteins could be wet-spun into macroscopic fibers, and structural analyses unveiled a high degree of crystallinity with the presence of β-nanocrystals that resembled the cross-β structure of amyloid nanofibrils. These polymeric amyloid fibers displayed strong and molecular-weight-dependent mechanical properties. Fibers made from the high molecular weight protein containing 128 repeats of the FGAILSS sequence exhibited an average ultimate tensile strength of 0.98 ± 0.08 GPa and an average toughness of 161 ± 26 MJ*m^{-3}, surpassing most recombinant protein fibers and even some natural spider silk fibers. The design strategy and the biosynthetic approach presented in this study can be generalized to create numerous functional materials, and the macroscopic amyloid fibers will enable a wide range of mechanically demanding applications.

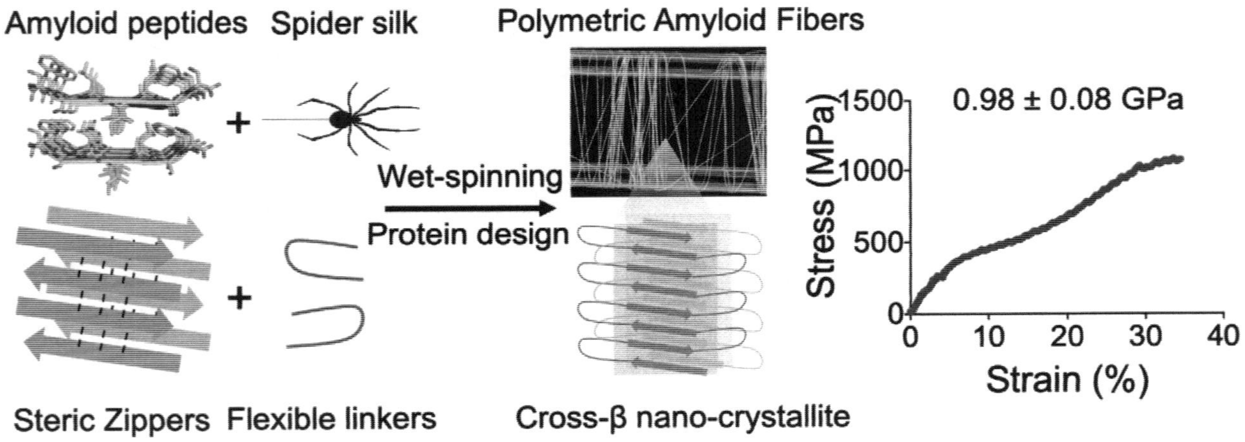

713g Design of Protein Nano-Architectures As Genetically Programmable, Functional Biomaterials

Claudia Schmidt-Dannert

In biological systems, proteins, nucleic acids and lipids are precisely organized to form higher ordered structures across multiple length scales. We believe that harnessing the principles and mechanisms underlying the assembly and organization of natural bionanomaterials offers tremendous opportunities for the design and scalable fabrication of functional biomaterials with emergent properties. Proteins and peptides provide the greatest versatility for the bottom-up design and low-cost production of such self-assembling supramolecular materials due to the chemical diversity of their amino acid building blocks. They are also genetically encoded, allowing for the genetically programmable production of self-organizing materials using cell factories or in the future, de novo via cell free expression systems. Proteins are also key players in the formation inorganic-organic composite materials with properties unmatched by synthetic properties. Inspired by the spatial organization of enzymes at the subcellular level via different types of protein nanostructures, we are taking advantage of these mechanisms for the design of self-assembling protein-based nano-architectures for different applications, including for *in vitro* biocatalysis and for the fabrication of new types of functional materials. Of key interest to us is the discovery and design of mechanism with which to interface protein-based materials with biomineralization processes for the production of innovative materials with unique mechanical and other properties. This presentation will discuss possibilities and examples from our work for the design of genetically encoded biomaterials as functional materials for diverse applications, including e.g. for biocatalysis and biosynthesis, as coatings and as living materials.

120a Are Genome Scale Metabolic Models (GeMs) Fit for Purpose? A Comprehensive Analysis into the Reliability of Chinese Hamster Ovary (CHO) Cell GeMs.

Benjamin Strain, James Morrissey, Cleo Kontoravdi

Background

Chinese hamster ovary (CHO) cells are the leading platform for the production of many therapeutic recombinant proteins, meaning increasing the efficiency and yield of these cell lines is of critical importance to meet demands and reduce costs. CHO cell genome scale metabolic models (GeMs) possess the power to revolutionise cell line efficiency and yield by their ability to predict and understand whole cell metabolism *in silico,* allowing for model directed metabolic and process engineering strategies to be implemented. Despite this, the effectiveness of CHO cell GeMs at capturing both intracellular flux distributions and extracellular phenotypes has yet to be comprehensively assessed.

Methods

In this work we present an evaluation of the extracellular and intracellular predictive power of CHO cell GeMs using a novel holistic assessment strategy. Firstly, we synthesise current state-of-the-art models to generate the most comprehensive CHO cell GeM to date, iCHO2441. From this model, we generate industrially relevant CHO-K1 and CHO-S cell line specific models via 'omics integration. We constrain the GeMs with 31 published metabolomic datasets and employ flux balance analysis (FBA) and flux sampling methodologies to compare these new models against the original CHO GeM, iCHO1766, as well as a small-scale CHO metabolic model, CHOmpact. Model predictions are assessed via comparison with experimentally measured growth rates, gene essentialities, amino acid auxotrophies, and 13C intracellular reaction rates.

Results

We demonstrate all CHO cell models are able to capture extracellular phenotypes, highlighting that cell line specific models show significant improvements in gene essentiality and amino acid auxotrophy predictions compared to other models tested. Secondly, we demonstrate that CHO GeMs are adept at predicting intracellular flux states across varying cell culture conditions, showing GeMs can predict qualitatively and quantitatively for many intracellular reactions using only simple metabolomic datasets to constrain models. Our in-depth interrogation identifies alternate pathways and infinite loops as key causes of poor predictive performance. We show that our updated GeM outperforms the original CHO GeM and highlight instances where the use of cell line specific and small-scale models is beneficial.

Summary

Overall, we have introduced an updated CHO GeM to the community and have employed the most intensive CHO cell model assessment to-date. Our results provide assurance that CHO GeMs can be used as an effective toolbox for interrogating cellular metabolism in many industrial settings. Ultimately, this work lays a foundation for the development and assessment of next-generation flux analysis techniques, highlighting areas for model improvements and providing recommendations for when different CHO cell metabolic models are best employed.

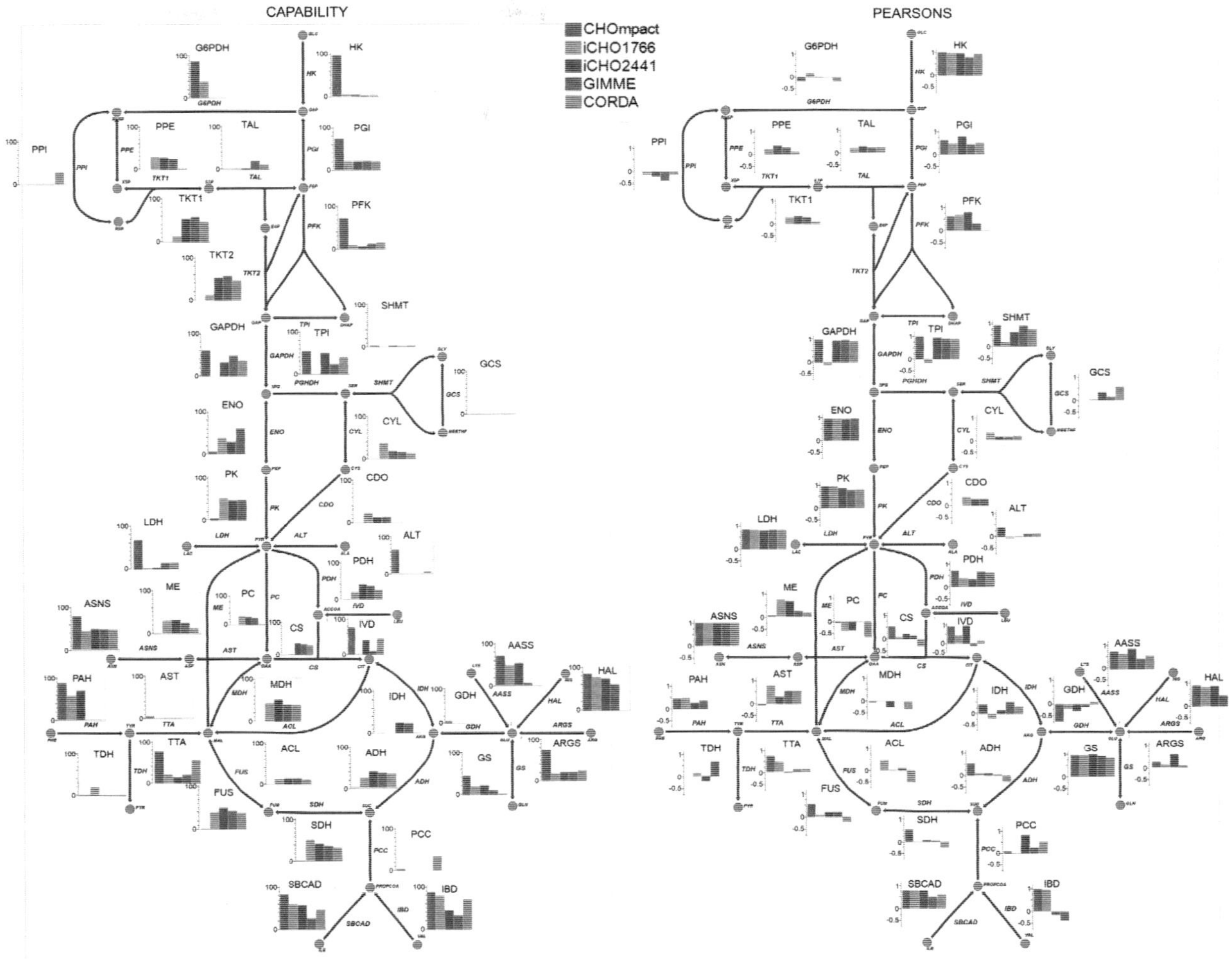

Figure 1. Capability and Pearsons correlation coefficient per reaction for each metabolic model tested. Capability is defined as the percentage of times a reaction is correctly predicted by the model within the experimental range.

120b Quantification of Metabolic Differences in Platelets Exposed to Estrogen Using Flux Balance Analysis

Samantha Siska, Cara Sake, Alexander Metcalf, Nanette Boyle, Keith B. Neeves

Platelets, while essential to human health, are implicated in severe health problems such as heart attacks, strokes, and venous thrombosis. Pre-menopausal women have lower incidences of these events compared to post-menopausal women and men, leading to a hypothesis that estrogen has a cardioprotective effect (Maccarrone et al 2015, Khetawat et al 2000). However, it has also been observed that women on hormonal oral contraceptives are at an increased risk (4-6 fold) of venous thrombosis (Vinogradova et al 2015). Conflicting studies exist on the effect of estrogen on platelets both *in vivo* and *ex vivo* (Grodstein et al 2000, Cushman et al 2004, Bar et al 2000, Moro et al 2005). Here, we apply metabolomic and fluxomic techniques to quantify changes in resting platelet metabolism upon acute exposure to physiologically relevant concentrations of estrogen with reaction level resolution. 124 extracellular metabolites and 175 intracellular metabolites were analyzed over 180 minutes for platelets exposed to 10 nM estradiol, 100 nM estradiol, or vehicle control. These experimental measurements were used to constrain a flux balance analysis model to identify differences in key energy and activation pathways: central carbon metabolism and arachidonic acid metabolism. Quantifying changes in resting metabolism upon estrogen treatment may help explain differences in activation between estrogen treated/untreated platelets, as well as explain sexual dimorphisms in platelet behavior *in vivo*.

References:

1. Bar, J., Lahav, J., Hod, M., Ben-Rafael, Z., Weinberger, I., andamp; Brosens, J. (2000). Regulation of Platelet Aggregation and Adenosine Triphosphate Release In Vitro by 17B-Estradiol and Medroxyprogesterone Acetate in Postmenopausal Women. *Thrombosis and Haemostasis*, *84*(4), 695–700.
2. Cushman, M., Kuller, L. H., Prentice, R., Rodabough, R. J., Psaty, B. M., Stafford, R. S., Sidney, S., andamp; Rosendaal, F. R. (2004). *Estrogen Plus Progestin and Risk of Venous Thrombosis*. http://jama.jamanetwork.com/pdfaccess.ashx?url=/data/journals/jama/4946/
3. Grodstein F, Manson JE, Colditz GA, Willett WC, Speizer FE, Stampfer MJ. A prospective, observational study of postmenopausal hormone therapy and primary prevention of cardiovascular disease. Ann Intern Med. 2000 Dec 19;133(12):933-41. doi: 10.7326/0003-4819-133-12-200012190-00008. PMID: 11119394.
4. Khetawat, G., Faraday, N., Nealen, M. L., Vinod Vijayan, K., Bolton, E., Noga, S. J., andamp; Bray, P. F. (2000). *Human megakaryocytes and platelets contain the estrogen receptor and androgen receptor (AR): testosterone regulates AR expression.*
5. Maccarrone, M., Bari, M., Battista, N., andamp; Finazzi-Agrò, A. (2002). Estrogen stimulates arachidonoylethanolamide release from human endothelial cells and platelet activation. *Blood*, *100*(12), 4040–4048. https://doi.org/10.1182/blood-2002-05-1444
6. Moro, L., Reineri, S., Piranda, D., Pietrapiana, D., Lova, P., Bertoni, A., Graziani, A., Defilippi, P., Canobbio, I., Torti, M., andamp; Sinigaglia, F. (2005). Nongenomic effects of 17β-estradiol in human platelets: Potentiation of thrombin-induced aggregation through estrogen receptor β and Src kinase. *Blood*, *105*(1), 115–121. https://doi.org/10.1182/blood-2003-11-3840
7. Vinogradova, Y., Coupland, C., andamp; Hippisley-Cox, J. (2015). Use of combined oral contraceptives and risk of venous thromboembolism: nested case-control studies using the QResearch and CPRD databases. *BMJ (Clinical Research Ed.)*, *350*, h2135. https://doi.org/10.1136/bmj.h2135

120c Metabolic Modeling of Eicosapentaenoic Acid (EPA) and Arachidonic Acid (AA) Fatty Acids Using a Cybernetic Framework

Sana Khanum, Shakti Gupta, Rubesh Raja, Lina Aboulmouna, Naseeha Mohammed, Mano Maurya, Shankar Subramaniam, Doraiswami Ramkrishna

Introduction

This research concentrates on studying the inflammatory response of the murine macrophage-like RAW cells upon external perturbation, such as a cut or an infection. In response to a disturbance from the environment, prostaglandins (PG2 series) arise from the cyclooxygenase (COX) oxidation of Arachidonic acid (AA), an ω-6 polyunsaturated fatty acid. They are the pain mediators, and are responsible for regulating the induction of inflammation. However, the pro-inflammatory PG2's heightened concentration contributes to widespread chronic inflammatory diseases, including asthma, and impaired wound healing. Non-steroidal anti-inflammatory drugs (NSAIDs), which work by inhibiting COX, are among the most consumed pharmacotherapeutic agents to treat these diseases. They exhibit anti-inflammatory roles and are beneficial in relieving and blocking acute inflammation but ineffective at terminating inflammation or promoting resolution and tissue repair.

Instead of controlling AA metabolism by COX inhibition, an alternate strategy is to activate the competitive substrate that binds COX, thereby inhibiting the PG2 series formation. The ω-3 fatty acid eicosapentaenoic acid (EPA) is a promising substitutable substrate. EPA competes with AA to bind with COX during the body's inflammatory response. The metabolism of EPA results in the PG3 series formation, thus reducing the levels of PG2-series metabolites. PG3 series is known to exhibit anti-inflammatory attributes and partake in inflammation resolution. Moreover, studies have confirmed that EPA-containing diets, fish oil being the primary source, assist prevent cardiovascular and other diseases.

Metabolic regulation forms an integral part of a biological system. All organisms control the reaction mechanism to achieve a biological goal, such as survival during a dreadful situation in unicellular organisms. Biological systems are challenging to model because of the vast intricacy of the regulatory processes due to unknown details. The cybernetic model is a mathematical framework for studying biochemical networks [1]. A distinguishing feature of the cybernetic model is that it successfully accounts for complicated metabolic regulation. Unlike other biological models that do not study regulation or can best incorporate known control, the cybernetic model excels in capturing known and unknown exhaustive control attributes. Thus, the cybernetic model has been remarkably successful in expounding the dynamics of a complex biological system.

Development of the Cybernetic Model

This work focuses on developing a cybernetic framework to study the AA and EPA metabolism in the presence of inflammatory markers, such as TNFα and IL-6. We model the metabolic network adopted by Gupta et al. [2], Fig. 1. Studies focused on modeling the competitive metabolism of AA and EPA are emerging; however, few studies are available. Gupta et al. [2] successfully captured the underlying mechanism. However, they accounted for transcriptomic regulation only and did not include cytokines in their system. In this work, we incorporate cytokines, well-known markers of inflammation, that exhibit a relationship with AA and EPA networks. Consequently, this study aims to gain insight into the fundamental mechanism ruling the competitive metabolism of AA and EPA in the presence of cytokines.

The cybernetic framework is based on defining a cybernetic goal and correspondingly computing the cybernetic control variables. For the network shown [1], an enzyme can bind with both substrates S_1 and S_2 to form products P_i and Pj as per $S_1 + e_i \rightarrow P_i$ and $S_2 + e_i \rightarrow$ Pj. The rate of formation of P_i and e_i follow eq. 1.

The cybernetic control variable u_{P_i} controls the enzyme synthesis rate; hence it appears in the enzyme synthesis rate formation equation. In eq. 1(b), α denotes the constitutive enzyme synthesis rate, β is the degradation rate and the second term denotes the enzyme synthesis due to Michaelis-Menten kinetics. The cybernetic control variable v_{P_i} controls the enzyme activity; thus, it appears with the P_i rate formation in 1(a), and γ_{P_i} is the decay rate of the product. Defining u_{P_i} and v_{P_i} requires formulating the cybernetic goal F. We define the cybernetic goal as eq. 1(c), which is the weighted combination of the inflammatory marker levels C_1 (TNFα) and C_2 (IL-6). The cybernetic goal is to maximize the production rate of the markers C_1 and C_2 which are associated with the AA and EPA branches, respectively. We will capture these associations using a data-driven approach. We define the control variables u_{P_i} and v_{P_i} using the cybernetic goal as shown in eq. 2(a) and 2(b).

Study outcomes

We lay the groundwork for studying the competitive metabolism of AA and EPA using the cybernetic model while including their interactions with different inflammatory markers. This study will provide insight into the underlying mechanism of AA and EPA formation under the influence of markers and ways to control their competition. Additionally, we expect to better understand how to modulate competitive metabolism of AA and EPA by tackling their marker levels.

Reference:

1. Aboulmouna, L., Raja, R., Khanum, S., Gupta, S., Maurya, M.R., Grama, A., Subramaniam, S. and Ramkrishna, D., 2020. Cybernetic modeling of biological processes in mammalian systems. *Current Opinion in Chemical Engineering*, *30*, pp.120-127

2. Gupta, S., Kihara, Y., Maurya, M.R., Norris, P.C., Dennis, E.A. and Subramaniam, S., 2016. Computational modeling of competitive metabolism between ω3-and ω6-polyunsaturated fatty acids in inflammatory macrophages. *The journal of physical chemistry B*, *120*(33), pp.8346-8353.

$$\frac{dP_i}{dt} = v_{P_i} k_{S\rightarrow P_i} S_1 e_i - \gamma_{P_i} P_i \quad \text{(a);} \quad \frac{de_i}{dt} = \alpha + \left[\frac{u_{P_i} k_{e_i S_1}[S_1]}{K_{e_i S_1}+[S_1]} + \frac{u_{P_j} k_{e_i S_2}[S_2]}{K_{e_i S_2}+[S_2]}\right] - \beta e_i \ \text{(b)} \qquad (1)$$

$$u_{P_i} = \frac{K_i e_i \sum_{k=1}^{n}\left(\frac{\partial F}{\partial C_k}\right)\left(\frac{\partial C_k}{\partial P_i}\right)}{\sum_{j=1}^{2}(K_j e_i)\sum_{k=1}^{n}\left(\frac{\partial F}{\partial C_k}\right)\left(\frac{\partial C_k}{\partial P_i}\right)} \ \text{(a)} \quad v_{P_i} = \frac{K_i e_i \sum_{k=1}^{n}\left(\frac{\partial F}{\partial C_k}\right)\left(\frac{\partial C_k}{\partial P_i}\right)}{\max_j\left(K_j e_i \sum_{k=1}^{n}\left(\frac{\partial F}{\partial C_k}\right)\left(\frac{\partial C_k}{\partial P_i}\right)\right)} \ \text{(b)} \quad F = a_1 C_1 + a_2 C_2 \ \text{(c)} \quad (2)$$

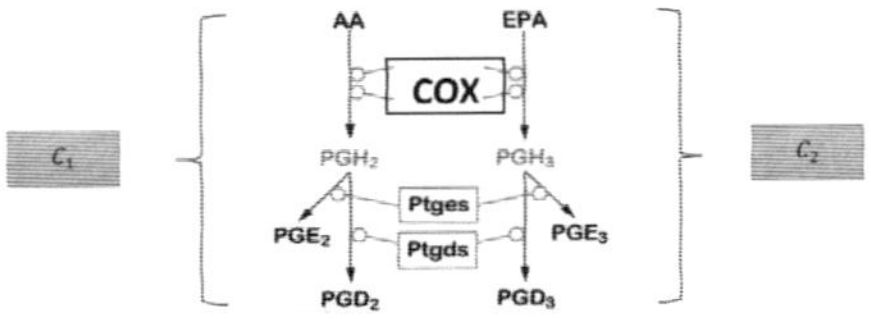

Figure: AA and EPA metabolic network. C_1 and C_2 denote the cytokines TNFα and IL-6 respectively.

120d Design of Diverse, Functional Mitochondrial Targeting Sequences across Eukaryotic Organisms Using Variational Autoencoders

Aashutosh Boob, Nilmani Singh, Xueyi Xue, Airah Zaidi, Li-Qing Chen, Huimin Zhao

A eukaryotic cell consists of a well-defined architecture of organelles carrying biochemically distinct cellular processes. These membrane-bound compartments provide nuclear encoded proteins a physiological context to carry out their function. Recently, utilizing the organelle environment and machinery has been of interest for applications in fabricating compartments for biochemical production and developing new methods of therapeutic intervention. The multifaceted contributions of mitochondria to cellular metabolism have made it an important target not only for metabolic engineering but also for disease treatment. However, only a few protein-localization tags have been characterized for mitochondrial localization even though the targeting efficiency is known to be influenced by the passenger protein. To address this limitation, we exploited generative models, an unsupervised machine learning framework to design a toolkit of mitochondrial targeting sequences (MTS) followed by in-vivo characterization in multiple organisms. A variational autoencoder was constructed to sample artificial sequences. A high fraction of generated peptide sequences (95%) were predicted to be functional by TargetP2.0. Furthermore, using UniRep model embeddings, clustering was implemented to assign an 'organism' label for in-vivo testing. The artificial sequences were added to the N-terminus of fluorescent proteins and confocal microscopy was performed for characterization. Success rate of 50%, 100% and 75% were obtained for *Saccharomyces cerevisiae*, HEK293 and *Nicotiana tabacum*, respectively. The functional sequences are highly diverse with normalized edit distance greater than 50% when compared to the MTS reported in UniProt database. Overall, these results suggest highly functional, diverse peptides capable of localizing proteins in mitochondrial matrix of diverse organisms can be designed. Currently, we are implementing this strategy for designing artificial peptides to perform protein localization in mitochondrial matrix of non-model organisms like *Rhodosporidium toruloides*. We also provide a list of potentially functional MTS for various organisms to expand the metabolic engineering toolbox for mitochondrial compartmentalization. This work should aid in enhancing titers of biochemicals, expressing functional heterologous proteins, and improving mitochondrial DNA editing and biomolecule delivery capabilities.

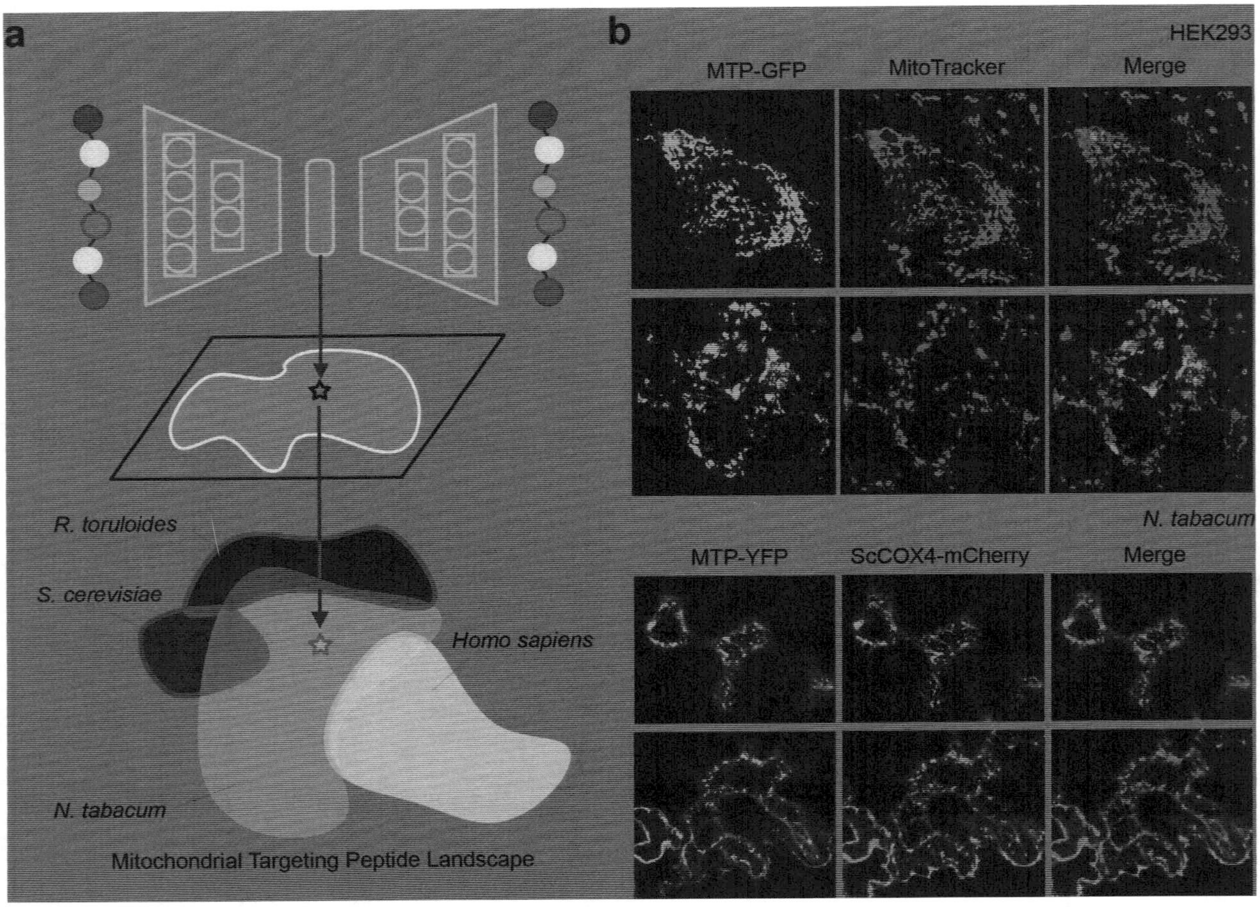
a
R. toruloides
S. cerevisiae
Homo sapiens
N. tabacum
Mitochondrial Targeting Peptide Landscape
b
HEK293
MTP-GFP
MitoTracker
Merge
N. tabacum
MTP-YFP
ScCOX4-mCherry
Merge

120e Next-Flux (Neural-net EXtracellular Trained Flux)

James Morrissey, Benjamin Strain, Gianmarco Barberi, Pierantonio Facco, Cleo Kontoravdi

Genome-scale modelling has emerged as an effective tool for the modelling of mammalian cell metabolism. Despite the acceleration of advancements and applications in the field of genome-scale modelling, there are many fundamental challenges to be addressed. Current mammalian genome-scale models (GeMs) are heavily underdetermined, and methods used to constrain the solution space are limited in depth and applicability. These constraints fail to guide flux in a biologically consistent manner, require expensive/unavailable data and are often based on a multitude of assumptions.

In this work we present a novel method for constraining Chinese hamster ovary (CHO) cell GeMs that exploits the link between extracellular and intracellular metabolomics to create biologically relevant bounds on intracellular fluxes. We train an artificial neural network (ANN) with a variety of published CHO cell extracellular metabolomics and corresponding ^{13}C-labelled intracellular fluxomic data to quantify the relationship between *easily measured* metabolites and *difficult to predict* fluxomics. The resulting model is used to predict intracellular flux distributions for unseen conditions, given a set of process-level experimental data (metabolite uptake rates, cell growth rate, and recombinant protein production rate). The model predicts the expected upper and lower bound of up to 47 intracellular fluxes across glycolysis, pentose phosphate pathway, TCA cycle and amino acid metabolism, which are then used to constrain a CHO cell GeM.

We show that this hybrid approach that combines stoichiometric modelling with data-driven constraints accurately predicts physiologically relevant metabolic states across a variety of validation conditions. Flux of key reactions is constrained within tight bounds, preventing infeasible solutions and guiding flux down appropriate pathways. The resulting intracellular flux distribution is more biologically consistent, improving the reliability of CHO cell GeMs in the application to industrial systems.

We present this method in an easy-to-implement package for the systems biology community to employ.

120f Dgpredictor: Automated Fragmentation Method for Gibbs Energy Change Prediction of Metabolic Reaction and *De Novo* Pathway Design

Vikas Upadhyay, Costas D. Maranas

Thermodynamic analysis of metabolic pathways is important in identifying novel pathways for biochemical synthesis. To this end, Group Contribution (GC) methods are widely applied for estimating the standard Gibbs energy change ($\Delta_r G'^{\circ}$) of enzymatic reactions using limited experimental measurements. However, drawbacks of such methods arise from their dependence on manually curated functional groups and inability to include stereochemical information; these obstacles lead to limited reaction coverage. Here, we present a moiety-based automated fragmentation method using molecular fingerprints to design a thermodynamic analysis tool called dGPredictor. It allows the ability to include stereochemistry within chemical structures and thus broadens biochemical reaction coverage. dGPredictor shows comparable accuracy as the current GC methods and, in particular, can capture Gibbs energy change for reactions that only undergo stereochemical changes, such as isomerase and transferase reactions, which show no overall group change. We apply dGPredictor to predict the Gibbs energy change for reactions involving novel structures and integration with *de novo* metabolic pathway design tools such as novoStoic to prohibit reactions steps for which the directionalities are thermodynamically infeasible. We also demonstrate a graphical user interface for dGPredictor that facilitates easy access for predicting Gibbs energy change of reactions at different pH and ionic strengths. dGPredictor allows customized user input of molecules as KEGG IDs or as InChI strings (for novel metabolites) and is open source (https://github.com/maranasgroup/dGPredictor). We find that dGPredictor improves the goodness of fit of the linear regression to find the moiety contribution over the widely used GC method by 78.8% (i.e., Mean Squared Error (MSE) over training data from the TECRDB database). Finally, dGPredictor also increases the coverage of $\Delta_f G'^{\circ}$ and $\Delta_r G'^{\circ}$ estimation for metabolites and reactions present in the KEGG database by 17.2% and 102%, respectively, over GC by allowing for stereochemical considerations not captured in previous expert-defined chemical groups.

120g Multidimensional Resource Allocation Predicts Contrarian Reverse Diauxie Phenotype in *Pseudomonas Aeruginosa*

S. Lee McGill, Yeni Yung, Kristopher A. Hunt, Luke Hanley, Ross P. Carlson

Pseudomonas aeruginosa is a globally-distributed bacterium with many industrial applications and often found in medical infections. The bacterium does not utilize a classic diauxie phenotype, as defined by model microorganisms, nor does it follow common systems biology assumptions including preferential consumption of glucose with an 'overflow' metabolism. Despite these contradictions, *P. aeruginosa* is competitive in many, disparate environments underscoring knowledge gaps in microbial ecology and systems biology. Physiological, omics, and *in silico* analyses were used to quantify the *P. aeruginosa* carbon catabolite repression strategy known as 'reverse diauxie'. An ecological basis of reverse diauxie was identified using a genome-scale, metabolic model interrogated with *in vitro* omics data. Reverse diauxie preference for lower energy, nonfermentable carbon sources, such as acetate or succinate over glucose, was predicted using a multidimensional strategy which minimized resource investment into central metabolism while completely oxidizing substrates. Application of a common, *in silico* optimization criterion, which maximizes growth rate, did not predict the reverse diauxie phenotypes. This study quantifies *P. aeruginosa*metabolic strategies foundational to its wide distribution, usefulness as a biocatalyst, and virulence including its potential for mutualistic interactions with microorganisms found commonly in the environment and in medical infections.

500a Cell-Free Biosynthesis to Inform Metabolic Engineering of Non-Model Bacteria

Blake J. Rasor, Ashty S. Karim, Michael Koepke, Michael C. Jewett

The accelerating climate crisis requires sustainable technologies, such as biological systems, to reduce greenhouse gas emissions across industrial sectors. A promising approach has emerged in pursuit of this goal using *Clostridium autoethanogenum* as a chassis for biological carbon capture. Although this anaerobic, non-model bacterium natively produces ethanol from carbon monoxide, efforts to expand the product scope have been limited due to slow growth and a lack of well-developed genetic tools. To facilitate metabolic engineering in *C. autoethanogenum* to produce value-added products from waste gases, we implemented cell-free biosynthesis for metabolic pathway prototyping. Cell-free systems comprising crude cell extracts enable the exploration of biochemical conversions in the absence of growth and viability constraints with greater throughput and flexibility than living cells, which facilitates rapid design-build-test cycles. We have demonstrated the production of diverse metabolites in cell-free systems derived from *E. coli* with enzymes expressed either during cell growth or directly in the cell extract, which enables prototyping of heterologous biosynthetic enzymes and competing endogenous pathways to maximize flux to the desired product. This cell-free prototyping approach accelerated the engineering of *C. autoethanogenum* by identifying multiple gene knockouts that resulted in highly productive strains converting greenhouse gases to acetone and isopropanol. Such iterative prototyping and strain optimization would have taken at least 12 months longer using *in vivo* methods alone. Continuous culture of engineered acetone production strains at industrial pilot scale produced up to 3 g/L-h while demonstrating significant greenhouse gas savings compared to conventional processes. Together, cellular and cell-free metabolism offer a promising path toward the rapid development of efficient and sustainable biosynthesis platforms.

500b Engineering *Vibrio Natriegens* for Degrading and Assimilating Poly(ethylene terephthalate)

Tianyu Li, Nathan Crook

Background

Poly(ethylene terephthalate) (PET), an extensively used and manufactured plastic, has become a major global concern due to its accumulation and resistance to degradation in the natural environment. Current strategies for handling waste PET, such as mechanical and chemical recycling, have been applied to alleviate PET pollution. However, some recycling processes can potentially cause secondary pollution by consuming massive amounts of energy. Moreover, the amount of recycled PET is quite limited and certainly much less than the amount of PET that ends up in the landfill. Consequently, the majority of PET waste is further broken down into small pieces during the landfilling process and then dispersed in the environment. These plastic pieces with sizes smaller than 5 mm are classified as microplastics. The leakage of microplastics into the ocean can severely threaten marine lives and ecosystems, and further jeopardize human health through the food web. Our project focuses on developing a remedy for PET microplastics by applying metabolic engineering to construct a whole-cell biocatalyst for degrading and assimilating waste PET *in situ*. *Vibrio natriegens*, a fast-growing, nonpathogenic, salt-tolerant marine bacteria, is the ideal chassis to achieve this bioremediation purpose.

Methods and Results

Firstly, we engineered *V. natriegens* to express PET-degrading enzymes. Previous studies revealed several microorganisms and enzymes that hydrolyze PET. Among them, a two-enzyme system composed of PETase and MHETase was discovered in *Ideonella sakaiensis* and was demonstrated to degrade PET efficiently at ambient temperature. We transferred the *I. sakaiensis* PETase and MHETase into *V. natriegens* after codon optimization, and fused PETase to a surface anchor, successfully constructing a *V. natriegens* that displays PETase on its outer membrane. We verified the surface display by using immunocytochemistry and demonstrated the ability of the displayed PETase to degrade the PET monomer bis(2-hydroxyethyl) terephthalate (BHET) using liquid chromatography.

After hydrolysis by PETase and MHETase, PET is broken down into terephthalic acid (TPA) and ethylene glycol (EG). Therefore, we constructed a metabolic pathway for assimilating TPA in *V. natriegens*. TPA is commonly metabolized via enzymes that convert TPA into protocatechuic acid (PCA), followed by further catabolism through the benzoate pathway. *V. natriegens* proved to be an ideal chassis for this project as it demonstrated cell growth in minimal media with PCA as the sole carbon source. We will present our efforts to enable TPA catabolism in *V. natriegens* by first prototyping the necessary transporters and metabolic enzymes in *E. coli,* demonstrating growth of *E. coli* using TPA as a sole carbon source for the first time.

Finally, we will describe our efforts to apply directed evolution for improving the surface display, PETase, and TPA degradation efficiency. For this purpose, we will describe a combination of error-prone PCR and *in vivo* mutagenesis (using temperate phage), followed by fluorescence-activated cell sorting and growth-based selections. The ability of these evolved strains to grow on PET as a sole carbon source in seawater media will also be presented.

Implications

Our project proposes a sustainable *in situ* bioremediation strategy for PET microplastic pollution. The engineered *V. natriegens* with the two-enzyme system may be able to perform as a whole-cell catalyst that can be easily separated or recycled by centrifugation and proliferated by cultivation so that it can overcome the major limitation of enzymatic PET recycling - product inhibition. Our research also contributes to the advancement of synthetic biology by expanding the genetic toolkit of *V. natriegens*, an under-characterized bacterium with superior growth rates compared with *E. coli* and yeast. Specifically, we have enabled surface display of heterologous proteins by *V. natriegens* for the first time. Taken together, this work expands the knowledge about engineering *V. natriegens*, and establishes a promising strategy for mitigating PET accumulation in seawater.

500e Integrative Omics Analysis of *Yarrowia Lipolytica* for the Bioconversion of Lipid-Based Substrates

Alyssa M. Worland, Zhenlin Han, Jeffrey Czajka, Yinjie Tang, Wei Wen Su

Motivation

Major cornerstones for modern sustainability are centered around the reduction of waste and the conservation of world resources. The development of microbial cell factories has been leveraged to provide a solution to both concerns through synthetic biology. *Yarrowia lipolytica* has emerged as an important host for many industrial applications, such as production of organic acids and terpenoid compounds. Equally, it is also desired for its ability to metabolize hydrophobic substrates, like alkanes and lipids, and for its high chemical tolerance. The US alone generates billions of pounds of waste oils each year which are often overlooked as substrates for microbial biorefineries, and offer a potentially cheaper alternative to carbohydrate based feedstock. While the bioconversion of sugars has been widely studied in *Y. lipolytica*, its potential for catabolizing lipids is less explored. Due to this, there are critical knowledge gaps in *Y. lipolytica* lipid catabolism, including NADPH generation routes and flux compartmentalization of key metabolites. This study utilizes integrative omics analysis to address the critical knowledge gaps in *Y. lipolytica* lipid metabolism in order to enable effective metabolic and process engineering strategies.

Methods

Omics analysis of *Y. lipolytica* W29 was performed to provide insights into the cellular regulation of lipid assimilation. Namely, transcriptomics, metabolomics, and fluxomics were used for this study. RNA sequencing data was collected and compared for cells grown on glucose, glycerol, and oleic acid. Free metabolite analysis evaluated the metabolic pool sizes of cells grown on different substrates. Moreover, steady-state ^{13}C metabolic flux analysis (MFA) was performed. Specifically, cells were grown in positionally labelled substrates (1,2-^{13}C glucose; 1,3-^{13}C glycerol; 1,2,3,7,8-^{13}C oleic acid) and the resulting amino acid and free metabolite labelling profiles were collected and used for flux calculation. MFA was conducted using WUFlux [1] to simulate mass isotopomer distribution (MID) values using the input carbon transition matrices of pathway reactions. The objective function minimized the sum of square residuals between the experimentally measured and computationally simulated MIDs. Compartment-specific amino acid generation was used to constrain fluxes for parallel metabolic reactions in multiple compartments. In addition, biomass composition analysis was conducted to inform the biomass equation used to fit the model and illuminate differences in bulk macromolecule distribution within the cell.

Results

^{13}C-MFA results showed the flux distribution response to oleic acid differs significantly from the flux results of glucose and glycerol metabolism. For instance, *Y. lipolytica* grown on oleic acid show higher activity in the TCA cycle than in the pentose phosphate or glycolysis pathways. Following β-oxidation, the generated acetyl-CoA enters into the TCA cycle or contributes to the process of lipogenesis. The flux results were used to constrain the genome scale model (GSM) for *Y. lipolytica* to provide an overall picture of the metabolism when oleic acid is used as the carbon source. Additionally, time-course dynamics of metabolite isotopologues were examined during lipid catabolism [2]. The isotopologue profiles suggests a segregated metabolic network in *Y. lipolytica* during growth on lipids (i.e. glycerol and fatty acid co-utilization). Biomass composition analysis shows protein content significantly reduced in oil and fatty acid conditions, however amino acid composition remains consistent. Transcription data analysis shows the differentially expressed genes between glucose and oleic acid-grown cells, pointing to unique genetic regulations in oleic acid-grown cells. Furthermore, metabolomics was performed on *Y. lipolytica* cultured on different substrates. Metabolite pool size data provide further insights into cell metabolic regulation. Based on metabolic analysis, strain engineering was performed to improve the productivity of β-carotene and astaxanthin producing strains.

Implications

Understanding the changes in cellular networks and regulation of *Y. lipolytica* during growth on lipid substrates provides insights to help establish this oleaginous yeast as an industrial host for the bioconversion of waste oils. This study evaluated the ability of *Y. lipolytica* to grow on lipid-based substrates and analyzed the cellular response at multiple levels. An important application of lipid catabolism in *Y. lipolytica* is for the biosynthesis of terpenoid compounds [3]. Terpenes are derived from the native mevalonate pathway, which is powered by acetyl-CoA. Catabolism of fatty acids can introduce an abundance of acetyl-CoA needed to power the mevalonate pathway and increase production of terpene compounds. This knowledge found from this work provides insights which can be applied to genetic engineering and process development strategies to use *Y. lipolytica* as a host for a microbial biorefinery generating value-added products from waste oils.

This work was supported by the USDA NIFA (award number 2020-67022-31146).

References

1. He L, Wu SG, Zhang M, Chen Y, Tang YJ. WUFlux: an open-source platform for ^{13}C metabolic flux analysis of bacterial metabolism. BMC Bioinformatics. 2016; 17(1):444
2. Worland AM, Czajka JJ, Xing Y, Harper WF, Moore A, Xiao Z, Han Z, Wang Y, Su WW, Tang YJ. Analysis of Yarrowia lipolytica growth, catabolism, and terpenoid biosynthesis during utilization of lipid-derived feedstock. Metabolic Engineering Communications, Volume 11, 2020, e00130
3. Worland AM, Czajka JJ, Li Y, Wang Y, Tang YJ, Su WW. Biosynthesis of terpene compounds using the non-model yeast Yarrowia lipolytica: grand challenges and a few perspectives. Current Opinion in Biotechnology, Volume 64, 2020, Pages 134-140

500f Investigating the Role of pBBR1's Mobilization Protein in Plasmid Maintenance in Non-Model Bacteria

Mark Kathol, Cheryl Immethun, Dianna Morris, Rajib Saha

Stable plasmid maintenance is crucial to accurate characterization and testing of biological devices. Model bacteria are often transformed with plasmids to utilize biochemical capabilities they do not naturally possess, and the mechanisms for replication of plasmids in these model bacteria are well understood. However, the plasmid maintenance mechanisms for non-model bacteria are not as well known. The mobilization gene found in the plasmid BBR1, Mob, has shown to be crucial for plasmid stability in *Rhodopseudomonas palustris CGA009* (hereafter *R. palustris*). *R. palustris* is an extremely metabolically versatile bacterium that is capable of all four modes of metabolism, catabolizes recalcitrant feedstocks such as lignin monomers, produces hydrogen and the industrially relevant bioplastic, polyhydroxybutyrate (PHB), and possesses both carbon dioxide and nitrogen fixation abilities. Strangely, BBR1 has been successfully employed in *Paraburkholderia sacchari* LMG 19450 LFM101 (hereafter *P. sacchari*), a pentose and hexose consuming bacterium, without the presence of Mob. Mobilization proteins facilitate horizontal gene transfer between two different bacteria through the Type IV secretion system by nicking DNA before conjugation and rejoining the DNA afterwards. A previously annotated mobilization protein from the streptococcal plasmid pMV158 has shown to have a very similar amino acid sequence to Mob. The nicking behavior of the pMV158 mobilization protein was severely impaired by replacing catalytic histidines and other active site residues. Using analogous methods, the effects of mutation on the active site residues of Mob will be explored in *R. palustris* and *P. sacchari*. This study will provide groundwork to engineering non-model bacteria and harnessing their unique biochemical capabilities.

500g Development of Non-Model Microbes As Chassis Organisms for Bioconversion

Adam Guss

Many non-model microbes have native phenotypic advantages over model organisms like *E. coli* and *S. cerevisiae*, such as the ability to catabolize polymeric biomass feedstocks, extreme tolerance to various stressors, or high flux through complex metabolic pathways. Unfortunately, these non-model microbes typically lack the genetic tools that would enable further modifications and metabolic engineering. We routinely improve transformation efficiency or demonstrate first-of-kind transformation in phylogenetically diverse bacteria by evading restriction systems and then use in-house libraries of genetic parts to enable gene deletion and heterologous expression. Beyond initial transformation technologies, we have also developed a high efficiency DNA integration tool called SAGE (Serine-recombinase Assisted Genome Engineering) using site-specific DNA recombinases. This system uses a suite of orthogonal site-specific DNA recombinases to enable rapid, stable, and iterative insertion of heterologous DNA into the chromosome of the target host. This allows rapid screening of both genetic parts (e.g., promoters, terminators, insulators, reporter genes) and metabolic pathways in diverse organisms. Because of the high efficiency, these recombinases also enable combinatorial libraries to be constructed in vivo for pathway optimization. Together, these approaches allow the rapid development of non-model microbes into bioengineering platforms for the conversion of plant biomass and waste feedstocks like plastics into the next generation of fuels and chemicals.

553a Complete Biosynthesis of Diverse Plant-Derived Styrylpyrones in *Saccharomyces Cerevisiae*

Yinan Wu, Maple Chen, Sijin Li

Plant styrylpyrones exerting well-established neuroprotective and other medicinal properties have attracted increasing attention in recent years. To meet the globally growing demand, microbial biomanufacturing bypassing the time-consuming agricultural processes provides a sustainable and efficient approach to produce the valuable plant styrylpyrones as daily supplements or potential drugs complementary. In this study, we firstly demonstrated the heterogenous biosynthesis of two 7,8-saturated styrylpyrones (7,8-dihydro-5,6-dehydrokavain (DDK) and 7,8-dihydroyangonin (DHY)) and two 7,8-unsaturated styrylpyrones (desmethoxyyangonin (DMY) and yangonin (Y)), in *Saccharomyces cerevisiae.* Although plant styrylpyrone biosynthetic pathways have not been fully elucidated, we functionally reconstructed the recently discovered core kava styrylpyrone biosynthetic pathway that has high substrate promiscuity and combined it with upstream hydroxycinnamic acid biosynthetic pathways to produce diverse plant-derived styrylpyrones without the native plant enzymes. We optimized the *de novo* pathways by engineering yeast endogenous aromatic amino acid metabolism and endogenous double bond reductases and by CRISPR-mediated δ-integration to overexpress the rate-limiting pathway genes. These combinatorial engineering efforts led to the first three yeast strains that can produce diverse plant-derived styrylpyrones *de novo*, with the titers of DDK, DMY and Y at 4.40 μM, 1.28 μM and 0.10 μM, respectively. This work will lay the foundation for larger-scale styrylpyrone biomanufacturing and the complete biosynthesis of more complicated plant styrylpyrones.

553b Optimizing Campesterol-Producing Yeast Strains for Enhanced Functional Reconstitution of Plant Membrane Enzyme

Shanhui Xu, Yanran Li

Campesterol is an important phytosterol regulating membrane properties, which is also the precursor of many phytosteroids of pharmaceutical values. In our previous work, we have established a set of engineered yeast strains accumulating free campesterol. However, the growth of the engineered campesterol-producing strain is deteriorated by the inactivation of the endogenous sterol acyltransferases, which impedes further application of the engineered strain to synthesize downstream phytosteroids of higher economic values. In this study, the growth of campesterol-producing yeast strain was optimized by restoring and tuning sterol esterification and engineering upstream FPP supply. The CRISPR-Cas9 library for inactivation and upregulation of genes of interests at genome-scale was establish and applied in the strain. Strains with enhanced growth was selected. Combination of optimization from different strategies resulted in improved growth and higher production of campesterol, which serves as the foundation to reconstitute the diverse phytosteroid structures. In addition, the campesterol-accumulating yeast strain is a preferred host for the heterologous reconstitution of plant cytochrome P450s, which exhibits higher enzymatic efficiency than expressed in wildtype yeast.

553c Metabolic and Epigenetic Engineering Enables Temporal Control of Taxane Metabolism in *Taxus Chinensis* Plant Cell Culture

Cassandra Brzycki Newton, Alexandra Harrison, Lauren Revene, Jay Gandhi, Sean Horton, Eric Young, Susan Roberts

Many medicinal plants have developed unique biosynthetic capabilities that make production of specialized metabolites in plant cell culture (PCC) an attractive alternative to extensive engineering of model hosts. However, rational engineering of PCC systems is limited by both a lack of robust metabolic engineering tools and temporal downregulation of secondary metabolism in continuously subcultured cell lines. To this end, we have developed two engineering strategies that can be utilized in parallel to achieve stable, high titers of the chemotherapeutic paclitaxel in *Taxus chinensis* PCC. First, using a pathway overexpression approach, four genes identified as potential rate-limiting steps in paclitaxel biosynthesis (TASY, DBAT, BAPT, and DBTNBT) were constitutively overexpressed in *Taxus chinensis* cell lines, resulting in up to a 9.9-fold increase in paclitaxel titer and differential accumulation of the pathway intermediates 10-deacetylbaccatin III (10-DAB) and baccatin III. These genes remained overexpressed and stably integrated in the genome even after 3 months of continuous subculturing. Second, in previous work we have demonstrated that increased DNA methylation of promoters, transcription factors, and other regulatory elements in continuously subcultured cell lines results in global downregulation of secondary metabolism, which can be rescued with DNA methyltransferase inhibitors. Thus, to enable finely tuned epigenetic control of secondary metabolism, we have developed CRISPR/dCas9-based tools for targeted methylation and demethylation of secondary metabolite biosynthesis genes using the catalytic domains of the *Nicotiana tabacum* DNA methyltransferase NtDRM1 and the human DNA demethylase TET1, respectively. These tools were applied to epigenetically downregulate PAL, which controls phenylpropanoid metabolism and competes with taxane biosynthesis, and upregulate BAPT, a gene in the taxane biosynthetic pathway hypothesized to be controlled by DNA methylation. This work is easily generalizable to other PCC systems and builds the foundation for use of rational epigenetic manipulation of metabolism as a tool for metabolic engineering.

553d Engineering Nucleotide Sugar Metabolism in *S. Cerevisiae* for the Glycosylation of Complex Natural Products and Heterologous Proteins

Samantha Crowe, Yuzhong Liu, Fei Gan, Jay Keasling

Glycosylation is an important modification step in cells that affects the bioactivity and stability of secondary metabolites, as well as the localization and function of proteins. *S. cerevisiae* has the proven ability to produce complex heterologous products and proteins but lacks the necessary precursors for glycosylation because yeast natively only produces a few of them including UDP-glucose and UDP-galactose. Nucleotide sugars in the form of UDP-sugars are the most common substrates for glycosylation in plants, the microbial biosynthesis of which can be achieved *via* the expression of nucleotide sugar interconversion enzymes (NSEs) that are native of plants. As such, UDP-glucose can also be converted to UDP-glucuronic acid, UDP-galactose, UDP-rhamnose, etc. in yeast. However, nucleotide sugar metabolism is highly regulated by many UDP-sugars inhibiting different NSEs, with UDP-xylose being a particularly important inhibitor of many enzymes involved in nucleotide sugar synthesis in order to maintain a high level of UDP-glucose in the cell. In this work, the heterologous production of various nucleotide sugars in *S. cerevisiae* as well as effective means to control their relative production levels with different promoters driving the expression of NSEs, as well as metabolic efforts to alleviate the inhibitory feedback caused by UDP-xylose. This work will pave the way for engineering *S. cerevisiae* to produce bioactive and complex glycosylated products with importance in food and pharmaceutical products.

553e Microbial Conversion of Vegetable Oils into Omega-3 Oils

Jiansong Qin, Na Liu, Ya-Hue V. Soong, Elif Kurt, Ashley Willan, Qiuyan Chen, Dongming Xie

Omega-3 fatty acids, including eicosapentaenoic acid (EPA C20:5) and docosahexaenoic acid (DHA C22:6), are long-chain unsaturated fatty acid with great benefits in heart health, improving immune function, and cognition development. Recent research also shows that omega-3 fatty acids can help patients recover from COVID-19. However, human body cannot directly synthesize omega-3 fatty acids, people need to obtain these long-chain unsaturated fatty acids from their daily diet. The traditional source of DHA and EPA are obtained from fish oils via an extraction process, whose supply is subject to a big challenge due to the uncertainty caused by overfishing and the concern of contaminations in ocean. Biomanufacturing of omega-3 fatty acids by microorganisms are considered a more sustainable and green approach to produce healthier and higher-quality omega-3 fatty acids. Recently, DuPont has developed a technology that uses metabolically engineered *Yarrowia lipolytica* to produce omega-3 EPA from glucose. However, the manufacturing cost needs to be further reduced due to the low conversion yield. Here we report a novel strategy for high-yield production of omega-3 fatty acids from vegetable oils. The metabolic pathways and fermentation strategies for direct conversion of canola oil into omega-3 EPA are investigated and optimized, which include (1) acceleration of oil substrate uptake by increasing lipase activity, (2) enhancing intracellular omega-3 EPA formation and accumulation *via* pathway engineering and optimization in fatty acid synthesis and metabolism, (3) minimizing lipid leaking from the yeast cells, and (4) fermentation engineering to avoid limitations in mass transfer and mixing. We have demonstrated more than 50% improvements in omega-3 EPA production when the canola oil is used to partially or completely replace glucose as the carbon source. To our best knowledge, this is the first research reporting a direct and efficient microbial conversion of a plant oil into a much more valuable omega-3 oil at significantly higher titer and yield.

Key words: *Y. lipolytica,* metabolic engineering, fermentation engineering, omega-3 fatty acids, eicosapentaenoic acid (EPA), vegetable oil.

553f Pathway Localization and Redox Engineering of Yarrowia Lipolytica for Fatty Alcohol Production

Siva Somasundarama, Michael Spagnuolo, Mark Blenner

Fatty alcohols currently find its use in areas such as surfactants, plasticizers, lubricants, fuels, and the cosmetics industry; however, traditional production methods rely on non-renewable petroleum-derived compounds or the low-yield harvest from plants. Recently, conventional hosts including *E. coli* and *S. cerevisiae* have been used for fatty alcohol production with some success. Oleaginous yeast, such as *Yarrowia lipolytica*, offer significant advantages for the production of oleochemicals, as its native metabolism evolved to high flux in fatty acid biosynthetic pathway. Fatty alcohol production in the cytosolic space is hindered by low availability of acyl-CoA and presence of competing metabolic pathways. As a result, we have targeted fatty alcohol production in the peroxisome, where acyl-CoA flux is directed during beta-oxidation, and potential fewer competing pathways. Several Fatty Acyl-CoA Reductases (FAR) from bacterial and mammalian sources were screened using canonical peroxisome targeting sequences, as well as used enzymes fusion strategy to physically localize FAR next to 3-ketoacyl-CoA thiolase (3KAT) in peroxisome. 3KAT fusion resulted in nearly double the titer of fatty alcohol irrespective of any FAR's overexpression. The nature of the linker attached between 3KAT and FAR had a significant impact on the titer. Engineering of peroxisome physiology had moderately improvement of titer. In addition, we addressed the cofactor availability in the peroxisome by overexpression of auxiliary enzyme IDP3 and NADH kinase to provide an additional co-factors supply for FAR, which had a large impact on fatty alcohol titer as well as free fatty acid concentration. The highest titer achieved was over 1.6 g/L of fatty alcohol in shake flask culture.

594a Engineering Medium Chain Fatty Acid Production in *Synechococcus Sp.* PCC7002

Joshua P. Abraham, Baltazar E. Zuniga, Miles N. Crockett, Hawkins Shepard, Jody C. May, John McLean, Jamey Young, Brian F. Pfleger

Sustainable production of target chemicals in microbes is limited by the need for plant-derived sugar feedstocks and freshwater requirements. These processes compete for arable land and resources that could otherwise be used for food production. The model cyanobacterium *Synechococcus sp.* PCC 7002 (henceforth PCC7002) is an attractive photoautotrophic host that grows in brackish waters, removing the requirement for an external sugar feedstock and freshwater. Medium chain fatty acids (MCFAs) are an important class of chemical precursors for fuels, pharmaceuticals, detergents, and personal care products but are typically derived from unsustainable coconut and palm oils. Competitive production of MCFAs in heterotrophic organisms like *Escherichia coli* has been enabled by protein engineering of thioesterase enzymes that terminate the cycle of fatty acid biosynthesis (FAB).

In this project, we generated analogous strains of PCC7002 that overexpress these thioesterase enzymes and produce up to 860 mg/L of octanoic acid (C8). We also characterize growth effects of various organic overlays on PCC7002 and demonstrate their functionality to partition free fatty acids from aqueous media and subsequently increase titers. Like other cyanobacteria, PCC7002 lacks functional β-oxidation and has a naturally low flux through its lipid pathways, making oleochemical pathway engineering in PCC7002 unique from other microbes. To enhance titers further, we seek to identify the flux and genetic bottlenecks of our FAB-enhanced strains using untargeted metabolomics, ^{13}C-based isotope nonstationary metabolic flux analysis (INST-MFA), and CRISPR interference. This work provides a foundation for engineering cyanobacterial lipid metabolism and can be extended to produce oleochemicals with other functionalities, such as fatty alcohols and triacylglycerols.

594b Bioproduction of Aromatic Amines Enabled By Stress-Resistant Bacteria

Cholpisit Kiattisewee, Ian Faulkner, Jesse Zalatan, James Carothers

Pseudomonas putida is a potential chassis for bioproduction due to its high tolerance against aromatic compounds, such as *p*-aminocinnamic acid (pACA) – a precursor for high-performance polymers synthesis. In this work, we demonstrated that pACA can be produced in the non-model bacteria *P. putida*. Biosynthesis of pACA was previously challenging in *E. coli* due to its inherent toxicity. With optimization of enzymes of choice, production of pACA directly from glucose can be improved from a trace-level to more than 100 μM. We then fine-tuned enzyme expression number using CRISPRa/i transcriptional regulation and plasmids with varying copy-number, and therefore improved the pACA titer to 500 μM. Further investigation at different cell density suggested that conversion of pACA peaked later at the stationary stage when the other competing aromatic analogs were consumed. This finding illustrates the potential of the transcriptional CRISPR tools for tuning of gene expression level, which could provide a new methodology to optimize metabolic engineering processes in non-model bacteria.

594c Genomic Engineering of *Escherichia coli* for Improved Aromatic Aldehyde Stability

Neil Butler, Shelby Anderson, Roman Dickey, Michaela Jones, Ishika Govil, Aditya Kunjapur

Aromatic aldehydes have notable applications as flavors, fragrances, reactive intermediates, and potential residues for click chemistry conjugations. Metabolic routes to aldehyde synthesis provide a green chemical alternative to petrochemical synthesis or costly native extraction methods, but microbial methods to aldehyde synthesis in *E. coli* have been limited by instability of aromatic aldehydes in a cellular context. In our prior work, we addressed the rapid reduction of most aromatic aldehydes in *E. coli* by creating the reduced aromatic aldehyde reduction (RARE) strain, which eliminated most of the reductive instability in aldehydes via genetic knockouts of aldo-keto reductases, alcohol dehydrogenases, and the regulator gene *yqhC*. However, for many aromatic aldehydes of commercial or synthetic interest, the RARE strain alone is insufficient for stability due to the oxidative activity of native aldehyde dehydrogenase enzymes. In this work, we have engineered an *E. coli* strain with knockouts of 6 key aldehyde dehydrogenase enzymes performed rapidly using multiplexed automated genome engineering (MAGE) resulting in further stability of a range of aromatic aldehydes over 24h in *E. coli* culture conditions. In addition, further genetic engineering using MAGE was utilized to expand aldehyde stabilization for specialized *E. coli* strains, namely the recoded C321.ΔA strain utilized for genetic code expansion. Here, we performed MAGE to create aldehyde stabilization knockouts, enabling the use of recoded C321.ΔA for use with normally unstable aldehydes. Using this strain, we incorporated an aldehyde-containing non-standard amino acid (nsAA) into proteins. Further use of this strain will enable incorporation of conjugatable, aldehyde-substituted nsAAs and stabilization of aldehydes for biosynthetic pathways relevant to genetic code expansion.

594d Screening of Norbaeocystin Methyltransferase Variants Enables Enhanced Psilocybin and Baeocystin Production in *E. coli*

Madeline McKinney, J. Andrew Jones

While psilocybin is currently under clinical evaluation for the treatment of severe depression, post-traumatic stress disorder (PTSD), and addiction, anecdotal evidence from recreational users has led some to postulate that the ratio of naturally occurring psychoactive metabolites in various mushroom species may greatly impact the experience and overall effect on the brain. This "Entourage Effect" may necessitate a specific production strain capable of high baeocystin production to meet necessary pharmaceutical demand. Norbaeocystin methyltransferase (psiM) is responsible for the iterative methylation of norbaeocystin, resulting in the production of psilocybin via a monomethylated intermediate, baeocystin. The genetic variation between mushroom species and resulting differences in their metabolite profile, have motivated this study to evaluate a range of psiM variants for their ability to selectively produce the intermediate product baeocystin. Here, we examine the effect of genetic variability of the psiM gene and attempt to create a biosynthetic production platform for producing high titers of baeocystin and higher titers of psilocybin than previously reported. Screening of genes from *Psilocybe cubensis*, *Psilocybe cyanescens*, *Panaeolus cyanescens* and *Gymnopilus dilepis* in a high throughput well plate assay yielded a production strain capable of producing more than 480 mg/L of psilocybin prior to fermentation optimization and scale up. This surpasses the current best performing psilocybin production strain by an increase of 245% under identical conditions. Additionally, initial screening produced a range of strains with enhanced baeocystin selectivity, with individual titers as high as 124 mg/L. Ongoing efforts to validate, optimize, and scaleup multiple lead strains is currently underway. This work further enhances the viability of industrial batch production of psilocybin using fermentation and provides a starting point for the development of strains capable of selectively producing its precursors for combined medicinal use.

594e Natural Product Synthesis from Lignin Aromatics Using *Acinetobacter Baylyi* ADP1

Bradley Biggs, Keith Tyo

Two major focuses in sustainable chemical synthesis are reducing the environmental impact of individual processes and minimizing or recovering waste materials. Biological upgrading of lignin biomass can advance both goals. Lignin has proved a recalcitrant feedstock, but biological approaches can leverage the native metabolism of certain microbial hosts to take complex and heterogeneous feedstocks and simplify them to tractable starting points for upgrading. Recently, we demonstrated that one of these hosts, *Acinetobacter baylyi* ADP1, is highly advantageous for these applications owing to engineerability and capacity undergo rapid design-build-test-learn cycles. Here, we use these traits and ADP1's native metabolism to covert mock alkali pretreated liquor (lignin) to two highly valuable natural products, vanillin-glucoside and resveratrol, creating strains with up to 22 genetic modifications including up to 8 heterologously expressed enzymes. In our approach we leverage preexisting aromatic species (vanillate, ferulate, and p-coumarate) to create shorter routes to these products, ultimately demonstrating this host's potential as a platform for upgrading lignin derived aromatics waste streams to valuable goods.

594f The Metabolic Engineering of *E. coli* for N-Glycolyl Chondroitin Production

Adeola Awofiranye, Ke Xia, Sultan N. Baytas, Robert J. Linhardt, Mattheos Koffas

In this study, we synthesized a derivative of chondroitin sulfate called N-glycolyl chondroitin sulfate (Gc-CS). This derivative is very valuable because it has medical and evolutionary applications. Gc-CS is hypothesized to be able to detect diseases like carcinomas, atherosclerosis, etc. This is because, according to recent studies, it was found to be a metabolite of one of the sialic acids that cause these diseases in humans. N-glycolylneuraminic acid (Neu5Gc) is a sialic acid biosynthesized from the hydroxylation of N-acetylneuraminic acid (Neu5Ac). The enzyme responsible for this hydroxylation is CMP-Neu5Ac hydroxylase encoded by the *CMAH* gene. During evolution, humans lost this gene and, therefore, cannot synthesize Neu5Gc. Although humans cannot synthesize Neu5Gc, it can be incorporated through exogenous sources such as eating red meat. Since humans don't synthesize this sialic acid, the body recognized this as an antigen, and it recruits antibodies. This interaction triggers an inflammation that affects epithelial and endothelial tissues. To test this correlation between red meat consumption and the development of diseases in humans, a more stable metabolite of Neu5Gc called N-glycolyl chondroitin sulfate (Gc-CS) can be used. We have been able to synthesize chondroitin with this glycan modification using metabolically engineered *E. coli* K4 adapted for chondroitin production. The bacteria were fed with a glucose carbon source supplemented with chemically synthesized N- glycolyl glucosamine, and this allowed the incorporation of N- glycolyl into chondroitin. More recently we are focusing on the metabolic engineering of the *E. coli* strain to increase N-glycolyl incorporation into the cell. After exploring the pathway for chondroitin sulfate synthesis, we explored two avenues to shift the metabolic flux towards N-glycolyl incorporation. We explored the effect of downregulating the bifunctional N-acetylglucosamine-1-phosphate uridyltransferase and glucosamine-1-phosphate acetyltransferase gene (GlmU) and the Glutamine-fructose-6-phosphate aminotransferase gene (GlmS).

594g Inducible Directed Evolution of Multigene Pathways in E. coli

Nathan Crook

Directed evolution is a powerful method for engineering biology in the absence of detailed sequence-function relationships. To enable directed evolution of complex phenotypes encoded by multigene pathways, we require large library sizes for DNA sequences >5-10kb in length, elimination of genomic hitchhiker mutations, and decoupling of diversification and screening steps. To meet these challenges, we developed Inducible Directed Evolution (IDE), which uses a temperate bacteriophage to package large plasmids and transfer them to naive cells after intracellular mutagenesis. Here, I will discuss the development of IDE, as well as its ability to quickly achieve improvements to several multigenic phenotypes, including carbon catabolism, pH tolerance, and biochemical production.

450a Synthetic Biology Tool Development Advances Predictable Gene Expression in the Metabolically Versatile Soil Bacterium *Rhodopseudomonas Palustris*

Mark Kathol, Cheryl Immethun, Taity Changa, Rajib Saha

Model organisms have been used as a biological, more sustainable alternative for manufacturing products such as pharmaceuticals, cosmetics, surfactants, and fragrances. However, these organisms are often metabolically restrained to only catabolizing simple sugars, and bioprocesses involving model organisms are often required to have low stress conditions. Non-model organisms possess many unique biochemical capabilities. Some of these organisms are capable of photosynthesis, can fix carbon dioxide and nitrogen, produce hydrogen and bioplastics, and can catabolize recalcitrant feedstocks such as lignin monomers and methane. Efforts to transfer metabolic pathways from non-model organisms to model organisms however have been fraught with difficulties. The cofactors, cellular conditions, as well as the energy and carbon sources needed for heterologous pathways to function may not be present in the host model bacterium. As a result, these pathways do not perform as well as in other microorganisms. The metabolically versatile non-model organism *Rhodopseudomonas palustris* is a purple non-sulfur soil bacterium which can produce hydrogen, is capable of all four modes of metabolism, and can catabolize lignin derived aromatic compounds. *R. palustris* is also capable of carbon dioxide fixation, as it possesses two forms of RuBisCo, and utilizes three nitrogenase isozymes to fix nitrogen while simultaneously producing hydrogen. This chassis benefits from an extreme metabolic versatility, yet stable heterologous gene expression from this bacterium has been problematic due to its high intrinsic resistance to several antibiotics. There is also a lack of synthetic biology parts that have been investigated in this microbe. We have provided a synthetic biology toolbox for *R. palustris* by including several synthetic biology parts such as origins of replication, terminators, and 5' untranslated regions, characterizing the fluorescence strengths of various fluorescent reporters, as well as the utilization of *R. palustris'* endogenous plasmid for heterologous protein production. This work determines the efficacy of a variety of synthetic biology tools for engineering *R. palustris* and harnessing its many unique biochemical processes. This study has also helped define the principles by which this microorganism can be engineered to express heterologous genes through a methodology that could be applied to other non-model microorganisms. This work has been published in the Journal of Frontiers in Bioengineering and Biotechnology, and can be found at https://doi.org/10.3389/fbioe.2022.800734.

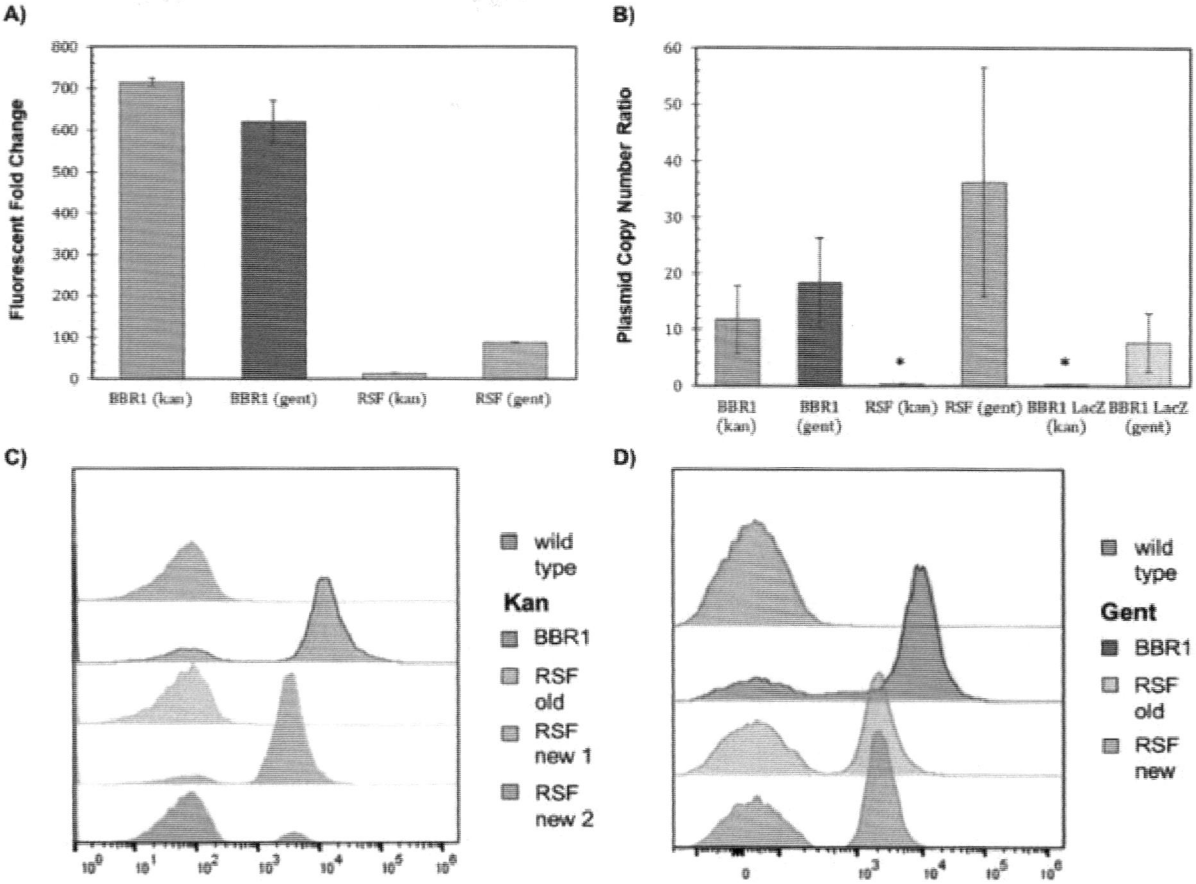

A)
Fluorescent Fold Change
BBR1 (kan)
BBR1 (gent)
RSF (kan)
RSF (gent)
B)
Plasmid Copy Number Ratio
BBR1 (kan)
BBR1 (gent)
RSF (kan)
RSF (gent)
BBR1 LacZ (kan)
BBR1 LacZ (gent)
C)
wild type
Kan
BBR1
RSF old
RSF new 1
RSF new 2
D)
wild type
Gent
BBR1
RSF old
RSF new

Activate W

450b Enhanced Synthetic Biochemistry Systems Enabled By Improvements in *Bacillus subtilis* Spore-Display

Trevor Nicks, Todd C. Chappell, Vikas D. Trivedi, Karishma Mohan, Nikhil Nair

Currently, there are two production platforms available to metabolic engineers: (1) cellular systems that utilize recombinant production organisms such as *Escherichia coli* and (2) cell-free systems which utilize isolated enzymatic pathways to produce compounds of interest. Both platforms have their own advantages and limitations. Cellular systems have proven scalability but have limited yield due to competing cellular metabolism. Meanwhile, cell-free systems can achieve near-theoretical yields but their scalability is hindered by a lack of enzyme stability. A synthetic biochemistry system that maintains the high yield and metabolic minimalism of cell-free systems while increasing enzyme stability would be highly advantageous. In this presentation, we present progress towards creating an enhanced synthetic biochemistry platform using *Bacillus subtilis* spore-display.

B. subtilis spore-display is an increasingly powerful enzyme immobilization strategy. For spore-display, a gene of interest is fused to a spore coat protein and, after sporulation, becomes anchored to the spore surface. Spores are a uniquely robust microbial phenotype and are metabolically inactive, meaning they have little intrinsic catalytic activity. When enzymes are displayed on spores they often show increased stability and can be easily purified by centrifugation. These qualities make spores particularly attractive for use as a self-assembling enzyme immobilization technology as they enable catalyst recycling. While previous studies have demonstrated spore-based cell-free metabolic engineering as a proof-of-concept, the technology is new, not well-developed, and has limited modularity.

There are at least 44 known spore-coat proteins that could potentially act as anchors for spore-display without affecting spore formation. But only 18 of these have previously been tested. To improve the modularity of spore-display for use in synthetic biochemistry systems we set out to identify optimal anchor proteins. In collaboration with the DOE Joint Genome Institute, we designed 88 synthetic DNA constructs that would allow us to test the suitability of each of the 44 spore coat proteins to act as either N or C-terminal fusion partners for the tetrameric enzyme beta-glucuronidase (GusA). From this screen, we have identified additional spore coat proteins that are capable of acting as enzyme fusion anchors and characterized their surface-availability. We are now completing work on increasing enzyme display density and building more complex synthetic biochemistry systems using spore-display.

450c A Modular Synthetic Biology Toolkit for Environmental Actinobacteria

Zach Jansen, Matan Lieber-Kotz, Ross Thyer

Actinobacteria are a broad group of Gram positive, high GC content bacteria, which are widely distributed in soils and aquatic environments. While genera such as *Streptomyces* are familiar as the source of many antibiotics, and *Mycobacterium* are well-studied as the causative agents of human and livestock disease, our knowledge of the wider phylum and the genetic tools available to interrogate them remain comparatively limited. However, recent advances in genome sequencing and several taxonomic revisions have shed new light on these species. In particular, the mycolata taxon, which is defined by the presence of mycolic acid as a component of the cell wall, is now thought to possess vast, untapped potential as new hosts for the bioproduction of commodity chemicals, biodegradation of natural and synthetic polymers, and a source of novel secondary metabolites including polyketide antimicrobials. Unfortunately, although this taxon includes *Mycobacterium* which has a large research community and established genetic tools, few of these have proven directly transferable to the other members of interest, particularly genera such as *Rhodococcus*, *Nocardia*, *Gordonia*, and *Dietzia*.

To facilitate the use of these bacteria as hosts for metabolic engineering, we have built a new, modular DNA assembly toolkit for constructing plasmids and genetic circuits which maintains broad compatibility across a wide range of environmental mycolata isolates. This toolkit comprises six plasmid origins of replication, five different selectable markers, a gradient series of synthetic promoters and universal, bicistronic ribosome binding sites, and various transcriptional logic. Collectively, these tools enable the construction of robust genetic circuits in multiple genera with previously unattainable control of gene expression. In addition, this toolkit has been designed to maximize the use of common parts between these genera and *E. coli*, where most plasmid assembly is expected to occur due to the ease of genetic manipulation. This dramatically reduces the size and complexity of the component DNA parts required and is expected to improve genetic stability during cloning. Using this toolkit, we demonstrate genetic circuits for metabolic engineering of high-value isoprenoid biosynthesis, protein engineering to enable novel carbon assimilation pathways, and modular assembly of transposons for genome-wide mutagenesis.

450d Development of a Synthetic Biology Toolbox for Soil Bacteria

Maya Venkataraman, Taylor Cook, Brian F. Pfleger

Soil microbes have the potential to greatly increase food yield and improve disease resistance in relevant crops; however, the ability to manipulate these non-model organisms is currently limited. In this study, we have developed tools for genetic manipulation and heterologous gene expression in two species of interest: *Pseudomonas putida* and *Azotobacter vinelandii. P. putida* is a common soil microorganism previously shown to associate with crops such as maize, whereas *A. vinelandii* is a robust nitrogen-fixing organism that has previously been engineered to supply additional ammonium to the soil. We characterized and optimized several plasmids, constitutive promoters, and different induction systems in *Azotobacter vinelandii* DJ and *Pseudomonas putida* KT2440. These characterized tools and induction systems were subsequently used to develop high-efficiency, λRed/CRISPR and I-SceI recombineering strategies for genetic modification of both microbes in order to create scarless knockouts and gene insertions.

We were able to identify stable plasmids for both organisms, characterizing their copy numbers using qPCR, as well as relative expression strength using a GFP reporter. Additionally, we identified reliable constitutive promoters for both *P. putida* and *A. vinelandii*, including induction systems that demonstrated approximately 100-fold induction for both organisms. We show that optimization of repressor systems from *E. coli* can greatly increase induction range. Additionally, the recombineering approaches highlighted here resulted in high efficiency knockouts in *Pseudomonas* (>90% efficiency of desired knockouts with optimized protocol), and genetic integrations were used to produce several heterologous natural products. Finally, we developed nitrogen-secretion strains of *A. vinelandii* using an I-SceI recombineering strategy. This study demonstrates synthetic biology tools for reliable gene expression in *P. putida* and *A. vinelandii*, two soil microbes which may have a role in developing robust synthetic communities for various crops in the future.

450f High Efficiency Multiplexed Cytosine Base Editing in Yarrowia Lipolytica

Vijaydev Ganesan, Dheeraj Pedada, Anthony Stohr, Jaya Gupta, Mark Blenner

Metabolic engineering involves manipulating gene functions such as knockouts, knock-in, and overexpression to produce the desired target molecule. Knockouts can be performed by CRISPR-Cas9. However, multiple dsDNA breaks are toxic to the host. To alleviate this problem, we have developed an efficient base editor guided by Cas9 nickase in the oleaginous yeast Yarrowia lipolytica. We observed indel-free base editing when we used dead Cas9 fused to PmCDA1 or the KU70 strain. We increased the editing efficiency by optimizing gRNA expression by removing intragenic nucleotides between tRNA and gRNA, using nCas9(H840A) to nick the non-edited strand of DNA, integrating base editor, and optimizing cell culture conditions. We observed that the editing window, PAM preference, efficiency, and indel percentage vary depending upon the type of Cas9, strain and DNA strand targeted. We extended the editing to multiplexed base editing of 5 genes simultaneously. An optimized golden gate assembly was used to assemble guide RNAs targeting multiple genes without pre-cloning. tRNA arrays were constructed to express multiple gRNAs without requiring individual promoters. tRNA arrays were optimized with one PolIII promoter driving two or three gRNA arrays. Finally, we achieved highest efficiency by using CAN1 co-selection mediated editing to disrupt multiple genes. We also observed cytosine base editing by fusing cytosine deaminase to LbCpf1 to target A/T rich sequences. To perform base editing, we used deaminase fused to LbCpf1 with truncated gRNA or dead Cpf1 fused to gRNA.

450g Onboarding Environmental Microbes for Efficient (re)Use of Renewable Feedstocks and Consumer Wastes in Biomanufacturing

Kevin Solomon

Microbial chemical factories are sustainable biomanufacturing platforms that complement traditional petrochemical industries by using renewable and/or waste alternative carbon sources. Environmental microbes that live in harsh environments with high resource competition, such as animal microbiomes, are particularly attractive due to their rich repertoire of enzymes to efficiently process diverse carbon sources and unique capacity to catalyze certain industrial chemistries at scale. However, these organisms frequently lack molecular tools to control these processes for societal applications. In this talk, I will describe our progress towards the creation of these tools in two systems. In the first example, I will highlight progress towards genetic and epigenetic tools for anaerobic fungi native to large herbivore digestive tracts and describe approaches to deploy anaerobic fungi today for direct production of fragrances and solvents from untreated feedstocks. In the second example I discuss our discovery of powerful enzymes for the degradation of post-consumer plastic wastes such as polypropylene and polystyrene from microbes native to the mealworm gut microbiome. I will also touch on emerging engineering tools that may be used to overcome some of the features that have long stymied the study and engineering of these powerful microbes.

556a Exploring *Streptococci* Cas9 Orthologs with Distinct PAM-Binding Motifs to Develop a Titratable Bacterial Gene Regulator

Yuxi Teng, Jian Wang, Xinyu Gong, Jianli Zhang, Yifei Wu, Lei Lou, Michelle Li, zhong-Ru Xie, Yajun Yan

The CRISPR-Cas9s have been developed as versatile tools for programmable genome editing and gene expression control. However, their targetable sequences and targeting efficacies are restricted by the strict requirement on a specific protospacer adjacent motif (PAM) sequence. The SpCas9 from *S. pyogenes*, representing most *Streptococci* Cas9s, requires a NGG PAM by its signature RxR PAM-binding motif. In this study, we reveal that three rarely reported PAM-binding motifs among *Streptococci* Cas9 orthologs generate distinguished PAM specificities. Rational site-directed mutagenesis of the SeCas9 from *S. equinus* with a different motif naturally preferring NAG PAM produces a variant with expanded PAM range. With this, we further develop a PAM-dependent titratable gene regulator by fusing an ω factor with the deactivated SedCas9 variant. Applying such a regulator for the metabolic control of endogenous genes in *Escherichia coli* leads to higher 4-hydroxycoumarin production. These results demonstrate the PAM diversity of closely-related *Streptococci* Cas9 orthologs and enrich the PAM-flexible Cas9 toolkits for broader applications.

556b Programmable Synthetic Biomolecular Condensates for Cellular Control

Yifan Dai, Ashutosh Chilkoti, Lingchong You

The formation of biomolecular condensates mediated by phase separation plays a critical role in controlling diverse cellular functions in nature. This has recently inspired the use of phase separation to design synthetic systems. While design rules of phase separation have been established for many synthetic intrinsically disordered proteins, most efforts have focused on investigating their phase transition behaviors in a test tube. It remains unclear whether and to what extent these engineered components can mediate phase separation in a living cell and drive specific cellular functions. In this talk, I will present a rational approach to enable programmability of the formation and the property of synthetic condensates toward realizing target cellular functions. This generalizable strategy lays the foundation for engineering designer biomolecular condensates for diverse synthetic-biology applications.

556c Engineered Prokaryotic Promoters Enable Multi-Input Processing Crispra/I Circuitry

Diego Alba, Ryan Cardiff, Benjamin Tickman, Cholpisit Kiattisewee, Jesse Zalatan, James Carothers

CRISPRa/i transcriptional regulatory networks hold great promise as a framework for engineering complex metabolic networks and information processing circuits. Recent analysis of CRISPRa/i circuit properties suggests a high degree of composability and scalability, highlighting the possibility of building more complex regulatory programs. However, to increase the complexity of the circuits and the range of their applications, an expandable set of high-performing CRISPRa/i promoters and guide RNAs is required. To engineer a set of high-performing activatable promoters, we characterize the effect of different promoter parts on the promoter expression profiles by using CRISPRa as a promoter interrogation tool. We design promoters with up to 1000-fold dynamic range in an E. coli-based cell-free expression system (CFS), constituting a 33-fold improvement from previous synthetic promoters. These new activatable promoters enable implementation of novel ligand-responsive CRISPR activation domains and circuit topologies not accessible with previous components, such as dense overlapping regulons and deep activation cascades. Additionally, we find a high degree of transferability of the insights gained with CRISPRa to other systems, such as the light-inducible transcriptional activator EL222. We demonstrate multi-input processing by integrating both ligand-responsive CRISPRa and EL222 into a tunable CRISPRa/i pulse generator. Overall, this work provides a toolbox of high-performing activatable promoters and orthogonal inducible transcriptional activators that expand the capabilities of the CRISPRa/i regulatory framework.

556d Programmable Control of Protein Activity Via Reversible Formation of Biomolecular Condensates in Bacteria

Mrugesh Parasa, Benjamin Rubino, Benjamin Howard, Kevin Solomon

Biomolecular condensates or membraneless organelles are phase separated proteins and/or other biomolecules that are ubiquitous in eukaryotic cells. While these condensates may be liquid with exchange and diffusion of their components with the rest of the cell (e.g. cytoplasm), they locally concentrate their constituent biomolecules altering their interactions in normal cellular processes. Here, we exploit this phenomena via reversible coacervate formation to control the activity of cellular proteins in *E. coli*. To induce liquid-liquid phase separation, we fuse proteins to elastin-like polypeptides (ELP) that reversibly aggregate in response to increases in temperature and/or changes in intracellular pH. In so doing, we sequester their fusion partners from the cytoplasm, limiting their ability to participate in cytoplasmic reactions. We have demonstrated this concept *in vivo* with enzymes and transcription factors for switchable control of protein activity with temperature and characterized its dynamics in cell-free systems. For example, I-SceI mediated cleavage of a host genome can be inhibited by increasing the cultivation temperature, creating a simple temperature-sensitive kill switch; accidental release will lower the culture temperature leading to cell death. Similarly, coupled transcription factors exhibit an order of magnitude reduction in GFP expression relative to unfused transcription factor controls at elevated temperatures. More importantly, the threshold for coacervate formation and control of protein activity may be tuned through appropriate design of the ELP fusion partner. We also examine the reversibility and dynamic response of these methods. Our results introduce a simple yet effective, rapid and tunable approach to control protein activity via induction of coacervate formation that may form a powerful new tool for synthetic biology.

556e Optogenetically-Controlled Bacterial Persistence

Yousr Dhaouadi, Dacheng Ren

Bacterial populations produce tolerant persister cells that survive lethal concentrations of antibiotics due to dormancy-related phenotypic changes. Recently, tolerance has been shown to promote the development of antimicrobial resistance based on accumulating genetic mutations. Persister research focuses on linking causality of persister formation to their recalcitrance towards antimicrobial agents. This is challenging, however, due to the low frequency of persisters within a population (typically <1%) and their rapid reversion to normal cells. Several methods have been developed to induce persister formation but require the addition of chemical or biological agents. In this study, we engineered an *Escherichia coli* strain to control the expression of the toxin gene *hipA* and thus persistence using blue light. The engineered strain demonstrated 100% persistence with overnight blue light induction, which is more than 30 times higher than the dark control. The blue light-controlled gene expression was validated using a fluorescent reporter strain and real-time quantitative PCR (RT-qPCR) analysis of *hipA* expression. Using blue light to control persistence provides a new method to study persistence without using chemical agents, which can have impacts on cell metabolism and pathways not related to persistence. Using light also has advantages as a non-invasive tool and reversible induction for studies that require spatial control of bacterial gene expression and persistence.

556f Mapping the Sequence-Function Landscape for Quorum Sensing Specificity: Mitigating Signal Crosstalk of the Lasr Quorum Sensor for Programmable Bacterial Consortia

Min Zeng, Vanessa Vu, Stephanie Call, Biprodev Sarker, Lauren B. Andrews

Programmable intercellular communication using quorum sensing components allows for coordinated and dynamic responses to be engineered in synthetic microbial consortia for biomanufacturing and other applications. Homoserine lactone-mediated quorum sensors are well-studied and widely applied, yet they commonly exhibit some degree of crosstalk with other LuxR-type quorum sensors natively. This signal interference is often detrimental to applications where specific intercellular signaling allows for more tightly controlled cell-cell communication. In this project, we apply targeted protein engineering strategies to elucidate the sequence-function relationship for a LuxR-type quorum sensing regulator and aim to identify variants with improved signaling specificity across a set of commonly used quorum sensors.

Here we target the LasR quorum sensor, which is among the highest level of crosstalk observed and for which the protein-ligand interactions and protein structure have been well studied. Mutations of ligand binding residues were generated for LasR using site directed mutagenesis. Among other mutations, a LasR mutation V76T has been identified that confers selectivity towards its cognate 3-oxo-C12-HSL signal relative to induction with the set of noncognate signals. A combinatorial pooled saturation mutagenesis library of LasR variants (9,486 mutants) was designed by mutating a ligand binding pocket region of LasR (L125 – L130) by custom scripts and then constructed using an oligo pool to build the LasR sensor library. LasR variants were designed to contain single, double, and triple mutations with all amino acid substitutions. Sequencing of the pooled library confirmed all 9,486 LasR protein variants (100% design coverage) present in the pool. Pooled sort-seq assays using the cognate and noncognate homoserine lactone (HSL) were performed to determine the activation of each LasR variant for both HSL ligands. Based on this dataset, analyses identified sensor variants with improved specificity and positions to maintain the function of the LasR sensor (e.g. G126). The elucidated sequence-function relationship reveals that single mutations can confer LasR with improved specificity, while double or triple mutations can give rise to variants with interesting properties, such as reversed signal specificity. This work informs how the LasR quorum sensor can be engineered to mitigate signal crosstalk and elucidates the tradeoff of sensitivity and specificity for this sensor. This approach can be used for further high-throughput engineering of orthogonal HSL-mediated quorum sensors necessary for programming bacterial consortia.

556g Approaching Sequence Completeness: Model-Predictive Design and Engineering of Synthetic Genetic Systems

Howard Salis

We introduce new biophysical models of gene expression to accurately predict transcription rates, translation rates, and mRNA decay rates from sequence information. We show how combining sequence-to-function model predictions leads to the accelerated design and engineering of genetic systems with valuable biotechnology applications. Our model building and testing efforts utilize the latest advances in oligopool synthesis, library-based cloning, and next-generation sequencing to carry out thousands of defined experiments per workflow. As a demonstrative application, we show how engineered organisms can detect explosive compounds inside the soil and report on their presence by producing pleasantly potent odorant molecules. Overall, our model-based design platform for engineering organisms is utilized by over 10,000 registered researchers who have engineered over 850,000 synthetic genetic systems with diverse biotech applications.

693c Plasmidmaker: A Versatile, Automated, and High Throughput End-to-End Platform for Plasmid Construction

Behnam Enghiad, Pu Xue, Nilmani Singh, Aashutosh Boob, Chengyou Shi, Vassily Andrew Petrov, Roy Liu, Siddhartha Suryanarayana Peri, Stephan Thomas Lane, Huimin Zhao

Plasmids are used extensively in basic and applied biology. However, design and construction of plasmids, specifically the ones carrying complex genetic information, remains one of the most time-consuming, labor-intensive, and rate-limiting steps in performing sophisticated biological experiments. We report a robust, versatile, and automated end-to-end platform for plasmid construction named PlasmidMaker that enables scarless and error-free construction of virtually any plasmids in a high throughput manner. To implement this platform, we first developed a versatile, scarless, parallel, robust, and accurate method for assembly of multiple DNA fragments using *Pyrococcus furiosus* Argonaute (*Pf*Ago) based artificial restriction enzymes (AREs). We then created a user-friendly frontend for plasmid design, and a backend that streamlines the workflow. Finally, we integrated the DNA assembly method and the software with a robotic system named Illinois Biological Foundry for Advanced Biomanufatcuring (iBioFAB) to create a nearly fully automated workflow. As a proof of concept, we used this platform to construct 101 plasmids across 6 different species (bacteria, yeast, plants, and mammals) which involves ~2×10^4 pipetting steps. Plasmids with sizes ranging from 5 to 18 kb were assembled from up to 11 DNA fragments efficiently with limited human intervention. PlasmidMaker should greatly expand the potential of synthetic biology.

693d Development and Assembly of a Synthetic Membrane-Less Enzymatic ATP Regeneration Cascade Powered By Fuel Oxidation

Scott Banta, Emma Willett

ATP regeneration is an important bioengineering capability that can be used to support for cell-free protein synthesis (CFPS), pseudocell development, and other *in vitro* applications. Methods for regenerating ATP include expensive phosphate donors and ATP-synthase enzymes that require a proton gradient across a membrane. In this talk, we propose a membrane-less ATP-generating synthetic enzyme cascade that is driven by NADH oxidation. We demonstrate the cascade using the oxidation of formate, a cheap fuel source, and polyphosphate, as a cheap phosphate donor. At the center of the proposed cascade is the ATP-NAD^+ kinase (NADK). All of the commonly investigated isoforms of NADK (human, *S. cerevisiae*, *Arabidopsis* and *E. coli*) have been measured, or assumed, to have irreversible reactions and can only accept ATP as a reactant. One group, in 1977, has reported NADK obtained from pigeon livers (*C. livia*) with reverse activity. First, we confirmed a reversible reaction for the recombinant pigeon NADK. Additionally, we identified 2 other isoforms (duck and cat) that can use ATP as a substrate and as a product. We determined that the human NADK has an affinity for NAD^+ that is ~600 times higher than the pigeon, duck and cat isoforms, and we concluded that NAD^+ serves as a potent product inhibitor in the reverse direction for the human NADK, which accounts for the observed irreversible behavior. Then, we constructed the membrane-less enzyme cascade with an NADK with reversible activity, a polyphosphate NAD^+ kinase, a formate dehydrogenase, and an NADPH oxidase. This cascade continuously produced ATP for several hours when coupled with firefly luciferase. To show the utility of the cascade, the purified enzymes were incorporated into a CFPS system and we demonstrate formate-powered recombinant protein expression of sfGFP and HRP. We believe this enzyme cascade is a valuable new biosynthetic system for regenerating ATP in emerging biotechnology applications.

693e Using the Cell-Free Gene Expression System for Highly Accurate Microrna Detection

Caroline Copeland, Yongchan Kwon

Early diagnosis and the monitoring of diseases and infections is a crucial medical practice to avoid aggressive and malignant disease progression and to retain the quality of one's health. However, many patients struggle to monitor their health regularly or there are no proper methods to detect certain illnesses early enough. Recently, numerous studies have confirmed that the dysregulation of microRNA (miRNA) can be correlated with the development and specific stage of major diseases. These miRNAs serve as the perfect biomarker, as they circulate in the blood during cell-to-cell communication in highly protective packaging. The current gold standard and other technologies that detect this biomarker suffer from high costs, high variability, and inaccuracies. These flaws stem from the same methods, resulting in low differentiation between the thousands of different forms of miRNA resulting in low accuracy.

The cell-free gene expression system is the remedy to these flawed methods, providing a cellular environment and finely tuned genetic circuitry capabilities programmed by the user. Here we have programmed an ultrasensitive and highly accurate miRNA sensing platform using the *E. coli*-based cell-free gene expression system. This platform can differentiate a 1 nucleotide difference between similar miRNA when the gold standard, RT-qPCR, could not. The genetic circuitry component allows for meticulous amplification steps to be added after the precise miRNA has been detected, removing false positives, and increasing sensitivity. Our circuit also has an extremely low leak compared to others that does not trigger the amplification cascade the same as the sample, allowing for low background noise. With the cell-free system's tolerance to toxic substances, the circuit can perform from a blood sample with minimal purification, lowering cost. The system is also able to be freeze-dried, making it inexpensive to transport and store. Given these advantages and our findings, the cell-free gene expression system is the optimal tool for miRNA sensing and profiling.

693f Profiling of Putative Threonine Transaldolases for Biosynthesis of Non-Standard Amino Acids

Michaela Jones, Neil Butler, Sean Wirt, Ishika Govil, Shelby Anderson, D'Jana Wyllis, Aditya Kunjapur

Beta-hydroxy non-standard amino acids (β-OH-nsAAs) have broad utility as small molecule drugs, precursors for beta-lactam antibiotics and building blocks for introducing novel functionality into proteins. Despite the valuable applications for these non-standard amino acids, organic synthesis routes are expensive and documented for only a limited number of β-OH-nsAA and existing biosynthesis approaches are limited by unfavorable equilibrium dynamics. A recently discovered enzyme class of threonine transaldolases can address the current limitations for β-OH-nsAA biosynthesis through a stereo-selective and thermodynamically favorable aldol condensation of an aromatic aldehyde and L-threonine. We have probed the substrate specificity of a TTA from *Pseudomonas fluorescens* to access novel β-OH-nsAAs. To explore the specificity of this enzyme class more broadly, we have expressed and characterized naturally occurring, putative TTAs. Through the screen of putative TTAs, we have identified different enzymes with broader substrate scope, lower threonine affinity, and faster initial reaction rates. To characterize the broad specificity of many TTAs, we optimized a high-throughput absorbance-based assay that couples an alcohol dehydrogenase with a TTA to produce a measurable output of enzyme activity. We have also produced β-OH-nsAAs in metabolically active, engineered *E. coli* cultures, a novel chassis for β-OH-nsAA production. Overall, our work helps identify new candidate enzymes that could enable biological synthesis of valuable small molecules and building blocks that expand protein functions.

23b Evaluating Proteome Allocation of *Saccharomyces Cerevisiae* Phenotypes with Resource Balance Analysis

Hoang Dinh, Costas D. Maranas

Saccharomyces cerevisiae is an important model organism and a workhorse in biochemical production. Using available molecular biology information and proteomics data we reconstructed and parameterized a genome-scale resource balance analysis (RBA) model (i.e., *sc*RBA) that captures both cellular metabolism and proteome allocation. Specifically, *sc*RBA contains the genome-scale metabolic network plus production pathways for protein, enzyme, and ribosome. The RBA system is constrained by reaction-enzyme and protein-ribosome coupling equations, and by experimentally observed rRNA and protein capacity. ATP maintenance rates and *in vivo* apparent turnover numbers (k_{app}) parameters were regressed from metabolic flux and protein concentration data. Our k_{app} parameterization workflow accounts for enzyme subunit stoichiometry and estimates minimally required intracellular fluxes that satisfy observed extracellular fluxes. Estimated parameters were found to differ when perturbing oxygen and nutrient availability. k_{app} values are higher *in vivo* compared to *in vitro* reported values for some enzymes in glycolysis, electron transport chain, and amino acyl-tRNA synthetase pathways reflecting their higher *in vivo* efficiency, matching literature-reported metabolons and post-translational modifications. We benchmark *sc*RBA by recapitulating yeast's aerobic fermentation phenotypes (i.e., the Crabtree effect) and proteome allocation, as well as demonstrate its applicability by predicting necessary proteome re-allocation from wild-type allocation to achieve optimal product yield.

23c Metabolomics and Isotope Tracing Reveal the Dynamic Role of Parallel Glycolytic Pathways

Richard Law, Junyoung Park

Glycolysis is a highly conserved pathway for the production of energy and biomass. However, some organisms simultaneously express the parallel Entner-Doudoroff (ED) pathway. The ED pathway yields only half as many ATP as textbook glycolysis, and the role of this seemingly less efficient pathway remains unclear. Here, we demonstrate that the ED pathway supports cell growth by rapid acceleration of glycolysis. We developed an isotope tracing strategy to quantify the distribution of glycolytic pathways during dynamic metabolic responses to changes in nutrient environment. While ED pathway utilization is low in the nutrient limited state, flux through the ED pathway increases disproportionately higher than textbook glycolysis upon nutrient upshifts. Consequently, the two pathways jointly increase total glycolytic flux to more rapidly increase cellular growth rate. Intermittent feeding of a limiting nutrient showed that the strain with both glycolytic pathways achieve higher cell densities than its ED-knockout counterpart. Thus, organisms with parallel yet specialized pathways can rapidly accelerate metabolism and growth, netting an edge over those with less metabolic flexibility. We surmise that the benefits of parallel pathways play a role in the holistic design of metabolic systems and that organisms may employ similar strategies elsewhere in metabolism.

INVESTIGATING THE NEURODEVELOPMENTAL EXPOSOME: A PROPOSED TRANSCRIPTOMIC AND METABOLOMIC ANALYSIS OF MOTHER-CHILD PAIRS IN PORTUGAL

D.R. Schultz[1,2], N. Papaioannou[1,2], C. Gabriel[1,2], T. Papageorgiou[1,2], I. S. Frydas[1,2], S. Karakitsios[1,2,4], D.A. Sarigiannis[1,2,3]*

[1] *HERACLES Health and Exposome Research Centre, Centre for Interdisciplinary Research and Innovation, Aristotle University of Thessaloniki, Greece*
[2] *Environmental Engineering Laboratory, Department of Chemical Engineering, Aristotle University of Thessaloniki, Greece*
[3] *Environmental Health Engineering, School for Advanced Study IUSS, Pavia, Italy*
[4] *ENVE.X, Thessaloniki, Greece https://enve-x.com/*

ABSTRACT

The exposome paradigm is the concept that the environment, lifestyle factors and behaviours, as well as internal biological factors of an individual can impact health outcomes. This concept has become an area of intense interest in recent years as an important tool in human health risk assessments. Particularly, there is an emerging and urgent need to understand the dynamic interactions occurring that may influence child neurodevelopment. The aim of this project is to investigate the link between prenatal exposure to environmental pollutants and infant neurodevelopment using integrated omics technologies. Mother-child cohort samples were collected from Portuguese subjects as part of the Health and Environment-wide Associations based on Large populations Surveys (HEALS) study and applied to the European Exposure and Health Examination Survey (EXHES) study. Together, multiomics analysis will be performed to identify genes and metabolites of interest. This data will then be linked questionnaire data to investigate the associated connections between genes and metabolites of interest and potential causal agents using environment-wide association studies (EWAS) approaches. The study also aims to generate adverse outcome pathways for neurodevelopmental disorders. This work will help to inform human health risk assessment efforts.

KEYWORDS: multi-omics, transcriptomics, metabolomics, risk assessment

1. INTRODUCTION

The human exposome is comprised of incredibly complex and dynamic exposure scenarios that includes every exposure that an individual is subjected to during their entire lifespan. As such, the exposome, and therefore the effect it may exert, are in constant flux, which makes quantifying and predicting the effects very difficult. Regardless, an untangling of the exposome is instrumental to understanding the connection between the environment and human health and further, the mechanisms behind many diseases that can lead to death, illness, and overall decreases in quality of life. One method to uncover the intricacies of the exposome is by using omics techniques.

Omics is the study of the complicated network of interactions that can influence the phenotype of an organism and involves various systems of biological organisation including genes (genomics), RNA (transcriptomics), proteins (proteomics), and metabolites (metabolomics). When integrated, omics have been helpful in multitude of applications including disease mechanism identification and biomarker discovery. As omics techniques expand, so do the attempts to integrate such data across biological pools. Multi-omics integration techniques are rapidly expanding as they allow for the exploration of the interconnections across the molecular groups without a priori knowledge of specific modes of action of a compound. One common technique for multiomics analysis is the Exposome-Wide Association Study (EWAS) approach, which was developed to systematically investigate and associate various exposure factors and modifiers in order to identify disease risk. Another method is enrichment analysis, which is used to determine whether a specified set of differentially expressed genes share biologically important attributes, such as the processes and functions that they are involved in on various levels. While both are important techniques in multiomics, they lack the ability to provide risk levels associated with the chemical itself. Therefore, we proposed to apply another method to integrate the data generated from the HEALS project.

Benchmark dose (BMD) is a point-of-departure (POD) method to derive human health guidance levels including reference doses, derived no-effect levels (DNEL), and acceptable daily intakes (ADI). BMD is defined as the concentration of a compound of interest that will elicit a predetermined change in the response rate for an adverse effect, also known as the benchmark response (BMR). Zhao et al (2022) used BMD in an assessment of metabolomic characterization of breast tissue exposed to BPA and found that most metabolites were dysregulated in a dose-response manner. While there are an increasing number of studies using POD for genomic (gPOD) and transcriptomic (tPOD) data, there are far less studies investigating metabolomic POD (mPOD) let alone a combination of the two. Moreover, in enrichment analysis, relying solely on common differentially expressed features between comparisons (i.e., identifying the common features that are differentially expressed across three treatments) ignores the fact that the POD for each feature that may be higher than the lowest dose used. This would mean that it may not be statistically detectable in the lowest treatments, but is in the other treatments and therefore, this feature is missed or overlooked. BMD therefore addresses this issue and therefore may be a more powerful option. Moreover, traditional BMD is generally associated with an apical outcome such as mortality (5%) applied to a population, although it is increasingly being repurposed and applied to high-throughput data. In this regard, many features may contribute to an apical outcome and so, identifying these contributing features separately may lend more insight into disease mechanisms and also may be able to detect dysregulation at much lower concentrations than apical-based BMDs. This may then allow for more informative and protective allowable limits

2. METHODOLOGY

In the present study and as a part of the HEALS European Exposure and Health Examination Survey (EXHES), 548 pregnant mothers were recruited and biosamples were collected at various timepoints (Figure 1). Biosamples (serum, plasma) were collected from pregnant women during delivery whereas serum from children were collected around one year of age. Questionnaire data was collected during pregnancy, during the first months post-parturition, and 1-year post-parturition. The questionnaires assessed various aspects of the maternal and infant exposome including personal health, environmental exposures, and socio-economic conditions.

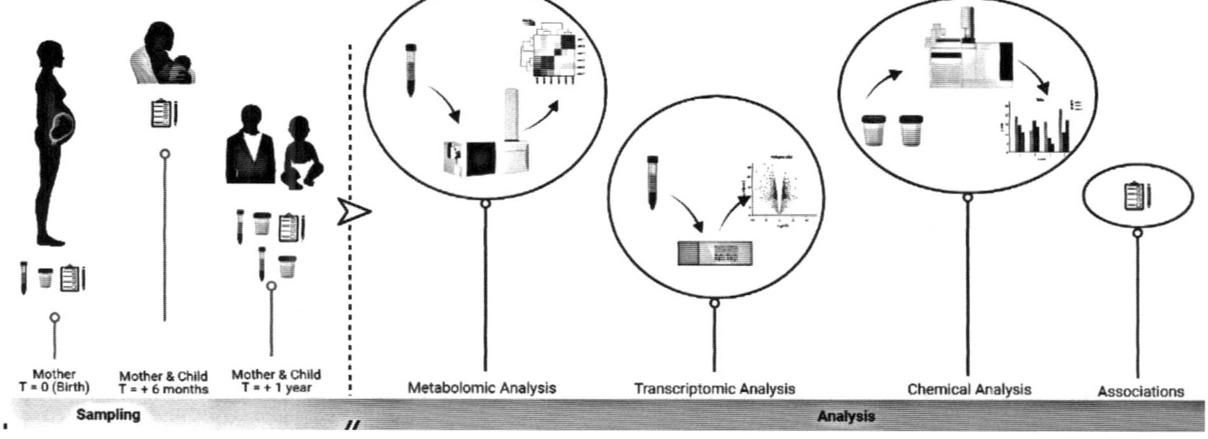

Figure 1: Phases of the HEALS-EXHES study, including sample collection and intended analysis (metabolomic, transcriptomic, chemical, and questionnaire).

Across the three biological sample types, chemical and omics analyses can be performed. Chemical analysis is ongoing and will identify contaminants of concern analysis using Thermo Scientific TSQ Vantage LC-MS/MS and/or Agilent 7500ce ICP-MS with ORS interference reduction system. Contaminants of interest may include phthalates, bisphenols, heavy metals, and perfluoronated compounds. Results will provide the independent concentration variable for BMD calculation. Untargeted metabolomic analysis has been completed on serum using an Agilent 1290 Infinity HPLC LC System coupled to an Agilent 6540 HRMS-QTOF using a reverse phase column in both positive and negative ionisation and followed in-house analytical procedures. Serum samples were stored in screwtop cryovials and placed at -80°C until later analysis. Samples were thawed in stable conditions overnight on the day before analysis at 4°C. Each sample was briefly vortexed to homogenize (5 s at 2000 rpm), 200 μl was transferred to a new 1.5 mL Eppendorf tube, and 50 μL was used to prepare a pooled quality control (QC) samples which followed the sample extraction protocol. After aliquoting of all samples, 600 μL of pre-chilled methanol was added to each sample for protein precipitation and metabolite extraction (1:3 sample to methanol ratio). Samples were stored at -20°C for 30 minutes to facilitate protein precipitation and then centrifuged (15000 rpm for 15 min at 4 °C). 300 μl of the supernatant was transferred to a new 1.5 mL Eppendorf tube and concentrated by nitrogen blown down evaporation at a gentle flow. The dry extracts were reconstituted with 100% LC-MS grade water with internal standards and vortexed (5 min at 2000 rpm). QC samples were reconstituted with 100% LC-MS grade water with 1 ppm internal standard additions. Samples were centrifuged (15000 rpm for 10 min at 4 °C) and 100 μL of the supernatant was transferred to autosampler vials. Sample randomization was performed to control for batch effects. Mobile Phase A was LC-MS grade water with 0.1% formic acid, and mobile phase B was methanol with 0.1% formic acid, both for positive and negative ionization mode. The chromatography technique applied was reverse phase HPLC, using a Fortis SpeedCore pH+ C18 column (2.1 x 100 mm, 2.6 μm, Fortis Technologies, United Kingdom). The flow rate was set at 0.250 mL/min at a constant temperature of 40°C. The following gradient elution program was the same for both positive and negative mode: 0% B at 2 min, 100% B at 17 min, 100% B at 22 min, 0% B at 24 min, and 0% B at 26 min. The injection volume was 5 μL. The data are acquired between 50 and 1000 m/z at a scan rate of 1.4 spectra/s in centroid mode at a resolution of 40,000 FWHM. The source conditions of the Q-TOF system are the following: gas temperature 300°C, drying gas 7 L/min, nebulizer 50 psig, fragmentor 150 V, skimmer 65 V, and capillary voltage 3500V or -3500V in positive or negative mode, respectively. Data acquisition is completed and resulting data will be preprocessed using R Bioconductor packages. Briefly, data preprocessing step also follow the in-house protocol and include chromatographic peak detection using the centWave algorithm, optimization of mass error and mzdiff parameters using IPO package, signal to noise estimation using in-house approaches, peak alignment using ObiWarp method, gap filling procedures, isotope and adduct annotation, filtering based on contamination and missing values, and relative standard deviation, and finally, normalisation by median, mean centering scaling, and log-transformation is performed. Transcriptomic samples will be selected after chemical analysis in order to select a subset of samples from across the measured chemical concentrations. Analysis will use Agilent microarrays and will follow standard pipelines provided in the R package limma, which includes background correction, quantile normalization methods, control probe removal and probe averaging, and batch correction or removal.

To derive PODs, a few pieces of information are required. Namely, it is important to obtain ranking metrics to characterize potential relationship of chemical compound and feature (metabolite/transcript) expression. Ranking metrics include compound concentrations, expression/count data from transcriptomic data, or chromatographic peak areas in metabolomic data. For chemical analysis, it will be necessary to divide the dataset into groups based on concentration ranges depending on the magnitude of spread of chemical concentrations. Groups based on ranges will include low, mid, and high, where the lowest concentration group acts as reference group.

However, it may be beneficial to include more ranges in order to better describe the dose-response. Moreover, it is important to assess other covariates to identify potential confounders, although these will be identified from the EWAS analysis.

3. DISCUSSION

In the present proposed study, we are interested in investigating three areas: (1) can cohort data be effectively used to generate mPODs, (2) can mPODs and tPODs be meaningfully combined in multiomics analysis, and (3) can the use of mPODs be used to decrease untargeted metabolomic preprocessing times and if so, how will this affect the EWAS and enrichment analysis results.

Recent years have shown an increasing number of studies being published using PODs to identify more protective indicators of toxicity (Thomas et al 2013a; Critzer et al 2021; Alcaraz et al. 2021). Moreover, they provide a quantitative result instead of a qualitative one, which is more useful for regulatory purposes. One major challenge of utilizing metabolomic data is finding software that can be effectively used for the data as metabolomic data is inherently different than genomic or transcriptomic data. However, a few recent studies have also attempted to apply the POD methodology (Crizer et al 2021; Zhao et al 2022). One difficulty that was found with metabolomic data is that model fitting issues may occur owing to the narrow dose ranges or low number of doses used in *in vitro* data sets that are frequently used in BMD studies. However, using cohort data may remedy this issue as there is generally a much wider dose range as well as significantly more replicates. Another issue that may arise is that using population-level samples may introduce a lot of natural variation to the analysis and may prove difficult to handle.

Recent evidence also suggests that omics-based dose-response (ODR) studies generally provided a good approximation of toxicological outcomes typically observed in apical outcome-based studies (Thomas et al 2013a; Critzer et al 2021; Alcaraz et al. 2021) and therefore, combining two omics may provide valuable insight into disease mechanisms. This would further be useful in biomarker discovery and disease mechanism identification as it would provide

Currently, untargeted metabolomic analysis generates hundreds to thousands of features across positive and negative ionisation modes and techniques (RP and HILIC). This is a massive barrier to the field as there is not yet effective and reliable software that can computationally annotate features across multiple databases and therefore requires manual annotation. This process is incredibly time-consuming and is inherently prone to researcher bias which can affect the replicability and confidence in certain results. It is also important to note that, with so many features, many features may be flagged as statistically significant due to chance, even with an False Discovery Rate cutoff. BMD could potentially be used to decrease the amount of time it takes annotating metabolites and increase confidence in results, by using it as a filter. By applying BMD, it could be possible to identify features that follow a dose-response type curve and filter out those that do not, even if they are identified as significant. Metabolites that exhibit a dose-response will be better candidate biomarkers as they should have a more quantifiable relationship between the independent and dependent variables. It is also uncertain whether features that do not exhibit a type of dose-response are even useful or informative as biomarkers or further downstream EWAS or enrichment analysis. It will be interesting to investigate whether removal of these features from the dataset will result in differences in EWAS and enrichment analysis and if so, to what extent.

There are various software available for generating PODs, including BMDExpress (https://www.sciome.com/bmdexpress/), BMDS (https://www.epa.gov/bmds/about-bmds), fastBMD (https://www.fastbmd.ca/), and TOXcms (http://pattilab.wustl.edu/software/toxcms/toxcms.php). However, these resources are largely geared towards genomic and transcriptomic data generated from *in vitro* datasets. Therefore, an effective method to preprocess metabolomic cohort data will be necessary. Regardless, using multiomics BMD has the potential to be a very useful tool in understanding the complex interactions of environmental pollutants and adverse health effects. Furthermore, it may be possible to decrease data processing time in untargeted metabolomics and increase biomarker discovery confidence by applying dose-response filtering to metabolomic data

REFERNCES:

[1] Alper James G. Alcaraz, Shaina Baraniuk, Kamil Mikulášek, Bradley Park, Taylor Lane, Connor Burbridge, Jessica Ewald, David Potěšil, Jianguo Xia, Zbyněk Zdráhal, David Schneider, Doug Crump, Niladri Basu, Natacha Hogan, Markus Brinkmann, Markus Hecker. Comparative analysis of transcriptomic points-of-departure (tPODs) and apical responses in embryo-larval fathead minnows exposed to fluoxetine. Environmental Pollution, Volume 295, 2022, 118667. doi.org/10.1016/j.envpol.2021.118667.

[2] Crizer, D. M., Ramaiahgari, S. C., Ferguson, S. S., Rice, J. R., Dunlap, P. E., Sipes, N. S., Auerbach, S. S., Merrick, B. A., & Devito, M. J. (2021). Benchmark Concentrations for Untargeted Metabolomics Versus Transcriptomics for Liver Injury Compounds in In Vitro Liver Models. *Toxicological Sciences*, *181*(2), 175–186. doi.org/10.1093/TOXSCI/KFAB036

[3] Thomas, R. S., Allen, B. C., Nong, A., Yang, L., Bermudez, E., Clewell, H. J., & Andersen, M. E. (2007). A Method to Integrate Benchmark Dose Estimates with Genomic Data to Assess the Functional Effects of Chemical Exposure. *Toxicological Sciences*, *98*(1), 240–248. doi.org/10.1093/TOXSCI/KFM092

[4] Sarigiannis, DA. Assessing the impact of hazardous waste on children's health: The exposome paradigm. 2017. Environmental Research, 158: 531 – 541. DOI: https://doi.org/10.1016/j.envres.2017.06.031.

[5] Sarigiannis, D.A., Papaioannou, N., Handakas, E., Anesti, O., Polanska, K., Hanke, W., Salifoglou, A., Gabriel, C., Karakitsios, S. 2021. Neurodevelopmental exposome: The effect of in utero co-exposure to heavy metals and phthalates on child neurodevelopment. Env. Res. 197: 110949. https://doi.org/10.1016/j.envres.2021.110949

[6] Yao C-H, Wang L, Stancliffe E, Sindelar M, Cho K, Weitong Y, Wang Y, and Patti GJ. Dose-Response Metabolomics to Understand Biochemical Mechanisms and Off-Target Drug Effects with the TOXcms Software. Analytical Chemistry, in press, 2019. doi:10.1021/acs.analchem.9b03811

[7] Zhao, H., Liu, M., Lv, Y., & Fang, M. (2022). Dose-response metabolomics and pathway sensitivity to map molecular cartography of bisphenol A exposure. *Environment International*, *158*, 106893. doi.org/10.1016/J.ENVINT.2021.106893

23e Multi-Omics Approaches to Decipher Human Melanoma Initiation

Vivek Bajpai, MD, PhD, Tomek Swigut, Joanna Wysocka

Melanoma is the most aggressive skin cancer that arises in skin pigment cells called melanocytes. Developmentally, melanocytes are derived from a short-lived embryonic stem cell population called neural crest stem cells. Recent evidence suggests that when melanocytes turn cancerous, they activate a developmental program similar to embryonic neural crest cells[1]. In addition, melanoma cells also utilize molecular machinery of neural crest cells to spread to different parts of the body (metastasis) and in acquiring resistance against currently used drugs[2]. Taken together, these observations suggest that melanocytes possibly de-differentiate into neural crest cell-like state during melanoma initiation and this neural crest-like state further drives melanoma progression. However, it is not clear, what is the epigenetic landscape of this dedifferentiated stage that melanocytes acquire in order to initiate melanoma. Moreover, it is not known whether these epigenetic signatures actually endow dedifferentiated melanocytes with melanomagenic potential and drive melanoma growth under permissible oncogenic environment. BRAF mutation where a valine is substituted by a glutamate at position 600 (BRAF (V600E)) is the most common mutation found in melanomas (in 80% primary and 68% metastatic melanomas)[3].

In this work, we identified global gene expression changes upon BRAF (V600E) overexpression in early melanocyte developmental context as well as in primary human melanocytes. Importantly, quantitative examination of cis-regulatory histone marks (H3K27Ac and H3K27me3) discovered over 700 enhancers that were differentially enriched between wild type and BRAF (V600E) mutant cells indicating their potential role in mediating transcriptional changes in early melanoma-genesis. Particularly, in mutant cells, the putative enhancers involved in cell proliferation (e.g., CDKN2A) displayed higher H3K27Ac marks in agreement with their more proliferative status while this active enhancer mark was depleted at melanocyte differentiation genes such as EDNRB suggesting BRAF(V600E) commits cells towards more melanoma-like cell state by simultaneously affecting both cell-cycle and cell differentiation programs. Our study identified earliest gene expression changes and associated epigenetic events that are involved in reprogramming of normal melanocytes into melanoma like-cells upon BRAF (V600E) overexpression. This study has generated a plethora of novel candidate enhancers that will fuel more studies towards identifying novel molecular mechanisms involved in the melanoma initiation and will potentially pave the way for development of novel diagnostics and therapeutics.

References:

1 Kaufman, C. K. *et al.* A zebrafish melanoma model reveals emergence of neural crest identity during melanoma initiation. *Science* 351, aad2197, doi:10.1126/science.aad2197 (2016).

2 Johannessen, C. M. *et al.* A melanocyte lineage program confers resistance to MAP kinase pathway inhibition. *Nature* 504, 138-142, doi:10.1038/nature12688 (2013).

3 Pollock, P. M. *et al.* High frequency of BRAF mutations in nevi. *Nature genetics* 33, 19-20, doi:10.1038/ng1054 (2003).

23f Comparative Genomics and Phenomics Reveal Genetic and Functional Diversity and Fitness within *Pseudomonas Aeruginosa* clinical Isolates

Mohammad Mazharul Islam, Glynis Kolling, Jason A. Papin

Pseudomonas aeruginosa is the leading cause of nosocomial infections and well-known for infecting patients with cystic fibrosis. These infections can be complicated by a variety of virulence mechanisms and a robust metabolic functionality repertoire. However, how key components of the environment modulate these metabolic pathways and virulence mechanisms during infection are poorly understood.

We hypothesize that mucin-driven modulations of the metabolism of *P. aeruginosa* are dependent on a complex combination of host and pathogen-specific factors which can be delineated using a combination of genomic, transcriptomic, and phenomic study. With that goal, we obtained 971 clinical isolates of *P. aeruginosa* from 590 patients from the UVA Health System Clinical Microbiology Laboratory, which collects samples from a range of outpatient and inpatient settings in the region. For each isolate, we recorded associated patient metadata, bacterial morphological phenotypes, and antimicrobial susceptibility profiles.

We selected a set of 25 phenotypically representative isolates from the entire isolate collection through stratified random sampling while guaranteeing the robustness of original phenotypic characteristics during repeat cultures. These 25 isolates were then cultured in LB medium with 5% Fetal Bovine Serum for whole genome sequencing. The genome sequence data was used for comparative genomic analysis using the PA14 strain as the reference genome. A dissimilarity matrix was enumerated from the output of multiple local alignment searches and was used to cluster the isolates according to their sequence similarities. The genotypic clustering was contrasted with the phenotypic clustering generated from a multi-parametric analysis to assess the genotype-phenotype correlation. In addition, we enumerated the fitness profile of the isolates across a vast library of Carbon substrates in defined media conditions. The fitness profile, combined with the genotypic variation of the isolates, explain their universal pattern of genetic diversity as well as the key metabolic and regulatory identifiers for phenotypic diversity. Overall, this study demonstrates the value in surveying clinical data in terms of pathogen isolate sources, patient metadata, morphological phenotypes, and correlating that data with genome sequence data and phenotype fitness to answer biologically relevant questions about genotype-phenotype relationships.

23g Multi-Omics Integrative Analysis Using Mutual Information-Based Machine Learning

Yan Tang, Mark P. Styczynski

Systems-scale analysis of multiple layers of molecular and cellular data has significant potential for both industrial and biomedical advances. While approaches have been developed and applied extensively for mining of a single type of genome-scale data at a time, the direct integration of multiple levels of these data types has remained challenging. Post hoc integration of individual analyses can provide some supporting insight, but direct integration of data types, whether driven by a priori knowledge or the data itself, could have significant impact. We present here a unique approach to the analysis of a longitudinal multi-omics dataset; while the specific application is in the context of a non-human primate model of malaria, the underlying approach could be used for arbitrary applications with available multi-comics datasets. We analyzed relationships across multiple biological layers using a mutual information-based machine learning approach to integrate heterogeneous longitudinal datasets and constructed an atlas of multi-omics relatedness networks (MORNs). Using this technique, we were able to detect signatures with substantial biomedical relevance. To our knowledge, this is the first report of large-scale, longitudinal multi-omics analysis of malaria in any system, as well as a novel approach for multi-omics data integration.

62a Predicting Antibody Titer Using Multi-Scale Dynamic Flux Balance Analysis Bioreactor Models

Saratram Gopalakrishnan, Jasmine Tat, Miguel Angel Valderrama-Gomez, Fabrice Schlegel, William Johnson, Pablo A. Rolandi, Cleo Kontoravdi, Nathan Lewis

The goal of pharmaceutical bioprocess optimization is the maximization of antibody titer at harvest. Common strategies to achieve this goal include cell line optimization for improved growth and antibody specific productivity, and process optimization for optimal media feeding and timing of hypothermic shift to induce antibody production. Mathematical models that can accurately predict cellular responses and antibody production under changing bioreactor conditions in the bioreactor are a valuable tool that inform interventions and strategies to accelerate cell line and process development. This talk presents a novel dynamic flux balance analysis (dFBA)-based framework to predict concentration profiles of cells, glucose, lactate, amino acids, and therapeutics in a perfusion bioreactor. Because process performance depends on the changing metabolic states over the course of the process, we first characterize the metabolic features of five CHO cell lines engineered to produce three types of antibodies. In four of five cell lines, we find that high producer clones typically grow to higher cell densities than the unselected clones, and that the final titer positively correlates with the final cell density at the end of the growth phase. Interestingly, the final titer negatively correlated with cell density for the fifth cell line, implying that this clone follows a different pathway to achieve optimal antibody titer. Exo-metabolomic characterization of the cell lines revealed two distinct metabolic states corresponding to the growth phase and the production phase, respectively. Upon performing a principal component analysis on the computed fluxes, we found that a shift from growth to production phase was marked by a reduced growth, reduced lactate secretion, and increased antibody production. Hierarchical clustering of gene expression data measured for each cell line revealed that differences in cell state were primarily based on process phase (i.e., growth vs production phase) and that the cell state varied minimally across different cell lines. Upon extracting context-specific models using mCADRE, we found that the growth phase models preferentially activated anabolic pathways such as lipid and nucleotide biosynthesis, whereas amino acid degradation pathways were preferentially activated in the production phase models. Next, we developed a novel framework to simulate the concentration profiles of all metabolites and cells in a perfusion bioreactor using the knowledge gained from the characterization of the five cell lines. The framework expands on the traditional dFBA framework by (i) incorporating nutrient uptake using kinetic rate laws, (ii) constraining fluxes through key metabolic tasks based on process phase, and (iii) limiting scope of the metabolic model based on the process phase. A machine-learning model is trained to predict the process phase based on reactor conditions. The concentration profiles predicted by this multiscale dFBA model are validated using experimental data generated across different bioreactor operating conditions.

62b Characterizing the Interplay between Rubisco and Nitrogenase Enzymes of *Rhodopseudomonas Palustris*

Rajib Saha, Adil Alsiyabi, Niaz Chowdhury

Rhodopseudomonas palustris CGA009 (*R. palustris*) is a gram-negative purple non-sulfur bacteria that can grow either phototrophically or chemotrophically. It does so by fixing or utilizing a diverse range of substrates, including one of the most abundant sources of carbon, lignin breakdown products, to meet its carbon and nitrogen requirements. It can grow aerobically or anaerobically and generate energy from light, inorganic, and organic compounds. Due to its ability to convert different carbon sources into useful products in anaerobic mode, this study, for the first time, reconstructed a metabolic and expression (ME-) model of *R. palustris* to investigate its anaerobic-photoheterotrophic growth. ME-models, as opposed to metabolic (M-) models, include transcription and translation reactions, as well as macromolecule synthesis, and couple these reactions with growth rate. This distinct feature of the ME-model resulted in nonlinear growth curve predictions that closely matched experimental growth rate data. The ME-model predicted malate dehydrogenase and glycerol-3 phosphate dehydrogenase as alternate electron sinks at the theoretical maximum growth rate. Furthermore, the ME-model identified ferredoxin as a key regulator in electron distribution between two major redox balancing pathways. Since ME-models include the turnover rate for each metabolic reaction, the model in this work was successfully used to capture the experimentally observed temperature regulation of various nitrogenases. Overall, these unique features of the ME-model demonstrated the influence of nitrogenases and rubiscos on *R. palustris* growth and predicted a key regulator in distributing electrons between major redox balancing pathways, thus establishing a platform for *in silico* investigation of *R. palustris* metabolism from a multi-omics perspective.

62c Flux Balance Analysis at Single-Cell Level

Shriramprasad Venkatesan, Rudiyanto Gunawan

Flux balance analysis (FBA) is a powerful method for simulating cellular metabolism and has been driving genome-scale metabolic network reconstructions. The FBA produces metabolic reaction rates, termed 'fluxes', such that these fluxes are stoichiometrically balanced and that they achieve an assumed cellular objective, such as biomass production (Orth, Thiele et al. 2010). FBA has been combined with omics data such as transcriptome (Machado and Herrgård 2014) and proteome (Sánchez, Zhang et al. 2017, Bekiaris and Klamt 2020), to contextualize, constrain, and improve metabolic flux predictions for specific cells. More recently, along with the torrent of single-cell omics data, notably single-cell RNA-sequencing (scRNA-seq) data, there has been a growing interest in applying FBA to simulate metabolic fluxes at single-cell level(Hrovatin, Fischer et al. 2022).

There are challenges in adapting FBA to single cell resolution. First, as it has been shown for bulk data, the expression of metabolic genes is only weakly connected to the flux of reactions that they are associated with based on the gene-protein-reaction (GPR) mapping (Zhang, Li et al. 2010). With the higher noise in scRNA-seq data, this correlation is likely to be lower. Further, scRNA datasets are known to have high sparsity and noise due to technical reasons such as dropouts, as well as biological reasons such as stochastic transcriptional bursting. Therefore, the integration of FBA with scRNA-seq need to account for the complexity of flux control by gene transcription (beyond metabolic genes) and the high level of noise at single cell level.

There have been a few recent developments for flux analysis at single-cell resolution. Yilmaz *et al.* adapted iMAT, a bulk expression integration method, and developed iMAT++ based on the idea that metabolic fluxes directly correlate with the expression of metabolic genes involved (Yilmaz, Li et al. 2020). Alghamdi *et al.* developed the algorithm scFEA to generate fluxes of metabolic modules in single cell level using a graph neural network model (Alghamdi, Chang et al. 2021). Here, flux balancing, metabolic gene expression, and cell metabolic activity are implemented in the loss function that is minimized during model training. scFBA is another single-cell FBA method, developed based on linear programming to minimize the sums of a penalty representing the inverse of a gene expression evidence that a reaction is active, while accounting for the stoichiometry within the network (Damiani, Maspero et al. 2019). In these methods, expression of metabolic genes are mapped to the specific reactions based on the GPR association, which ignores the possible roles of non-metabolic genes in (indirectly) regulating metabolic reaction.

In this study, we developed a novel deep learning-based FBA method that leverages the graphical structure of the metabolic network to predict reaction activity from the expression of both metabolic and non-metabolic genes and stoichiometrically-balanced metabolic fluxes. We built on the concept of a reaction flow graph (Beguerisse-Díaz, Bosque et al. 2018) to represent metabolic networks as reaction graphs. Contrary to the typical depiction of metabolic networks where metabolites are nodes and directed edges are reactions occurring between them, in reaction graphs the nodes represent reactions and the edges describe product-substrate relationship. A directed edge between two nodes indicates that the product(s) of the originating reaction (node) is consumed by the destination reaction. The edge weights of the graph capture the probabilistic flow of metabolites between these reactions. The gene expression of a cell is mapped to reaction activity using a single layer perceptron as illustrated in Figure 1A. The discrepancy between the weights in the perceptron and the GPR association is minimized via the loss function. The output of this perceptron is a vector of reaction activity, which serves as the input to the graph neural network layer of the model.

A graph neural network model based on metabolic reaction graph is used to convolve the activity of reactions to predict metabolic fluxes and state (see Figure 1A). By utilizing multiple rounds of graph convolution, the activity of each reaction is passed through the reaction graph using attention-based rule. The number of layers of graph convolution is decided relative to the diameter of the graph (maximum distance between any two nodes in the graph) and through model tuning to allow maximum classification accuracy between the phenotypes. Our rationale is that the flux of a reaction in a cell is determined not only by its noisy gene expression, but also by those of its parent and ancestors. The convolved reaction activity values are used for two tasks. The first is to classify metabolic phenotype that a cell belongs to (*e.g.* cell type or culture condition). This is to ensure that during the training, the convolved reaction activities are more similar among cells that are of the same or similar phenotype. The second task is to generate stoichiometrically-balanced fluxes. At the single cell level, not all cells may be functioning optimally to satisfy the assumed cellular objective. Thus, a relaxation of flux bounds is done based on the convolved reaction activity. Initially, flux variability analysis (FVA) is performed to identify lower and upper flux bounds that are possible near the optimal flux distribution. During model training, the convolved reaction activities are used to adjust the bounds for reaction fluxes, where reactions with higher (lower) activity are allowed to adopt higher (lower) flux values.

We applied our method to single cell dataset generated by Jariani *et al.* for *Saccharomyces cerevisiae* under different growth conditions: glucose vs. maltose media (Jariani, Vermeersch et al. 2020). In this study, yeast cells were initially grown with maltose substrate prior to experiment. At the start of the experiment, it was shifted to a glucose based media for 12 hours. After 12 hours, the substrate was reverted to maltose and grown for another 20 hours. Samples were periodically taken during 5 timepoints and single cell RNA sequencing analysis was made to get single cell transcriptome of the yeast cells. In the study, the authors found that there was a period of time after the substrate was shifted to maltose that growth was arrested. This was a period where the cells were beginning to adapt their metabolism to the new substrate. They termed this period lag-phase. For our study, we utilized the genome-scale metabolic model iMM904 and gene expression data for two time points - glucose at 6 hours(glu-6h) and maltose in lag-phase for 3 hours (Figure 1B) were chosen for determining fluxes. Figure 1B shows a selected set of fluxes which showed notable shifts in fluxes before and after the switch as predicted by our model. We found that hexokinase and ethanol transport and exchange were significantly altered by the switch in media. Indeed, hexokinase (HXK1) gene was reported as being suppressed in the presence of glucose (Jariani, Vermeersch et al. 2020). Further, as were reported by the authors, we do see that biomass flux is nearly the same between the two time points, but the inlag biomass fluxes show a higher variance which indicate that the cells are beginning to adapt to the new substrate and resume growth.

In summary, our multilayered neural network model is able to generate metabolic fluxes that are both stoichiometrically balanced as well as consistent with gene expression at a single cell resolution. Further improvements of this method are currently underway to elucidate novel metabolic signatures for more complex phenotypes and cell subtypes in human cells.

Alghamdi, N., et al. (2021). 'A graph neural network model to estimate cell-wise metabolic flux using single-cell RNA-seq data.' Genome Research **31**(10): 1867-1884.

Beguerisse-Díaz, M., et al. (2018). 'Flux-dependent graphs for metabolic networks.' npj Systems Biology and Applications **4**(1).

Bekiaris, P. S. and S. Klamt (2020). 'Automatic construction of metabolic models with enzyme constraints.' Bmc Bioinformatics **21**(1).

Damiani, C., et al. (2019). 'Integration of single-cell RNA-seq data into population models to characterize cancer metabolism.' Plos Computational Biology **15**(2): e1006733.

Hrovatin, K., et al. (2022). 'Toward modeling metabolic state from single-cell transcriptomics.' Molecular Metabolism **57**: 101396.

Jariani, A., et al. (2020). 'A new protocol for single-cell RNA-seq reveals stochastic gene expression during lag phase in budding yeast.' eLife **9**.

Machado, D. and M. Herrgård (2014). 'Systematic Evaluation of Methods for Integration of Transcriptomic Data into Constraint-Based Models of Metabolism.' Plos Computational Biology **10**(4): e1003580.

Orth, J. D., et al. (2010). 'What is flux balance analysis?' Nature Biotechnology **28**(3): 245-248.

Sánchez, B. J., et al. (2017). 'Improving the phenotype predictions of a yeast genome-scale metabolic model by incorporating enzymatic constraints.' Molecular Systems Biology **13**(8): 935.

Yilmaz, L. S., et al. (2020). 'Modeling tissue-relevant Caenorhabditis *elegans* metabolism at network, pathway, reaction, and metabolite levels.' Molecular Systems Biology **16**(10).

Zhang, W., et al. (2010). 'Integrating multiple 'omics' analysis for microbial biology: application and methodologies.' Microbiology **156**(2): 287-301.

A

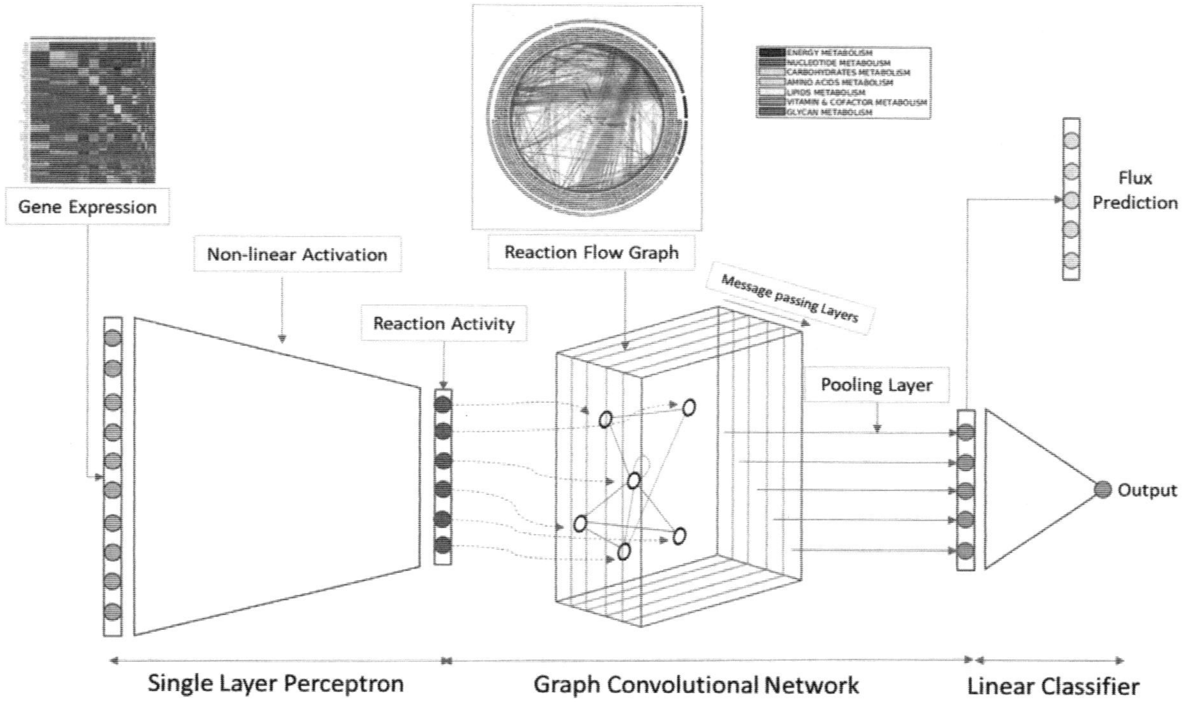

$$loss_{total} = Classification.loss + GPR.loss + S.v$$

B

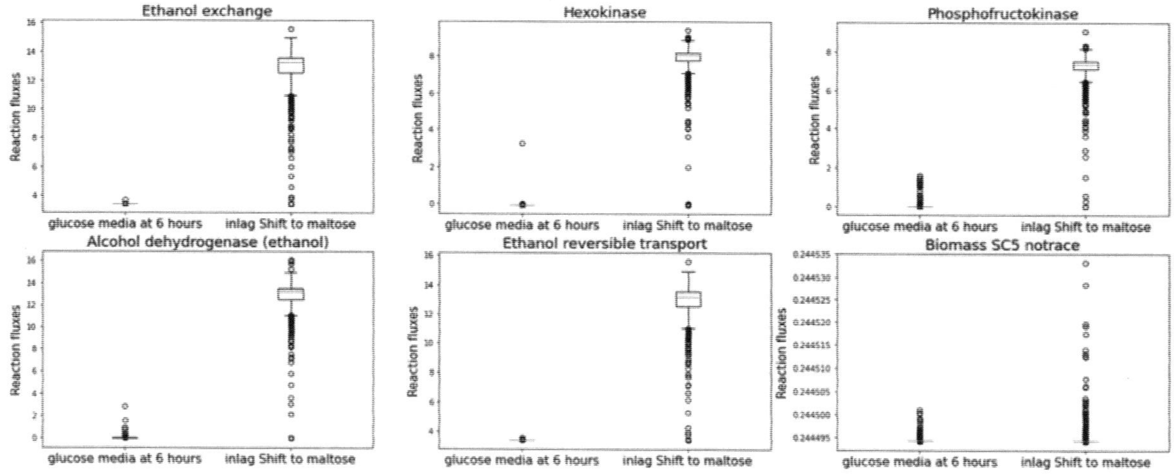

Figure 1. (A) Multi-layer neural network model of cell metabolism. The input to the model is single-cell metabolism. The perceptron layer describes the GPR mapping from gene expressions to reaction activity. The graph convolutional network convolves the single-cell reaction activity using metabolic reaction graph, which are subsequently pooled to produce stoichiometrically balanced fluxes and to classify cell phenotype (e.g. cell type, growth condition). (B) Flux prediction for yeast at single-cell resolutionat two time points – glucose at 6 hours and maltose in lag phase for 3 hours are shown.

62d Nonlinear Modeling of the Estrogen Receptor Activity of Environmental Chemicals Using Classification Algorithms and Single-Cell Level Data

Zahir Aghayev, Adam T. Szafran, Anh Tran, Hari S. Ganesh, Fabio Stossi, Lan Zhou, Michael A. Mancini, Efstratios N. Pistikopoulos, Burcu Beykal

Environmental emergencies such as floods and hurricanes increase the potential risk for contact, inhalation, and ingestion of the released toxic household, medical, and industrial chemicals [1]. Environmental chemicals, both synthetic and natural, have the potential to adversely affect the endocrine system posing long-term risks to human and wildlife reproductive and metabolic health by binding to the endocrine hormone receptors, like estrogen (ER) and androgen (AR) receptors [2]. Modeling the estrogenic activity of benchmark endocrine-disrupting chemicals will allow us to predict agonistic and antagonistic behavior of novel chemicals and complex mixtures and facilitate the decision-making process for mitigative actions.

Recent advancements in big-data analytics and machine learning algorithms aid the computational characterization of estrogenic activity of chemicals using cell-based high throughput assays [3-4]. In this work, high-throughput microscopy and single-cell image analysis followed by machine learning methods were used to construct a classification model of estrogen receptor activity by environmental chemicals based on the single-cell level data without any data reduction strategy. We used 55 different compounds in the experiments which were defined as agonists, antagonists, or inactive chemicals of the ER by the US Environmental Protection Agency (EPA) [5]. After preprocessing the data obtained from 70 size, shape, and intensity features from cell and nuclear compartments after treatment with each chemical, supervised machine learning analyses – Support Vector Machine (SVM), Random Forest (RF), Neural Network (NN) classifiers were trained to ultimately predict the estrogen receptor-α agonists or antagonists. The results show that the two-class non-linear SVM model predicts the outcome of the test data with more than 92% accuracy.

References

[1] Cooper, C.M. and Wardropper, C.B., 2021. Environmental, public health, and economic development perspectives at a Superfund site: AQ methodology approach. *Journal of Environmental Management*, *279*, p.111571.

[2] EPA, 2022. What is Endocrine Disruption? https://www.epa.gov/endocrine-disruption/what-endocrine-disruption (Accessed on April 5, 2022).

[3] Ganesh, H.S., Beykal, B., Szafran, A.T., Stossi, F., Zhou, L., Mancini, M.A. and Pistikopoulos, E.N., 2021. Predicting the Estrogen Receptor Activity of Environmental Chemicals by Single-Cell Image Analysis and Data-driven Modeling. In *Computer Aided Chemical Engineering* (Vol. 50, pp. 481-486). Elsevier.

[4] Mukherjee, R., Beykal, B., Szafran, A.T., Onel, M., Stossi, F., Mancini, M.G., Lloyd, D., Wright, F.A., Zhou, L., Mancini, M.A. and Pistikopoulos, E.N., 2020. Classification of estrogenic compounds by coupling high content analysis and machine learning algorithms. *PLoS computational biology*, *16*(9), p.e1008191.

[5] Judson, R.S., Magpantay, F.M., Chickarmane, V., Haskell, C., Tania, N., Taylor, J., Xia, M., Huang, R., Rotroff, D.M., Filer, D.L. and Houck, K.A., 2015. Integrated model of chemical

perturbations of a biological pathway using 18 in vitro high-throughput screening assays for the estrogen receptor. *Toxicological Sciences*, *148*(1), pp.137-154.

62e Measurement and Modeling of Heterogeneous Single-Cell PI3K Inhibition Dose Responses

Patrick C. Kinnunen, Gary D. Luker, Kathryn E. Luker, Jennifer J. Linderman

Precision health care in cancer increasingly focuses on treating patients with drugs targeted against specific driver mutations present in malignant cells. Targeted drugs typically produce transient responses with subsequent recurrence fueled by drug resistant cancer cells. Drug resistance arises from multiple cell intrinsic and/or environmental causes, including secondary genetic mutations, compensatory signaling pathways, intercellular signaling, and mass transfer limitations for drugs in spatially heterogeneous, disorganized tumor environments. Standard models for drug development and testing in cell-based assays typically rely on population-scale metrics defined by concentration-response curves. These metrics include maximum effect (E_{max}) and the concentrations of a compound required to reach 50% of the maximum effect (EC_{50}). However, population-scale metrics of efficacy mask heterogeneity among single cells. Drug resistance and tumor recurrence commonly arise from small subsets of cancer cells, motivating new approaches to analyze drug effects in single cells.

In this work, we first simulated concentration-response curves from heterogeneous populations of breast cancer cells. Our simulations reveal that arbitrary distributions of concentration-response parameters, including bimodal distributions or distributions with small subpopulations, can yield identical population level concentration-response curves. Next, we used live, single cell fluorescence microscopy to measure concentration-dependent responses to clinically-approved inhibitors (alpelisib and omipalisib) of the oncogenic phosphatidylinositol-3-kinase (PI3K) - Akt pathway. We quantified effects of these compounds on activation of Akt and ERK, two major kinases driving proliferation and survival in breast cancer and multiple other malignancies, with multiplexed kinase translocation reporters (KTRs). KTRs reversibly move between the nucleus and cytoplasm of cells, providing a dynamic, quantitative readout of kinase activity and inhibition. We developed automated image processing methods to quantify KTRs in thousands of cells, tracking activities of Akt and ERK in single cells over time. We stably expressed KTRs for Akt and ERK in multiple breast cancer cell lines with mutations that activate the Akt pathway. Our studies revealed that cells respond to PI3K inhibitors rapidly and heterogeneously, with different magnitudes and durations of inhibition. We next subjected cells to continuously escalating doses of PI3Ki and extracted dose-dependent kinase activity measurements from individual cells over time. We determined that both breast cancer cell lines displayed heterogeneous EC_{50} and Emax values for each tested PI3K inhibitor. Approximately 10% of cells displaying EC_{50} values 10 times higher than the population EC_{50} and approximately 10% of the cells showing an E_{max} close to zero, indicating a lack of kinase inhibition.

We next investigated the impact of measured cell-intrinsic heterogeneity on drug resistance relative to variations in pharmacokinetic and mass transport dynamics of small molecule drugs that govern tissue distribution in vivo. We integrated a two-compartment pharmacokinetic model with a Krogh Cylinder model of drug distribution in tissue around a blood vessel. We calibrated this model based on measured PK data for one of our PI3K inhibitors and then calculated different distributions of pharmacokinetic parameters at various locations around a blood vessel. For instance, we compared random spatial distributions of EC_{50} and E_{max} values for single cells or distributions that were a function of radius from the blood vessel against the baseline case where all cells had identical concentration-dependent responses. We found that intrinsic variation in cellular dose-response can accentuate tissue heterogeneity, increasing the likelihood of resistance to an administered drug.

In summary, our systems biology approach based on single-cell imaging data provides an innovative framework to analyze susceptibility or resistance to cancer therapies. Predicting how cell-intrinsic and environmental factors regulate efficacy of drugs promises a new pathway to overcome drug resistance and improve outcomes for patients with cancer.

62f Nuclear Organization and Genome Regulation By RNA-Dependent Phase Separation

Krishna Shrinivas, Jonathan E. Henninger, Ozgur Oksuz, Pradeep Natarajan, Phillip A. Sharp, Richard A. Young, Arup K. Chakraborty

Precise gene expression is essential for cell-type specification, development, and physiology. In multi-cellular organisms such as mammals, this regulation is dependent on the coordinated organization of many DNA, RNA, and protein species, which are often present at far distances from each other in a crowded nuclear milieu. Recently, phase separation has emerged as a key mechanism contributing to organization of multiple nuclear compartments such as the nucleolus, transcriptional condensates, and nuclear speckles. The interplay between these nuclear compartments that form by phase separation, gene activity, and newly synthesized RNA is not well understood. To investigate this interplay, we develop computer simulations based on statistical physics models, which are tested and refined in collaboration with experiments *in vitro* and in cells. We first find that different RNA species and transcriptional components undergo phase separation in a manner consistent with complex coacervation of charged poly-ions. We identify that diffusion is limiting under cellular conditions, and thus synthesis of RNA can either assemble (low rates, such as at non-coding or regulatory elements) or dissolve (high rates, such as during productive gene expression) phase-separated compartments. We argue that this provides a model of dual feedback by RNA levels and may contribute to self-limited regulation of gene expression. Model predictions that connect the effect of perturbations in RNA synthesis to variations in condensate size and lifetime are verified by through live-cell imaging as well as gene-reporter assays in cells. We then extend the model to investigate how spatially localized gene activity (for e.g. through clustering on the genome) influences condensate properties. Through extensive simulations, we identify a unified framework that describes diverse phenomena in nuclear organization that often lack direct mechanistic bases - including condensate flow, nucleation by RNA synthesis, nucleolar vacuoles, and aspherical morphologies. Overall, our model implicates gene activity (RNA synthesis) as a potent regulatory axis of gene expression and nuclear organization.

62g Metabolic Modeling Tools in Kbase to Integrate Multi-Omics Data, Understand Microbiome Interactions, and Explore Energy Biosynthesis Mechanisms

Christopher S. Henry

Increasingly multi-omics data is becoming more accessible for the study of a wide range of complex biological systems. Today, large-scale metagenomes can be readily obtained from soil microbiome systems, while the instruments and protocols surrounding the collection of metabolomic and proteomic data are constantly improving. Yet analysis methods still struggle to annotate and integrate these individual datasets, let alone combine them to discover new biological principles. Many of the species observed within microbiome systems have never been cultured and it is a challenge to assemble their genomes from metagenomic data; many of the genes observed within microbiomes are not annotated or misannotated; and many of the metabolites observed cannot be identified or associated with known biochemical pathways. Here we will discuss approaches we are applying with KBase and ModelSEED to address all three of these problems.

First, I'll be discussing recent advances in our ModelSEED reconstruction pipeline to dramatically improve the quantitative accuracy of our draft models in predicting ATP yields across the tree of life. We apply this pipeline to 5000 reference and representative genomes, showing dramatic improvements in model size, accuracy, and completeness compared with previous versions of ModelSEED. In comparing to competing tools like CarveMe, we show improved accuracy. By correcting ATP predictions in draft reconstructions, we empower a broad family of model analysis algorithms like E-matrix analysis that rely on accurate ATP predictions to work properly.

We worked in collaboration with Mikayla Borton and Kelly Wrighton to apply our improved pipeline to analyzing multi-omics data from the Genome Resolved Open Watersheds (GROW) project. In this project, the GROW team has already sequenced 161 river microbiomes, from which they have assembled 2093 metagenome assembled genomes (MAGs) dereplicated across all samples. We applied the annotation and modeling systems in KBase to construct and characterize metabolic models for all of these MAGs, exposing insights into variation in metabolic pathways and energy biosynthesis mechanisms across these strains.

Unfortunately, draft metabolic models still struggle to accurately predict phenotypes for new genomes and MAGs, as they are extremely vulnerable to errors propagated from annotation gaps in metabolic pathways. Machine learning (ML) is more robust to incomplete data and generally more accurate than models when sufficient training data is available, but ML lacks mechanistic detail in output. Clearly there is great potential synergy in combining models and ML. We explore this idea in collaboration with Jim Davis by developing machine learning classifiers to predict growth on 60 distinct carbon substrates based on a training set of 178 diverse genomes. We similarly generated draft models for these genomes, demonstrating that machine learning predictions significantly outperform draft models. Thus, by fitting models to ML predictions for new genomes, we can rapidly identify missing annotations and raise the quality of models, even for incomplete genomes.

Another significant challenge is that many metabolic pathways remain unknown, while the genes associated with these pathways remain unannotated or misannotated. To address this challenge, I'll be discussing the gene and pathway function discovery pipeline in KBase. We apply this pipeline to predict the pyridine degradation pathway and its associated genes in *Arthrobacter luteus*. This pipeline integrates data from genomes, models, cheminformatics, transcriptomes, and protein structure to identify the most probable candidate gene, which was subsequently validated in the lab. We also applied this pipeline to predict novel metabolic pathways in the minimal genome developed by JCVI, boosting our ability to explain the metabolome of this species from 10% of observed metabolites to 50% of the observed metabolites.

664a A Low-Cost Paper-Based Sample Preparation Module to Lyse Bacterial Cells and Extract Genomic DNA Using Isotachophoresis

Shruti Soni, Bhushan Toley

ABSTRACT

In recent years, much work and effort have been put into developing affordable and simple POC-nucleic acid amplification tests (NAATs) for diagnosing infectious diseases like tuberculosis. While the primary focus has been on nucleic acid (NA) amplification and detection strategies, many POC-NAAT assays still rely on complex sample preparation methods. To overcome this limitation, we demonstrate a POC-compatible paper-based device capable of purifying and concentrating *Mycobacterium tuberculosis* genomic DNA (gDNA) from viscous mock-sputum using isotachophoresis. We also demonstrate the device's capability to electrochemically lysis *M. Smegmatis* bacteria, along with purification and concentration of the *Mycobacterium smegmatis* gDNA. This is the first demonstration of sputum DNA preparation using paper-based ITP.

KEYWORDS: Paperfluidics, sample preparation, NAAT, isotachophoresis (ITP), point-of-care (POC) diagnostics, bacterial lysis

INTRODUCTION

Dependence on off-line sample preparation methods involving complex protocols, instrumentation, and trained technicians has proved to be one of the major roadblocks in the progression of POC-NAATs. The presence of several amplification-inhibitory molecules within complex biological samples such as sputum makes sample preparation a critical pre-analytical step to obtain more clinically relevant results for TB diagnosis. Sample preparation generally involves sample collection, storage, NA isolation (cell lysis), purification, and concentration.

ITP is an electrokinetic separation and concentration technique that utilizes an electric field and a discontinuous buffer system consisting of a fast-moving leading electrolyte (LE) and slow-moving trailing electrolyte (TE). It is used to concentrate the charged species (DNA) at the LE/TE interface. The electrophoretic mobility of the DNA must lie between LE and TE. Paper-based ITP formats have been shown to work well in focusing dyes, short-ssDNA, dsDNA ladder, and extract/amplify DNA (200 bp) spiked in serum and blood samples.[1,2] Moving towards the goal of an integrated sample-to-answer POC-NAAT device for TB, we present an inexpensive paper-based ITP platform capable of lysing bacterial cells and purifying and concentrating gDNA from viscous mock-sputum within 15 min.

EXPERIMENTAL

The paper-based ITP (**Fig. 1**) device consists of an 11-layered foldable structure with circular disks of Whatman filter paper sandwiched between the two reservoirs (containing TE and LE buffer) that forms a paper channel for the liquid to flow between the two reservoirs. gDNA (10^2-10^0 copies/μL) mixed with 900 μL of TE buffer (2 mM tris-taurine, TTD) was added to the TE reservoir along with simultaneous addition of 900 μL of LE buffer (1 M tris-HCl, TH) to the LE reservoir. A voltage bias of 18 V was applied between the platinum electrodes in the reservoirs. Post-ITP, DNA gets concentrated in some of the initial paper disks. Each paper disk can be used as a PCR/real-time PCR template, or the concentrated DNA can be eluted out from the paper disk for DNA amplification. The device's capability to purify gDNA in the presence of inhibitory substances was also tested.

In another set of initial experiments, we also demonstrate the device's capability to electrochemically lysis M. *Smegmatis* bacteria, spiked in mock sputum (TE reservoir), and simultaneous purification and concentration of the *Msm* gDNA into the paper disks.

RESULTS AND DISCUSSION

Gel electrophoresis images of the gDNA concentrated within the paper layers (L1-L11) after ITP, along with some positive (P: 230 bp target; red arrow) and negative controls (N, NP, TH, TT), are shown in **Fig. 2.** The concentration of 10^2 copies gDNA/μL and 1-copy (10^0) gDNA/μL is depicted in **Fig. 2A** and **2B**, respectively. gDNA mixed with TE buffer (TTD) before undergoing ITP concentration is represented by the green arrow, whereas the bands highlighted within the yellow rectangle represent the gDNA concentrated within the paper layers. The higher intensity of the bands within some of the paper layers compared to the intensity of the band corresponding to their respective TTD signifies the concentration and focusing of gDNA within the paper layers. The number of paper layers with the gDNA after ITP depended on the initial copy number of the sample. The device was able to concentrate and purify the DNA even in the presence of inhibitory molecules present in the mock-sputum (egg yolk emulsion + methylcellulose) (**Fig. 2C**). qPCR results showed approx. 13x concentration of purified DNA, post-ITP (**Fig. 3**). **Fig. 3 A-C** shows the normalized DNA concentration (w.r.t. the liquid sample before performing ITP, TTD) for different gDNA copy numbers ranging from 10^3-10^1 copies/μL. PTTD (in Fig. 3) refers to the control in which a blank paper disk was dipped in the DNA sample (TTD) before performing ITP. The qPCR was performed on the concentrated gDNA eluted out of the paper disk (using 1X TE buffer) after completing the ITP experiment. **Fig. 3D.** depicts the normalized DNA concentration for 3X-diluted mock sputum spiked with 10^2 gDNA/μL, suggesting that the paper-based ITP device can purify and concentrate the gDNA even in the presence of inhibitory molecules. ITP was performed for 20 minutes at 18 V for mock sputum samples.

Using this paper-based ITP device, we were also able to electrochemically lyse (via electrochemical generation of hydroxide) *M. Smegmatis* bacteria cells, spiked in mock sputum (TE reservoir), and simultaneously purify and concentrate M*sm* gDNA into the paper disks. (results not shown here)

CONCLUSION

This low-voltage paper-based ITP platform is the first demonstration of sputum DNA preparation using paper-based ITP. One of the advantages of this device is that it can accommodate large sample volumes (900 μL). It offers a critical pre-analytical step involving bacterial lysis along with purification and preconcentration of gDNA from complex sputum samples and complements paper-based isothermal nucleic acid amplification TB tests.[3] Ultimately, we aim that these tools would enable molecular diagnostics of TB in low-resource settings.

REFERENCES

1 A. T. Bender, M. D. Borysiak, A. M. Levenson, L. Lillis, D. S. Boyle and J. D. Posner, *Anal. Chem.*, 2018, **90**, 7221–7229.

2 X. Li, L. Luo and R. M. Crooks, *Lab Chip*, 2015, **15**, 4090–4098.

3 N. Kaur, J. S. Michael and B. J. Toley, *Sci. Rep.*, 2019, **9**, 15367.

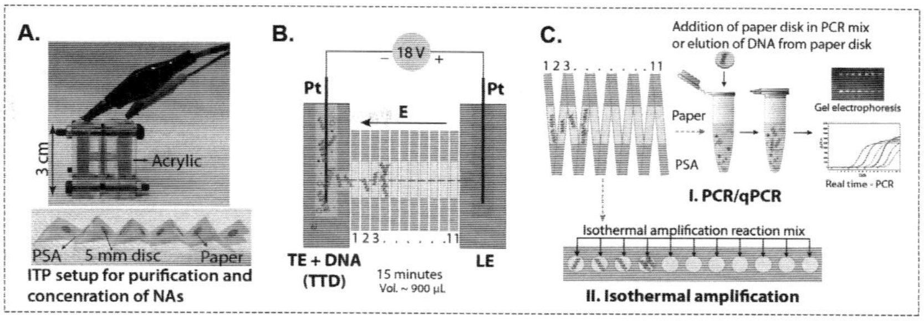

Figure 1: ***A.*** *Paper-based ITP platform (below: Whatman filter paper containing a blue dye for visibility secured in PSA)* ***B.*** *Schematic of a paper-based ITP device depicting LE and TE reservoirs made from acrylic sheets, 11-layer paper structure, Pt electrodes, and a voltage source.* ***C.*** *After performing ITP, the paper disks could directly be used as a template for PCR in tube or in-situ isothermal amplification.*

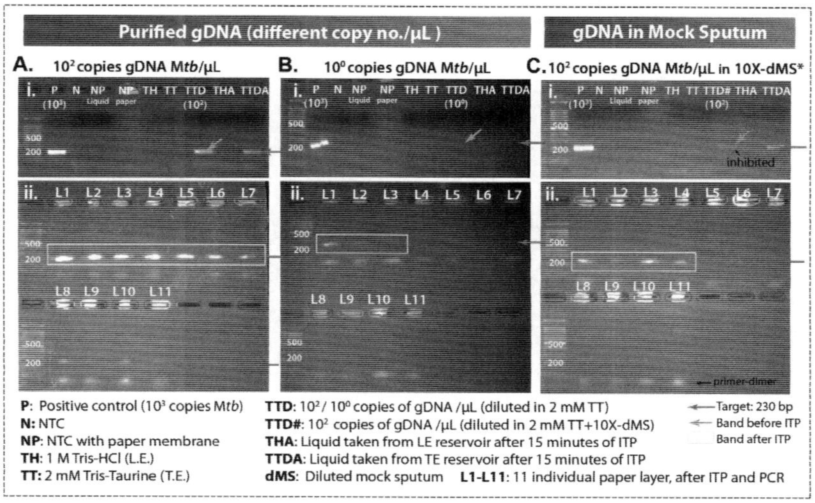

Figure 2: *Gel electrophoresis of positive and negative control along with different layers (L1 to L11) of the paper disks after performing ITP for 15 minutes at 18 V followed by PCR for three cases:* ***A.*** *10^2 copies gDNA Mtb/µL and* ***B.*** *10^0 copies gDNA Mtb/µL diluted in 2 mM tris-taurine buffer (TE), and* ***C.*** *10^2 copies gDNA Mtb/µL in 10X-diluted mock sputum.*

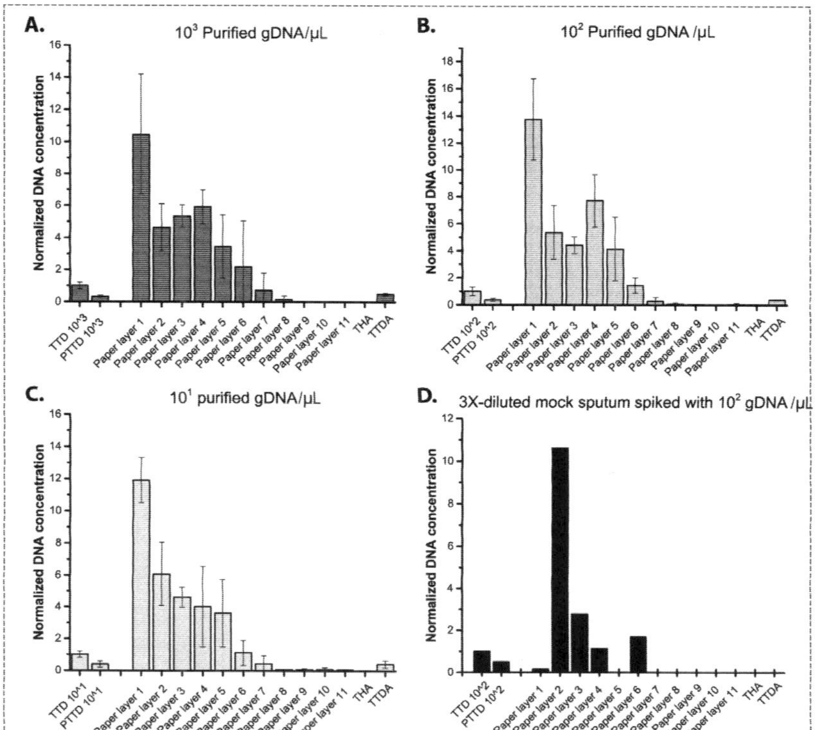

Figure 3: *Normalized DNA concentration w.r.t. liquid sample TTD, post ITP for different gDNA copy number* ***A.*** *10^3 copies gDNA Mtb/µL and* ***B.*** *10^2 copies gDNA Mtb/µL and* ***C.*** *10^1 copies gDNA Mtb/µL diluted in 2 mM tris-taurine buffer (TE),* ***D.*** *10^2 copies gDNA Mtb/µL spiked in 3X-diluted mock sputum.*

664d Leveraging the Ordered Packing of Spherocytes As a Potential Diagnostic

Logan Piegols, Tobias Dwyer, Sharon Glotzer, Omolola Eniola-Adefeso

Introduction: Hereditary spherocytosis is a genetic disorder that causes red blood cells (RBCs) to become more rigid and spherical in shape. This apparent change in RBC morphology leads to lifelong health complications such as anemia, jaundice, gallstones, and spleen damage. Current clinical diagnostic methods can be prohibitively expensive and nevertheless, due to the mildness of most cases the disorder is not diagnosed. By leveraging hexactic packing of spheres, it may be possible to identify this disorder using a minimal cost sedimentation method. This study seeks to quantify the granular structures for mixtures of discocytes (biconcave disc RBCs) and chemically altered spherocytes (spherical RBCs) in a sedimentation system. We describe the radial distribution functions of each cell type, hexactic order, clustering of cell mixtures using advanced analysis computational methods. We find that the presence of spherocytes can cause differences in the ordering and disordering of the sedimentation system. Furthermore, we investigate the role of rigidity in the differential hexactic ordering of pure discocytes and spherocytes, and find significantly higher hexactic ordering of spherocytes compared to discocytes.

Methods: Blood draw protocols have been approved by the University of Michigan Internal Review Board (IRB-MED). RBCs were separated from whole blood samples using centrifugation with the plasma and white blood cells being discarded. Spherocytes were made from healthy RBC samples via exposure to sodium salicylate and exposed to tert-butyl hydroperoxide for rigidification as needed. Alteration of RBC shape was confirmed with confocal imaging of fluorescently stained RBC membranes via Wheat Germ Agglutinin, Alexa Fluor™ 488 (Thermo Fisher). Fractions or pure samples of discocytes and spherocytes were then reconstituted in phosphate buffer solution at less than 1% hematocrit. The blood solutions were then pipetted onto the bottom side of a suspended glass slide and allowed to sediment for 10 minutes. The monolayer of RBCs formed at the air-water boundary where then imaged using light microscopy. Centroid locations of each cell were selected via ImageJ. Centroids were analyzed using OVITO software to quantify average hexactic ordering, radial distribution functions, and clustering of cells.

Result and Conclusions: This work highlights how an increasing fraction of spherocytes can cause fundamental changes in how both spherocytes and discocytes hexatically order in sedimentation (Figure 1A). A significant jump in hexatic order from low to high following a bulk fraction of spherocytes of 0.6 may be indicative of a phase transition. Having spherocytes does not always mean increased hexatic ordering but instead regions of varying order are produced through the introduction of spherocytes. Investigations into the role of rigidity on hexatic order for pure samples, highlight shape playing a significant role in hexatic order while rigidity being not significant (Figure 1B). Thus, shape likely plays a larger role in the hexactic ordering of the sedimented RBCs. Radial distribution functions show spherocytes having longer range order than that of discocytes. Undoubtedly, spherocytes produce differential granular crystallization from that of just discocytes. Further investigations into the structures spherocytes can produce via sedimentation may provide key insights into developing a cheap and effective method for identifying hereditary spherocytosis.

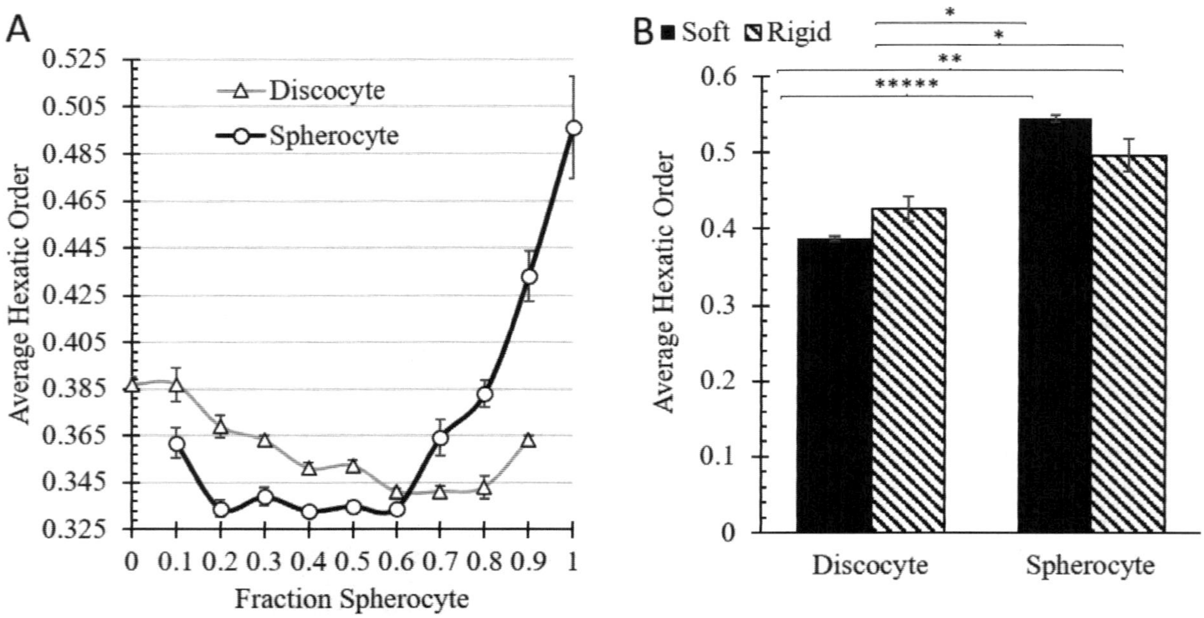

Figure 1: Average hexactic ordering of RBCs in sedimentation. A) Average hexatic order of each type of RBC as bulk fraction of spherocytes is increased. B) Average hexatic order of pure samples of both rigidified and soft (non-rigidified) discocytes and spherocytes.

664e Excitation-Scanning Hyperspectral Imaging Technologies for Multilabel Cellular Imaging

Silas Leavesley, Naga Srilakshmi Annamdevula, Madison Howard, Craig Browning, Marina Parker, William Oswald, Andrea Britain, C. Michael Francis, Na Gong, Thomas Rich

Hyperspectral imaging technologies were originally developed by NASA and the Department of Defense for satellite imagery and remote sensing applications. More recently, these technologies have found a wide range of potential uses in the biomedical field, from basic research studies to clinical translation. One area in which hyperspectral imaging technologies have been highly utilized is fluorescence microscopy. Typical hyperspectral imaging fluorescence microscopy approaches involve optical filtering or dispersion of the emitted fluorescence in order to measure the fluorescence emission spectrum (emission-scanning). However, an alternative technological approach can be used to allow measurement of the fluorescence excitation spectrum (excitation-scanning). When implemented using broad-band or long-pass fluorescence emission filters, we have previously shown that the technique of excitation-scanning hyperspectral imaging may allow for significantly improved signal strength and/or imaging speed. Here, we report on the capabilities of excitation-scanning hyperspectral imaging microscopy for detection of many (5+) fluorescent signals in cell and tissue preparations for two application areas: live cell imaging (cell signaling) and tissue diagnostics.

Cell signaling studies were performed utilizing both pulmonary microvascular endothelial cells and human airway smooth muscle cells. Cells were labeled to allow detection of either cAMP (using the H188 Turquoise-Epac-Venus FRET reporter) or Ca^{2+} (using the calcium indicator Cal-520 AM). Time-lapse cell signaling spectral image data were analyzed using custom spectral unmixing and image post-processing algorithms implemented in MATLAB (MathWorks), as well as a custom software package for automated ROI-based cell signal quantitation called S8. Tissue imaging studies were performed using de-identified pairs of resected cancerous and non-involved colorectal tissues, obtained in accordance with the University of South Alabama Institutional Review Board (IRB exempt). Tissue spectral image data were analyzed using convolutional neural network-based approaches to differentiate cancerous and non-involved tissues. For both live cell imaging studies and tissue imaging studies, images were acquired using a custom excitation-scanning spectral imaging microscope that allowed for sequential excitation of fluorescence across a range of wavelengths, using either a 300 W Xe arc lamp (Titan X300, Sunoptic Surgical) and tunable filter assembly (VF-5, Sutter Instruments) or a custom LED-based spectral illumination module, with detection provided by a high-speed sCMOS camera (Prime 95B, Photometrics).

Results from both cell and tissue imaging studies indicate that sufficient information is present in excitation-scanning spectral image data to allow separation of 5+ fluorescent labels with sufficient signal strength so as to visualize and automatically quantify fluorescent molecule levels within subcellular regions. Additionally, results from tissue imaging studies indicate that excitation-scanning spectral image data can be used for automatic classification and differentiation of cancerous and non-involved tissue types. Finally, our results indicate that excitation-scanning spectral imaging can be implemented using a range of technological approaches, including broad-band illumination sources (arc lamps or supercontinuum laser) with tunable filters or through combination of many discrete wavelength sources (LEDs). These results suggest that excitation-scanning spectral imaging is a versatile alternative to emission-scanning spectral imaging that may be implemented on a range of microscope platforms. This work was supported by NIH awards P01HL066299, R01HL58506, and R01HL137030, and NSF award MRI1725937. Drs. Leavesley and Rich disclose financial interest in a start-up company, SpectraCyte LLC, that was formed to commercialize spectral imaging technologies.

664f Vapor Biomarker Analysis for Precision Healthcare: From Benchtop to Wearable

Xudong Fan

In this presentation, I will first discuss vapor detection technologies that are commonly used, especially, gas chromatography (GC). Then I will discuss various micro-GC technologies developed in our lab, including both components like micro-thermal injectors, micro-columns, and micro-photoionization detectors, etc., and systems such as one-dimensional GC and two-dimensional GC. Finally, I will talk about how to move vapor analysis from benchtop to portable and even wearable for breath and body odor analysis. A few biomedical applications that use breath analysis will be described, such as acute respiratory distress syndrome (ARDS), asthma, and COVID-19.

447a Delivery of Immunostimulatory Biologics for Skin Tissue Repair

Muhammad Raisul Abedin, Shubham Pallod, Jordan Yaron, Mallikarjun Gosangi, Jacquelyn Kilbourne, Kaushal Rege

The management of dermal wound healing at a chronic stage is a major medical burden. Dysregulation of wound healing processes can lead to chronic wounds and increased morbidity and mortality. An impaired immune system significantly delays the wound healing process. Timely delivery of the relevant therapeutics to stimulate the immune response in the wound area is crucial for an orchestrated response to injury needed for a faster healing process. In this work, we explored the role of a natural biopolymer along with an immunostimulatory nucleic acid (iNA) to induce a robust innate immune response aiding in skin wound healing. A splinted dermal wound healing mouse model was used to investigate wound healing in immunocompetent Balb/c mice to better reproduce the granulation and re-epithelialization characteristics of human wound healing. The iNA was delivered at the wound site alone or in combination with the natural biopolymer. On day 7, mice were euthanized, and tissue was collected for immunohistochemical analysis. Trans-epidermal water loss (TEWL) data, indicative of skin barrier function, were recorded with a VapoMeter (Delfin). On each day, including days without tissue collection, photographs were collected for planimetric analysis of the wound area. Individual treatment of the immunomodulatory natural biopolymer and the therapeutic biologic produced from the iNA resulted in significant wound closure. In the case of the delivered drug without the biopolymer, we found significantly high localization of the delivered therapeutic at day 1 which was transiently expressed at the wound area and maintained the expression level till day 7. However, in the case of the drug delivered from the biopolymer, the availability of the localized therapeutics was reduced and tended to last for shorter days compared to the individual drug treatment alone. The TEWL data for the natural biopolymer and iNA alone showed significantly higher barrier function recovery compared to the untreated open wounds. In summary, delivery of immunomodulatory biopolymers and iNA derived therapeutics resulted in significant enhancement of wound healing in immunocompetent mice, thereby opening new avenues for tissue repair.

447b Engineering a Logic-Gated Drug Delivery Platform for Cancer Immunotherapy

Blaise Kimmel, John Wilson

Immune checkpoint inhibitors (ICIs) are a recently developed immunotherapy for an increasing number of cancer types. The activation of pattern recognition receptors (PRRs), such as the stimulator of interferon genes (STING), can produce type-I interferons (IFN-Is) and other cytokines to induce and amplify a coordinated immune response within the tumor microenvironment (TME). This is increasingly important for the engineering of immunostimulatory molecules that can be delivered intratumorally to "cold" tumor sites where the number of active T cells are low in order to create a more highly immunogenic "hot" TME. Recent work in our group has demonstrated that engineered nanoparticles containing cyclic-dinucleotide (CDN) STING agonists (STING-NP) offer enhanced intracellular delivery of CDNs, leading to increased tumor immunogenicity and response to ICI. However, the STING-NP strategy currently lacks tumor-specific targeting, and its efficacy and safety are limited due modest tumor distribution. Thus, controlling the spatial delivery of STING agonists to the TME is necessary to systemically treat metastatic disease safely. In this talk, I will discuss our engineering of a logic-based immunotherapy for STING activation. Here, we utilize the design/build/test cycle from synthetic biology to construct "smart" STING agonist with logic-gated activation within solid tumors. We used advanced molecular engineering strategies to design and construct a logic-gated drug delivery agent with precise drug loading of the STING agonist. Within our drug delivery agent is a single domain antibody fragment (nanobody) for selective targeting of EGFR+ tissue. Further, the loaded STING agonist features environmentally restrictive activation at the TME. The combined logic-gated activity of tumor binding with the nanobody and site-specific activation of the STING agonist is demonstrated in vitro to confirm biological activity and improve upon design features for maximum efficacy within EGFR+ cell lines. We next evaluated our logic-gated drug delivery agent in breast cancer mouse models to quantify STING activation in the TME, quantify therapeutic efficacy/toxicity, and evaluate biodistribution. The engineering strategies identified within this talk provide fundamental insights into the role of logic-gated immune activation in vivo.

447c Localized Delivery of Tumor-Specific Antigen and Immune Checkpoint Antibodies to Induce Systemic Antitumor Immunity

Rong Tong, Eungyo Jang

The development of cancer immunotherapeutics to enhance anti-tumor immune responses has led to significant clinical advances. However, the rate of complete response to single immunotherapy drugs remains low, due in part to the complex immune network of the tumor microenvironment. Combinations of immunotherapies improve the efficacy but result in exacerbating toxicities. None of the current delivery strategies, including local administration that still leaks potent drugs to the circulation, can avoid debilitating toxicities that prevent immunotherapies from reaching their full curative potential. To overcome such delivery problems, we developed a localized drug delivery platform using tumor antigens and immune-checkpoint antibodies with prolonged intratumoral retention to enable sustainable delivery of combinations of immunotherapeutics and elicit durable antitumor immune responses. This was supported by our preliminary studies that intratumorally administered OVA antigen and anti-PD1 significantly delayed mouse lymphoma EG7-OVA tumor outgrowth without exacerbating toxicity, compared with the local treatment of individual drugs. Additionally, we also demonstrated that the sustainable intratumoral delivery of immune checkpoint antibodies and OVA antigens could elicit durable anti-tumor responses with robust immune memories in such tumor models. We therefore envision that high efficacy and high safety can be achieved together using such localized delivery technology that allows for specific tumor eradication without systemic toxicities.

447d Composite Deoxycholic Acid-Based Microparticles for Localized Adipose Tissue Reduction with Reduced Inflammation

Daniel Kupor, Michael L. Felder, Hanieh Safari, Omolola Eniola-Adefeso

Excess adipose tissue can lead to various physical health concerns and body image issues. Deoxycholic acid is an FDA-approved treatment for removing adipose tissue, initially proposed as a safer injectable solution than standard yet risky surgical procedures such as liposuction. However, limited efficacy, excessive doses, and severe injection site inflammation have impeded the success of this drug formulation. Recently, microparticles composed of bile salts and gold nanoparticles have been shown to lyse adipocytes *in vitro* and *in vivo* with lower inflammation than the clinically used deoxycholic acid treatment. Here, we describe the long-term local reduction of adipose tissue by subcutaneous injection of deoxycholic acid-based composite microparticles using an optimized dosing scheme. These composite microparticles release the active drug and gold nanoparticles over time through surface erosion to curb inflammation and cause fewer side effects at the injection site. We demonstrate that these particles significantly reduce adipose tissue mass over 4 weeks post-injection in an obese mouse model with a subcutaneous injection and contralateral control. Additionally, the particle therapy exhibits reduced inflammation and associated immune cell infiltration compared to deoxycholic acid solution, observed through histology and cytokine levels within the adipose tissue. Any increase in inflammation for the particle treatment is completely resolved by 4 weeks post-injection, as shown in **Figure 1**. Finally, we show that a second dose of particle therapy after 4 weeks can further reduce adipose tissue mass. Overall, deoxycholic acid-based composite microparticles promise to deliver an effective and safer injectable therapy for local adipose tissue reduction with fewer treatments.

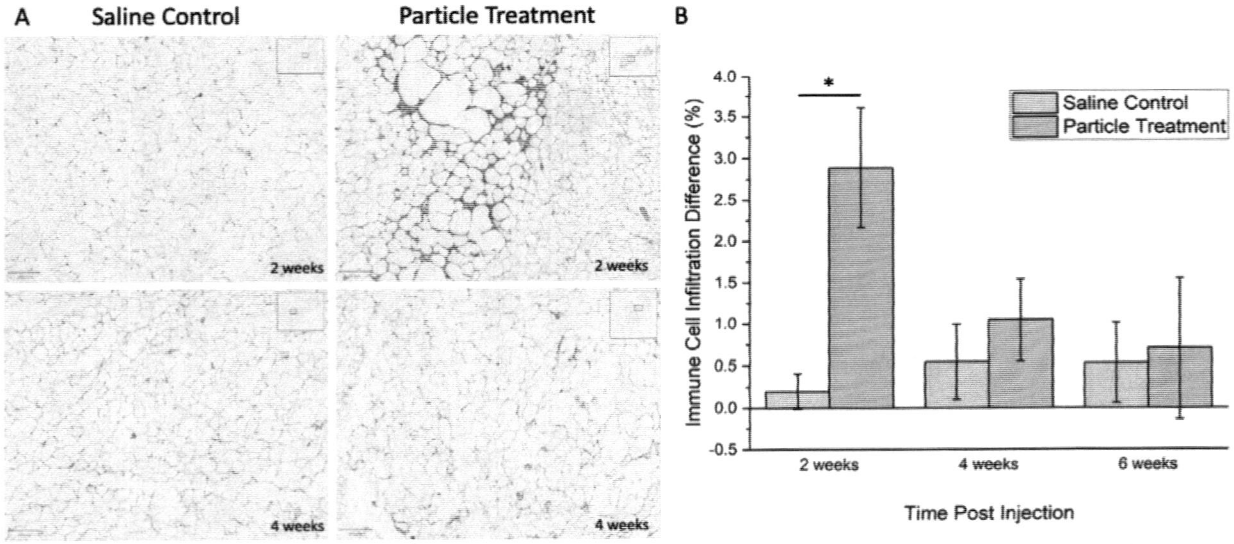

Figure 1: (A) Histology images of the adipose tissue post-treatment with H&E staining showing immune cell infiltration in purple. (B) Quantification of immune cell infiltration in both the saline and particle-treated fat pads compared to the contralateral control showing the resolution of particle-caused inflammation after 4 weeks (* p<0.05).

447e Nanoparticle Delivery of Small Molecule Inhibitors to Treat Sting-Associated Inflammatory Diseases

Lucinda Pastora, Kyle Becker, Lihong Wang-Bishop, Plamen Christov, John Wilson

Problem: The cyclic GMP-AMP synthase (cGAS)/Stimulator of Interferon Genes (STING) pathway is implicated in the development and progression of a myriad of inflammatory diseases including nonalcoholic steatohepatitis, colitis, sepsis, age-related macular degeneration, and cellular senescence. Thus, STING pathway inhibitors (in isolation or in combination with other therapies) could have wide therapeutic application in inflammatory conditions. The cGAS inhibitor RU.521 and the STING inhibitor H-151 have improved outcomes in mouse models of colitis, AKI, and ALS; however, these studies required frequent high-dose i.p. injections of the inhibitors, which limits translatability. Furthermore, long-term use of systemically administered cGAS or STING inhibitors may leave patients vulnerable to viral infections and tumor development. Thus, targeted inhibition of the cGAS/STING pathway may be an attractive, broadly applicable treatment for a variety of cGAS/STING-driven ailments. We hypothesize that targeted, nanoparticle-mediated delivery of RU.521 and H-151 to tissue-resident macrophages can enhance and sustain inhibition of cGAS/STING signaling in the inflamed tissue to slow or reverse progression of cGAS/STING-driven inflammatory conditions, while increasing the translatability of cGAS/STING inhibitors as therapeutics by decreasing the time between dosing and improving safety. Poly(propylene sulfide) (PPS) and poly(lactic-co-glycolic acid) (PLGA) nanoparticles (with and without added excipient polymer) will be used to deliver these inhibitors.

Methods: RU.521 and H-151 were encapsulated in PPS and PLGA nanoparticles using oil-in-water (o/w) emulsions stabilized with PVA. The size of these particles was measured using dynamic light scattering, drug loading and encapsulation efficiency were quantified using high performance liquid chromatography, and *in vitro* drug release at 37°C PBS was monitored over 7 days. Type-I interferon reporter cell assays were used to assess the capacity of the inhibitors and particle formulations to suppress cGAS/STING signaling in RAW-Dual cells and THP1-Dual cells stimulated with the cGAS agonist herring testes (HT) DNA. RT-qPCR and Western blots were used to assess cGAS/STING pathway activation in primary macrophages treated with a cGAS agonist and inhibitor-containing nanoparticlek. Flow cytometry was utilized to observe changes in M1 and M2-associated surface markers after cGAS/STING activation or inhibition. Additionally, the free inhibitors and loaded nanoparticles were tested in a model of liver interferonopathy and colitis.

Results: All NPs fabricated by o/w emulsion are ~160 nm in diameter. Inclusion of an excipient polymer improved drug loading by up to >7-fold and increased the drug release from 3 days to more than 7 days. All nanoparticle formulations tested were as potent, if not more potent, in blocking cGAS/STING-driven inflammatory responses than dose-matched free drug in both human monocyte and murine macrophage cell lines and primary macrophages in vitro. Treating M0 BMDMs with inhibitor-loaded NPs decreased M1-like gene expression and cell surface markers when co-administered with a cGAS agonist. In vivo, NPs accumulated in the target tissues, and the inhibitors decreased cGAS/STING signaling in an *in vivo* model of artificial model of liver interferonopathy. The therapeutic efficacy of our nanoparticle formulations is currently being evaluated in a murine model of colitis, which is a STING-associated inflammatory disease.

Conclusions and Implications: All nanoparticle formulations were as potent or more potent in inhibiting cGAS/STING signaling compared to dose-matched free drug in both human and murine macrophages in an *in vitro* model of interferonopathy. Treating M0 BMDMs with inhibitor decreased M1-like gene expression and cell surface markers when co-administered with a cGAS agonist. NPs reached their target tissues, and artificial models of liver interferonopathy and colitis were established. Finally, the inhibitors decreased cGAS/STING signaling in vivo. Chronic, aberrant overactivation of the cGAS/STING has been implicated in a variety of inflammatory conditions, and the targeted delivery and sustained release of STING pathway inhibitors may help resolve inflammation and improve patient outcomes in STING-associated autoimmune and inflammatory diseases.

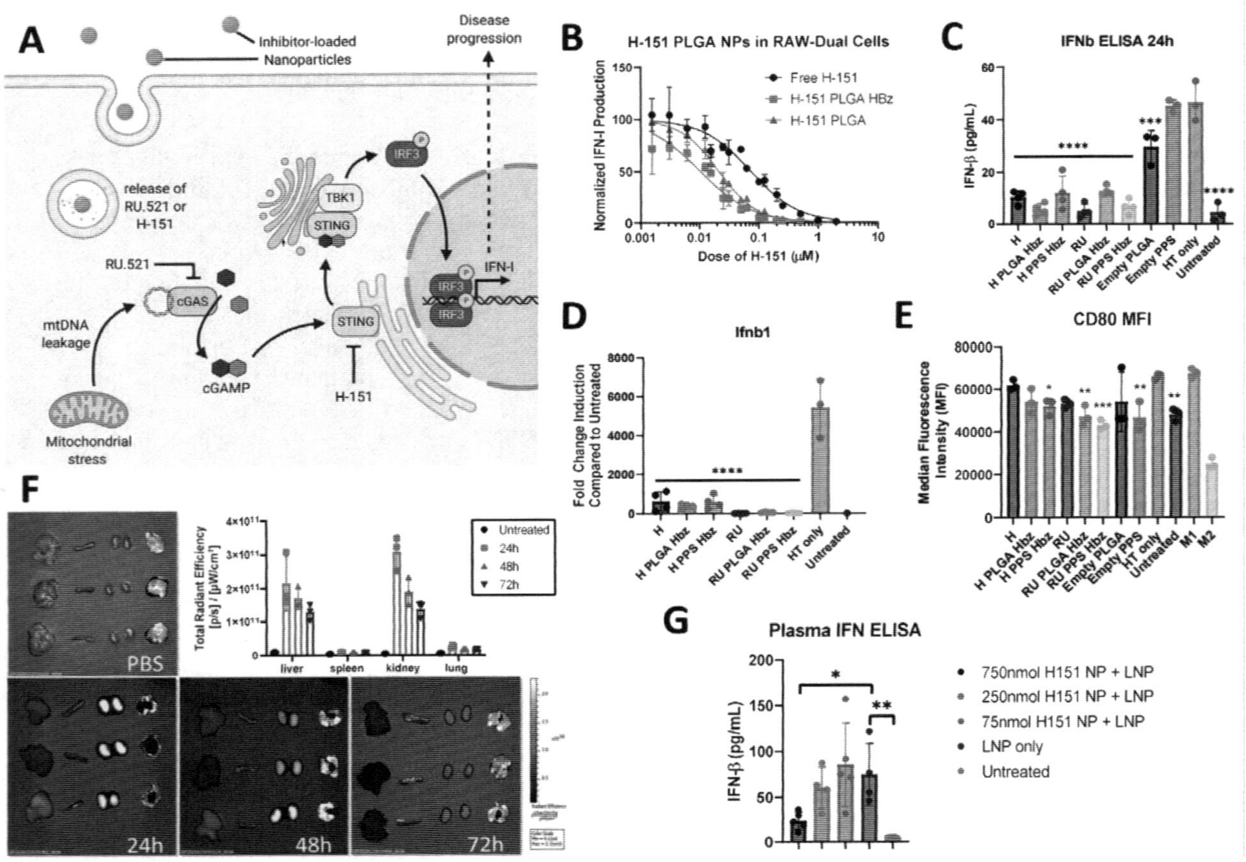

Figure 1 – Nanoparticle Delivery of Small Molecule Inhibitors to Treat STING-associated Inflammatory Diseases

Inhibitor-loaded nanoparticles block cGAS/STING signaling, preventing IFN-I production and disease progression (A). NP formulations containing H-151 outperformed dose-matched free drug in RAW-Dual cells stimulated with 1μg/mL herring testes (HT) DNA (B). BMDMs were isolated from mice and polarized for 8 days before being stimulated with 1μg/mL HT DNA and co-treated with the designated inhibitor or NP formulation (C-E). BMDMs were co-treated with HT DNA and 2μM H-151 or 25 μM RU.521, and supernatants were collected after 24h; IFN-β secretion was assessed by conducting an IFN-β ELISA (C). BMDMs co-treated with HT DNA and 2 μM H-151 or 25 μM RU.521 were lysed after 6h, and RT-qPCR was performed to assess transcription of the gene *Ifnb1* (D). Flow cytometry was conducted 24 hours after co-treating BMDMs with HT DNA and 0.1 μM H-151 or 5 μM RU.521 to assess expression of CD80 on the cell surface (E). PLGA NPs were loaded with Cyanine-7 and injected into the tail veins of mice. Organs were excised 24, 48, and 72h post-injection, and fluorescence was measured using an IVIS Imaging System (F). Mice were injected with H-151 NPs via tail vein injection, and 24h later, mice were injected with MC3 lipid nanoparticles containing 10ug 95bp dsDNA. After 6h, blood was collected, and plasma IFN-β was quantified using an IFN-β ELISA (G). The data represent mean ± S.D., n=3-4, *p<0.05, **p < 0.01, ***p<0.001, ****p<0.0001 using one-way ANOVA.

447f Optimized Lipid Nanoparticles and Delivery Conditions for State-of-the-Art Nebulized mRNA Delivery to the Lungs

Allen Jiang, Jacob Witten, Idris Raji, Sayo Eweje, Corina MacIsaac, Favour Oladimeji, Robert Langer, Daniel G. Anderson

Safe, efficient nebulized delivery of mRNA using lipid nanoparticles (LNPs) has the potential to treat, cure, or prevent a wide variety of diseases. While nebulized mRNA delivery has excellent potential, safe and effective delivery of mRNA to the lung epithelium remains a challenge. This is primarily due to two barriers: first, nebulization itself induces powerful shear forces that can disrupt nanoparticle structure and induce aggregation and second, the lung epithelium itself is a challenging target because NP uptake is most efficient on the basolateral side of polarized lung epithelial cells, but the basolateral side is rendered inaccessible to apically delivered LNPs by tight junctions. Here, we addressed each of these barriers to yield an optimized delivery approach. We first found that incorporation of cationic helper lipid into LNPs improved stability to nebulization and subsequent *in vivo* delivery. Next, we further enhanced stability by altering the nebulization buffer and adding additional excipients to prevent nebulization-induced aggregation which increased nebulized delivery efficiency *in vivo*. Lastly, we used fully differentiated air-liquid interface cultured primary lung epithelial cells as an *ex vivo* model for the lung to screen a large, novel library of biodegradable ionizable lipids to identify several excellent lung-transfecting hits. Our final combination of novel ionizable lipid, charge-stabilized formulation, and stability-enhancing excipient yielded a significant improvement over state-of-the-art delivery vehicles for nebulization in terms of nebulized stability and *in vivo* delivery efficacy. Furthermore, these optimized, nebulized LNPs also drive protein expression for multiple days, allow for multiple dosing, and have no apparent toxicity. Thus, these LNP formulations are excellent candidates for applications from pulmonary vaccination to protein replacement therapy.

447g Delivery of RNA Editing Machinery Enhances Immunogenicity of Cold Tumors

Lisa Volpatti

Immune checkpoint inhibitors (ICIs) have revolutionized cancer immunotherapy but remain effective in only a subset of patients and tumor types. The successful treatment of tumors that lack T cell infiltration, known as immunologically cold tumors, remains a major roadblock in cancer immunotherapy. Tumors with high mutational burden consistently respond better to ICI due to the abundance of new proteins or "neoantigens" that become recognizable by the immune system. Here, we deliver RNA editing machinery to alter codons and ultimately protein expression to induce the formation of neoantigens in tumor cells and unmask them to the immune system. We test this approach in a highly aggressive, immunologically cold model of murine melanoma and show significant reduction in tumor growth and increase in overall survival compared to ICI alone. This approach has the potential to expand the use of ICIs in the clinic through combination therapy and enhance our understanding of the role of neoantigens in ICI treatment.

497a Branched-Tail Lipid Nanoparticles Potently Deliver mRNA across Multiple Murine Models

Mariah Arral, Pooja Pandya, Namit Chaudhary, Saigopalakrishna Yerneni, Kathryn Whitehead

Messenger RNA (mRNA) has the potential to revolutionize modern medicine. We have already seen the beginning of this with the mRNA based COVID-19 vaccines by Moderna and Pfizer/BioNTech. Therapeutic mRNA can be used for protein replacement therapies, immunotherapies, gene editing, and many other applications. To deliver the fragile mRNA cargo to its target, a delivery vehicle is needed. Lipid nanoparticles are the most clinically advanced mRNA delivery vehicle. The preclinical screening of mRNA lipid nanoparticles is critical for their advancement from benchtop to bedside. Most preclinical studies use C57BL/6 and BALB/c mice, which are genetically identical (inbred). These mice are often selected as they tend to reduce within-group variance. Resulting in lowering the number of animals needed. However, outbred mice have been shown to more accurately mimic the genetic diversity of humans, and some strains (e.g. CD-1) are more cost-effective than inbred mice. To-date a direct comparison of mRNA lipid nanoparticles efficacy in inbred and outbred murine models has not been done. This study aims to investigate the effect that chemistry has on delivery efficacy and immunogenicity in inbred and outbred murine models. We have designed 20 novel branch-tail lipid nanoparticles to screen in C57BL/6, BALB/c, and CD-1 mice. Our work provides a direct comparison of mRNA lipid nanoparticle efficacy in inbred and outbred mouse strains for the first time. Of the 20 novel lipid nanoparticles screened we have identified nine promising candidates for potent mRNA delivery. Additionally, we also screened lipid nanoparticles in outbred Sprague Dawley rats. Furthermore, we will investigate the acute immune response observed after delivery of lipid nanoparticles. Future work will evaluate if there are efficacy, specificity, or immunological changes when the route of delivery is altered.

497b Image-Guided, Bioactive Gas Delivery with Ultrasound-Activatable Microbubbles for Treatment of Ischemic and Traumatic Injuries

Rajarshi Chattaraj, Misun Hwang, Chandra Sehgal, Daniel A. Hammer, Daeyeon Lee

Therapeutic gases like xenon (Xe), argon (Ar), nitric oxide (NO), oxygen, have been shown to be an effective treatment for acute onset conditions of hypoxia and ischemia reperfusion injuries. Xenon is especially promising due to its otherwise inert nature thus precluding harmful side effects. Xenon is postulated to block the N-methyl-d-aspartate (NMDA) subtype of glutamate receptors on cell surfaces, preventing a cascade that results from an injury and ultimately leads to apoptosis. Argon, another noble gas is postulated to stop an injury cascade by interfering with the PI3k/Akt apoptotic pathway and can serve as a less expensive and more widely available alternative to xenon. Clinical trials with xenon use inhalation for administration: this method is systemic, wasteful, and not cost effective.

Microbubbles containing a gas core offer an excellent option to local, safe, image-guided delivery of bioactive gases instead of systemic administration. Microbubbles, commercially used for diagnostic imaging, usually encapsulate a water-insoluble fluorous gas like perfluorocarbons or sulfur hexafluoride to prevent efflux from the bubble in aqueous media. We have developed microbubbles (MBs) encapsulating pure xenon gas as a first in the field. While there are limited examples of Xe MBs in literature, these have used a stabilizing fluorocarbon gas in the core to keep the otherwise water-soluble gas trapped inside the bubble, since therapeutic gases are tens of time more water-soluble than fluorocarbon gases. Through optimization of the phospholipid composition of the microbubble shell, we demonstrated that a long-chain phospholipid dibehenoylphosphatidylcholine (DBPC, C22:0), can stabilize 1-10 μm Xe and Ar microbubbles and keep them stable for several days in storage. This is likely due to the greater intermolecular van der waals cohesive energy for long-chain lipids which provides rigid packing to reduce gas efflux. We have then showed that both Xe and Argon microbubbles Xe and Ar microbubbles elicit significant and persistent non-linear, bubble-specific, ultrasound signal, in both a tissue-mimicking phantom and *in vivo* in a small (mouse) and large (pig) animal model. On intravenous injection, both as bolus of $\sim 10^9$MBs/m and as an infusion at 0.6 mL/min of $\sim 10^8$MBs/mL, we saw sustained ultrasound contrast for pure Xe MBs in a pig heart and carotid artery.

To establish the therapeutic effects of these Xe MBs in injury protection, we carried out trials for neuroprotection using Xe MBs in a large animal model of traumatic brain injury (TBI). A well-established, focal controlled cortical impact porcine model of TBI was used with 1 month old, ~10 kg pigs, whereupon Xe MBs were injected through infusion for ~20 min at 1, 3, and 24 hrs after injury onset. Fluorocarbon bubbles were used as Control. MBs were visualized using a commercial clinical scanner in the carotid artery followed by an increase in ultrasound power to burst the bubbles and release the gas near the brain. After 1 and 5 days, magnetic resonance imaging (MRI) scans of the pig brain revealed significantly reduced perilesional edema in the xenon-treated group. In the Control group treated with fluorocarbon bubbles with the same shell composition,core/hemorrhage size was seen to increase. Histopathological Handamp;E staining studies carried out with the 5 mm coronal brain sections showed lower endothelial proliferation and perivascular inflammation (lower migration of lymphocytes and macrophages) near the injury site for the xenon treated group. In summary, this study presents a first demonstration of pure xenon and argon microbubbles that are echogenic *in vivo*. It also shows the feasibility of neuroprotection using Xe MBs, thus providing a highly promising theranostic agent for local, non-systemic, gas delivery for ultrasound image-guided treatment of different kinds of injury. Ongoing work is exploring the lowered migration of cells near injury site *in vitro* in the presence of Xe MBs. Additionally, we are testing argon and nitric oxide bubbles for treating ischemia reperfusion injuries *in vitro*, where cells have been exposed to oxygen-glucose deprivation. Together, this presentation will focus on a variety of microbubbles with bioactive gases like xenon, argon, and nitric oxide as image-guided gas delivery agents for non-invasive, non-systemic, cytoprotective treatments of acute ischemic/traumatic injuries.

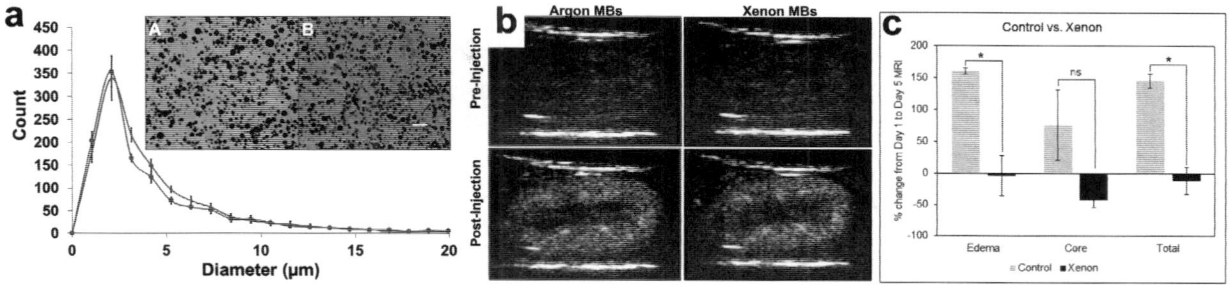

a: Xenon microbubble size distribution and corresponding images after 0 (*A*) and 5 (B) days. **b**: Microbubble signal from ultrasound *in vivo* in a mouse kidney after IV injection of 0.03 mL of ~ 10^8 bubbles/mL. **c:** Neuroprotective efficacy of xenon microbubbles in a porcine model of traumatic brain injury. There is marked decrease in edema from Day 1 magnetic resonance imaging to Day 5 in the xenon- treated group. Graph is showing percentage change in volumes (cm^3) from Day 1 MRI to Day 5 MRI of edema, core, and total volume in the perfluorobutane (PFB)-treated (gray) and xenon-treated groups (n = 3 each).

497c PLG Microparticles for Small Molecule Delivery to Muscle Cells

Candice Cheung, Kidochukwu Atube, Michael Gower

Statement of Purpose: Older adults experience a reduced capacity to regain muscle mass after atrophy that results from short bouts of disuse (illness, injury, surgery), which occurs often for this age group. Reduced responsiveness to therapy in aged muscle occurs due to low expression of growth factors, such as Insulin-like Growth Factor-1 (IGF1), which is produced by muscle-resident macrophages[1,2]. Transcriptomics data indicates that activation of retinoic acid receptors (RAR) in macrophages enhances IGF1 expression[3], which can be achieved with all-*trans*-Retinoic acid (ATRA). In addition to ATRA being a potential IGF1-inducer in muscle macrophages, ATRA has also been established to induce muscle differentiation and maintain muscle stem cells[4]. Administration of ATRA to muscle is limiting as it has a short half-life in vivo when administered orally[5] or intravenously[6], has poor solubility in aqueous solution[7], and is sensitive to light and oxidation[8]. Moreover, high systemic doses have side effects, such as hypertriglyceridemia, mucocutaneous dryness, as well as ATRA resistance[9]. Thus, what is needed is a method to deliver and release ATRA to the skeletal muscle that extends its presence in the tissue and limits off-organ side effects. This study aims to: (1) synthesize ATRA-loaded microparticles that remain bioactive after encapsulation and (2) test the myogenic ability of particle-treated macrophages in an *in vitro* model of muscle regeneration.

Methods: To confirm ATRA's muscle-formative ability, ATRA was used to treat C2C12 myoblasts daily for 8 days. After performing immunofluorescence microscopy for myosin heavy chain, image analysis was used to quantify myotube formation and growth. ATRA-loaded PLG particles were synthesized using a single emulsion/solvent-evaporation technique by adding 5 mg/mL of ATRA to the organic phase for ATRA-PLG particles. Drug loading was determined using UV-vis spectroscopy. To confirm the bioactivity of ATRA after encapsulation, RAW264.7 macrophages were incubated with ATRA-PLG particles for 24 hours, then media were collected for IGF1 ELISA analysis. To investigate the impact of ATRA-PLG particles on macrophage-induced myoblast fusion into myotubes (an *in vitro* model of muscle regeneration), macrophages were treated with ATRA-PLG particles for 24 hours, after which macrophage conditioned media was collected and used to treat C2C12 myoblasts daily for 3 days. Control groups included conditioned media from macrophages without particle treatment and media conditioned with ATRA particles alone. After performing immunofluorescence microscopy for myosin heavy chain, image analysis was used to quantify myotube formation and growth.

Results and Discussion: In a myotube formation assay using C2C12 murine myoblasts, ATRA was able to successfully increase myotube formation as shown by myotube width and number of myotubes (Figure 1). ATRA was successfully incorporated into PLG microparticles, producing spherical and monodisperse particles with a diameter of 2.9 μm (Figure 2A). Drug loading for the ATRA-PLG particles were 78 μg of ATRA per milligram of particle with a linear drug release of 2.88% released per day (Figure 2B). When incubated with macrophages, ATRA particles increase IGF1 secretion into the media compared to blank PLG microparticles (Figure 2C), indicating that the ATRA remained bioactive after encapsulation. In the myotube formation assay, media collected from macrophages treated with ATRA-PLG particles induced myoblasts to fuse more readily into myotubes (Figure 3A) compared to all other treatments, as quantified by the number of myotubes per image field (Figure 3B) and number of incorporated nuclei per myotube (Figure 3C). The data indicate that while untreated macrophages and ATRA released from the particles increase myotube formation, there is an enhanced effect when macrophages are pre-treated with the ATRA-PLG particles. Since macrophages are present in skeletal muscle, we expect that there are opportunities to modulate muscle growth with ATRA-PLG particles indirectly through macrophages and directly through ATRA acting on muscle.

Conclusion: ATRA can be encapsulated into PLG microparticles while maintaining bioactivity. Treatment of macrophages with ATRA-PLG particles changes the secretome in a way that enhances myotube formation compared to media from untreated macrophages or media containing ATRA, but no macrophage-derived factors. The data suggest that elevated levels of IGF1 play a role, but changes in other myogenic or myostatic factors cannot be ruled out. Future studies will assess the efficacy of ATRA-PLG particle delivery on muscle growth in a mouse model of disuse muscle atrophy.

(1) Zhang, Y.; Pan, X.; Sun, Y.; Geng, Y. J.; Yu, X. Y.; Li, Y. The Molecular Mechanisms and Prevention Principles of Muscle Atrophy in Aging. **2018**, *1088*, 347–368. https://doi.org/10.1007/978-981-13-1435-3_16.

(2) Tidball, J. G.; Welc, S. S. Macrophage-Derived IGF-1 Is a Potent Coordinator of Myogenesis and Inflammation in Regenerating Muscle. **2015**, *23* (7), 1134–1135. https://doi.org/10.1038/mt.2015.97.

(3) Széles, L.; Póliska, S.; Nagy, G.; Szatmari, I.; Szanto, A.; Pap, A.; Lindstedt, M.; Santegoets, S. J. A. M.; Rühl, R.; Dezsö, B.; Nagy, L. Research Resource: Transcriptome Profiling of Genes Regulated by RXR and Its Permissive and Nonpermissive Partners in Differentiating Monocyte-Derived Dendritic Cells. **2010**, *24* (11), 2218–2231. https://doi.org/10.1210/me.2010-0215.

(4) Mezquita, B.; Mezquita, C. Two Opposing Faces of Retinoic Acid: Induction of Stemness or Induction of Differentiation Depending on Cell-Type. *Biomolecules* **2019**, *9* (10). https://doi.org/10.3390/biom9100567.

(5) Ozpolat, B.; Lopez-Berestein, G.; Adamson, P.; Fu, C. J.; Williams, A. H. Pharmacokinetics. **2003**, 10.

(6) Kawakami, S.; Opanasopit, P.; Yokoyama, M.; Chansri, N.; Yamamoto, T.; Okano, T.; Yamashita, F.; Hashida, M. Biodistribution Characteristics of All-Trans Retinoic Acid Incorporated in Liposomes and Polymeric Micelles Following Intravenous Administration. *J. Pharm. Sci.* **2005**, *94* (12), 2606–2615. https://doi.org/10.1002/jps.20487.

(7) Szuts, E. Z.; Harosi, F. I. Solubility of Retinoids in Water. *Arch Biochem Biophys* **1991**, *287* (2), 297–304. https://doi.org/10.1016/0003-9861(91)90482-x.

(8) Tolleson, W.; Cherng, S.-H.; Xia, Q.; Boudreau, M.; Yin, J.; Wamer, W.; Howard, P.; Yu, H.; Fu, P. Photodecomposition and Phototoxicity of Natural Retinoids. *Int. J. Environ. Res. Public. Health* **2005**, *2* (1), 147–155. https://doi.org/10.3390/ijerph2005010147.

(9) David, M.; Hodak, E.; Lowe, N. J. Adverse Effects of Retinoids. *Med. Toxicol.* **1988**, *3* (4), 273–288. https://doi.org/10.1007/BF03259940.

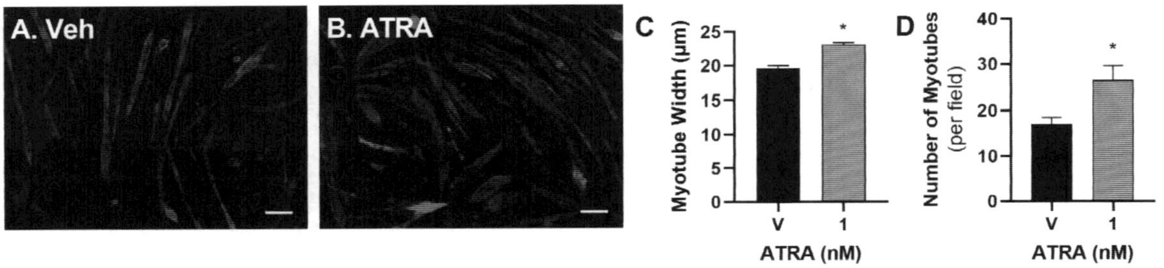

Figure 1: ATRA enhancing myotube formation in C2C12 Myoblasts. After daily administration of Vehicle (A) or ATRA (B) for 8 days, myotubes were stained for myosin heavy chain. Image analysis was conducted for (C) myotube width and (D) number of myotubes per image field. Student's t-test was conducted. Scale bar = 100μm

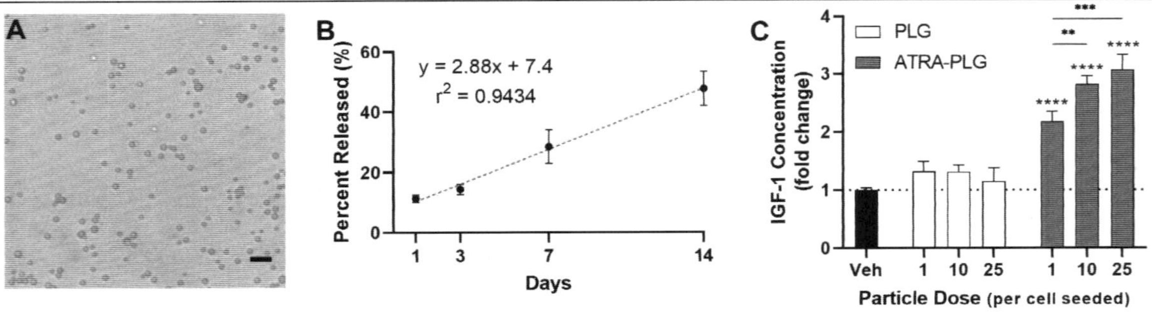

Figure 2: ATRA-loaded PLG microparticles. (A) ATRA-PLG microparticles with drug loading of 78 μg ATRA per mg of particle. (B) Drug release of ATRA from ATRA-PLG. (C) Impact of ATRA-PLG on RAW264.7 macrophage IGF1 response. Two-way ANOVA with Tukey's multiple comparison was conducted between Blank PLG and ATRA-PLG with particle type and dose as sources of variation. **, ***, and ****indicate p < 0.01, 0.001, and 0.0001 respectively. Scale bar =10 μm

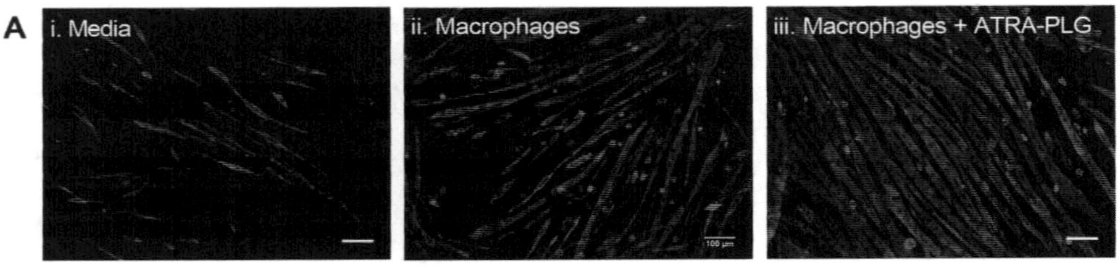

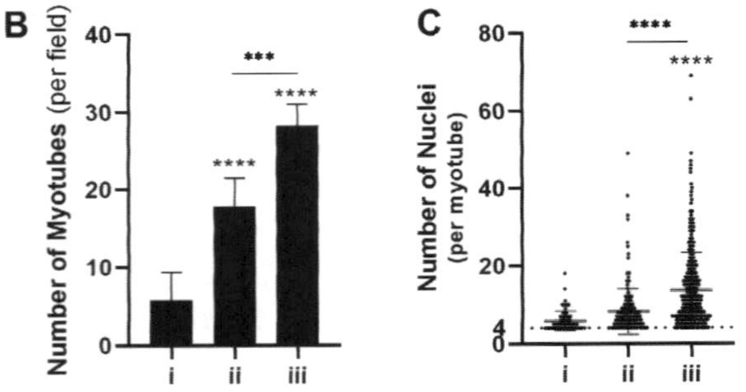

Figure 3: Myotube formation assay. (A) Immunofluorescent images of formed myotubes after conditioned media treatment for 4 days. Scale bar = 100 μm. (B) Number of myotubes per field. (C) Number of nuclei per field. One-way ANOVA was conducted between means where *** and **** indicates , < 0.001 and 0.0001 compared to (i), respectively, uinless otherwise stated.

497d Highly Tunable Fluorescent Nanoparticle Library for Optimizing Neuronal Drug Delivery

Rachel Pollard, Parker Lewis, Dane Jensen, Rocco Latorre, Shlok Paul, Brian Schmidt, Nigel Bunnett, Nathalie M. Pinkerton

Nanomedicine is a promising avenue for neuronal drug delivery, due to 1) favorable pharmacokinetics of encapsulated drugs, and 2) the potential for targeting of specific tissue and cell types. However, in order to optimize these nanotherapeutic systems, nanoparticle-neuron interplay needs to be fully characterized. Previous studies have indicated that size and surface properties, such as charge and functionalization, play an important role in nanoparticle uptake by different cell populations. However, robust studies of these interactions in neurons remain to be seen. In order to address this gap, we have developed a novel, tunable nanoprobe that allows for the characterization of uptake kinetics while recapitulating the characteristics of drug-loaded nanoparticles.

In this study, a quantum-dot nanoparticle (qNP) library was assembled by encapsulating CdSe-based quantum dots into polymeric nanoparticles using Flash Nanoprecipitation (FNP). Size and surface characteristics were tuned respectively by varying the ratio of core to stabilizing polymer and altering stabilizing polymer chemistry. Spatiotemporal qNP uptake was examined *in vitro* and *in vivo*. The resulting library of qNP probes demonstrated tunability in size (50-250 nm diameter) and surface charge (-30 to 0mV). *In vitro* studies showed size-dependent uptake kinetics of PEGylated qNPs into Schwann cells and colocalization with endosomes. In dorsal root ganglia, particles were visualized in both soma and neurites. In vivo studies show that qNPs were taken up by mouse intestinal epithelium and enteric nerves.

In conclusion, we demonstrate the versatility of FNP as a scalable single-step process for formulating nanocarriers, and begin to elucidate the assembly mechanics of the novel polymer-nanocrystal composites. Preliminary biological studies confirm the promise of polymer nanoparticles to act as delivery systems to nerves via intrathecal and rectal delivery. Additionally, we present foundational work elucidating the relationship between nanoparticle size, surface charge, and neuronal internalization efficiency. Overall, the qNP library is a novel tool that can be used for optimizing drug delivery systems and studying nano-bio interactions in any tissue system.

497e Brain-Derived Extracellular Vesicles: Molecular Probes and Therapeutic Vehicles in the Neonatal Ischemic Brain

Nam Phuong Nguyen, Hawley Helmbrecht, Elizabeth Nance

EVs as a Promising Nanomedicine Platform

Extracellular vesicles (EVs) are unique bio-derived nanoparticles that are produced in every cell type and serve as critical cellular mediators within the body. Surrounded by a lipid bilayer, EVs are capable of carrying a wide repertoire of biomolecules involved in critical physiological pathways. EVs are categorized by their size and pathway of biogenesis: exosomes are small vesicles (30-100nm) that originate from endosomal multivesicular bodies fusing with the plasma membrane, microvesicles (100-1000nm) are shed directly from the plasma membrane, and apoptotic bodies (>1000nm) bleb from the cell during apoptosis. EVs have many strong advantages for the design of therapeutic platforms, including: 1) biocompatibility, 2) small size, 3) low immunogenicity, 4) stability, and 5) inherent targeting capacity.

Though the mechanisms of cell-specific uptake of EVs remain unclear, EVs contain cargo that are horizontally transferred to neighboring cells, acting as an important mode of cellular communication. This unique ability to transfer information between cells in the body elevate EVs as a promising therapeutic platform for areas that have been traditionally difficult to target with other nanomedicine platforms, such as the central nervous system (CNS). While the blood-brain barrier (BBB) serves as a significant obstacle to many nanotherapeutic platforms, EVs have a large advantage in that they are capable of crossing this barrier. Due to the advantages of EVs as native therapeutic vehicles, our group has successfully sequestered brain tissue-derived EVs to study their therapeutic potential in neonatal brain injury. We evaluated the therapeutic potential of EVs using a holistic approach – quantitative data collected from cellular cytotoxicity, gene expression, and bio-barcoding analysis paired with qualitative confocal imaging data tracking changes in microglial morphology and EV fate with quantum dots. Our research goal is to understand the role and potential therapeutic effect of EVs in neonatal Hypoxic-Ischemic Encephalopathy (HIE), which is the leading cause of morbidity and mortality in neonates.

Therapeutic Motivation

HIE is characterized as a reduction of adequate blood flow to the brain that can lead to stroke or permanent brain damage if left untreated. The impact of this brain condition is on a global scale: for every 1000 live births, 1-8 neonates are affected by HIE in high resource countries and this number is as high as 26 neonates in low resource countries. Currently, the standard treatment option for HIE patients is therapeutic hypothermia (TH) by continually cooling newborn infants to 33.5-35°C for 72h. However, TH only offers a 15% absolute reduction in the risk of death and disability and is resource-intensive. Thus, investigating key mediators of HIE response to identify points of therapeutic intervention has significant clinical potential within the neonatal population.

A hallmark of HIE is the immediate and sustained inflammation that occurs as a physiological response to the injured tissue. At this stage, cells are simultaneously releasing pro-inflammatory cytokines that lead to oxidative stress and cell death as well as anti-inflammatory cytokines that help clear debris and rebuild tissue. Though inflammation plays both beneficial and detrimental roles in the injury outcome of HIE, it serves as a vital time period for brain repair and is a major target for the development of therapies. However, a critical obstacle to the development of new therapeutics is a lack of understanding about the EV-mediated responses following HIE. Due to their ability to regulate communication between immune cells, EVs have emerged as important players in injury response, neuronal development, and neuronal proliferation following brain injury in multiple *in vitro* and *in vivo* models. Despite recent interest in studying the role of EVs in cerebral ischemia, most studies and all ongoing clinical trials in this field use adult models administered with stem cell culture-derived EVs, which do not recapitulate native EVs found in a 3D tissue environment.

Results andamp; Significance: EV Therapeutic Potential

In contrast to standard *in vitro* cell culture models, we successfully isolated and characterized endogenous EVs from neonatal postnatal (P)10 rat brain tissue (BEV) using a combination of ultracentrifugation, ultrafiltration, and size exclusion chromatography sequestering techniques. Following successful BEV isolation and characterization, we performed various dose- and time- dependent therapeutic efficacy studies on *ex vivo* brain slice models of HI. Our work demonstrated that BEVs elicit a significant therapeutic response by decreasing cytotoxicity within oxygen glucose deprivation (OGD) conditioned brain slices. The minimum therapeutic dosage was determined to be 25mg BEVs , and the therapeutic exposure time is 48h across all dosages. The therapeutic application window for treatment with BEVs was determined to be 4-24h post-injury. This result is clinically relevant because it suggests that BEVs remain therapeutically viable even when administered as early as 4 hours and up to 24 hours post-injury onset.

At the minimum therapeutic dosage, BEV treatment increased the gene expression levels of IL-10, a prominent anti-inflammatory cytokine. In particular, IL-10 expression after BEV treatment was comparable to the healthy control, pointing to the role that BEVs play in regulating injury attenuation. Extensive confocal imaging analysis of microglia supplemented our experimental results. Microglia are the resident immune cell of the brain, whose changes in morphology are connected to inflammation, disease onset and progression, and stimuli from the local environment. To supplement experimental results, we performed quantitative computer-aided image analysis on microglia over several key morphologic features on both an individual and population-based level. We observed a considerable morphology shift from an amoeboid, inflammatory phenotype to a restorative, anti-inflammatory phenotype between 24-48h of BEV exposure following OGD injury. We concluded that BEVs significantly attenuate inflammation and cytotoxicity in HIE neonatal models through increases in anti-inflammatory cytokine expression and microglial morphology shifts. From a technology perspective, studying the brain's response to endogenous EVs following ischemic injury is a first step to understanding how EVs can improve therapeutic payloads for neonatal HIE patients.

Research Innovation: EV Cellular Fate

Following our confirmation of the therapeutic efficacy of BEVs, we next sought to evaluate BEV fate and localization using our *ex vivo* platform. Understanding the cell-specific uptake and regional distribution of BEVs can inform us of therapeutic targets within the neonatal HIE brain. A two-pronged approach is used to probe our BEVs: 1) tagging BEVs with quantum dots (QD-BEVs) that can be qualitatively tracked using confocal microscopy, and 2) conjugating BEVs with unique oligobarcode sequences (oligo-BEVs) to quantitatively access and compare uptake across different cell types.

Quantum Dot Tagging of BEVs

Though current techniques for labeling EVs exist such as commonly used lipophilic dyes, a considerable disadvantage of these methods is dye aggregation and non-specific labeling that can give false-positive signals. While fluorescent reporter proteins provide signal specificity, this process requires genetic engineering of cells to produce specific proteins, thereby limiting the translational relevancy of this work. Quantum dots (QDs) are excellent candidates for labeling and tracking EVs due to their tunable excitation/emission spectra, photostability, versatile surface chemistry and resistance to photobleaching. Our lab has tagged EVs with QDs through a simple "click chemistry" coupling reaction between 4-formylbenzoate (4FB) and hydrazinoni-cotinate acetone hydrazone (HyNic). We have shown that our QD-EV formulations are effective at tracking BEV localization within microglial BV-2 cells *in vitro* as well as in *ex vivo* brain tissue slices. Confocal imaging of QD-BEV conjugates revealed that BEVs preferentially localize to microglia in the corpus collosum and striatal regions of OGD-conditioned *ex vivo* brain slices. This result supports the role BEVs may play in shifting microglial morphology following BEV treatment.

Oligobarcode Labeling of BEVs

Supplementing qualitative image analysis using QD tagging techniques, we also utilized oligobarcode labeling as a quantitative approach to determining BEV fate. Oligobarcodes are short oligonucleotide strands that can be designed with unique sequences to differentiate between different markers. We conjugated oligobarcodes to BEVs (oligo-BEVs) using a similar click chemistry technique and quantified their localization and internalization within different cell types in the CNS. We used oligobarcodes to determine preferential cell-specific localization of BEVs, giving us insight into the fate of these nanoparticles.

Conclusions

Together, our therapeutic evaluations of BEVs along with our two-pronged labeling approach to determine and quantify BEV fate reflect the important role that BEVs play in attenuating inflammation and cell death in HIE. Future research in BEVs can inform the design and administration of EV-based therapies for improving outcomes for neonatal HIE patients.

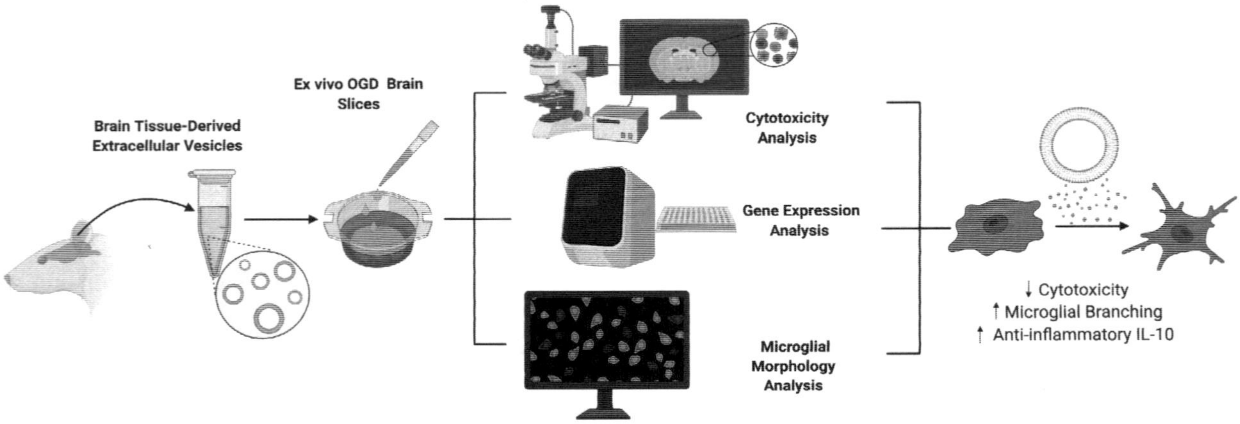

497f Nano-Hitchhikers for Effective Delivery of Alpha-Particle Radiotherapy to Glioblastomas

Rajiv Ranjit Nair, Tony Wu, Anjali Sharma, Rangaramanujam Kannan, Stavroula Sofou

Introduction: Brain cancer is the leading cause of cancer death in children and adolescents, surpassing leukemia. Pediatric glioblastoma, a type of brain cancer in children, is an aggressive disease with no defined standard therapy. Pediatric glioblastoma, is quite distinct from its adult counterpart. Despite aggressive treatment with multimodal therapy (surgery, combined with even highly toxic combination of chemotherapies and/or radiotherapy) most children fail to reach a long-lasting response and do not survive. The reported median survival in pediatric glioblastoma, despite multi-modal therapies ranges from 13 to 73 months with a 5-year survival of less than 20 %. We propose a nanotechnology-based delivery strategy that 'hitchhikes' an existing biological mechanism: this mechanism enables selective and uniform delivery of a cytotoxic agent (an α-particle emitter) in brain tumors; through our work proposed we expect to demonstrate that it is more effective at controlling pediatric glioblastoma cancers than currently available approved and alternative methods of treating this disease.

Current therapeutic approaches following surgery necessarily resort to combinations of therapeutics and/or external radiotherapy but, unfortunately, even with highly toxic combination regimens, most pediatric patients fail to reach a long-lasting response partly due to resistance to therapies. Additionally, since (external) radiation therapies in pediatrics have severe consequences to the developing brain, there are further limitations to the therapeutic options that are available for adults.

Alpha-particle radiopharmaceutical therapy (α-RPT) is a form of internal radiation and has already been shown to be impervious to most resistance mechanisms. Due to the massive amount of energy deposited as they traverse tissue, α-particles produce orders of magnitude more DNA damage (complex double strand breaks – the most difficult to repair type of DNA damage) than all other forms of cytotoxic therapy. The complexity and level of rapidly induced DNA damage overwhelms cellular repair mechanisms; this inability to repair lethal damage is the reason that α-particles, if optimally delivered, are impervious to resistance. However, the short range of α-particles in tissue (up to 5 to 10 cell diameters) which makes them ideal for precision killing – especially needed in the brain - also limits penetration within the volume of brain tumors. This is because the diffusion-limited penetration depths of traditionally used carriers (antibodies, molecular targeting vectors, liposomes) in tumors, combined with the short range of α-particles, result in only partial tumor irradiation. This is the root cause of current treatment failure of glioblastoma with α-particle therapies. The limitations lowering efficacy are only exacerbated by the additional challenge of the traditional carriers' crossing of the Blood Brain-Tumor Barrier.

We propose a novel delivery strategy that is grounded in a strikingly simple, powerful, and clinically implementable approach: the systemic administration of α-emitters loaded on unique hydroxyl PAMAM dendrimer-nanoparticles, i.e. dendrimer NPs (sugar- and peptide-functionalized generation-6 hydroxyl PAMAM dendrimers) which are shown (1) to effortlessly and quickly cross the Blood Brain-Tumor Barrier and with minimal transport across the Blood Brain Barrier of the healthy brain, upon systemic administration, (2) to be effectively taken up by the tumor-associated macrophages (non-cancerous cells but strongly correlated with tumor aggressiveness), and (3) to relatively quickly clear from the body and the brain when outside the tumor-associated macrophages which (4) most importantly, retain the dendrimer-nanoparticles for days, uniformly distributed within the brain tumors, and would uniformly irradiate the brain tumors resulting in maximum killing efficacy of tumor cancer cells.

Materials and Methods:

In vitro, the sensitivity of the glioblastoma cells, GL-261 to the free form of drug (in this case Actinium-225 chelated to DOTA, ^{225}Ac-DOTA) and the ^{225}Ac-DOTA radiolabeled onto dendrimer nanoparticles was evaluated using radio-sensitivity assays. The survival of immune-competent C57BL/6 mice bearing intracranially GL-261 glioma tumors was evaluated upon intravenous administration of α-particle therapy delivered by the dendrimer carriers for varying total injected radioactivities and was compared to the no treatment condition. The maximum tolerated dose (MTD) and long-term toxicities on tumor-free mice were also studied.

Results and Discussion: A colony survival assay demonstrated the sensitivity of GL-261 cells, in monolayers, to α-particle irradiation (Figure 1A), which were preserved upon dendrimer chelation. On mice with intracranially implanted GL-261 tumors, a single injection of 600 nCi and/or of 700 nCi total administered radioactivity, delivered by hydroxyl PAMAM dendrimers, showed good inhibition of the growth of GL-261 glioma tumors, as indicated by the statistically significant prolonged survival of these tumor bearing mice (p-values <0.05) (Figure 1B). The MTD has still not been reached at 800 nCi per animal. No long-term toxicities (6 months post injection) have yet been observed.

Conclusions: The use of hydroxyl PAMAM dendrimer nanoparticles as a delivery carrier for α-RPT to combat glioblastoma is promising as indicated by the improved survival of tumor-bearing mice and the well-tolerated profile of the radiolabeled carrier.

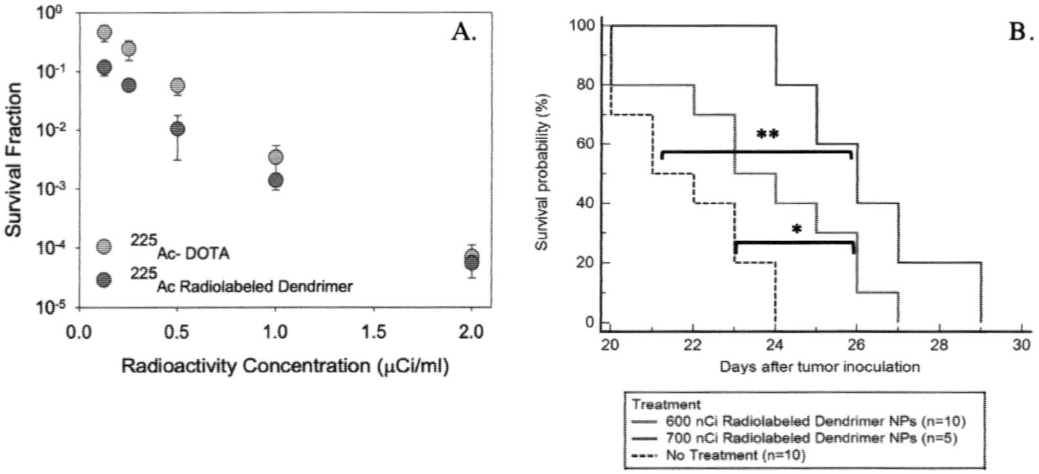

Figure 1: Glioblastoma cells are sensitive to alpha-particle radiation in vitro, and the delivery of alpha-emitters by dendrimers in vivo prolongs survival of animals with intracranially implanted glioblastoma tumors
*(A) Radiosensitivity assay on GL-261 cells as a monolayer; (B) Kaplan-Meier survival plots showing the increased survival of C57BL/6 mice bearing GL-261 glioma tumors when injected intravenously with 600 and/or 700 nCi total radioactivity delivered by the dendrimers which were administered intravenously only once on day 10 post tumor inoculation. *p-value<0.05, **p-value<0.01*

595a Effectiveness of Engineered Nanoceria at Inhibiting Thermally-Induced Oxidative Stress in Symbiodinium.

Liza M. Roger, Joseph A. Russo, Robert Jinkerson, Juan Pablo Giraldo, Nastassja Lewinski

Coral bleaching has been linked to an immune-like response to thermal stress. In this study we tested the antioxidant effects of engineered poly(acrylic acid)-coated cerium oxide nanoparticles (nanoceria) on free-living Symbiodinium, *Breviolum minutum*. The synthesized nanoceria with primary particle diameters ~4 nm were internalized by *B. minutum* within 30 minutes and subsequently localized in the cytosol. Nanoceria exposure did not inhibit cell growth over time, and an exponential growth phase was reached by 30 days of exposure. When the cells were held at an elevated temperature of 34°C for 1 hour, increased intracellular reactive oxygen species (ROS) concentrations were measured over time. A ROS scavenging effect of the nanoceria was observed with a 5 fold decrease in intracellular ROS levels in the *B. minutum* during thermal stress. The presence of nanoceria did not appear to influence nitric oxide concentrations. As a proof of concept, aposymbiotic anemones *(Exaiptasia pallida*) were incubated with nanoceria-loaded *B. minutum*. The successful reinfection observed suggests that nanoceria-loaded Symbiodinium could be used to inoculate partially bleached corals to reduce oxidative stress and provide corals with more time to recover and survive bleaching events.

595b Soft Poly-Ethylene-Glycol (PEG) Micelles and Lipid Nanoparticles (LNPs) Resolve Autophagy and Reduce Pro-Inflammatory Cytokine Secretion in RAW264.7 Murine Macrophages

Monireh Asoudeh, Nicole Nguyen, Paul Dalhaimer

The effects of soft poly-ethylene-glycol (PEG)-based nanoparticles (NPs) on macrophage autophagy and inflammatory signaling are largely unknown. With the prevalent use of PEG NPs that are lipid nanoparticles (LNPs) as vaccine adjuvants, drug delivery vehicles, and carriers of nucleic acid reagents, it is crucial to understand how they affect these two pathways. Here, we show that soft PEG-poly-butadiene (PEG-PBD) cylindrical and spherical micelle nanoparticles (CNPs and SNPs), and lipid nanoparticles (LNPs) that have the lipid makeup of the chemotherapy carrier DOXIL lower murine RAW264.7 macrophage autophagosome abundance over 24 hours of incubation. CNPs, SNPs, and LNPs mostly resolve autophagosome formation triggered by starvation, rapamycin, and lipopolysaccharide (LPS). All three PEG NPs lower autophagosome abundance when macrophages are cultured in serum from obese humans. CNPs, SNPs, and LNPs lower reactive oxygen species (ROS) in macrophages. Surprisingly, all the three PEG NPs – especially LNPs - decreased the levels of most of the macrophage cytokines compared to controls over 24-hour incubations. Our findings show that soft PEG NPs do not augment autophagy or inflammation like many hard NPs. Instead, they resolve macrophage autophagy and reduce inflammation. Also, macrophages seem to treat PEG NPs like lipoproteins, not foreign objects. Thus, the terminology that macrophages "clear" PEG NPs, which has been used over many decades, is misleading.

595c *Membrane Protein Cryo-EM and Assays in Dynamically Loadable Intact Lipid Bilayers on Nanoporous Membranes*

Malcolm Lane Gilchrist, Yueming Li

The objective of this research is the development of a nanoporous membrane-supported biomembrane platform to enable cryo-EM studies of membrane proteins in intact biomembranes that are not currently possible. This architecture is open to the loading of relevant membrane proteins (MPs), substrates and binding partners by lateral diffusion and can be combined with functional protease assays in the same intact biomembrane under cryo-EM study. Thus far, cryo-EM of MPs in intact biomembrane systems are prepared in batch and are not accessible via lateral diffusion, this includes MPs in liposomes, detergent micelles, and other recent work in nanodiscs. Nanoporous supported systems provide a unique solid structure to present MPs in for cryo-EM investigation in controlled orientations and in-membrane concentrations. Through this proposed technology we build open systems accessible to lateral diffusion-based loading where MP-MP and MP-soluble protein interactions can conceivably be staged in the cryo-EM observation volume to look at complex formation in intact lipid bilayers. And furthermore, phase separation is staged in the same volume allowing for studies of the direct effect of biomembrane thickness on MP structure and MP-MP complex formation *in situ*, and both of these types of studies are currently very difficult to stage in the cryo-EM observation region. We have coupled the cryo-EM studies in the observation region with functional assays and imaging microscopy at both the micro- and nanoscale in these systems, using the imaging to examine to bottom-up construction of the supported lipid bilayers. These measurements included atomic-force microscopy (AFM) which we have recently used to image MPs such as gamma-secretase at the single molecule level in intact lipid bilayers with biomembrane phase separation(Barros et al. (2020) Langmuir 36, 23, 6569–6579). This method for direct imaging studies of membrane proteins in intact biomembranes of controlled composition can significantly enable a new window of structural biology and assays of these drug targets.

595d Recombinant Protein Nanostructures for the Inhibition of Sars-Cov-2

Rajarshi Chattaraj, Christina Kim, Daeyeon Lee, Daniel A. Hammer

The emergence of the highly infectious SARS-CoV-2 virus, which is responsible for the Covid-19 pandemic, has led to an urgent need to develop therapeutics, treatments, and vaccines, both against the immediate threat, as well as developing a versatile platform for future treatments. It is known that binding between the receptor binding domain (RBD) of the spike protein (S) on SARS-CoV-2 and the human ACE2 receptor is essential for entry of the virus into the epithelium. SARS-CoV-2 can enter via endocytosis by binding to the ACE2 receptor, and/or via fusion, where, after ACE2-binding, it gets activated by the human proteases like furin, TMPRSS2, and cathepsins,allowing it to fuse at the plasma membrane. A powerful strategy to block viral infection is to use inhibitors such as short chain peptides and mini-proteins that bind to the S1 protein with high affinity. Such moieties have been designed based on either ACE2 receptor binding epitopes or designed *de novo*to have very low dissociation constants (K_d) with the S1 RBD. However, free peptide formulation can be difficult and non-uniform, especially with long chain lengths; additionally, free peptides and mini-proteins have a short *in vivo* half-life. In this work, we leverage the advantages of functional nanoparticles containing peptides/mini-proteins instead as blocking agents for the SARS-CoV-2 virion. We have developed micellar nanostructures with a functionalized recombinant protein called oleosin that will bind and inactivate the Spike S1 protein. Oleosin, found in nature as a stabilizer for fat bodies in plant seeds, is a rare protein that can act as a free-chain triblock surfactant, capable of assembling into micelles and vesicles when engineered in certain ways, as shown previously in our lab. Micelles were formed by spontaneous self-assembly of oleosin that was genetically modified with anti-S1 mini-protein and peptide sequence. Because oleosin is a protein, we can precisely incorporate into it a versatile range of functional moieties at the gene level, eliminating the need for complex post-formulation functionalization chemistry and washing steps involved in other varieties of synthetic micelles and vesicles. Unlike therapeutic antibodies, recombinant proteins can be formulated relatively easily, and unlike free peptides and mini-proteins in many cases, recombinant proteins are uniform and can be equipped with multiple functions with precision through genetic engineering. Furthermore, micelles can have a much greater in vivo half-life than free proteins.

We cloned S1-binding mini-protein genes called LCB*x*previously designed by David Baker's laboratory (at UW Seattle) to the oleosin N-terminus, which we call Oleo-LCB*x*. Additionally, several ACE2-mimicking peptides were also cloned to the parent oleosin gene in the same way using In-Fusion cloning. Proteins were expressed in E.*Coli*(BL21) and purified using Nickel-immobilized metal affinity chromatography.These proteins formed micelles in the 10-100 nm range as verified by dynamic light scattering. To assess the inhibitory activity of these micelles, we measured their effect on the "infection" of Reporter Virus Particles (RVPs) into 293T-hsACE2 cells; the RVPs contain a gene for green fluorescent protein (GFP) and "infection" is reported by GFP-associated fluorescence. Two proteins, Oleo-LCB1 and Oleo-LCB3 were seen to have the capacity to completely block of RVP infection into cells at 10 μM, as determined from the total lack of fluorescence in the cells after co-incubation with RVPs and the protein. Based on a therapeutic dose response, these proteins were able to reduce RVP infection at a concentration as low as 5 nM, where the functional Oleo-LCB*x* is present in a blend of Oleo-LCB*x* and the non-functionalized oleosin at a total final concentration of 10 μM (above the critical micelle concentration). This is a highly promising result showing that the LCB1 conformation and affinity for the spike protein is preserved in the oleosin conjugate. Another advantage of nanoparticles instead of unassembled/free peptides/proteins is that multiple functionalities can be incorporated in the same structure by blending different oleosins, leading to a multifunctional therapeutic. To validate the potential multifunctional nature of a micellar therapeutic, two different concentrations of Oleo-LCB1 and Oleo-LCB3 were co-mixed in the same sample. We found that certain combinations performed much better than individual molecules at the same concentration, with one combination completely eliminating RVP infection. Furthermore, micelles of Oleo-LCB1 were equally effective against the Delta variant of SARS-CoV-2. In summary, we present a molecular bioengineering approach to make multifunctional nanostructures (micelles or potentially vesicles) of precise chemistry to competitively bind to virus-mimicking particles. Ongoing work involves addition of other functionalities to micellar oleosin resulting in synergistic approach to block virus entry through the ACE2 receptor. For instance, we have developed, expressed, and purified oleosin variants with anti-fusogenic peptides that will inhibit the stage of viral interaction with the cell membrane that involves the activation of membrane proteases resulting in virus-cell membrane fusion. These variants also form 10-100 nm micelles, thus allowing blending of such functionalities with Oleo-LCB*x*to form composite micelles. Although the immediate concern is to combat SARS-CoV-2 infection, the strategy is modular, in that domains of proteins embedded in the nanostructurescan be swapped with other bioactive domains of different specificity as required as new infectious agents, such as mutant strains, emerge.

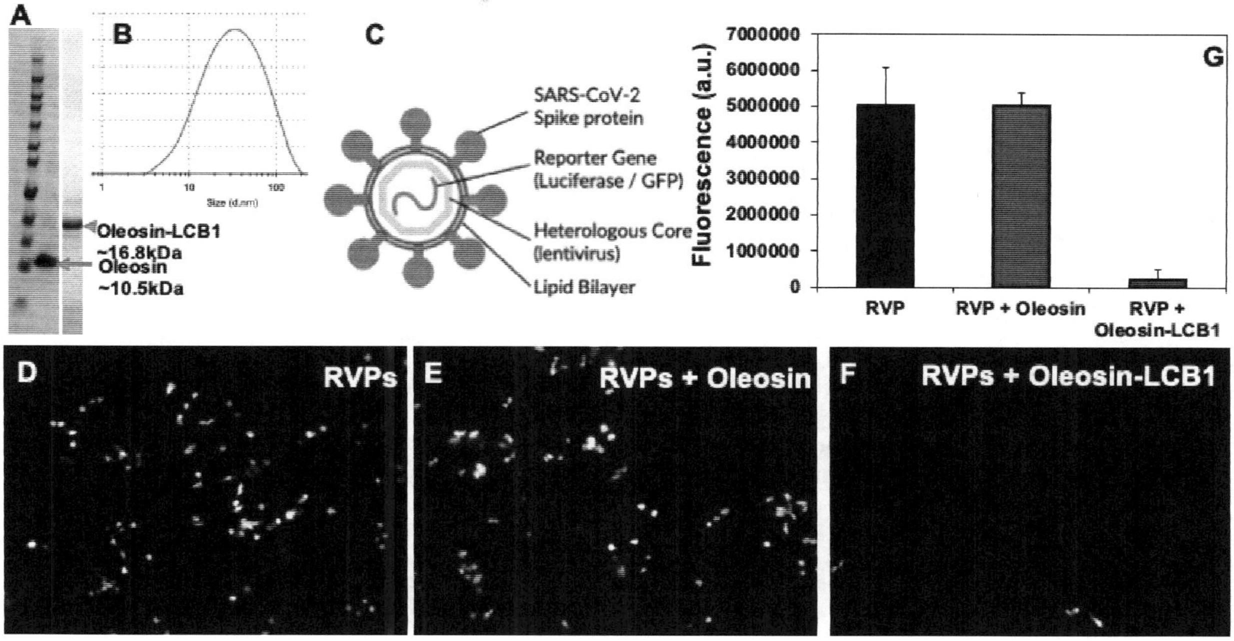

595e Characterizing Enzyme Cooperativity with Imaging Samdi-MS

Blaise Kimmel, Jennifer Grant, Milan Mrksich

Cooperative binding interactions are important in metabolic regulation, cellular signaling, and disease progression. Binding of a metal or ligand to multiple sites in a protein is described by the Hill coefficient *(n)* which describes the degree of cooperativity and where $K_{0.5}$ describes the ligand concentration required for half maximal activity. However, there remains a need for high-throughput strategies to quantitatively measure cooperative binding interactions. Our group has have previously shown that we can overcome this challenge by using iSAMDI-MS (imaging self-assembled monolayers for matrix-assisted laser desorption/ionization mass spectrometry), which can perform and analyze thousands of reactions in a single experiment for calculating the Michaelis constant and rate of a chemical reaction. This talk describes the utility of iSAMDI for calculating the cooperativity in binding of calcium ions to peptidylarginine deiminase type 2 (PAD2). Here, we will evaluate the design elements for our microfluidic device, focusing on the incorporation of a self-assembled monolayer that is functionalized with a peptide substrate for PAD2. The enzyme and different concentrations of calcium ions were flowed through each of eight channels, where the position along the channel corresponds to reaction time and position across the channel corresponds to the concentration of Ca^{2+}. Imaging SAMDI (iSAMDI) is then used to determine the yield for the enzyme reaction at each 200 μm pixel on the monolayer, providing a time course for the reactions. After extracting the imaging mass spectrometry data, we utilized numerical calculations to quantify the spatiotemporal conversion of the peptide substrate into the citrullinated product as a function of concentration, position, and flow rate. Ultimately, analysis of the peptide conversion as a function of position/time gives the degree of cooperativity (*n*) and the concentration of ligand required for half maximal activity ($K_{0.5}$) for the Ca^{2+} – dependent activation of PAD2. This work establishes a high-throughput and label-free method for studying enzyme-ligand binding interactions and widens the applicability of microfluidics and matrix-assisted laser desorption/ionization mass spectrometry (MALDI) imaging mass spectrometry.

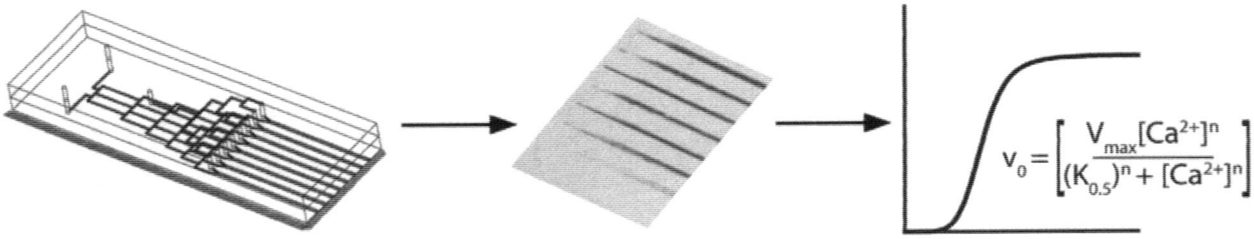

595f Delivery and Actuation of Aerosolized Microbots

Coy Zimmermann, Tyler Schraeder, Brandon Reynolds, Emily DeBoer, Keith B. Neeves, David W. M. Marr

Pulmonary drug delivery is commonly performed using inhalers which disperse μm-scale droplets throughout the lungs. However, once delivered, drug transport can be diffusion-limited and impeded by mucous build-up, especially in cystic fibrosis patients. In addition, inhalation-based therapies cannot selectively target within the lungs to treat localized diseases, such as cancer. Microbots show good promise for overcoming these issues; however, microbots are typically fabricated in sizes greater than the optimal droplet size for inhalation-based deposition in the lung. Here, we present a novel microbot delivery system that uncouples size from structure and function by delivering colloidal paramagnetic building blocks into the lungs for assembly *in situ* to form microwheels capable of translation, selective delivery, and mechanical work deep within lung mimics. To verify beads are successfully loaded, aerosolized bead-laden droplets are captured in oil to determine loading and size distributions. Aerosolized microwheel velocity distributions are measured and are not significantly different after delivery. Clear 3D-printed lung mimics are fabricated to demonstrate two targeting methods; one which generally drives microwheels deeper, and one which selectively targets specific areas. Together, this work opens the previously inaccessible pulmonary system for inhalation-based microbot therapies.

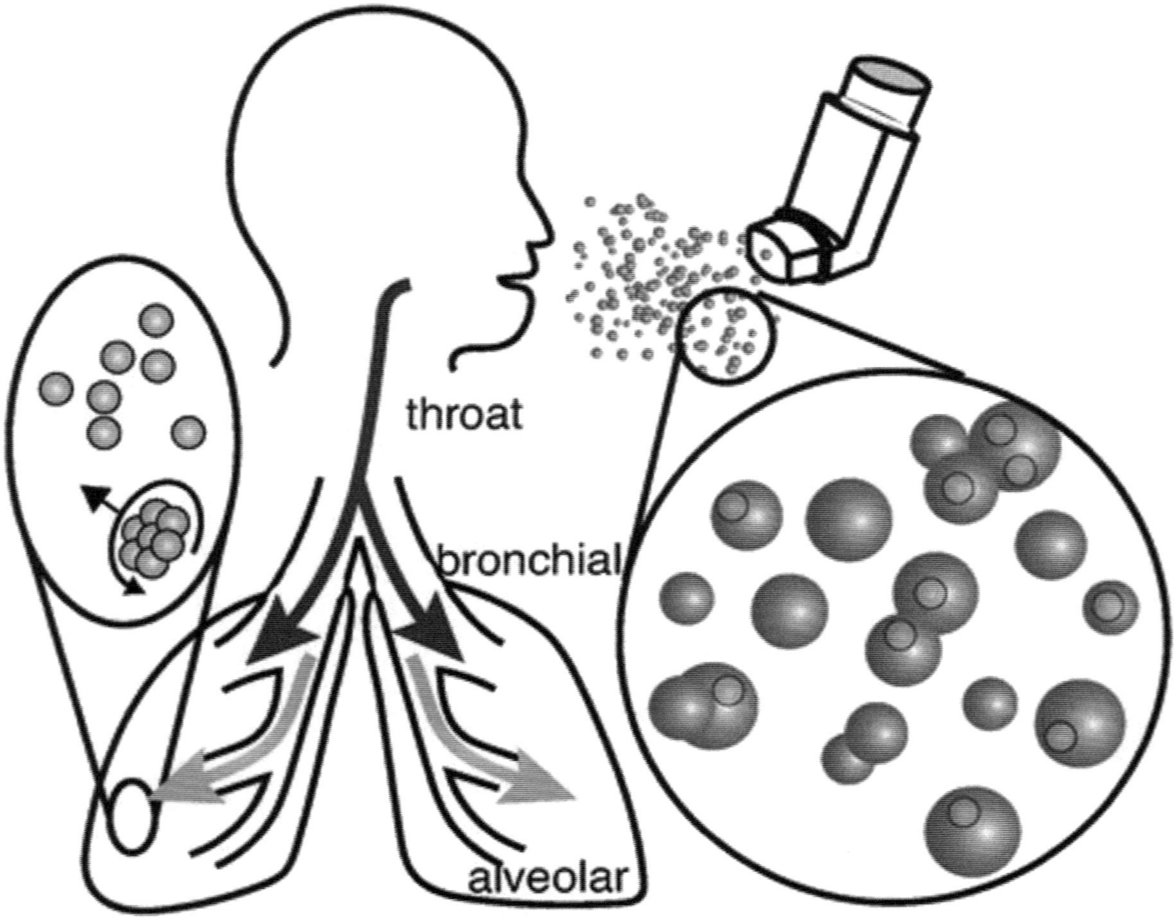
throat
bronchial
alveolar

595g Invited Talk: Characterizing Sugar Transporters with Biosensors and Cheminformatics

Lily Cheung

SWEETs are membrane sugar transporters that play critical roles in plant physiology, crop yields, and biotechnology. We recently reported the design of a biosensor consisting of a plant SWEET and a conformation-sensitive fluorescent protein that translates sugar binding to the transporter into a fluorescence response. We named this biosensor SweetTrac1 and showed with a mass action kinetics model that changes in fluorescence intensity correspond to particular molecular events in the transport cycle. In this talk we demonstrate how biosensors combined with cheminformatics can be used to decipher the variety of molecules that a transporter can recognize and provide suggestions on how to use biosensors for protein engineering.

633a A Combined Analysis of Metabolic Networks and Transcriptomic Data to Predict the Impacts of Copper Deficiency on the Liver Metabolism

Naeun Lee, Joseph Brake, Heejeong Kim, Jaekwon Lee, Hyun-Seob Song

Metabolic dysfunction-associated fatty liver disease is a widespread public health concern. While high-calorie intake and obesity are known risk factors for fatty liver disease, other contributors and modifiers remain characterized. Intriguingly, inadequate uptake or distribution of copper (Cu) leads to lipid and cholesterol metabolic disorders, and fatty liver disease and obesity dysregulate Cu levels. Cu is a vital mineral required for enzymes in several biological processes, including energy generation and redox balance; nevertheless, how Cu deficiency and toxicity induce metabolic disorders and the underlying mechanisms remain to be elucidated. For these reasons, identifying the metabolic pathways affected by Cu bioavailability to display macronutrient metabolic disorder followed by fatty liver disease is an important challenge. To fill in this knowledge gap, we aim to predict the impacts of Cu deficiency on metabolism in the liver, a central organ for Cu homeostasis and macronutrient metabolism, via a combined analysis of metabolic networks and transcriptomic data. We first experimentally characterized alterations in gene expression by Cu deficiency based on a comparative analysis of transcriptomic data collected from control and Cu-deficient mouse liver. Cu deficiency, specifically in hepatocytes of mice, was induced by knocking out the Ctr1 gene that encodes a high-affinity Cu importer. We subsequently performed flux balance analysis using a genome-scale hepatocyte network by constraining the actions of Cu-containing enzymes (cytochrome oxidase and superoxide dismutase) singly and in combination. This analysis predicted the changes in mitochondrial net activity from the Ctr1 gene knockout liver. Moreover, our data indicate that Cu deficiency leads to the regulation of several lipid metabolic pathways. The simulation shows an increase in a series of reactions for fatty acyl-CoA, acetyl-CoA, cholesterol, and steroid hormone synthesis; on the other hand, citric acid cycle activity is decreased with a less acetyl-CoA synthesis from citrate. These results identify metabolic pathways promoting lipid synthesis under Cu deficiency. For validation, we compare predicted flux distributions with transcriptomic data. Next, we integrate transcriptomic data into metabolic networks. As a follow-up step, we will examine the impacts of Cu deficiency on liver metabolism under different diet conditions. The results are expected to shed new light on the roles of Cu homeostasis in liver functions, fat metabolism, and the mechanisms behind fatty liver disease, which provides the fundamental knowledge needed for better treatment or prevention of metabolic dysfunction and fatty liver disease.

633b High-Performance Bioinformatics Workflow Incorporating Metabolomics Data Analysis, Processing, and Integration within the Exposome Concept.

Nafsika Papaioannou, Catherine Gabriel, Spyros Karakitsios, Dimosthenis Sarigiannis

In 2005, Prof. Wild introduced a new term; the exposome. The 'exposome' encompass the totality of human environmental exposures from conception onwards. The study of the exposome opens a new window to disease etiology, but there is a lack of standardised bioinformatics workflow for its analysis and visualisation. We provide a robust workflow developed in the R/Bioconductor architecture that includes methods for the data pre-processing, processing of metabolomics data, and integration with the environmental exposures. We validated the workflow with 64 serum samples from pregnant women. The samples were part of the European Exposure and Health Examination Survey (EXHES) study.

The sample preparation and data acquisition protocols for the metabolomics analysis of serum samples have been previously described. Briefly, serum samples were thawed at 4°C and immediately vortexed. A protein removal step of 200μL was followed using 1:3 methanol. The samples were centrifuged at 15000rpm for 15 min, and the supernatant was transferred to a new Eppendorf to dryness. We recontructed the dry extracts with LC-MS grade water containing internal standards, and we vortexed them vigorous. A final centrifugation at 15000rpm for 10 min and were transferred to autosampler vials using inserts for analysis. The analysis was performed with an Agilent 1290 Infinity HPLC LC System coupled to an Agilent 6540 HRMS-QTOF/ LCMS system in two ionisation mode; positive and negative. Spectra were acquired in full MS scan mode between 50 and 1000 m/z at a scan rate of 1.5 spectra/s in centroid mode at a resolution of 40,000 FWHM. The source conditions are as follows: gas temperature 300 oC, drying gas 7 l/min, nebuliser 50psig, fragmentor 250 V, skimmer 65 V, and capillary voltage 3500V or -3500V in positive or negative mode, respectively. A Fortis Speed Core pH+ C18 (2.1x 100 mm, 2.6 μm) preceded by a filter column was used. The column was thermostat at 40 °C, and the flow rate was 250 μl/min. The mobile phases were water (solvent A) and methanol (solvent B), amended with 0.01% formic acid. The optimised gradient elution program was the same for both ionisation modes: 0% B for 2 min, 100 % B for 17 min, 100 % B for 5 min, and restore the initial conditions with 0 % B for 4 min. The injection volume was 5 μL.

Spectral processing was based on the R packages XCMS and CAMERA. Briefly, immediatly after raw data import in .mzML format, we plotted the base peak chromatograms of the samples, boxplots representing the distribution of total ion currents per file, and heatmaps, as a first evaluation of experiment performance, thus to detect any failing run. QC samples were used to evaluate the performance of the selected parameters for the peaks detection, alignment, and grouping. Chromatographic peak detection was performed based on the centWave algorithm. We optimised the mass error (ppm) and mzdiff parameters using the IPO R-based package. The signal to noise (snthresh) and the noise level parameters were estimated based on an in-house developed approach. Obiwarp method was used for alignment. We performed gaps-filling to minimise the false negatives. The PerformPeakAnnotation function was used for isotope and adduct annotation. The list of the detected metabolites was filtered to minimise the false positives and obtain consistent variables. We excluded the metabolites having 100% presence in solvents and the variables with more than 80% missing values in QCs. The resulted matrix was further reduced by excluding the metabolites with RSD >30% in QCs. The instrument and overall process variability were then determined by calculating the median RSD for authentic internal standards and all endogenous metabolites. Normalisation by median, mean centering scaling, and log transformation was performed to transform the data matrix into a more Gaussian-type distribution, thus reducing systematic error in experimental conditions. A major concern in exposome studies that include metabolomics analysis is the systematic signal deviations between batches and within batches caused mostly by the build-up of dirt in the interface of an LC-MS system or imperfect column regeneration in the gradient program. We performed intensity drift correction to improve the signal/noise ratio and reduce various types of overfitting using a combination of the algorithms included in the batchCorr, and the DBnorm R packages. Finally, we run an in-house developed algorithm to annotate features by comparing the MS/MS from the analysed samples and the MS/MS from the online databases (HMDB, KEGG, and LipidMaps) and in-house build database. We stricked the query to metabolites that can be detected in blood. We searched based on the following adducts: 'M+2H', 'M+H+NH4', 'M+ACN+2H', 'M+2ACN+2H', 'M+H', 'M+NH4', 'M+Na', 'M+ACN+H' in positive ionization mode, and the following in negative ionization mode: 'M+ACN+Na', 'M+2ACN+H', '2M+H', '2M+Na', '2M+ACN+H' 'M-2H', 'M-3H', 'M-H2O-H', 'M+Na-2H', 'M-H', 'M+Cl', 'M+FA-H', 'M+K-2H', '2M-H'.For database matching, mass error in ppm was set as in the peaks detection step. Individual enrichment pathway analysis was using the Fisher's method. The databases used were the following: KEGG, WikiPathways, Reactome, HumanCyc, EHMN, PharmGKB, SMPDB, BioCart, INOH, and PID. The metabolites names, KEGG IDs, and PubChem IDs were used as identifiers. For the size of the population, we consider the metabolites that have been detected in human samples even if they have not been quantified according to the HMDB database. We did not consider the predicted metabolites to avoid the false positives results. The Exposome-Wide Association Study (EWAS) approach was adopted to comprehensively and systematically explore and associate multiple exposure factors and modifiers, discovering and replicating robust correlations with metabolites levels and dysregulated pathways. The 'X-Wide Association Analyses (XWAS)' R package was implemented to explore and associate the detected metabollite, and pathways with multiple exposure factors. The latest is called Exposome-Wide Association Study (EWAS) analysis, and it is based on the Genome-Wide Association Study (GWAS). The results were visualised using volcano plots and correlation globes.

As exposure factors, we considered 14 phthalate metabolites (MEP, MBzP, MiBP, MnBP, MCHP, MnPeP, MEHP, 5OH-MEHP, 5oxo-MEHP, 5cx-MEHP, MnOP, OH-MiNP, cx-MiNP, and OH-MiDP), and 2 Hexamoll® DINCH® metabolites (OH-MINCH and cx-MINCH) determined by online high-performance liquid chromatography coupled to tandem mass spectrometry (HPLC–MS/MS) using internal isotope-labelled standards.

Most perturbed metabolic pathways were related to oxidative stress and stress-activated signalling pathways, including the urea cycle, indicating an imbalance between the cellular reactive oxygen species (ROS), which may be an effect of exposure to phthalates, and the inability of the cell to detoxify them.

The list of the detected metabolites included the glutamate, which is utilised as a transamination source for aspartate and either pyruvate or 3-hydroxybutyrate as a source of acetyl-CoA. The acetyl-CoA is one of the fundamental units for metabolic energy production (ATP) through the tricarboxylic acid (TCA) cycle, followed by electron-transport chain mediated oxidative phosphorylation. In parallel, it is a fundamental unit for energy storage via gluconeogenesis and lipogenesis.

In addition, the integrated analysis revealed perturbation in the metabolism of biogenic amino acids (Biogenic Amine Synthesis/ Dopamine metabolism). Exposure to phthalates can disrupt the synthesis, transport, and release of biogenic amines, including dopamine, serotonin, norepinephrine, and glutamate, which modulate behaviour, cognition, learning, and memory. Thus, perturbations of the main pathways of dopamine, serotonin, norepinephrine, and glutamate metabolism could identify potential neurodevelopment biomarkers.

Overall, we present the usefulness of our advanced workflow in the R programming language by analysing a real dataset, including metabolomics and exposure data.

633c A Computational Framework for Studying the Survival/Inhibition Mechanism of *Vibrio Cholera El Tor* with Evolving Metabolic Capability in a Glucose-Rich Environment

Faiz Khan Mohammad, Raghunathan Rengaswamy, Swagatika Sahoo

Vibrio cholera El Tor (VC El Tor), causative bacteria of cholera disease, has a unique fermentation ability to consume glucose, produce neutral fermentation by-products and complete respiration end-products in an overflow strategy of metabolism. The secretion products in the medium of growth, in turn, can modulate the bacteria's metabolic capabilities as a feedback system and can determine the fate of bacteria. To predict the metabolic behaviour of such bacterial systems with a feedback loop (changes in the environmental condition modulate the metabolic capabilities), we have developed the Dynamic evolving species metabolic modelling (DevoMM) framework based on Pirt's law [1]. The framework considers, however, the Non-growth associated maintenance parameter in a genome-scale metabolic constraint-based model (COBRA model) represents the energy requirement for the cell to sustain its biomass which is fixed under specific extracellular conditions, deviations from these reference conditions lead to metabolic shifts and adjustments of the cellular nutrient uptake or maintenance requirements (NGAM). The major challenge in this framework is estimating the change in the maintenance requirement (NGAM) based on the presence of un-dissociated acid and pH of the growth medium. To address this, we use the relationship developed by Tack I.L.M et al., (2018) [2] for the influence of environmental conditions on maintenance requirements to estimate NGAM at specific environmental condition.

Furthermore, this evolution in the metabolic capability of bacteria with the environmental condition can form a basis for the cross-talk in a microbial community. Cross-feeding of acidic fermentation by-products from E.coli can modulate the survival of VC El Tor when co-cultured in a glucose-rich environment (observation from Sengupta et al., 2017 experiment [3]). DevoMM is extended to a multi-species, dynamically evolving Multi-species Metabolic Modelling Framework to predict the cross-talk in the microbial community and evolution in the metabolic capabilities of individual bacteria in the community. Together, these frameworks predict the survival benefits of VC El Tor in a glucose-rich medium and, when grown with E.coli in the same glucose-rich setting, death of VC El Tor. In Conclusion, the study highlights the importance of modelling evolving metabolic capability in the COBRA model.

The schematic of the modelling framework is shown below:

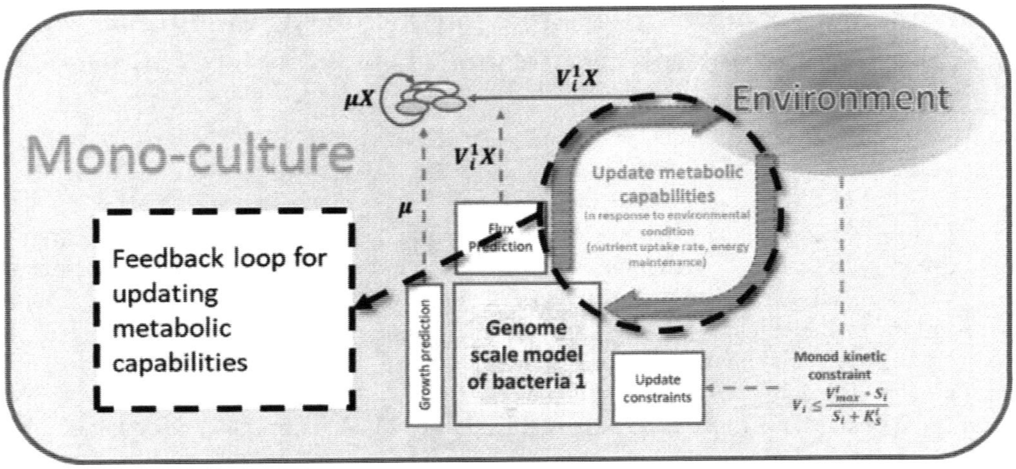

Figure 1: Dynamically evolving Metabolic modelling (DevoMM)

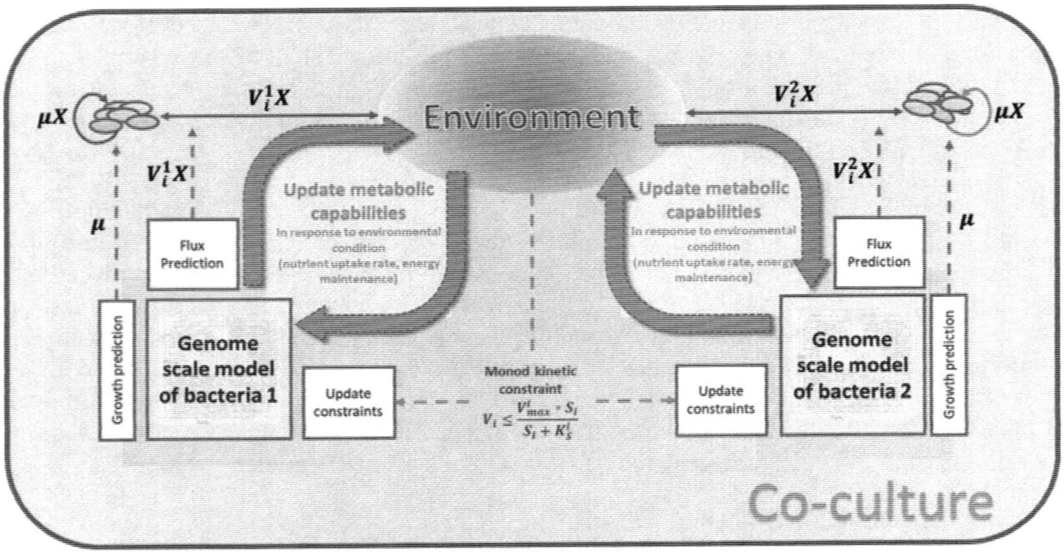

Figure 2: Dynamically evolving Multi-species Metabolic modelling (DevoMMM)

633d Developing a Machine Learning Model for Fitness Prediction in Multiplex Knockout *S. Cerevisiae* mutants

Michael Volk, Huimin Zhao

Metabolic engineering of yeasts allows for modification of industrial relevant phenotypes including robustness phenotypes like improved growth rates and resistance to feedstock broth inhibitors. Traditional in silico design tools relying on mechanistic representations of the cell have shown some success, but they typically represent only a fraction of the total genome. The most updated genome scale model for *Saccharomyces cerevisiae* covers ~20% of the genome. The most common approach for engineering robustness phenotypes is through adaptive laboratory evolution (ALE), where successive passages of mutating cells are selected for improved growth phenotypes. While ALE has proven its reliability in engineering growth phenotypes, it requires lengthy experimentation that can last months and whole genome next generation sequencing (NGS) analysis of checkpoint mutants to identify mutations that must then be reintroduced to the parent strain to deduce causal mutations. Additionally, ALE is restricted to evolving phenotypes that are dependent on growth and it can only take advantage of single nucleotide polymorphisms and small insertions or deletions. Recent activity in machine learning for biosystems design has demonstrated that a data driven approach can be used in making improved predictions for engineering desirable phenotypes. We aimed to construct a genome scale machine learning model that accounts for nearly the entire genome (~5,845 genes in *Saccharomyces cerevisiae*) which can predict targets for improved fitness without ALE experimentation or NGS analysis. To do so, we developed a multi-relational graph neural network (GNN), to predict high fitness *Saccharomyces cerevisiae* multiplexed knockout mutants, and an associated experimental pipeline for constructing and profiling mutants. Specifically, we have combined gene-gene interactions, protein-protein interactions, the regulome, gene ontology, and mined gene features, in a gene-graph to predict growth phenotypes for yeast with *n*-order gene deletions. The GNN models genes as nodes and their corresponding interactions as edges. The GNN has been generalized to construct representations at node, edge, and global levels of the gene-graph. Growth prediction corresponds to a global prediction on the gene-graph. The machine learning problem is formatted as a ranking problem where the GNN provides a rank-ordered list of gene deletions for high fitness mutants. When compared against the known top 200 fitness mutants, preliminary results achieved a breakeven precision of 0.59. High ranked mutants from the GNN model are then profiled with a liquid growth assay and compared against colony size measurements from synthetic genetic array data. In addition to global fitness prediction, the node representations from the GNN are used to suggest gene expression modifications, unachievable by ALE mechanisms, to confer tolerance against furfural, a common chemical inhibitor found in lignocellulosic hydrolysates. We envision this GNN serving as an intermediate design tool that can warn against genetic designs with high-order synthetic lethality when constructing multiplexed knockout mutants and for suggesting tuned genetic expression to maintain mutant fitness under environmental stress. Importantly, the GNN feature representation is generalized to be transferable to other system biology and metabolic engineering prediction tasks that are unrelated to growth phenotypes. Some of these prediction tasks could include, but are not limited to, gene function classification for under characterized genes, existence of novel regulation interactions, and other global cell states like changes in cell morphology resulting from gene knockouts.

633e Tracking Pyrophosphate Metabolism and Evaluating Its Significance in the Bioprocessing of Lignocellulosic Biomass By *Clostridium Thermocellum*

Wheaton Schroeder, Tuen Kuil, Daniel Olson, Costas D. Maranas

Lignocellulosic biomass is an abundant and renewable source of carbon for chemical manufacturing, yet it is difficult (and cost prohibitive) to use in conventional processes. A promising candidate for lignocellulose bioprocessing is the thermophilic anaerobe *Clostridium thermocellum* given its natural competence to produce ethanol, amino acids, and hydrogen gas as well as its ability to grow at high substrate loading. This organism has an atypical glycolytic pathway in that pyrophosphate drives it, being used as an energy source in phosphofructokinase (pfk) and pyruvate phosphate dikinase (ppdk) reactions. The identification of pyrophosphatase sources is key to the engineering of *C. thermocellum*. Multiple *in vivo* efforts to identity the primary pyrophosphate source have resulted only in the exclusion of various processes from being the primary pyrophosphate source. To better understand these experiments and closely track pyrophosphate metabolism, an existing Genome-Scale Model (GSM) of metabolism was carefully curated for pyrophosphate metabolism into a new model, iCTH659. Particularly focuses included transport, nucleotide, energy, vitamin, and amino acid metabolism in addition to addressing reaction imbalances, infeasible cycling, and removing the soluble pyrophosphatase assumption from biomass synthesis. The iCTH659 model agrees with fermentation and biomass yield metrics from the most recent investigations of *C. thermocellum* metabolism. From this model, stoichiometrically feasible pyrophosphate sources are explored. In addition, alternatives to pyrophosphate-utilizing reactions are proposed through the model curation process and are evaluated for feasibility. This shows stoichiometrically viable candidates as yet uninvestigated for providing the driving *C. thermocellum* glycolysis, including alternative pfk phosphate donors.

633f Multi-Faceted Analysis Reveals the Metabolic Underpinnings of Overflow Metabolism in the Pathogenic *Staphylococcus Aureus*

Adil Alsiyabi, Rajib Saha

Staphylococcus aureus is a dangerous human pathogen that infects approximately 20% of the world population. Furthermore, it has evolved multiple antibiotic-resistant strains making it difficult to treat with conventional medications. Experimental evidence has shown that unlike other pathogens, *S. aureus* performs overflow metabolism even during relatively low growth rates, indicating a potential bottleneck in the respiratory pathway. In this work, a multi-faceted modeling approach was used to investigate *S. aureus*'s metabolism. First, a thermodynamic analysis of the respiratory pathway was conducted to identify metabolic bottlenecks that compromise the pathway's efficiency. Incorporation of these findings into a genome-scale metabolic model of *S. aureus* revealed that this leads to significant changes in central metabolism and has important implications for overflow metabolism. Furthermore, transcriptomic data from 7 distinct strains revealed that activity through the electron transport chain is affected my membrane-crowding effects during exponential growth. Finally, a coarse-grained analysis based on optimal resource allocation revealed that metabolically induced changes in the efficiency of the respiratory pathway has direct effects on overflow behavior.

633g Invited Talk: Metabolic Engineering of Non-model Yeasts for Production of Chemicals and Fuels

Zengyi Shao

Many non-model yeasts exhibit desirable metabolic traits for industrial applications. However, they often lack genetic engineering tools that are essential for further research and development. Recently, we developed several genetic tools for rapid engineering of two non-model yeasts including low pH tolerant *Issatchenkia orientalis* and high lipid producing *Rhodosporidium toruloides*. Using these tools, we engineered an *I. orientalis* strain capable of producing 116 g/L succinic acid (0.63 g/L/hr) from sugarcane juice and an *R. toruloides* strain capable of producing 28 g/L triacetic acid lactone (0.24 g/L/h) from glucose in fed-batch fermentation. Succinic acid is one of the top 12 bio-based platform chemicals identified by the Department of Energy and has diverse applications in food, pharmaceutical, and agriculture industries. Triacetic acid lactone is also a platform chemical for the production of commercially valuable bifunctional chemical intermediates and end products, including phloroglucinol, acetylacetone, and sorbic acid. Preliminary technoeconomic analysis suggests that these fermentative production processes are not far from being economically viable. Therefore, these studies have demonstrated the potential of non-model yeasts as production hosts for industrial applications.

In the second part, I plan to highlight one technology we recently developed to address the issue of low efficiency for achieving precision genome editing in non-model yeasts. A common technology hurdle for engineering nonconventional species is that majority of them preferentially employ non-homologous end joining (NHEJ) to repair DNA double-strand breaks (DSB), which prohibits precise CRISPR/Cas9 genome editing. Most recently, we identified the shortcomings of the conventional genome editing strategies and developed a new CRISPR platform, named Lowered Indel Nuclease system Enabling Accurate Repair (LINEAR), which enabled precision editing without NHEJ disruption with efficiencies of 67-100% in four industrially relevant yeasts. With NHEJ preserved, we demonstrated its ability to survey genomic landscapes, identifying loci whose spatiotemporal genomic architectures yielded favorable expression dynamics for heterologous pathways.

597a Roughening Instability of Growing 3D Bacterial Colonies

Alejandro Martinez-Calvo, Tapomoy Bhattacharjee, R. Konane Bay, Anna Hancock, Ned Wingreen, Sujit Datta

While colony morphogenesis is well-studied in 2D, many bacteria grow as large colonies in 3D environments, such as gels and tissues in the body, or soils, sediments, and subsurface porous media. Here, we describe a morphological instability exhibited by large colonies of bacteria growing in 3D. Using experiments in transparent 3D granular hydrogel media, we show that dense colonies of different species of bacteria, including *Escherichia coli* and *Vibrio cholera*, generically roughen as they consume nutrients and grow beyond a critical size, eventually adopting a characteristic branched, broccoli-like self-affine morphology independent of variations in the cell type and environmental conditions. This behavior reflects a key difference between 2D and 3D colonies: while a 2D colony may access the nutrients needed for growth from the third dimension, a 3D colony inevitably becomes nutrient-limited in its interior, driving a transition to rough growth at its surface. In support of this picture, we elucidate the onset of roughening using linear stability analysis and numerical simulations of a continuum model that treats the colony as an 'active fluid' whose dynamics are driven by nutrient-dependent cellular growth. We find that when all dimensions of the growing colony substantially exceed the nutrient penetration length, nutrient-limited growth drives a 3D morphological instability that recapitulates essential features of the experimental observations. Our work thus provides a framework to predict and control the organization of growing colonies—as well as other forms of growing active matter, such as tumors and engineered living materials—in 3D environments.

597b The Instabilities in Bacterial Colony Patterns

Uttam Kumar, Subramaniam Pushpavanam

In this work, we use a modeling and experimental approach to study the self-organization of bacterial cells expanding on a solid agar. Growing microbes form a biofilm by secreting extracellular polymeric substances (EPSs) that form an integral part of biofilm-like structures. In these structures, interaction among mechanical and physical forces result in different morphological patterns such as concentric rings, dendrites, Eden like structures. Surfactant molecules that lower the surface tension of the colony are secreted by these growing cells. Lowering the surface tension helps spreading of bacteria on solid surfaces. the difference in the surface concentration of surfactant molecules causes Marangoni effect that facilitates spreading.

In this work, we study the effect of bacterial surfactant secretion, diffusional flux of nutrients, substrate thickness and wettability on colony growth and shape within the framework of a hydrodynamic thin film model. We observed the different colony patterns in *Peudomonas* and *Bacillus Subtilis* under similar condition. Our framework accounts interplay among mechanical stress from bacterial migration, growth, surface stress and Marangony effects. Our mathematical model is able to explain front instability in bacterial colony pattern and various experimental observations of microbial colony patterns in literatures.

597c Multi-Scale Modeling of Mucin-Driven Modulations of Microbial Phenotypes

Mohammad Mazharul Islam, Tracy Kuper, Glynis Kolling, Roseanne Ford, Jason A. Papin

Pseudomonas aeruginosa is the leading cause of nosocomial infections and well-known for infecting patients with cystic fibrosis. These infections can be complicated by a variety of virulence mechanisms and a robust metabolic functionality repertoire. This study focuses on the microbial interactions with mucin, which is a key component in the mucus layer, and is known to modulate the pathogen's metabolic traits as well as the proliferation and development of biofilms. Understanding the mechanistic mucin-driven modulations of microbial phenotypes is of paramount importance in multiple diseases including cystic fibrosis, a disease characterized by defective clearance of mucus. While it is appreciated that mucus provides a critical barrier against microbial pathogens in the lung, gastrointestinal tract, vagina, and several other body surfaces, the specific molecular interaction between mucins (the primary component of mucus) and microbes is poorly understood.

Genome-scale metabolic reconstructions are powerful tools to simulate microbial metabolism and interactions with molecular modulators in infection environment for in silico hypothesis testing of regulatory mechanisms and potentially devising treatment strategies. In this work, to improve our understanding of mucin-driven metabolic phenomena in infectious biofilm, we use a computational framework involving a genome-scale metabolic reconstruction of Pseudomonas aeruginosa. The model identified key modulators of metabolic behavior of the clinically relevant phenotypes of Pseudomonas, specifically key genes and pathways regulated by different mucin molecules. This research will guide us to devise innovative solutions for combating this pathogen, which can then be translated to many other infectious diseases.

597d Data-Informed Catastrophic and Harmonious Genetic Codon Bias for Multi-Gene Expression

Aaron Love, Nikhil Nair

Expression of genes in microbial hosts forms the foundation of basic research and applied biotechnology. A significant amount of recent work has focused on de-coupling non-native gene expression from host resources to improve the outcome of synthetic biology and metabolic engineering efforts. So far, most research has focused on transcription, translation initiation, or global sequence parameters, but fewer studies have investigated the mechanistic underpinnings of resource allocation during translational elongation. The degenerate genetic code provides an opportunity to allocate cellular tRNA and ribosomal resources optimally between host and heterologous protein expression. 18 of the 20 natural amino acids can be encoded with multiple codons, which can result in significant codon usage bias due to translational selection for codons with faster elongation times, or that are tuned to host tRNA supply. Through our research, we aim to improve the predictability and robustness of genetic engineering in microbes by systematically determining optimal codon bias schemes.

We investigate how the partitioning of microbial translational resources, specifically through allocation of tRNA by incorporating dissimilar codon usage bias (CUB), can drastically alter expression of proteins and reduce the burden on gene expression systems. Utilizing nearly identical fluorescent reporters (CFP and YFP), we assayed genetic resource competition both in vitro and in vivo using novel designs that isolate translation elongation from other variables. We find that alternative CUB designs can trans-regulate gene expression of competing heterologous and endogenous genes, yielding profitable or catastrophic design options. By isolating individual codons experimentally, we correlate specific codon usage patterns with genetic burden, and derive novel coding schemes for multi-gene expression. These empirically derived coding schemes based on a new codon adaptation index enable the design of harmonious multi-gene expression systems while avoiding catastrophic cellular burden.

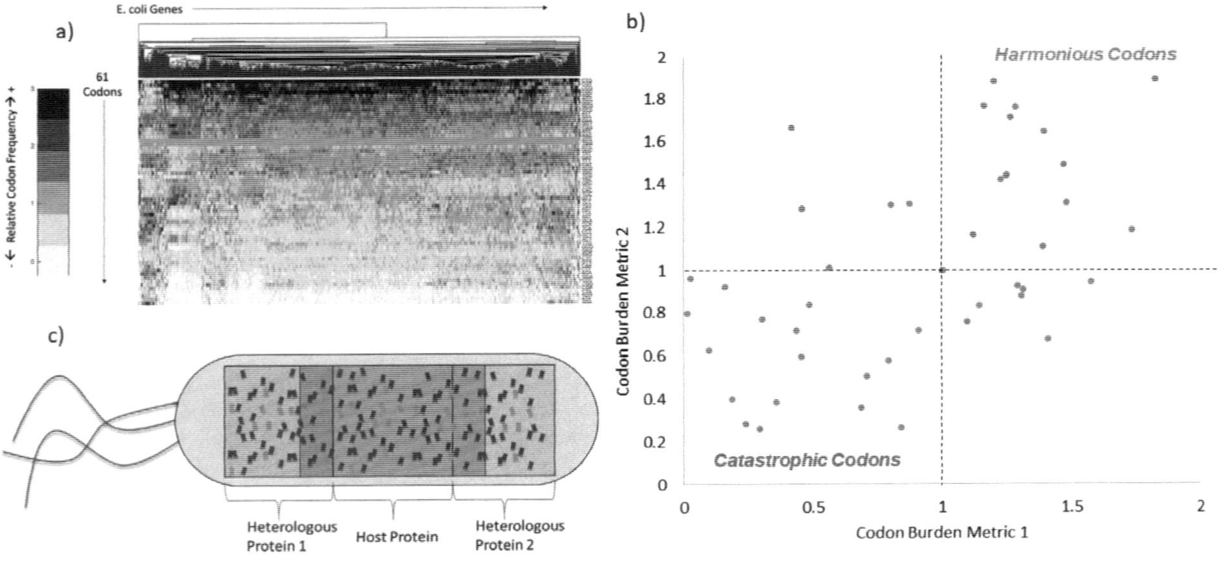

Figure 1 – Determination of coding schemes for catastrophic and harmonious gene expression.
a) Assessing coding scheme of all E. coli genes, including highly expressed genes.
b) Binning of codons based on experimental outcomes.
c) Experimental evaluation of phenotypes associated with different coding schemes (catastrophic, harmonious, or neutral).

597e High-Throughput Screening of a Promoter Library Reveals New Persister Mechanisms in *Escherichia coli*

Sayed Golam Mohiuddin, Aslan Massahi, Mehmet Orman

Abstract

Persisters are generally nongrowing cells that are transiently tolerant to bactericidal antibiotics. This transient state enables persister cells to recover and grow into an antibiotic-sensitive population. Persister cells can be formed stochastically or triggered by environmental stresses such as antibiotic treatment. Although the quinolone-induced SOS response is known to play a critical role in persistence (1, 2), other mechanisms induced by antibiotics are largely unknown. Using promoter and knockout cell libraries in *Escherichia coli*, we have identified several genes, such as *waaG*, *guaA*, and *guaB,* whose expression was considerably up-regulated in antibiotic treated cultures. These genes were found to be important for persister formation in *E. coli* as their deletion significantly enhanced the sensitivity of cells to various antibiotics (**Fig. 1A**). The GuaA and GuaB are the essential enzymes for the biosynthesis of ppGpp (a global persister molecule) and their deletion significantly reduces the ppGpp levels in the cells. The WaaG glucosyltransferase plays a vital role in the outer membrane by providing the first outer core D-glucose of lipopolysaccharides. The deletion of *waaG* selectively disrupted the transmembrane proton gradient component of the cellular proton motive force, impaired the intracellular ATP production, and significantly reduced the formation of type I persisters in stationary phase (**Fig. 1BC**). Previously, it has been shown that active respiration in the stationary phase aids in the arrest of growth in some cells, and inhibition of stationary phase respiration, which drives the proton motive force (PMF), reduces the number of non-growing cells (3). Here, our findings associated with the Δ*waaG* strain further validates this notion. Overall, given that persistence is a critical survival strategy that evolved in many bacteria, our study will enhance the current molecular-level understanding of this conserved mechanism.

Note: This study has been published in "*Microbiology Spectrum*" (4).

References

1. Völzing KG, Brynildsen MP. 2015. Stationary-phase persisters to ofloxacin sustain DNA damage and require repair systems only during recovery. MBio 6:e00731-15.
2. Goormaghtigh F, Van Melderen L. 2019. Single-cell imaging and characterization of Escherichia coli persister cells to ofloxacin in exponential cultures. Sci Adv 5:eaav9462.
3. Orman MA, Brynildsen MP. 2015. Inhibition of stationary phase respiration impairs persister formation in E. coli. Nat Commun 2015 61 6:1–13.
4. Mohiuddin SG, Massahi A, Orman MA. 2022. High-Throughput Screening of a Promoter Library Reveals New Persister Mechanisms in Escherichia Coli. Microbiol Spectr 10:e02253-21.

Figure Legend

Fig. 1. Deletion of the *waaG*, *guaA*, and *guaB* genes reduced persister levels in *E. coli*. (**A**) Bi-phasic kill curves of WT, Δ*waaG*, Δ*guaA*, and Δ*guaB* were generated by treating exponential phase cultures of these strains with ofloxacin for 5 h. Colony formation units (CFU) was determined by plating cells on the agar plates at indicated time points. (**B**) Late stationary phase (t=24 h) cells of WT and ΔwaaG strains were collected, washed, and stained with $DiSC_3$ probe. Once the fluorescence levels reached equilibrium state the stained cells were treated with polymyxin B (PMB), chlorpromazine (CPZ) at the specified concentrations. The fluorescence levels were measured at the indicated time points using a plate reader. (**C**) The intracellular ATP levels of WT and ΔwaaG strains were measured at late stationary phase using a BacTiter-Glo Microbial Cell Viability assay kit. A two-tailed Student's t test with unequal variance was used to determine the statistical significance where, $^{**}P<0.01$. The number of independent biological replicates, N=4. Data points represent mean ± standard deviation (SD).

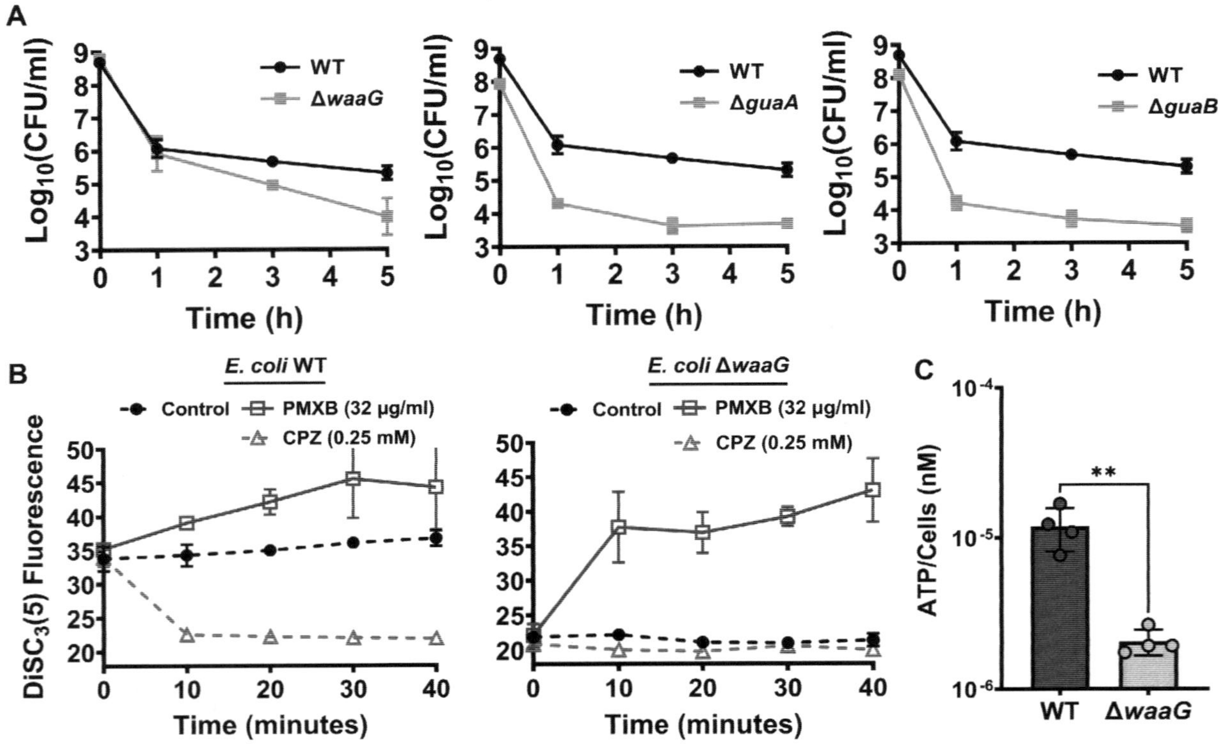
A
Log10(CFU/ml)
Time (h)
WT
ΔwaaG
ΔguaA
ΔguaB
B
E. coli WT
E. coli ΔwaaG
Control
PMXB (32 μg/ml)
CPZ (0.25 mM)
DiSC3(5) Fluorescence
Time (minutes)
C
ATP/Cells (nM)
**
WT
ΔwaaG

597i Combined feedforward/feedback strategies in engineering transcriptional motifs

Gregory Reeves

Background: Regulation of gene expression is of paramount importance in all living systems. In the past two decades, it has been discovered that certain motifs, such as the feedforward motif, are overrepresented in gene regulatory circuits. Feedforward loops are also ubiquitous in process control engineering, and are nearly always structured so that one branch has the opposite effect of the other, which is a structure known as an "incoherent" feedforward loop in biology. In engineered systems, feedforward control loops are subject to several engineering constraints, including that (1) they are finely-tuned so that the system returns to the original steady state after a disturbance occurs (perfect adaptation), (2) they are typically only implemented in the combination with negative feedback, and (3) they can greatly improve the stability and dynamical characteristics of the conjoined negative feedback loop. On the other hand, in biology, incoherent feedforward loops can serve many purposes, one of which may be perfect adaptation. It is an open question as to whether those that achieve perfect adaptation are subject to the above engineering principles.

Results: We analyzed an incoherent feedforward gene regulatory motif from the standpoint of the above engineering principles. In particular, we showed that an incoherent feedforward loop Type 1 (I1-FFL), from within a gene regulatory circuit, can be finely-tuned for perfect adaptation after a stimulus, and that the robustness of this behavior is increased by the presence of moderate negative feedback. In addition, we analyzed the advantages of adding a feedforward loop to a system that already operated under negative feedback, and found that the dynamical properties of the combined feedforward/feedback system were superior.

Conclusions: Our analysis shows that many of the engineering principles used in engineering design of feedforward control are also applicable to feedforward loops in biological systems. We speculate that principles found in other domains of engineering may also be applicable to analogous structures in biology.

597j Invited Talk: Correlating microbiota structure with the exo-metabolome

Pushkar Lele

A soup of biomolecules and metabolites characterizes microbial niches –the exon-metabolome. The exon-metabolome powerfully regulates the physiology of the native bacterial species in an existing niche, but perhaps more crucially, it can determine the transport properties of invading foreign species. The ability to influence cell transport lends defensive functions to specific prominent metabolites, enabling them to shape the microbiota structure. I will discuss our work that reveals the molecular mechanisms by which one key component of the exon-metabolome, indole, selectively controls bacterial access to niches rich in indole. We find that the metabolite attracts one type of bacteria while repelling another from the niche. It does so by activating major bacterial receptors in a signal transduction network that biases cell motility. In Escherichia coli, we show that the activation of one type of receptor causes the cell to swim toward indole-rich niches. In contrast, activating a different receptor causes the cell to swim away from those niches. Net transport outcomes are dominated by the kinetics of the biphasic receptor activation [1]. Next, I will discuss our FRET-based experiments in live cells that helped identify molecular determinants in the receptors of this unique biphasic behavior. Biophysical experiments on single cells further revealed an additional mode of action: indole influences bacterial transport by controlling the energy source that powers motility [2]. By differentially affecting cell energetics, which modulates growth, indole's impact on communities appears to extend beyond transport. This work offers a basis for understanding the complex functions of metabolites in community formation and subsequent growth. The data is expected to guide models for predicting microbiota structure from metabolite profiles and designing novel communities.

[1] Yang et al, PNAS, 2020

[2] Gupta et al, PNAS Nexus, 2022

20a Matrix Remodeling Modulates Human Neural Maturation

Michelle Huang, Julien G. Roth, Renato Navarro, Bauer L. LeSavage, Sarah C. Heilshorn

Human induced pluripotent stem cells (hiPSCs) have emerged as a promising *in vitro* model system for studying neurodevelopment. However, current models remain limited in their ability to efficiently incorporate biochemical and biomechanical signaling cues imparted by the neural extracellular matrix (ECM) and achieve limited degrees of maturation. The native brain ECM is viscoelastic and stress relaxing, exhibiting a time-dependent response to an applied force, yet most conventional biomaterials are formed via static covalent crosslinks and are therefore highly elastic. Here, we develop a family of recombinant materials, consisting of aldehyde-modified hyaluronic acid and hydrazine-modified elastin-like protein (ELP), which are crosslinked with dynamic covalent bonds and exhibit independently tunable stiffness and stress relaxation rate. This platform facilitates the creation of hyaluronan-elastin-like protein (HELP) hydrogels with consistent brain-mimetic stiffnesses, a range of stress relaxation rates, and modular cell-interactive domains. hiPSC-derived neural progenitor cells (NPCs) encapsulated within these gels exhibit high viability and undergo relaxation rate dependent maturation. Specifically, NPCs within faster relaxing, more remodelable HELP matrices extend longer, more complex neuritic projections, adopt decreased metabolic activity, and express higher levels of genes and proteins associated with neural maturation. In addition to relaxation rate, a cell's ability to remodel its microenvironment is also influenced by degradation of the local biopolymer scaffold. We observe heightened matrix degradation and a concomitant increase in protease and hyaluronidase secretion in faster relaxing gels. To elucidate the cell-matrix interactions underlying these observations, we replace the integrin-binding peptide domain (RGD) within HELP with a scrambled analog and introduce inhibitors of myosin contractility and actin polymerization. In these gels, we observe that both the loss of integrin ligands and the inability to polymerize actin preclude NPCs from extending neurites. Taken together, these results suggest that cell-mediated strain and enzyme secretion within remodelable matrices drive neural maturation and that tuning biomechanical signaling cues within engineered microenvironments may advance current models of human neurodevelopment.

20b Stiffness Promotes Metabolic Reprogramming in Primary Hepatocytes

Michael Moeller, Senthilkumar Thulasingam, Kimberly M Stanke, Madhusudhanan Narasimhan, Srivatsan Kidambi

Metabolic-associated fatty liver disease (MAFLD) is the most common liver disease globally and can progress from simple steatosis to nonalcoholic steatohepatitis (NASH) and potentially liver cirrhosis and failure. Hepatocytes are the primary functional liver cell and accumulate excess lipids in MAFLD. Liver stiffness is established as an indicator of disease progression and useful for prognosis. However, the role of liver stiffness in MAFLD and hepatocyte metabolic dysfunction is not well understood. To study the effect of matrix stiffness on hepatocyte function, we cultured primary hepatocytes on our novel biomimetic platform, BEASTS (Bio-Engineered Adhesive Siloxane substrate with Tunable Stiffness). Unlike other common methods for investigating the role of matrix stiffness, such as polyacrylamide gels, BEASTS is able to facilitate cell attachment with a protein-free surface and avoid confounding interactions between the gel's mechanical properties and adhesive proteins. Using BEASTS, hepatocytes were cultured on a range of stiffnesses representative of healthy (2 kPa) to fibrotic (25, 55 kPa) liver tissue. We observed altered energy metabolism including increased glycolysis in hepatocytes cultured on fibrotic stiffness. Hepatocytes cultured on fibrotic stiffness displayed greater intracellular lipid accumulation with concurrent changes to lipogenic gene expression. We also observed disrupted redox balance in hepatocytes cultured on fibrotic stiffness, with decreased intracellular glutathione and altered inflammatory markers, indicating a pro-inflammatory environment. Our data together implicate increased tissue stiffness in promoting MAFLD-like metabolic changes to hepatocytes in fibrotic liver. This work helps to address a gap in knowledge of how matrix stiffness influences hepatocyte metabolism and to build a more nuanced understanding of disease progression for therapeutic development.

20c Directing Multicellular Organization By Varying the Aspect Ratio of Soft Hydrogel Microwells

Gayatri Pahapale, Jiaxiang Tao, Milos Nikolic, Sammy Gao, Giuliano Scarcelli, Sean X. Sun, Lewis H. Romer, David Gracias

Multicellular self-organization with precise spatial definition is essential to various biological processes, including morphogenesis, development, and healing in vascular and other tissues. This self-organization of cells is a complex spatiotemporally phenomenon regulated by the interplay of diverse environmental cues, including biochemicals such as signaling molecules and transcription or growth factors and physical parameters like matrix stiffness and geometry. Gradients and patterns of chemoattractants (biochemical cues) as guides of multicellular self-organization are well-described, but the influence of three-dimensional geometry at physiological stiffnesses is less well defined. We present the discovery of a new mode of endothelial cell self-organization guided by a combinatorial effect of stiffness and three-dimensional (3D) geometry (microwells), independent of protein or chemical patterning.

To investigate the effect of combinatorial cues, we microfabricated geometrically and mechanically tunable 3D microwells in gelatin hydrogels using photo- and soft lithography. We used scanning electron microscopy, Brillouin microscopy, rheometry, and atomic force microscopy to characterize the uniformity of these micropatterned hydrogels. We utilized live-cell time-lapse microscopy and quantitative epifluorescence/confocal microscopy to examine the self-organization of cells seeded to the hydrogel microwells.

We discovered that endothelial cells in soft 2 kPa microwells are approximately 30 times more likely to migrate to the microwell edge to organize in ring-like patterns than in stiff 35 kPa microwells (**Fig 1A and B**). We systematically examined this striking observation by varying the geometric and mechanical properties of the substrate (or the cell). We observed that the self-organization is independent of curvature and cross-sectional shape. Instead, we observe a quantitatively striking dependence on the microwell aspect ratio (perimeter/depth) when cells are growing on soft 2 kPa microwells with ring-like self-organization significantly more pronounced in microwells with aspect ratio < 25 (**Fig 1C**). Further, pharmacologically decreasing cytoskeletal tension for cells in stiff 35 kPa microwells led to self-organization of cells on the edges of the stiff 35 kPa microwells, similar to that in soft 2 kPa microwells, indicating that cells with lower cytoskeletal tension are sensitive to geometry. Following a discussion of these observations, this presentation will explain the mechanism leading to the striking differences in multicellular organization in response to combinatorial physical cues and a mathematical model which reveals that the ring-like self-organization results from the balance between tangential cytoskeletal tension, cell-cell, and cell-substrate adhesion. Finally, the presentation will show the technological implications of this new mode of self-organization which can be used to program reproducible cell patterning in customized CAD-designed microwell shapes, such as the letters of the word "CELL" simply by providing appropriate geometric and mechanical cues.

Our findings highlight the importance of the combinatorial effects of geometry and stiffness in complex cellular self-organization and provide a new method to guide the patterning of endothelial cells. We anticipate that this discovery will further facilitate the investigation of environmental cues on cell behavior and can be leveraged to design microenvironments that direct multicellular organization for complex devices, mimetic tissues, and organoid designs in the absence of protein patterning.

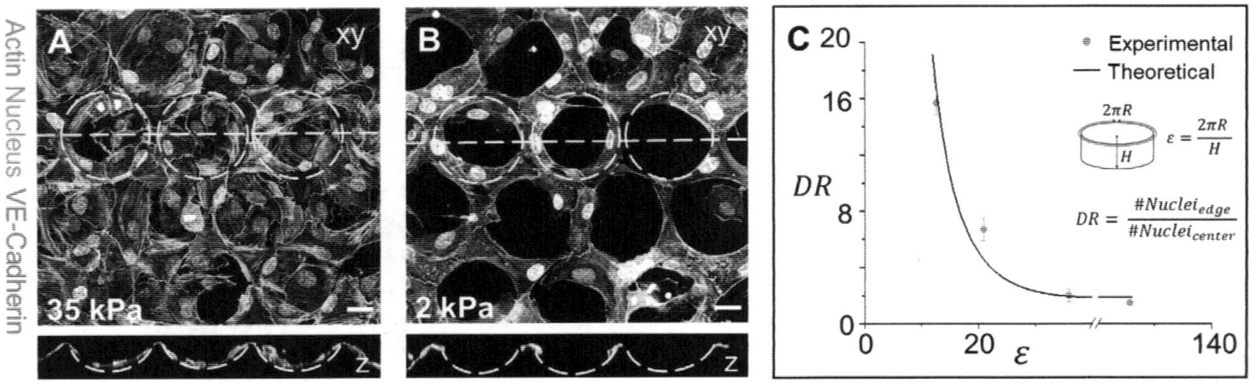

Figure 1: Multicellular self-organization in response to a combinatorial mechanical and geometrical cue. (A, B) Confocal imaging showing the top (xy) and side-views (z) of actin (green), nucleus (gray), and VE-cadherin (magenta) stained cells 24 h after seeding in hemispherical microwells. The images show uniform multicellular distribution all over the stiff 35 kPa hydrogel microwell area (A). In contrast, we observe a distinct ring-like multicellular self-organization around the edge of the soft 2 kPa hydrogel microwell (B). Yellow dashed lines indicate the positions of the microwells and the axis along which the side view is shown. Scale bar = 25 μm. **(C)** Plot depicting the relation between multicellular organization as quantified by cell distribution ratio (DR) and the aspect ratio ε, for the soft 2 kPa microwell. Data presented are mean±SD and are quantified from 100 cells for each aspect ratio from three experiments. Solid line represents data from the mathematical model.

20d Fluid Viscosity Enhances Breast Cancer Metastasis through Coordinated Activity between Volume-Regulatory and Mechanosensitive Ion Channels

Alexander Kiepas, Kaustav Bera, Inês Godet, Selma Serra, Anindya Sen, Se Jong Lee, Yuqi Zhang, Daniele Gilkes, Miguel Valverde, Konstantinos Konstantopoulos

Many physiological processes, including angiogenesis, neurodevelopment and wound healing, rely on the directed movement of cells through the extracellular matrix (ECM). Cellular migratory programs are also engaged in pathological conditions, such as cancer progression to metastatic disease, where they contribute to tumor cell dissemination. Cells must navigate a complex network of proteoglycans and fibrous proteins, such as collagens, elastins, fibronectins and laminins. Fiber architecture and cross-linking density regulate ECM porosity, stiffness and viscoelasticity, whereas extracellular fluid, which hydrates the ECM, imparts hydrostatic pressure and viscosity on migrating cells. Previous studies have established that ECM ligands, substrate stiffness (durotaxis), and the local accumulation of chemical moieties (chemotaxis) can affect migration mechanisms engaged by cells; however, it remains elusive how cells sense and respond to physiologically relevant values of fluid viscosity.

To investigate the effect of fluid viscosity on cells, we designed a microfluidic device with prescribed dimensions that mimics the confining architecture of tissue microenvironments (3.5 x 10 μm^2). We found that elevated fluid viscosity (8 cP) counterintuitively increased the motility of MDA-MB-231 breast cancer cells in confinement and on 2D surfaces (**Fig. 1a**). Live-cell imaging of myosin IIA-GFP followed by immunofluorescence staining of phosphorylated (Ser19) myosin light chain 2 (pMLC) revealed increased myosin II activity at the leading edge of cells migrating at 8 cP. MIIA activity is regulated by RhoA and Rho-associated protein kinase (ROCK). Therefore, we performed FLIM/FRET imaging of the RhoA2G biosensor and found significantly elevated RhoA activity in cells exposed to 8 cP (**Fig. 1b**). Myosin IIA (MIIA), but not myosin IIB, knockdown, reduced cell motility at 8 cP down to basal levels. Similarly, Y27632 (10 µM), which inhibits ROCK, or blebbistatin (50 µM), which inhibits myosin ATPase activity, abrogated viscosity-induced increases in cell migration.

Actomyosin contractility is regulated by intracellular Ca^{2+} levels via Ca^{2+} channels in the endoplasmic reticulum and plasma membrane. For this reason, we performed whole-cell patch clamp recordings and live-cell imaging of the genetically encoded fluorescent Ca^{2+} indicator GCaMP6s to quantify Ca^{2+} influx. We found elevated calcium activity in breast cancer cells exposed to 8 cP (**Fig. 1c**). Accordingly, both cell impermeable (BAPTA) and permeable (BAPTA AM) calcium chelating agents abolished the viscosity-induced enhancement of cell migration. Given that fluid viscosity applies mechanical forces to the cell exterior, we hypothesized that mechanosensitive ion channels in the plasma membrane facilitate elevated Ca^{2+} influx in response to viscosity. Indeed, Ca^{2+} activity in response to 8 cP was abolished in cells with TRPV4 knockdown (**Fig. 1c**), or upon treatment with the TRPV4 inhibitor HC067047, preventing an increase in cell migration speed.

Changes in cell membrane tension can lead to membrane channel activation. Interestingly, we observed that cells exposed to elevated viscosity experienced a rapid increase in cell volume which was dependent on the expression of Na^+/H^+ exchanger 1 (NHE1) (**Fig. 1d**). FLIM imaging of the membrane tension probe Flipper-TR revealed significantly elevated plasma membrane tension in cells exposed to 8 cP (**Fig. 1e**). Consequently, NHE1 knockdown abrogated viscosity-induced increases in Ca^{2+} activity (**Fig. 1c**) and migration speed. To determine the physiological significance of fluid viscosity sensing through the coordinated activity of NHE1 and TRPV4, we performed tail-vein injection experiments in mice with pre-conditioned (8 cP) or naïve (0.77 cP) breast cancer cells and quantified lung metastatic burden. Pre-treatment with 8 cP media significantly increased metastatic burden after 3 weeks. In contrast, TRPV4 depletion reduced metastatic burden in both naïve and pre-conditioned cells after 3 weeks (**Fig. 1f**).

Cumulatively, our data reveal that extracellular viscosity is a novel physical cue, which regulates migration and metastasis *in vivo*. In particular, we highlight the cooperation of volume-regulatory and mechanosensitive ion channels in this process. This finding is especially important in the field of cell biology as most in vitro functional assays are currently performed in medium with a viscosity close to water (0.7 cP), which is far below the viscosity of interstitial fluid (~2-3 cP) and whole blood (4.5-6 cP).

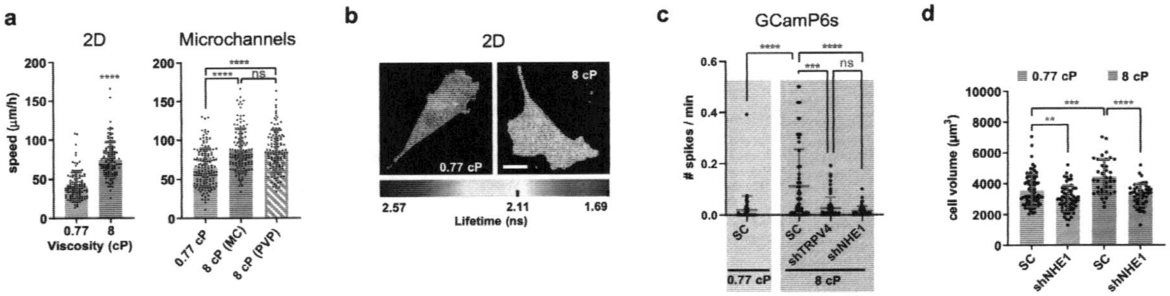

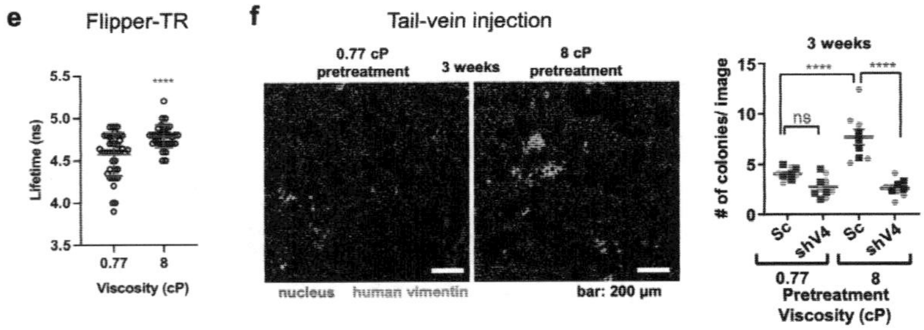

(a) Migration speed of MDA-MB-231 cells in 3.5x10 μm² microchannels or on 2D collagen I-coated surfaces. **(b)** Heat map based upon lifetime (nsec) of the RhoA activity biosensor in representative MDA-MB-231 cells on 2D at 0.77 cP and 8 cP, as determined by FLIM-FRET. Scale bar: 20 μm **(c)** Number of calcium flashes observed in ≥52 SC, shTRPV4 or shNHE1 MDA-MB-231 cells on 2D at 0.77 or 8 cP from 3 independent experiments. **(d)** Cell volume as measured from confocal z-stacks of Lifeact-GFP-expressing MDA-MB-231 cells at prescribed viscosities. **(e)** Lifetime (nsec) of Flipper-TR membrane tension sensor probe in MDA-MB-231 cells at 0.77 cP and 8 cP. **(f)** Number of metastatic colonies in the lungs 3 weeks post injection. 10 images of lung section per mouse (1.62 mm²/image) were acquired. Red lines represent mean ± SEM for ≥9 total mice in each group from 2 independent experiments.

20e Counterintuitive Effect of Fluid Viscosity on Enhancing Cell Motility Via Dynamic Load Response of Actin Network

Kaustav Bera, Alexander Kiepas, Brent Ifemembi, Konstantin Stoletov, Yizeng Li, Colin D. Paul, Anindya Sen, Se Jong Lee, Panagiotis Mistriotis, Kandice Tanner, John Lewis, Sean X. Sun, Konstantinos Konstantopoulos

It is well known that cancer cells respond to physical stimuli, such as stiffness, fluid shear stress and hydraulic pressure. The viscosity of the interstitial fluid is a key physical cue that varies under both physiological and pathological conditions. However, its impact on cancer biology and the mechanism by which cells sense and respond to changes in extracellular viscosity is unknown. Till date, most *in vitro* cell functional assays, including motility, are performed in medium with a viscosity close to that of water (0.7 cP at 37°C). However, the viscosity of the interstitial fluid varies up to 3.5 cP, and can be further increased by the presence of macromolecules, such as mucins, secreted by tumor cells. Elevated degradation of extracellular matrix (ECM) at tumor sites also exacerbates macromolecular crowding, which can further increase the viscosity of the interstitial fluid. To investigate this, we employed an interdisciplinary approach integrating sophisticated microfluidic and molecular biology techniques, super resolution imaging and combined mathematical modeling with *in vivo* zebrafish and chick embryo models. We demonstrate that elevated viscosity counterintuitively increases the motility of breast cancer cells on two-dimensional (2D) surfaces, in confinement (**Fig. 1a**), and cell dissemination from 3D tumor spheroids. Augmented cell motility was also observed with other cancerous (human osteosarcoma, glioblastoma) and non-cancerous (human aortic smooth muscle cells, fibroblasts) cells under elevated fluid viscosity. Increased mechanical loading imposed on the cell leading edge by elevated viscosity induces an Actin Related Protein 2/3 (Arp2/3) complex-dependent dense actin network (**Fig. 1b**) and promotes a mesenchymal cell phenotype which leads to faster cell motility. Moreover, breast cancer cells exposed to elevated viscosity exhibit increased migration in zebrafish (**Fig. 1c**) and increased extravasation in chicken embryo. This presentation will also discuss the role of Osmotic Engine Model in regulating cancer cell migration at physiologically relevant elevated viscosities. Cumulatively, we exhibit that extracellular viscosity is a physical cue that can regulate key (patho)physiological processes like cell migration and cancer metastasis. Our research will also reveal possibilities of investigating if extracellular viscosity affects other physiologically-relevant cellular processes like morphogenesis.

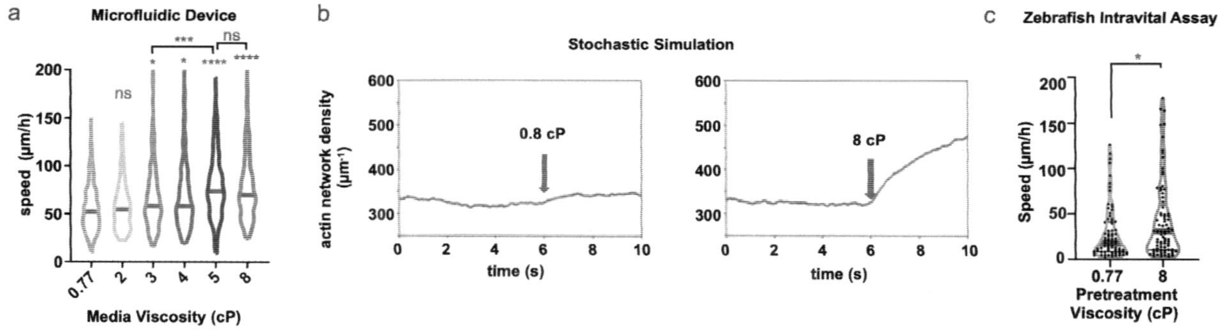

Fig. 1: a, Migration speeds of MDA-MB-231 cells inside 3.5x10 μm^2 channels at prescribed viscosities. b, Average actin density over time from 20 stochastic simulation runs. Viscous forces were applied at t=6 sec (indicated by red arrow) and maintained until the end of the simulation. c, Average speeds of ≥77 MDA-MB-231 cells inside zebrafish intersegmental vessels. Thick red line is median with quartiles marked by thin red lines.

20f A Modular Microfluidic Platform to Study How Fluid Shear Stress Alters Estrogen Receptor Phenotype in Single Breast Cancer Cells

Braulio Ortega Quesada, Jonathan Cuccia, Rachel Coates, Elizabeth C. Martin, Adam Melvin

Approximately 70% of breast cancer patients present with an estrogen receptor-positive (ER+) subtype. Many of these patients respond to endocrine therapy which is designed to specifically target the estrogen receptor; however, following metastatic spread many of these patients develop a resistance to endocrine therapy resulting in a poorer prognosis. Cancer cells are exposed to a range of chemical and physical cues during metastasis that can result in both phenotypic and genotypic changes; however, the underlying cause of endocrine resistance is currently unknown. During metastasis, cancer cells are exposed to fluid shear stress (FSS) while traversing the circulatory system which can result in both deformation and DNA damage. FSS can be understood as the internal frictional force between moving layers in laminar flow. The goal of this work is to further study the effects of fluid shear stress on its role in the developed resistance to endocrine therapies. The most common approach to study FSS is to flow cells through tubing using either a peristaltic pump or syringe pump. This approach has been effective; however, the individual cells can experience a range of FSS magnitudes due to the larger diameter of the tubing and parabolic velocity profile associated with laminar flow. To overcome this limitation, a micron-scale fluidic channel was developed to expose single ER+ breast cancer cells (MCF-7) to similar FSS magnitudes and durations. 'This shearing module' was then paired with a single cell 'trapping module' containing an array of 3000 traps to allow for on-chip immunostaining of MCF-7 cells for biomarkers associated with endocrine resistance. A strength of the modular approach is that the shearing module could be used by itself to shear a large number of cells for off-chip analysis using standard techniques including PCR and Western blot for both population analysis and single-cell resolution. In this way, for bulk analysis, the cells can be collected and frozen after shearing in a microtube for subsequent analysis. For single-cell resolution, they are trapped in the wells by pairing the shearing module. MCF-7 cells were found to exhibit both enhanced proliferation and increased protein phosphorylation following exposure to 10 dyn/cm^2 FSS. Interestingly, the bulk analysis showed no change in the phosphorylation of Akt; however, the single-cell resolution showed a significant change in pAkt levels. Analyzing this data showed that only a small subpopulation, which could correspond to an endocrine resistance population, exhibited enhanced pAkt levels validating the need for single-cell analysis. Phosphorylated Akt plays an important role in colonization and tumorigenesis since it promotes cell survival by inhibiting apoptosis and mediating cells proliferation. Finally, immunostaining for phosphorylation of the estrogen receptor showed diminished levels in cells exposed to shear when compared to the non-sheared control. This finding supports a potential mechanism for endocrine resistance by FSS-induced loss of the estrogen receptor at the metastatic site.

Figure 1. Experimental set-up diagram, COMSOL modeling, and devices pictures. A) Top view of the shearing device showing some dimensions, the inlet, and the outlet. B) Top view of the trapping array showing the usual distribution of cells once they are trapped. C) Cross-section of the shearing device showing the dimensions and the velocity profile inside. D) Side view of the trapping array. E) Brightfield image of MCF-7 cells after being sheared and trapped. F) Picture of the shearing device with red dye inside. G) Picture of the trapping array with red dye inside.

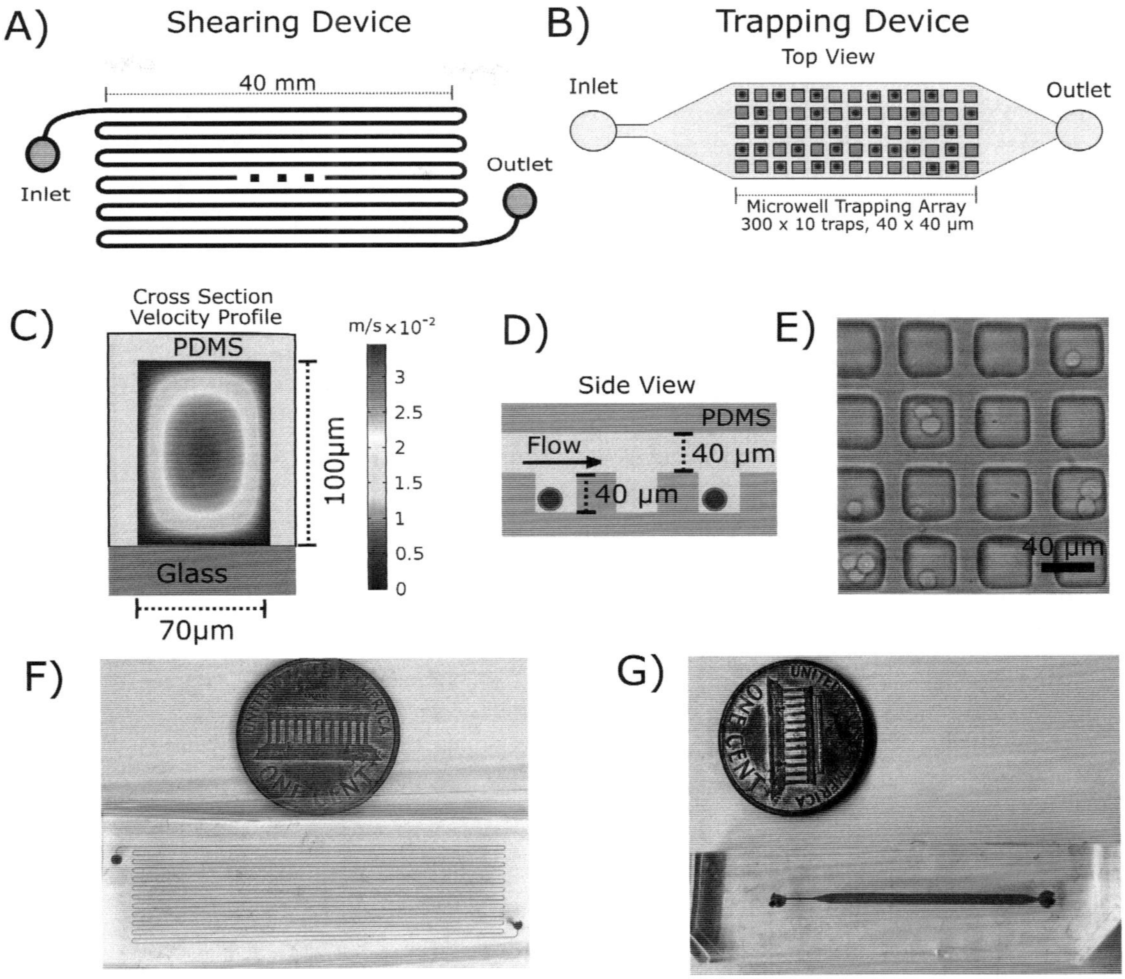
A)
Shearing Device
40 mm
Inlet
Outlet
B)
Trapping Device
Top View
Inlet
Outlet
Microwell Trapping Array
300 x 10 traps, 40 x 40 μm
C)
Cross Section
Velocity Profile
PDMS
Glass
100μm
70μm
m/s x 10^-2
3
2.5
2
1.5
1
0.5
0
D)
Side View
PDMS
Flow
40 μm
40 μm
E)
40 μm
F)
G)

20h Breaking through the physical armor of cancer cells

Matthew Paszek

Cancer cells deploy multiple strategies to circumvent or suppress immune-mediated targeting and killing. Identification of biochemical weaknesses in these defenses have led to new immunotherapies that can provide durable clinical responses for a subset of cancers. While cancer cells also may employ physical mechanisms for immune evasion, what these physical defenses are and how to exploit their vulnerabilities remain largely unresolved. In this talk, I will discuss how cancer cells can physically protect themselves from immune attack through modulation of the glycocalyx, the protein- and sugar-rich coating that encases eukaryotic cells. We find that cancer cells can construct a formidable glycocalyx "armor" that shields them from immune cell recognition and killing. We accurately measure the nanoscale thickness of the cellular glycocalyx using an imaging technique called Scanning Angle Interference Microscopy (SAIM). Combining SAIM with powerful genetic strategies to engineer the glycocalyx structure, we show that immune cell-mediated cytotoxicity exhibits a near perfect inverse correlation with the glycocalyx thickness of target cancer cells. Common cancer cell types construct a glycocalyx that is thick enough to provide substantial protection against immune cells, as well as engineered immune cell therapies. In the final part of my talk, I will discuss potential vulnerabilities in the glycocalyx armor and how to capitalize on them in immuno-oncology through cellular engineering strategies and pharmacological interventions.

59a Controlling the Immune Response through Controlling the Aggregation Kinetics of Engineered Peptides

Gokhan Gunay, Seren Hamsici, Handan Acar

Regardless of the organism in which it occurs, the immunogenic cell death process follows a similar trend: aggregation of proteins (or peptides) at a cell membrane causes damage to the cell membrane, and this damage subsequently releases natural adjuvants called damage-associated molecular patterns (DAMPs) which are natural adjuvants for the immune system. For example, Alzheimer's Disease is a multifactorial disease of unknown etiology, but the central hypothesis is that it is triggered by an accumulation of the pathogenic peptide amyloid beta $A\beta_{1\text{-}42}$ in the brain. $A\beta_{1\text{-}42}$ spontaneously aggregates into soluble fibrils and oligomers, which are the primary toxic forms of $A\beta_{1\text{-}42}$, and create immunogenic cell death and DAMPs release. We recently designed peptides inspired by the intermolecular interactions as in $A\beta_{1\text{-}42}$ peptides and aggregate in the cellular environment, inducing a local and transient immunogenic cell death. The results of the process akin to $A\beta_{1\text{-}42}$ oligomers in terms of inducing the immune response through cell membrane rupture, which we called "peptide-aggregation induced immunogenic rupture (PAIIR)".[1]

In this study, we showed the control over peptide assembly, as in $A\beta_{1\text{-}42}$ oligomerization, to control the immune response effect of PAIIR, as vaccine adjuvants.[2] In this study, we created aggregates of the PAIIR that can assemble in different kinetics (shown as 0.002% fastest and 0.2% slowest). The aggregation kinetics affected their cell membrane rupture kinetics, thus their immunogenicity. We showed detailed characterization of the peptide aggregation, membrane damage, and cell death profiles of these aggregates on several cell lines. We used ovalbumin (OVA) as the antigen and prepared cocktails of different aggregation PAIIRs. Antibody formation in mice followed the same aggregation kinetics profile **(Fig.1)**; the slowest aggregation provided the highest number of antibodies. In this study, we showed controlling the effects of biomaterials on the cells and overall organism, through molecular engineering principles.

[1] Hamsici, S. *et al.* Peptide framework for screening the effects of amino acids on assembly. *Science Advances* (**2022**)

[2] Gunay, G. *et al.* Peptide Aggregation Induced Immunogenic Rupture (PAIIR). *BioRxiv* (**2022**)

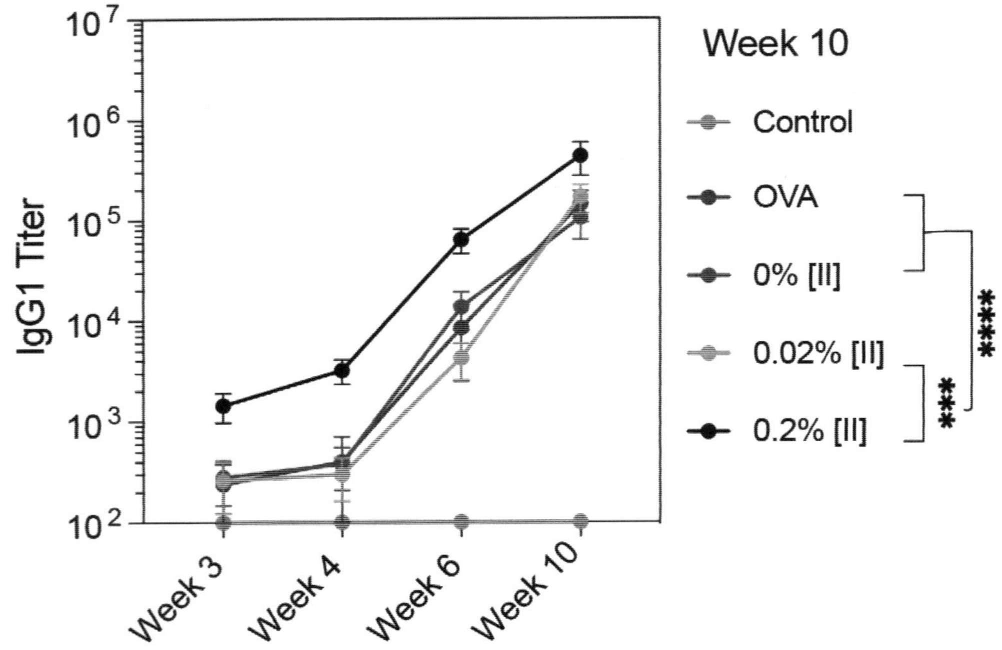

Week 10
Control
OVA
0% [II]
0.02% [II]
0.2% [II]
****
***
IgG1 Titer
10^7
10^6
10^5
10^4
10^3
10^2
Week 3
Week 4
Week 6
Week 10

59b Analysis of Microglia Morphology across Different Neuroinflammatory Rat Models

Hawley Helmbrecht, Elizabeth Nance, Kaleb Decker, Teng-Jui Lin, Sanjana Janakiraman, Mia Onodera

Background:

Animal models are common pre-clinical methods for studying pathological processes of brain diseases. Since no animal model matches the exact human presentation of any specific disease, most diseases are represented by multiple animal models that capture different aspects of cellular response. To combine the insights of different animal models, we propose an image processing-based cell quantification approach that enables comparison of pathological processes across models. However, effective combinatorial experimental modeling is complex, so we focus only on microglia - the immune cells of the brain - for the first application of our image processing pipeline. Microglia are of particular interest for animal models of neurological disease since they are a common measure of neuroinflammation. Neuroinflammation plays pathological roles in the ongoing impact of various neurological diseases such as by acting as the resident macrophage and by maintaining neural circuits[1]. We apply a microglia morphology quantification pipeline that analyzes individual cell features and common populational cell shapes to images acquired from three separate models of neuroinflammation. Two imaging data sets are obtained from cultured organotypic whole hemisphere brain slice models from rats: (1) an oxygen-glucose deprivation (OGD) model and (2) a lipopolysaccharide (LPS)-induced neuroinflammation model. Our third data set is obtained from the Vannucci model of HI in rats[2]. By analyzing the microglia morphology of all three models using the same image processing pipeline, we can study the reactivity of microglia across models and compare specific benefits or features of each model. Our microglia morphology analysis pipeline quantifies microglia at the individual level through geometric features including perimeter, circularity, and area and at the population level through principal components analysis and k-means clustering that allows us to group microglia into the most common shapes. Our multi-level microglia morphology analysis across models of neuroinflammation provides a basis for a combinatorial model approach to study unique aspects of disease models. The combinatorial approach will improve pre-clinical disease characterization of pathological processes by increasing the number of features of microglia morphology that can be quantified and then exploring the extent to which those features change.

Methods:

Image Processing

We begin with immunofluorescent neural images of various brain regions across the three models. After image acquisition, the steps of the image processing pipeline are to (a) segment every cell from all images using a threshold-based method, (b) compute geometric features of every cell – i.e. perimeter, circularity, aspect ratio – using the Sci-Kit Image package in Python, and (c) apply the Visually Aided MorphoPhenotyping Image REcognition (VAMPIRE)[3] package to apply principal components analysis and k-means clustering to determine the five most common shapes and (d) visualize the data as scatter plots of every cell for all geometric features and heat maps of population trends (Figure 1).

Figure 1. Microglia morphology analysis pipeline showing how cells are analyzed from multiple regional images. Cells go through two levels of analysis: (top row) individual: geometric feature analysis and (bottom row) population: group clustering and then are visualized as individual results and as population trends. The data is compared by either considering the whole slice as one group or the regions of each slice as different groups. After all slices have been quantified, inter-slice comparisons for every experimental group are computed.

Image Segmentation

Within each experimental group and each subset – i.e. image magnification, region, and sex differentiated groups – 1/3 of the data was randomly selected for manual segmentation. Manual segmentation and manual counting were carried out by three different researchers using ImageJ. After manual segmentation, ImageJ was also used to calculate the total cell area coverage of each image. Alongside manual segmentation, a set of thresholds – isodata, Li, mean, minimum, Otsu, triangle, Yen - were applied automatically using the Sci-Kit Image package in Python with the same set of data. After the thresholds were applied, cell count and total cell area were calculated using Python. The automated threshold-based segmentation scores were compared to the manual scores and the best threshold-based segmentation method was chosen for all data from each of the three HI models.

Data Splitting for VAMPIRE

For the train:test data split for the VAMPIRE package, we approached the data split in three ways: (1) all data together, (2) each model separately, and (3) cross model comparison. In all models, all data sets were bundled together and split into 80:20 train:test split and then all data sets were run using the compiled model. For each model separately, every disease model was used to create its own VAMPIRE model and then run on only that model's data. For the cross-comparison model, we ran each disease model through the VAMPIRE shape modes created by the other two disease models to compare clustered shapes and reactivity in a multi-level way.

Data Visualization

Data was visualized in two overarching ways - within each model and across each model - and at two separate levels - individual features and population trends. Individual feature maps included scatter plots of geometric features of every single cell analyzed. Meanwhile, population trends included heatmaps of shape mode frequencies and heatmaps of changes both in respect to other models and in respect to the non-treated control.

Results:

Results from the two *ex vivo* models show that both models change microglial morphological features after treatment in regards to perimeter, area, circularity in comparison to the non-treated control groups. However, when comparing models, we see that the change shows regional differences and different extents of reactivity. Additionally, the models show relatively similar overall clustering shape modes with each model returning a shape that is clearly circular and other shapes that are different extents of branching as supported by previous studies in literature[4, 5]. In population analysis, the models both show increased heterogeneity with treatment, but different extent of heterogeneity and shape mode reactivity in comparison to the non-treated control groups. Additionally, we completed the segmentation steps on different magnifications of the LPS model images that qualitatively show similar trends across magnification with increased insight to individual geometric features – especially branching - but decreased insight in population trends as magnification increases. Meanwhile, the Vannucci *in vivo* model shows decreased microglial count in the injured model with regional variations in microglia circularity[6].

Conclusions:

Different etiologies of disease can affect cellular response in unique ways. For example, in response to neuroinflammation microglia contract branches and swell therefore increasing circularity. We can capture microglial geometric changes with our image processing pipeline. In all models of neuroinflammation, we observed different extent of changes in microglial shape features and different population reactivity when comparing the injured groups to both non-treated control groups and each other. Since changes in microglia morphology provide insight into neuroinflammatory processes, comparing microglia across models may be a way to explore combinatorial model approaches for better modeling of human disease processes. By characterizing unique aspects of microglia morphology both across brain regions and across models, we learn specific features of microglia that can be teased out with future experiments. For example, by quantifying extent of circularity change of microglial in injured groups and treatment, we establish a method for studying disease severity and treatment impact across models. By establishing methods of characterizing and comparing cellular response across different etiologies that result in similar pathological processes, we show promise in using combinatorial model approaches to better match human pathological processing.

References:

1. Muzio, L.; Viotti, A.; Martino, G., Microglia in Neuroinflammation and Neurodegeneration: From Understanding to Therapy. *Front Neurosci* **2021,** *15*, 742065.
2. Rumajogee, P.; Bregman, T.; Miller, S. P.; Yager, J. Y.; Fehlings, M. G., Rodent Hypoxia-Ischemia Models for Cerebral Palsy Research: A Systematic Review. *Front Neurol* **2016,** *7*, 57.
3. Phillip, J. M.; Han, K.-S.; Chen, W.-C.; Wirtz, D.; Wu, P.-H., A robust unsupervised machine-learning method to quantify the morphological heterogeneity of cells and nuclei. *Nature Protocols* **2021**.
4. Nguyen, N. P.; Helmbrecht, H.; Ye, Z.; Adebayo, T.; Hashi, N.; Doan, M.-A.; Nance, E., Brain Tissue-Derived Extracellular Vesicle Mediated Therapy in the Neonatal Ischemic Brain. *International Journal of Molecular Sciences* **2022,** *23* (2), 620.
5. Wood, T. R.; Hildahl, K.; Helmbrecht, H.; Corry, K. A.; Moralejo, D. H.; Kolnik, S. E.; Prater, K. E.; Juul, S. E.; Nance, E., A ferret brain slice model of oxygen–glucose deprivation captures regional responses to perinatal injury and treatment associated with specific microglial phenotypes. *Bioengineering andamp; Translational Medicine* **2021**.
6. Joseph, A.; Wood, T.; Chen, C.-C.; Corry, K.; Snyder, J. M.; Juul, S. E.; Parikh, P.; Nance, E., Curcumin-loaded polymeric nanoparticles for neuroprotection in neonatal rats with hypoxic-ischemic encephalopathy. *Nano Research* **2018,** *11* (10), 5670®C5688.

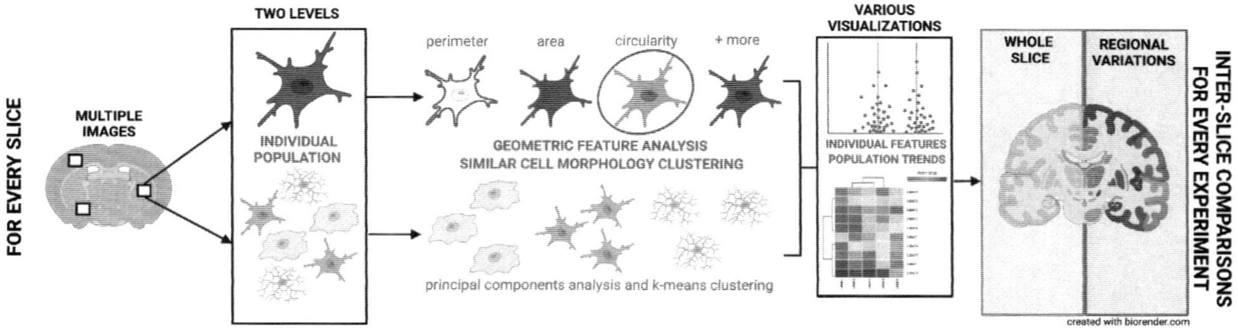

59c Regulation of Circadian Disruption through Inhibition of NOX2 in Microglia: A Potential Route to Target Neuroinflammation

Iswarya Muthukumarasamy, Jennifer M. Hurley, Jonathan Dordick

Sustained neuroinflammation has been shown to be a major contributor to the progression of various neurodegenerative diseases such as Alzheimer's (AD) and Parkinson's (PD) diseases, among others. Neuroinflammation, like other cellular processes, is affected by the circadian clock. Microglia, the resident immune cells in the brain, act as major contributors to neuroinflammation and are under the influence of the circadian clock. Microglial responses such as activation, recruitment and cytokine expression show rhythmic nature in their response to different stimuli including LPS, Aβ, etc. Mouse models have shown that impairment of the microglial clock results in increased levels of neuroinflammation and worsening of AD disease symptoms. Although the field has grown exponentially in recent decades, there remain significant gaps in our understanding of the complex relationship between circadian rhythms and neuroinflammation. Gaining a mechanistic insight of this intricate relationship will advance the development of interventional therapies that mitigate circadian disruption, thereby preventing neuroinflammation associated with various neurodegenerative diseases. In this study, we focused on a specific, yet under-evaluated, aspect of this complex relationship by examining the interaction between the microglial circadian clock and the enzyme NADPH Oxidase Isoform 2 (NOX2) in BV2 microglial cells. NOX2 activation in microglia results in the production of reactive oxygen species (ROS) leading to increased oxidative stress, an integral characteristic of neuroinflammation. We sought to understand the level of control the microglial circadian clock has on NOX2 expression. We examined the nature of oscillation of the clock gene *Per2* and the NOX2 subunits $gp91^{phox}$ and $p47^{phox}$ under distinct types of microglial activation. We also evaluated the impact of NOX2 inhibition on the microglial circadian clock. We discovered that the BV2 microglial clock exerted significant control over NOX2 expression and inhibiting NOX2, in-turn, enabled the microglia to retain a functional circadian clock and reduce levels of ROS and inflammatory cytokines. These findings make NOX2 an attractive therapeutic target to combat circadian disruption and neuroinflammation associated with neurodegenerative diseases.

59d The Role of Myeloid Derived Suppressor Cells in the Immune Response to Infection

Hannah Weppner, Laurel Hind

Sepsis is responsible for 5-6 million deaths per year making it one of the leading causes of preventable death worldwide and its incidence is increasing. Sepsis is caused by a dysregulated immune response to infection; however, our understanding of what causes this dysfunction remains limited. Myeloid-derived suppressor cells (MDSCs), an immature subpopulation of innate immune cells have recently been found in high numbers in septic patients. This population of cells constitute a unique component of the immune system, defined by their ability to suppress the response of T cells, yet the functions of MDSCs in the context of infection, are not well understood. MDSCs are generally detrimental during infection, likely due to their suppression of the adaptive immune response. However, in some patients with chronic inflammation the immunosuppressive functions of MDSCs can be beneficial, suggesting that MDSCs could play a protective role in other infections with hyperinflammation. Controlling the recruitment and function of MDSCs could, therefore, be an innovative strategy for treating infections, including sepsis. Unfortunately, very few studies have investigated MDSCs in the infectious microenvironment due to a) their low abundance in healthy patients, making them challenging to isolate and study, and b) the lack of relevant models of the human infectious microenvironment. In this project, we have overcome both challenges by inducing primary human MDSCs from primary human monocytes and studying their response to infection using our novel "infection-on-a-chip" microfluidic model. This device incorporates key aspects of the human infectious microenvironment including a model endothelial blood vessel, an extracellular matrix, primary human immune cells, and a live pathogen. Using this device, we discovered that the model endothelial blood vessel secretes both IL-6 and GM-CSF in response to the bacterial pathogen *Pseudomonas aeruginosa*. Interestingly, IL-6 and GM-CSF are both known to drive MDSC expansion and activation. Therefore, we are also investigating the effect of endothelial activation on MDSC recruitment.

To generate primary human MDSCs, we incubated healthy human monocytes, isolated from peripheral blood, with GM-CSF and IL-6 for 5 days. We then determined their phenotypic identity using flow cytometry to confirm expression of MDSC markers (HLA-$DR^{lo}CD14^{+}CD33^{+}$). Finally, we confirmed their functional identity by evaluating their ability to suppress T cell expansion using a T cell suppression assay. We found a significant reduction in proliferating T cells in the presence of MDSCs compared to stimulated T cells alone (Fig. 1), indicating the presence of functional MDSCs. We are now investigating MDSC recruitment to inflammatory proteins (IL-8) and bacterial pathogens (*P. aeruginosa*) using our novel infection-on-a-chip device. Finally, we are examining the role of the vascular endothelium in activating MDSC recruitment in the infectious microenvironment using a combination of our device, multiplexed ELISAs, and blocking antibodies to identify key signals in the MDSC response.

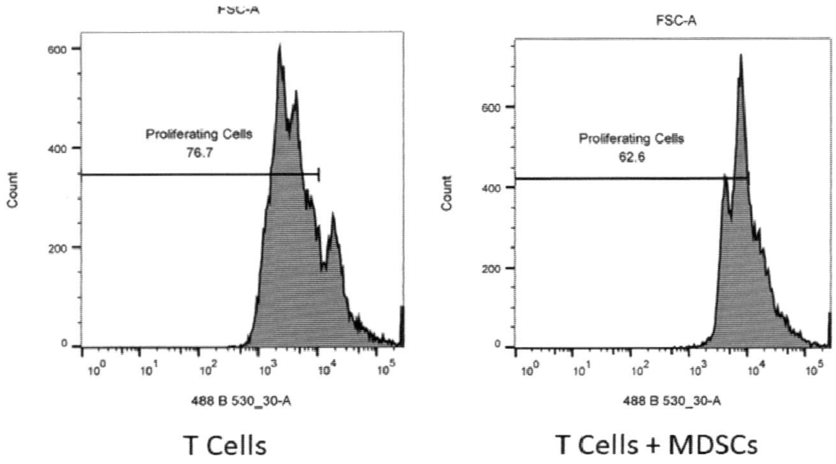

Figure 1. T Cell Suppression Assay. Primary human T cells were stained with carboxyfluorescein succinimidyl ester (CFSE), a fluorescent indicator used to trace cell generations using dye dilution, then stimulated with antibodies that induce proliferation (anti-CD3/anti-CD28) alone or in the presence of MDSCs. Flow cytometry showed almost 77% of T cells proliferating alone compared to only 63% with MDSCs.

59e Engineer CAR-Neutrophils from Hpscs for Targeted Cancer Immunotherapy

Xiaoping Bao

The major focus of our research is to engineer human pluripotent stem cells (hPSCs) for targeted cell and cancer therapies. In this talk, I will focus on our recent development of a simplified, chemically-defined platform for producing definitive hematopoietic stem and progenitor cells (HSPCs) and neutrophils for targeted cancer immunotherapy. Neutrophils, the most abundant white blood cells in the circulation, are closely related to cancer development and progression. Primary neutrophils from healthy donors present potent cytotoxicity against different human cancer cell lines through direct contact and via the generation of reactive oxygen species (ROS). However, due to their short half-life and resistance to genetic modification, neutrophils have not yet been engineered with chimeric antigen receptors (CARs) to enhance their anti-tumor cytotoxicity for targeted immunotherapy. Here, we genetically engineered human pluripotent stem cells with different synthetic CARs and successfully differentiated them into functional neutrophils by implementing an innovative chemically-defined differentiation platform. The resulting CAR-neutrophils presented superior and specific cytotoxicity against tumor cells both in vitro and in vivo. Collectively, we established a new platform for the production of CAR-neutrophils, paving the way to myeloid cell-based therapeutic strategies that would complement and boost current cancer treatment approaches. By reducing the complexity of hematopoietic induction and constructing stable CAR-expressing hPSCs, our approach will offer enhanced understanding of human embryonic hematopoiesis and expedite the development of hPSC-derived off-the-shelf hematopoietic and immune cell products for therapeutic purpose.

59f Temporal *ID2* transcription Factor Expression Improves Natural Killer Cell Differentiation from Human Pluripotent Stem Cells

Juhyung Jung, Yun Chang, Xiaoping Bao

Natural killer (NK) cells are one type of innate lymphoid cells, and NK cell-based immunotherapy serves as a potentially curative therapy for cancers. However, the lack of reliable resources for large amount of NK cells required for clinical infusion has limited the broader application of NK cells in targeted immunotherapy. Substantial effort has thus been made to generate NK-like cells from human pluripotent stem cells (hPSCs), but detailed molecular mechanisms regulating NK cell differentiation remain elusive, preventing us from developing robust strategies for NK cell production. Here, we genetically engineered hPSCs with inducible overexpression of transcription factors *NFIL3*, *ID2*, or *SPI1* via CRISPR/Cas9-mediated gene knockin and investigated their temporal roles during NK cell differentiation. Our results demonstrated *ID2* overexpression significantly promoted NK cell generation compared with *NFIL3* and *SPI1* under a chemically-defined, feeder-free culture condition. The resulting ID2 hPSC-derived NK cells exhibited various mature NK-specific markers and displayed effective tumor-killing activities, comparable to NK cells derived from wildtype hPSCs. Collectively, our study provides a new platform for efficient NK cells production, severing as a realistic off-the-shelf cell source for targeted cancer immunotherapy.

59h Engineering Immunity: A biomaterials and biomanufacturing perspective

Krishnendu Roy

In this presentation, we will discuss two emerging areas of ImmunoEngineering. First, we will present strategies to use materials-based modulation of vaccines and Immunotherapies, specifically in developing robust mucosal vaccines against SARS-Cov2. Second, we will present our work on combining materials, data science, and process control concepts for scalable biomanufacturing of functional immune cell subsets, especially in T and B cell manufacturing.

119a Developing an Ex Vivo Mesolimbic Pathway Model for Exploring Phenotypes Linked with Addiction

Thomas Rudibaugh, Albert J. Keung

The mesolimbic pathway, the connection between dopaminergic (DA) neurons of the midbrain and medium spiny neurons (MSNs) of the ventral forebrain, is of critical importance because it has been linked with reward behaviors such as addiction. Repeated exposures to drugs of abuse lead to long-term neuroplastic alterations, with ample evidence of molecular and cellular level changes in rodent models. To study these mechanisms in human backgrounds, researchers need a model system which can be used to manipulate and dynamically monitor both the short term dynamic electrical and molecular response and the long-term alterations produced by chronic drug use. Stem cell derived human cerebral organoids are good model candidates as they can be differentiated into specific neural subtypes that can mimic cellular responses seen *in vivo* and can be dynamically perturbed. Previous work has generated ventral forebrain and midbrain organoids, however they used different cell lines and medias to generate their organoids. Here we modified a whole brain organoid protocol to generate both DA+ midbrain and MSN+ ventral forebrain organoids while ensuring their media components are the same after 25 days of culture, allowing for organoid fusion. Cellular identity and functional activity were confirmed before exposing ventral forebrain organoids to acute and chronic dosing regiments of dopamine and ventral forebrain-midbrain organoid fusions to acute and chronic cocaine dosing. Through a combination of calcium imaging, and single cell sequencing we have produced a comprehensive characterization of this organoid system's electrical and cell type specific transcriptional responses to acute and chronic stimuli. The results demonstrate cerebral organoids' ability to recreate known phenotypes including increased expression of early genes FosB and cFos and pathway genes linked with increased synaptogenesis and complex phenotypes including crosstalk between drugs of abuse and molecular and cellular desensitization. We are advancing these models further by developing dynamic reporters of target genes and pathways to better understand underlying mechanisms of and provide platforms for screening potential therapeutics.

119b Investigating Changes in Extracellular Matrix Architecture in a 3D Glioblastoma Model

Rosalyn Hatlen, Padmavathy Rajagopalan

Introduction

Glioblastoma (GBM) is the most common type of brain cancer. Despite current treatment methods including resection, chemotherapy, and radiotherapy, the median survival time of patients has not been significantly altered in the last 10 years. In GBM tumors, the microenvironment includes extracellular matrix (ECM) components, cancer cells, stromal cells, and various growth factors. The GBM tumor ECM is unique in that it differs from surrounding brain tissue. It has been reported that a GBM tumor can be up to 12-fold stiffer than normal brain tissue. Additionally, the composition of the ECM is altered due to the increased collagen deposition from stromal and cancer cells. Understanding structural changes to the ECM in GBM can aid in understanding how these cells form and maintain the tumor environment.

Methods

We have assembled a 3D model of the perivascular niche in GBM. We assembled hydrogels composed of Type 1 collagen or HyStem-C® (Sigma), a gelin and HA-based biomaterial. A human GBM cell line LN229 (ATCC) was utilized since these cells display a mesenchymal subtype, and are moderately invasive. Human umbilical vein endothelial cells (HUVECs) were used as a vasculature mimic. While the vasculature is comprised of different cell types, endothelial cells are a principal component and are commonly used as an *in vitro* model.

Layered 3D hydrogels were prepared in glass-bottom tissue culture well plates. First, a thin gel of collagen (either 1.1 mg/mL or 2.2 mg/mL) was deposited to prevent cells from adhering to the glass surface beneath. This was followed by a layer of collagen I or HyStem-C®. After gelation, a fibronectin-coated coverslip of HUVECs was placed atop the stacked gels. For co-cultures, the second cell type was seeded around the coverslip on day 7. For LN229 only cultures, a blank coverslip was utilized to mimic HUVEC co-culture, and the LN229 cells were seeded around the coverslip as in co-cultures.

Results

The storage moduli of the hydrogels were measured to be 0.336 ± 0.073 Pa, 1.593 ± 0.111 Pa, and 241.373 ± 43.299 Pa for 1.1 mg/mL collagen, 2.2 mg/mL collagen, and HyStem-C®, respectively. HUVECs, LN229 cells, and co-cultures were cultured for up to 14 days. Gel contraction was observed in all cultures. Contraction was approximately 14% and 34 % in HUVEC and LN229 monocultures, respectively, at day 14. Gel contraction in co-cultures reached up to 68% in softer gels and 34% for cultures in stiffer gels. Cellular proliferation was also significantly increased in co-cultures by 10 – 16-fold from day 8 to day 14. Monocultures of each cell type showed statistically insignificant increases in proliferation of 1 – 4-fold. Physical colocalization of HUVECs and LN229 cells was also observed. Colocalization was found in up to 30% of total LN229 cells, and increased with higher collagen content. This indicates a role for the chemical composition of a biomaterial scaffold. Both gel contraction and colocalization data indicate that LN229-HUVEC interactions are influenced by the surrounding ECM.

Future and Ongoing Work

Currently we are investigating additional aspects of ECM remodeling, including the endocytosis of the biomaterial scaffold, matrix metalloproteinase (MMP) secretion, and other physical properties such as porosity and changes to collagen architecture. The increased stiffness of the ECM in GBM tumors can result in a barrier to drug diffusion. In the future, we plan to test the administration of chemotherapy treatments to investigate how ECM remodeling can affect drug efficacy *in vitro*.

119c Chemo-Therapeutic Screening on NOVEL Scaffold Assisted Dynamic *in Vitro* multicellular Models of Pancreatic Cancer

Priyanka Gupta, Melina Kitsiou, Pedro Perez-Mancera, Hemant Kocher, Andrew Nisbet, Giuseppe Schettino, Eirini Velliou

Introduction: With a 5-year survival rate of only 9%, Pancreatic Ductal Adenocarcinoma (PDAC) is the 7th leading cause of cancer related death worldwide[1]. The aggressive nature and high mortality rate of PDAC are attributed to its late diagnosis, heterogeneity in the tumour and the tumour microenvironment and its resistance to currently available treatment methods[2]. An in-depth study of PDAC's resistance to current therapeutic methods requires the development *of biomimetic, niche mimicking in vitro tumour models*. Current research focuses on the development of 3D *in vitro* tumour models to replace 2D culture systems and animal models in order to tide over limitations associated with such systems. 3D *in vitro* models are considered to have better *in vivo* niche mimicking capabilities in comparison to 2D culture systems while mitigating the cost and reproducibility problems associated with animal models. Additionally, as in all tissues, the PDAC tumour microenvironment (TME) is heterogeneous in cellular nature consisting, additionally to cancer cells, different cell types , e.g., stellate cells and endothelial cells, all contributing to the tumour formation, metastasis as well as its response and resistance to treatment. Thus, recent studies have focused on generating multicellular pancreatic cancer models, which are primarily spheroid based[3,4]. Spheroids are useful 3D models due to their ease of development, ability to allow for fast analysis and studies. However, it is difficult to maintain spheroid cultures for long time without requiring resuspension, the latter inevitably affecting the formed cell-cell and cell-ECM interactions. As an advancement of our previous works[5,6,7,8], we have recently developed a poly urethane (PU) based 3D hybrid multicellular model of pancreatic cancer model using pancreatic cancer cells, endothelial cells and pancreatic cancer cells, wherein we were able to show long term maintenance of the *in vitro* model (4 weeks), feasibility of extracellular matrix (ECM) mimicry through scaffold coating *via* passive absorption, formation of dense cellular masses, secretion of ECM proteins[9]. *The current work focusses on screening our multicellular scaffold based model for the purposes of therapeutic assessment (chemotherapy). We also look at the effects of a zonal vs a single scaffold cellular and biochemical architecture on chemotherapy assessment in vitro.*

METHODS: PU scaffolds were prepared using the Thermal Induced Phase Separation (TIPS) method. Absorption based surface modification of the scaffolds enabled coating with ECM proteins (collagen and fibronectin) for enhancement of ECM mimicry[4]. A *zonal cellular and biochemical structure* was achieved with (i) endothelial and stellate cells seeded on the outer side of the polymeric scaffold which were coated with collagen I and (ii) pancreatic cancer cells seeded in the inner scaffold coated which were coared with fibronectin. Further to the zonal architecture, a single scaffold based simplistic multicellular model was designed and compared to the zonal model[8]. More specifically, for the simplistic model, a cocktail of the three cell types (cancer cells, stellate cells and endothelial cells) were added to a singel 5x5x5 mm^3 scaffold and cultured for the relevant time period. For the chemotherapy assessment, 50μM Gemcitabine was applied to both models after 4 weeks of culture, both in static and dynamic (perfusion bioreactor assisted) conditions, followed by 7 days post-treatment monitoring. Various *in situ* assays for monitoring the cell viability, spatial organisation, ECM production and q-PCR analysis were carried out at specific time points throughout the culture period.

RESULTS andamp; CONCLUSION: We report here the effect of cellular and biochemical factors of our 3D models on the chemotherapy response of pancreatic cancer cells *in vitro*. Effects of therapeutic agents on cell viability, apoptosis and ECM secretion in our zonal and single scaffold were observed and compared. The importance of a dynamic culture system was also highlighted in this work. Furthermore, the spatial biochemical and cellular configurations of the model affect the response to chemotherapy. Our developed model is a low cost high throughput tool that can be used for personalized studies and treatment screening of pancreatic cancer.

ACKNOWLEDGMENT: The project was supported financially by the 3DBioNet (UKRI). E.V is grateful to the Royal Academy of Engineering for an Industrial Fellowship and to the Medical Research Council UK for a New Investigator Research Grant (MR/V028553/1), which also financially supports P.G.

REFERENCES:

1.Rawla et al. World J Oncol, 2019; 10(1):10-27.

2.H. S. Lee and S. W. Park, Gut and Liver, 2016;10:340-347.

3. Lazzari, et al. Acta Biomaterialia.2018; 78: 296- 307.

4. Di Maggio, et al. Pancreatology. 2016; 16, 995–1004.

5. Totti, S. et al. Drug discovery today. 2017; 22(4): 690-701.

6. Totti, S. et al. RSC Advances. 2018; 8(37): 20928-20940.

7.Gupta, et al. RSC Advances.2019; 9 (71): 41649-41663

8. Wishart , G. *et al.,* Cancers. 2021 13(23):6080

9. Gupta, et al. Frontiers in Bioengineering andamp; Biotechnology. 2020; 8 (290).

119d Secretome of Engineered Neuron-Innervated Muscle

Kai-Yu Huang, Gelson Pagan-Diaz, Young-Hak Cho, Sung Gap Im, Rashid Bashir, Hyunjoon Kong

Skeletal muscles can secrete cytokines and small proteins, so-called myokines, in response to the contraction. Myokines act to retain the homeostasis of muscle and other organs through autocrine, paracrine, or endocrine effects. Efforts have been made to create in vitro platforms to recapitulate skeletal muscle physiology and use it to study myokine secretion mechanisms. Most approaches expose an engineered muscle sheet or 3D muscle strips to electrical pulse stimulation. However, these studies miss a crucial element that regulates muscle contraction physiologically, the neuromuscular junction (NMJ).

The NMJ is a chemical synapse at the junction of a motor neuron and a muscle fiber. At the junction site, neurons transmit a stimulatory or inhibitory signal to the muscle by releasing neural transmitters and, in turn, controlling the muscular contraction. Recently, neuromuscular system engineered in vitro has gained attention because of their potential as neuromuscular disorder drug screening platforms and autonomous bio-actuators. However, these platforms have not been used to address the extent to which neural innervation affects myokine secretion.

In this work, we reproduced the neuron-innervated muscles to address the extent to which neural innervation affects myokine secretion and subsequently modulates the physiological activities of muscle and other organs. The muscles were cultured on a grooved substrate with a myofibril-like pattern and co-cultured with neural stem cell-derived motor neurons to recreate the high functional NMJs. We examined the contraction capability of resulting neuron-innervated muscles with the addition of excitatory neurotransmitters compared to neuron-free muscles. Then, we assessed the change in myokine gene expression and secretion between neuron-innervated muscles and neuron-free muscles. We found that the substrate with grooved pattern served to enhance the differentiation and alignment of myotubes, resulting better neural innervation. The neuron-innervated muscles contracted more actively in responded to glutamate. In addition, the neuron-innervation upregulated the myokine gene expression and increased the myokine secretion of the muscles. The results of this study will be useful for better understanding the crosstalk between muscles and neurons and provide a potential treatment for neural disorders.

119e Engineering an iPSC-Based 3D Liver Organoid for Hepatotoxicity Studies

Neeti Gandhi, Padmavathy Rajagopalan

Introduction

Approximately 23% of Americans take five or more prescription drugs every day. Adverse reactions to medications can cause drug-induced liver injury (DILI), which is the leading cause of acute liver failure in the United States. The majority of DILI cases are idiosyncratic, where the effects of drug- related injury can vary amongst individuals and result in very different responses. When hepatic function deteriorates, the entire body is affected due to the array of functions that the liver performs, such as metabolism, filtration, synthesis, and biotransformation.

The four major cell types in the liver sinusoid, the region where biotransformation of toxicants and xenobiotics occurs, include hepatocytes (parenchymal cells), and three major non-parenchymal cells (NPCs) including liver sinusoidal endothelial cells (LSECs), Kupffer cells (KCs), and hepatic stellate cells (HSCs). Hepatocytes secrete cytochrome P450 (CYP450) enzymes, which enable the biotransformation of drugs and other xenobiotics. Genetic variations in the transcriptome contribute to differences in the expression of various proteins that are involved in metabolism.

Currently, in vitro liver models use primary human hepatocytes (PHHs) or hepatocyte cell lines. While PHHs are ideal for such studies, they are difficult to obtain and are limited in availability. In efforts to address the issues surrounding PHHs, induced pluripotent stem cells (iPSCs) exhibit significant potential for organoid development. iPSCs can be retrieved non-invasively from dermal fibroblasts. iPSCs exhibit infinite proliferation, maintenance of the host genotype, and differentiation into various cell types, such as hepatocyte-like cells (iHLCs). One major concern when working with iHLCs is their fetal phenotype. iHLCs have been shown to exhibit similar morphology to PHHs, but express lower hepato-specific functions. Therefore, our goal is to investigate whether interactions with hepatic NPCs may improve liver functions in iHLCs. We have assembled 3D liver organoids that recapitulate the in vivo environment and architecture of the liver sinusoid for hepatotoxicity studies.

Methods

The 3D organoids were assembled using hepatic parenchymal cells and NPCs. The functions of iHLC-based organoids were compared to cultures containing PHHs. In general, iHLCs were seeded on Type 1 collagen hydrogels and allowed to mature for an additional seven days. To assemble organoids, LSECs were encapsulated in a hydrogel composed of Type 1 collagen and 1% (v/v) fibronectin above the iHLCs or PHHs. The control consisted of hepatocytes cultured as a collagen sandwich (CS). The initial ratio of hepatocytes : LSECs was approximately 5:1. Two prototypic hepatotoxicants, acetaminophen (APAP) and ethanol (EtOH), were administered 24h later at input concentrations of 2.5 mM (1/2 LC50, lethal concentration 50) and 5 mM (LC50) for APAP and 80 mM (1/2 LC50) and 160 mM (LC50) or EtOH. Cultures were ended either at 24h or 72h post-toxicant administration.

Results

Initial results show iHLC-based organoids produce approximately 76% lower urea than PHH-based organoids. PHH organoids exhibited a statistically significant 40% decrease in urea 24h after administering APAP at 1/2 LC50 APAP. Additionally, PHH organoids exhibited a 20% decrease in urea 24h after administering EtOH at LC50. The depletion of glutathione in iHLCs organoids was 1.6-fold and 1.7-fold after administration of EtOH at 1/2 LC50 and LC50, respectively, compared to iHLC organoid controls. PHH organoids also exhibited a 12% decrease in the secretion of a liver-specific enzyme, alanine aminotransferase, compared to PHHs in CS after administering EtOH at LC50.

Conclusions and Future Work

Initial conclusions suggest iHLCs may require more maturation prior to incorporation in liver organoids. Investigation of how the addition of NPCs can help enhance iHLC function through cell signaling mechanisms is ongoing. Future work will include the addition of KCs and HSCs to further recapitulate the in vivo architecture and environment of the liver.

119f Combined Cellular and Tissue Engineering Approaches for Functional Vascularization of Biomaterials

Mai T. Ngo, Ahmad S. Khalil, Christopher S. Chen

Over the past several decades, the field of tissue engineering has advanced the development of biomaterials as scaffolds for regenerative medicine. However, clinical translation continues to be hampered by the inability form functional vasculature within these materials to ensure survival and function post-transplantation. While several biomaterial platforms have been designed to regulate the delivery of angiogenic growth factors spatially and temporally, their utilization in clinical trials has been limited, potentially due to the high cost of short shelf-life of growth factors. In contrast, multiple clinical trials have investigated the delivery of cells or genes to stimulate therapeutic angiogenesis, but the inability to control cell behavior or the magnitude and timing of gene expression results in vasculature that is non-perfused and eventually regresses. Furthermore, no approach thus far has demonstrated the ability to control the extent of vascularization, which is critical for meeting the needs of different tissues and patients.

During development, the human body uses the process of vascular morphogenesis to create functional vasculature. Temporally regulated and concentration-dependent cascades of paracrine signaling between endothelial and stromal cells orchestrate the initial organization of endothelial cells into cords and direct their assembly into perfused vascular networks. Inspired by the process of vascular morphogenesis, we hypothesize that biomaterials can be functionally vascularized by the incorporation of endothelial and stromal cells that are engineered with user-controlled genetic circuits to recapitulate the paracrine signals necessary for vascularization.

In this work, we demonstrate the functional vascularization of fibrin hydrogels by encapsulation of genetically engineered endothelial and stromal cells. The engineered cells express inducible synthetic zinc finger transcription regulators (synZiFTRs) to enable user-controlled expression of paracrine factors responsible for vascular morphogenesis, such as VEGF. Using a library of small molecules (e.g. gravoprevir, 4-hydroxytamoxifen, etc.), we report orthogonal and dose-dependent expression of paracrine factors across a wide concentration of small molecule induction. Using microfluidic devices to model vascular morphogenesis, we show that vascular density can be tuned by varying the extent of paracrine signaling and explore the impact of varied ratios of paracrine factors on vascular architecture (e.g. network length and branching). Finally, we demonstrate that early- and late-stage paracrine factors can be sequentially expressed, and that the timing of late-stage paracrine factor induction influences vascular architecture and perfusion. Overall, we describe a technology for achieving dose- and time-dependent control over gene expression, which can be combined with biomaterials to achieve vascularized artificial tissues. Broadly, we envision that the combination of cellular and tissue engineering can be used not only for vascularization, but for directing tissue assembly in general.

119g Microfabrication Approaches to Control Stem Cell Fate and Function

Quinton Smith

The promise of stem cell based tissue engineering hinges upon the generation of functional derivatives that can act to augment, replace, or regenerate diseased tissues. A critical bottleneck towards this goal is efficiently maturing pluripotent derived populations. However, the inability to recapitulate aspects of the complex developmental niche limits the success of deriving cells that match their mature counterparts. Here we describe a range of approaches that enable molecular testbeds, for elucidating the role of microenvironment cues in differentiation, morphogenesis, and function. These approaches, which include leveraging biomaterial scaffolds, micropatterned substrates, and microfluidic devices, will be discussed in the context of vascular and hepatic lineage specification.

252a Organotypic Whole Hemisphere Brain Slice Models to Study the Effects of Age and Oxygen-Glucose-Deprivation on the Extracellular Properties of Cortical and Striatal Tissue

Jeremy Filteau, Elizabeth Nance

Introduction: Chronic diseases are the leading cause of death and disability, affecting 6 in 10 adults in the United States. Developing novel therapeutics is necessary to combat chronic disease and reduce this burden, but the process to take a drug from discovery to marketplace is slow and costly. Failure to translate outcomes in animal models to success in human clinical trials is a major contributor to the time and cost associated with bringing drugs to market and is in part due to the inability of animal models to recapitulate the physiology of the disease in question. 2D *in vitro* cell culture models are simple and inexpensive, but lack contact between neighboring cells and native tissue architecture. Organotypic slice models have advantages over primary cell culture, as multicellular interactions are retained and bias in cell type is eliminated. Additionally, tissue slices retain 3D architecture and the slices mature *in vitro* in a similar manner to native conditions, allowing slices to be used long term in culture without sacrificing physiological relevance. Relative to *in vivo* models, the extracellular environment can be better controlled, which allows for high throughput screening of various therapeutics or to study the biological response to any stimulus, including toxins and injury/disease models. The brain extracellular environment is involved in many critical processes associated with neurodevelopment, neural function, and repair following injury. Organization of the extracellular matrix and properties of the extracellular space vary throughout development and across different brain regions, motivating the need for platforms that provide access to multiple brain regions at different stages of development. Herein, we demonstrate the utility of organotypic whole hemisphere brain slices and establish them as a platform that can help satisfy this need. Whole hemisphere brain slices were then leveraged to better characterize how hypoxia impacts different regions of brain tissue.

Methods: Whole hemisphere brain slices taken from postnatal day 10 and postnatal day 17 rats were cultured up to two weeks. At set days in vitro (DIV), slices were either 1) screened for viability and metabolic activity using a lactate dehydrogenase assay (LDH), propidium iodide stain (PI), and Alamar Blue assay (AB), 2) fixed for antibody staining and imaging to study cellular response to culturing, or 3) used in particle tracking experiments to study changes in the extracellular . LDH and AB are colorimetric assays, the former quantifying toxicity via release of lactate dehydrogenase via damaged cell walls and the latter quantifying metabolic activity via conversion of a fluorophore (blue to red shift) by live cells. PI staining (only penetrates damaged cell membranes) allows quantification of viability with regional detail when imaged with a confocal microscope. Slices were fixed and antibody markers for neurons (Neun) and glial cells (Iba1, S100B) were applied prior to imaging by confocal microscopy. Cell distribution can be determined on a regional basis. Multiple particle tracking (MPT) is the tracking of fluorescent polystyrene nanoparticles in live tissue sections, using diffusion and other extracted parameters to better understand the microenvironment around particles. Oxygen glucose deprivation (OGD) is a model for hypoxia achieved by limiting oxygen (via nitrogen chamber) and glucose exposure (glucose-free media) to brain slices. Following 2h of OGD exposure, slices are rested for 24h or 72h until assays or imaging is performed at endpoints.

Results: Whole hemisphere brain slices taken from postnatal day 10 and postnatal day 17 rats retained viable, metabolically active cells through 14 days *in vitro*. Tissue extraction followed by oxygen-glucose-deprivation, used to model a hypoxic event *ex vivo*, resulted in reduced slice metabolic activity and elevated cytotoxicity, regardless of slice age. Multiple particle tracking in oxygen-glucose-deprived brain slices revealed that hypoxia-ischemia impacts the extracellular environment of brain tissue differently depending on brain age and brain region. In most instances, the extracellular space was most difficult to navigate immediately following insult, then gradually provided less hindrance to extracellular nanoparticle diffusion as time progressed. This response was not universal across all brain regions and . Slices from postnatal day 10 rats were highly viable for at least two weeks, while slices from postnatal day 17 rats had poor viability. In both ages, viability was lowered after slicing, but only recovered in the postnatal day 10 slices. There is a strong microglial response (change in morphology to proinflammatory state) initially after slicing and in response to OGD. OGD resulted in cell death in both neuronal and glial populations. Postnatal day 17 slices had higher cell death response to OGD and did not recover to the same extent as postnatal day 10 slices, which had a viability level return to baselines after 72h.

Conclusions: In addition to *in vivo* work being highly costly, which makes age-based studies resource intensive, experimental control/access is limited in these models. As a result, quantifying spatiotemporal response to brain disease is more difficult to achieve. We culture whole hemisphere brain slices taken from postnatal day 10 and postnatal day 17 rats to demonstrate our experimental approach. As onset of some neurological diseases is age-related, the ability to culture slices from older animals is advantageous to developing accurate ex vivo models. Our work demonstrates that brain slices can be used to probe extracellular and cellular responses to OGD with the ability to resolve the response by donor rat age, time after insult, and brain region. Importantly, this workflow is not limited to the OGD model. We anticipate that the combination of multiple particle tracking and antibody staining can be applied to other disease models, such as for Alzheimer's, to quantify extracellular and cellular effects of disease in tandem on a time, region, and age basis.

252b Defining Functional Subtypes of Senescence at Single-Cell Resolution

Pratik Kamat, Yukang Li, Bart Starich, Aaron Winston, Anshika Agrawal, Jeremy Walston, Denis Wirtz, Jude Phillip

Background. Aging is characterized by the accumulation of damage and physiological dysfunctions that drive age-associated diseases[1,2]. According to US census data, over the next 30 years, an estimated 1 in 5 persons will be above the age of 65[3]. Unfortunately, living longer does not always equate to living healthier. As such, there is a dire need to identify key contributors of age associated pathologies that could reduce disease burden. Recently, cellular senescence has emerged as a key mechanistic driver of age-associated diseases, connecting molecular and cellular changes to physiological dysfunctions [4,5].

Cellular senescence is a state of cell cycle arrest that is characterized by the secretion of pro-inflammatory factors, termed senescence-associated secretory phenotype (SASP)[6]. In humans, senescent cells (SnC) accumulate with age, and can be induced in response to stresses, such as DNA damage. However, it is unclear when senescence is beneficial vs. unfavorable, and how we can accurately identify and classify senescent cells [4]. Currently, senescent cells are identified based on protein biomarkers that include p16 and beta-galactosidase (BGAL), but there remains no universal biomarker of senescence [6]. Recent studies have indicated that based on the type of senescent induction, the resulting senescent cells secrete differential profiles of SASP and respond differently to senolytics—a new class of drugs that eliminate senescent cells [7]. *As such, we hypothesize that functional subtypes of senescent cells exist, and is encoded within the morphological phenotypes of single cells.*

Methods. To elucidate the notion of functional subtypes of senescent cells, we will develop a high-throughput morphological reference map that connects distinct morphological properties of single cells with protein biomarkers. Using a panel of primary dermal fibroblasts taken from healthy aging individuals, we induced senescence using four standard approaches, e.g. DNA damage, reactive oxygen species, and replication. Following induction, cells were stained for multiple senescence-associated biomarkers (p21, p16, Lamin B1, HMGB1, and BGAL) and imaged using confocal microscopy. Cells/Nuclei were segmented by a combination of computational and deep learning approaches, and cells were computationally clustered on individual morphological parameters. This analysis was performed on ~45,000 single cells.

Results. From these experiments we have developed approaches to: (1) Identify and classify senescent from non-senescent cells based on a combination of morphological and protein-based senescence biomarkers. (2) Classify functional subtypes of senescence based on a morphological reference map that enables the prediction of protein-based biomarker expression. (3) Predict senescence-burden as a function of induction kinetics and aging conditions. (4) Quantify the responses of senescence subtypes to senolytic treatments.

Conclusion. We have developed a framework to classify functional subtypes of senescence with potential applications in *precision aging in* health and disease.

[1] Phillip, et al., Annual Review BME (2015) [2] Otín, et al., Cell (2013) [3] Ortman, et al., US Census (2014) [4] Di Micco, et al., Nature Rev (2020) [5] Baker, et. al., JCB (2018 (2014) [6] Lozano, et al. Nature Rev (2019). [7] Campisi, et al., Annual Rev Path. (2014)

252c Age-Dependent Stiffening of the Extracellular Matrix Increases the Rate of Cellular Senescence and Cancer Metastasis

Bartholomew Starich, Joanne Baek, Derin Tanrioven, Pratik Kamat, Jude Phillip, Denis Wirtz

Background: Melanoma is the fifth leading cause of cancer related death in the United States. It is often associated with years of excess sun damage and as such disproportionately affects older generations. While the 5-year survival rate for localized melanoma is exceptionally high at 99%, the metastatic disease is much more devastating with survival dropping to ~30%[1]. Like many age-associated pathologies, melanoma likely arises from a combination of intrinsic (*e.g.*, genetic, molecular, extracellular) and extrinsic (*e.g.*, environmental) factors. While significant body of work focuses on targeting genetic aberrations associated with melanoma. The mechano-structural changes of skin with age, and the corresponding impact on age-related phenotypes such as cellular senescence have been largely overlooked.

Cellular senescence reflects an irreversible state of cell cycle arrest that is often associated with normal aging. Senescent cell populations increase with age, and the presence of senescent cells, particularly senescent fibroblasts, within the tissue microenvironment can stimulate pro-growth effects on cancer cells through the secretion of pro-inflammatory factors (*i.e.,* SASP). A key SASP factor, interlukin-6, can promote proliferation in a paracrine manner through activation of STAT3[2].

During aging, skin tissues exhibit increased collagen bundling and, as a result, increased ECM stiffness[3]. In homeostasis, this crosslinking of collagen with other ECM proteins such as fibrillin and elastin enhances the structural stability of the skin[4]–[6]. In solid tumors, the current consensus is that this age-associated stiffening of the matrix drives tumor progression through increased proliferation and migration of cancer cells. **Here, we hypothesize that age-dependent matrix stiffening promotes accelerated senescence and associated increased Interlukin-6-containing SASP that drive melanoma cell migration and metastasis.**

Methods: Using a gender-balanced cohort of primary dermal fibroblast cell lines from the Baltimore Longitudinal Study of Aging (BLSA), we assess the rate at which fibroblasts enter a state of cellular senescence in the presence of pro-senescence inducing agents. Dermal fibroblasts from four individuals aged 25, 35, 65, and 75 are plated on poly-acrylamide substrates of varying mechanical stiffness between 0.5 kPa and 25 kPa, corresponding to low stiffness of normal/young skin and high stiffness of solid melanoma tumors. The extent of senescence is assessed based on proliferation arrest, hallmark biomarkers of senescence (*e.g.* SA-βgal, p16, p21, and HMGB1), abundance of secreted factors (*e.g.,* IL-6 and IL-8) in cell culture media, and morphological assessment. Metastatic potential of melanoma cells is characterized by functional assessments of single cell motility in a 3D extracellular matrix.

Results andamp; Conclusion: Our results suggest that dermal fibroblasts cultured on stiff substrates are more susceptible to enter cellular senescence compared to soft. These results suggest 1) that senescence may be regulated via mechanotranduction, and 2) that there is a feedback loop by which increasing matrix stiffness contributes to the onset of cellular senescence, which in turn promotes further matrix stiffening and cancer metastasis.

252d Metabolic Rewiring of Aged Myoblasts and Restores Regenerative Potential of Progeric Skeletal Muscle

Nika Rajabian, Izuagie Ikhapoh, Shahryar Shahini, Ramkumar Thiyagarajan, Aref Shahini, Joseph Kulczyk, Susan Udin, Aimee Stablewski, Kenneth Seldeen, Bruce Troen, Kirkwood Personius, Stelios Andreadis

Skeletal muscle (SkM) comprises approximately 40% of total body mass and plays essential physiological roles in the body such as enabling skeletal movements and regulating metabolism. Age-related muscle loss, sarcopenia is a major medical problem facing the elderly and correlates with loss of metabolic function, falls leading to cranial fractures, type II diabetes, and cardiac insufficiency. Here we investigate the age-related metabolic rewiring that occurs in myoblasts using in vitro and in vivo models of aging and rejuvenation. RNA sequencing and pathway analysis data revealed that several metabolic pathways changed significantly upon senescence. We also provided evidence aged myoblasts have less glycolysis and insulin sensitivity, which leads to utilize a different source of energy to generate ATP. Our results suggested that aged myoblasts preferred to catabolize amino acids mostly methionine for ATP production and this came at the expense of accumulation of ammonium that leads to DNA damage and impaired cellular function, compromising regenerative capacity and myotube formation. Interestingly, we found that expression of the embryonic transcription factor, NANOG, in senescent cells restored insulin sensitivity, Akt2 signaling, glucose uptake and utilization of glucose for ATP production. In addition, NANOG decreased expression of methionine adenosyl-transferase (MAT) 2A and ammonium levels. Interestingly, inhibiting MAT2A using shRNA showed similar results as NANOG, including restoration of insulin sensitivity, Akt2 signaling and increased glycolysis. Most notably, decreasing methionine catabolism by NANOG expression or MAT2A inhibition led to dramatic improvements in skeletal muscle strength in a mouse model of premature aging.

252e A Microfluidic Approach for the Isolation of a Highly Migratory Presenescent Stem Cell Subpopulation

Farshad Amiri, Farnaz Hemmati, Panagiotis Mistriotis

Cellular senescence refers to the permanent cell cycle withdrawal (aka cell cycle arrest) as a result of prolonged cell expansion and it triggers a variety of phenotypic and structural alterations that diminishes stem cell potential. Ultimately, senescent cells contribute to disease progression and hinder tissue regeneration. The aim of this study is to devise a novel approach to eliminate senescent cells from adult stem cell populations in order to improve their therapeutic efficacy. To achieve this, we exploit the innate potential of stem cells to migrate toward damaged tissues *in vivo*. Since senescent stem cells display increased size and reduced motility and plasticity, we hypothesize that a Y-shaped microfluidic device that has been previously optimized to isolate highly migratory cells effectively distinguishes between presenescent and senescent stem cells. This device is composed of two parallel seeding and collection channels connected by a ladder-like arrangement of Y-shaped micron-sized channels, the dimensions of which recapitulate the size of pores and longitudinal channel-like tracks encountered by migrating *cells in vivo*.

To test our hypothesis, early, intermediate, or late passage human dermal fibroblasts or human adipose tissue-derived mesenchymal stem cells (hMSCs) were seeded close to the entrances of Y-shaped, polydimethylsiloxane (PDMS)-based microchannels and their migration was monitored for 13 hours using time-lapse microscopy. Highly migratory and non-migratory cells were isolated by the device and their morphological, molecular and functional properties were assessed. The results showed that highly migratory cells were significantly faster and smaller than non-migratory cells, especially at earlier passages. Irrespective of cell passage number, migratory cells also exhibited higher proliferative capacity (assessed via Ki-67 immunocytochemistry) and less DNA damage (assessed via γH2AX immunocytochemistry) compared to non-migratory cells. To determine the ability of our assay to detect presenescent cells within heterogeneous populations, we mixed early passage cells with RFP-labelled, senescent cells at a 1:1.5 ratio. The device identified ~70% of early passage cells and only ~30% of senescent cells, thereby enriching the % of control (presenescent) cells within the migratory cell population. In the future, we will assess whether the clonogenic, differentiation, immunosuppressive and proangiogenic potential of highly motile hMSCs is elevated compared to the parental stem cell population.

Taken together, our assay enables the physical isolation and characterization of a highly motile presenescent cell subpopulation in real-time and at the single-cell level. Such technology can improve the standardization of cell preparations, enhance stem cell potency, and facilitate the discovery of signature markers for highly motile, presenescent stem cells.

252f Vitamin K2 Modulates Mitochondrial Dysfunction Induced By 6-Hydroxydopamine in SH-SY5Y Cells Via Mitochondrial Quality Control Loop

Hengfang Tang Sr., Zhiming Zheng

Vitamin K2, a natural fat-soluble vitamin, is a potent neuroprotective molecule owing to its antioxidant effect, but its mechanism has not been fully elucidated. Therefore, we treated SH-SY5Y cells with 6-hydroxydopamine (6-OHDA) in a dose-dependent manner, followed by vitamin K2. 6-OHDA had reduced cell viability, decreased mitochondrial membrane potential, increased the accumulation of reactive oxygen species (ROS), and induced mitochondria-mediated apoptosis and abnormal mitochondrial fission and fusion compared with untreated SH-SY5Y cells. However, vitamin K2 significantly attenuated 6-OHDA-induced changes. Vitamin K2 plays a key role in apoptosis by upregulating and downregulating Bcl-2 and Bax protein expressions, respectively, thereby inhibiting mitochondrial depolarisation and ROS accumulation and maintaining the stability of mitochondrial structure and function. Additionally, vitamin K2 significantly inhibited the 6-OHDA-induced decrease and increase in MFN1/2 and DRP1 expressions, respectively, thereby maintaining the dynamic balance of mitochondrial fusion and fission. Furthermore, vitamin K2 downregulates the expression of p62 and upregulates the expression of LC3A in 6-OHDA-treated cells through the PINK1/Parkin signalling pathway, thereby promoting mitophagy. Moreover, it induced mitochondrial biogenesis in 6-OHDA damaged cells by promoting the expression of PGC-1α, NRF1, and TFAM. These indicate that vitamin K2 can alleviate mitochondrial damage, and that this effect is related to the participation of vitamin K2 in the regulation of the mitochondrial quality control loop, through the maintenance of the mitochondrial quality control system, and repair mitochondrial dysfunction, thereby alleviating neuronal cell death mediated by mitochondrial damage.

252g Human aging and senescence at single-cell resolution

Jude Phillip

Most older adults are affected by multiple chronic diseases and have an increased risk of developing physical and cognitive disabilities. However, the inherent variability observed among older individuals indicate that this risk of disability and disease is not uniform. As such, there is a critical need to develop new approaches to capture the underlying biological mechanisms of aging that describe disease susceptibility and resilience in older adults. A growing body of evidence shows that the interactions of intrinsic and extrinsic factors, across length scales from molecules to organisms, contribute to the rates of ageing and age-associated diseases in humans. However, it is unclear how underlying molecular states of an individual relate to their clinical outlook. We postulate that studying age-related changes at intermediate length scales of cells may provide a key link to understanding ageing patterns in humans. As integrators of molecular signals, cells offer a sensitive mesoscale view of ageing, with cellular dysfunctions likely occurring prior to the manifestation of age-related diseases. My lab is interested in developing new technologies and approaches to map aging trajectories based on the properties of cells. Our hypothesis is that if we can map healthy aging, then deviations from healthy aging trajectories can inform on disease states.
In my presentation, I will focus on how biophysical properties of cells constitute a framework to study aging and senescence in humans, the role of cellular heterogeneity in propagating aging and aging-related phenotypes, and a generalizable framework to understand emergent patterns of single-cell motility with applications beyond aging in health and disease. Through this work, we are seeking to develop a better understanding of why humans age differently, with the goal of developing precision aging strategies to eventually impact the health of aging individuals.

632a Expanding the Scope of Bacterial CRISPR Activation with PAM-Flexible dCas9 Variants

Ava Karanjia, Cholpisit Kiattisewee, Mateusz Legut, Samantha Koplik, Benjamin Kleinstiver, Neville Sanjana, James Carothers, Jesse Zalatan

CRISPR-Cas transcriptional tools have been widely applied for programmable regulation of complex biological networks. In comparison to eukaryotic systems, bacterial CRISPR activation (CRISPRa) has stringent target site requirements for effective modulation. We demonstrate that expanded PAM dCas9 variants can increase the number of targetable CRISPRa sites in bacteria and systematically evaluate the capabilities of different dCas9 variants. In particular, we observed that dxCas9-NG provided a high dynamic range of gene activation for NGN PAMs while dSpRY permitted modest activity across almost any PAM sequence. However, this improved PAM-flexibility comes with the tradeoff that CRISPR interference (CRISPRi) becomes less effective. Weaker CRISPRi gene repression can be partially rescued by targeting multiple sites in the gene of interest. By systematically characterizing the properties of expanded PAM dCas9 variants in bacterial CRISPRa/i, we provide a framework to choose the most effective variant for a given gene target. This toolbox of dCas9 variants will further expand the utility of CRISPRa/i in bacterial systems for a broad range of applications including genome-wide screens and metabolic engineering.

632b Next Generation CRISPR-Cas Tools for Post-Transcriptional Gene Regulation in Bacteria

Ryan Cardiff, Cholpisit Kiattisewee, Jesse Zalatan, James Carothers

The CRISPR-Cas system provides powerful tools to regulate gene expression in bacteria. Cas9-based systems typically act at the DNA level to activate, repress, or knockout gene targets. However, transcriptional regulation in bacteria has important limitations in its ability to precisely target individual genes for up- or down-regulation. Specifically, CRISPR-Cas9 is unable to regulate single genes in a transcriptional operon. We have shown that targeting of one gene for activation or repression in a multigene operon leads to simultaneous regulation of all genes in the same unit. These limitations greatly restrict our ability to efficiently direct metabolic flux to a desired product for biosynthesis. We have recently engineered Cas13-based RNA targeting tools for use in bacteria. These tools function post-transcriptionally to overcome the limitations of DNA-targeting CRISPR-Cas systems. Here, we characterize new tools leveraging CRISPR-Cas13 that allow for individual targeting of genes in a transcriptional operon for up- and down- regulation to improve microbial biosynthesis.

632c Development of a CRISPR Diversifying Base Editor in *Saccharomyces Cerevisiae* for Rapid Antibody Enhancement and Selection.

Andrew Cazier, John Blazeck

Therapeutic antibodies have received widespread interest and investment in recent years due to their ability to treat a variety of conditions including cancer, autoimmune diseases, and viral infections. Developing high-affinity antibodies often includes many laborious and time-consuming molecular cloning steps. Surface display using *Saccharomyces cerevisiae* is frequently employed to screen antibody candidates, but a diversified library must first be generated and transferred into yeast. To address this limitation, we are building a yeast-based platform that is capable of generating antibody diversity by mimicking how human antibodies are refined in B cells – by altering DNA sequences through somatic hypermutation. We have developed a method to recapitulate somatic hypermutation in yeast *in situ* using a CRISPR-based, "diversifying base editor" (DBE). Our system creates intense, localized mutations on a targeted gene (e.g., an antibody sequence) while minimizing off-target effects. Our DBE consists of a nuclease-dead Cas9 protein that can be complexed with multiple human deaminase protein fusions recruited via guide RNA scaffold sequences, allowed precise mutation of a targeted locus (**Figure 1**).

We will highlight three key results from our efforts: (1) Fluorescence-based, mutational assays that have allowed us to estimate the rate and breadth of mutations produced by the DBE, (2) Protein engineering efforts that enhanced the mutation rate of the DBE 5-fold, and (3) Optimization of the guide RNA scaffold sequences to further enhance mutation rate up to 15-fold (**Figure 1**). Finally, we will describe ongoing experiments employing our optimized DBE and yeast display for the *in situ* enhancement of antibody affinity and isolation of these sequences in a manner that mimics B cell somatic hypermutation. Ultimately, we foresee our system becoming a rapid, inexpensive, and easily implemented tool for antibody therapeutic development pipelines, as antibodies can be improved by applying multiple rounds of DBE-mediated mutagenesis and FACS selection.

Figure 1. A CRISPR diversifying base editing (DBE) causes localized mutations in a targeted gene. The DBE is composed of a nuclease dead Cas9 (dCas9), a deaminase fusion, and a guide RNA with aptamers. By engineering the guide RNA and deaminase, we have increased the mutation rate of the DBE.

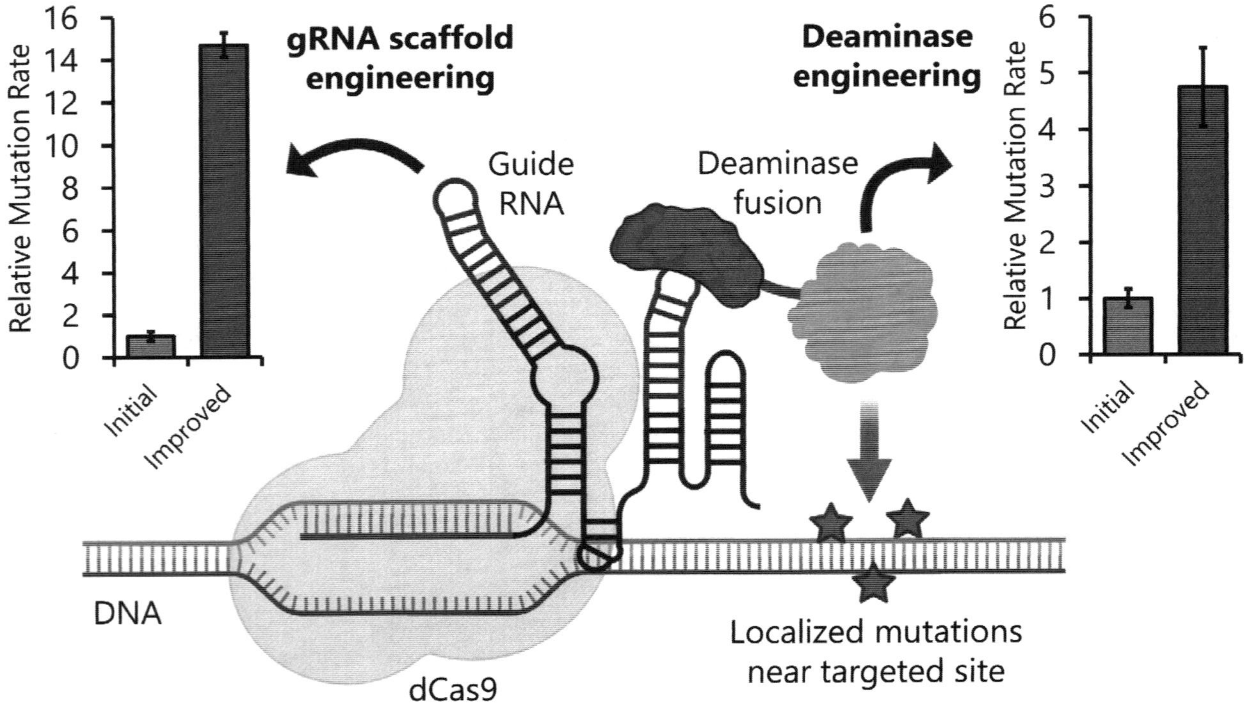

16
14
12
10
8
6
4
2
0
Relative Mutation Rate
Initial
Improved
gRNA scaffold
engineering
Guide
RNA
Deaminase
engineering
Deaminase
fusion
6
5
4
3
2
1
0
Relative Mutation Rate
Initial
Improved
DNA
dCas9
Localized mutations
near targeted site

632d Establishment of an Efficient Bacterium-Yeast Consortium for Biosynthesis and Discovery of Strigolactones

Anqi Zhou, Sheng Wu, Kang Zhou, Yanran Li

Strigolactones (SLs) are a class of valuable phytohormones with great agricultural significance in plant growth and development, yet the limited access to SLs largely hampers the SL-based foundational investigations and large-scale applications. Here, we developed an efficient *Escherichia coli-Saccharomyces cerevisiae* consortia for the high yield production of various SLs, including carlactone, carlactonoic acid, 5-deoxystrigol (5DS), 4-deoxyorobanchol (4DO), and orobanchol (OB), as well as for the discovery of novel SLs. The SL-producing microbial biosynthetic system enabled us to functionally characterize and compare unknown SL-related CYP450s from diverse plant species. We also demonstrated the great potential of engineering a 2-strain co-culture system. Through multiple metabolic engineering strategies, SL titers were enhanced from original μg/L to low mg/L, which enables the identification of novel SLs and biosynthetic enzymes. Our work provides a robust and promising platform for investigating the SL biosynthesis and evolution and lays the foundation for expanding the repertoire of new SLs in microbial hosts.

632e Towards Programmable Interspecies Cell-Cell Communication in Bacterial Consortia: Synthetic Homoserine Lactone-Based Quorum Sensors for Gram-Positive *B. Subtilis*

Min Zeng, Biprodev Sarker, Lauren B. Andrews

Bacteria use biochemical cell-to-cell communication via diffusible molecules and gene regulatory elements to bring about large-scale responses. The ability to predictably design multicellular responses in diverse bacterial communities would have a wide range of applications including efficient biomanufacturing, bioenergy conversion, novel therapeutic strategies, sustainable agriculture, and bioremediation. According to the paradigm for bacterial quorum sensing, gram-negative and gram-positive bacteria utilize different chemical classes of quorum signaling molecules, specifically homoserine lactone (HSL) versus oligopeptide signals, respectively. A universal biochemical signal for cell-cell communication would facilitate the programming of dynamic responses in diverse bacterial consortia.

Here, we present a strategy to develop synthetic HSL-mediated quorum sensors for gram-positive bacteria. In this work, we use the model gram-positive *Bacillus subtilis*. Challenging the bacterial quorum sensing paradigm, we show that synthetic LuxR-type quorum sensors can be engineered in *B. subtilis* 168. These synthetic HSL-mediated quorum sensors are comprised of the LuxR-type regulator and a synthetic cognate promoter. Starting from the inactive gram-negative bacterial promoter sequence, we show the promoter requirements to achieve HSL-mediated activation in the model gram-positive using a set of LuxR-type quorum sensors. To engineer the gram-positive RpaR quorum sensor for p-coumaric HSL, we constructed a combinatorial library of synthetic P_{Rpa} promoters by combining different -35 elements, -10 elements, -16 elements and transcription start sites. Further, we applied statistical design of experiments to dissect the effect of each promoter element and their interactions that contribute to quorum sensor activity in *B. subtilis*. The dynamic range for synthetic RpaR quorum sensor variants can exceed 300-fold in *B. subtilis*. Following the promoter structure of selected variants in the P_{Rpa} library, we engineered synthetic LuxR (~20 fold) and CinR quorum sensors (~30-fold) in *B. subtilis* that can achieve 20-fold and 30-fold activation, respectively. These HSL-based quorum sensing systems can be used for programmable multicellular genetic circuits between gram-negative and gram-positive bacteria and will facilitate the predictive design of customizable bacterial consortia with specified dynamic responses.

632f Modular and Syntrophic Cocultures of *Clostridia* to Produce Isopropanol and C6-C8 Alcohols and Carboxylic Acids

Jonathan Otten, Hyeongmin Seo, Noah Willis, John Hill, Eleftherios Papoutsakis

Synthetic syntrophic cocultures provide several advantages for the renewable production of target biofuels and chemicals. Engineering all desired metabolic pathways into one organism is difficult, but in a coculture, each organism can specialize according to their natural and engineered capabilities while sharing metabolites and proteins. This work explores cocultures of *C. acetobutylicum* (*Cac*), an industrial solventogen that produces ethanol, butanol, and acetate; *C. kluyveri* (*Ckl*), which can elongate acetate into longer-chain carboxylic acids; *C. ljungdahlii* (*Clj*), an acetogen that can capture carbon dioxide to produce ethanol and acetate; and *C. saccharolyticum* (*Csac*), which can quickly produce large amounts of ethanol and acetate. Butanol, hexanol, and octanol, as well as their respective carboxylic acids, are common industrial chemicals, but they are currently produced from petroleum-based processes. Isopropanol is an additional valuable industrial chemical that can be produced biologically, but current modes of production are inefficient. Syntrophic cocultures of *Clostridia* promise a sustainable and green replacement.

A key benefit of cocultures is their modularity. If the objective is high production of C6-C8 chemicals, *Ckl* will be needed to perform chain elongation. In these cocultures, an engineered high-ethanol-producing strain of *Cac* consumes glucose and produces ethanol, acetate, carbon dioxide, butyrate, and butanol. The ethanol and acetate are elongated by *Ckl* to create valuable hexanoate and octanoate, which can be converted by either *Cac* or *Clj* into their respective alcohols, which are easier to extract from the media. Cocultures of *Cac* and *Ckl* and *Csac* and *Ckl* have been studied. *Clj* can be added to these co-cultures to capture the carbon dioxide produced by the coculture organisms, which raises carbon efficiency and lowers costs. Notably, this work has resulted in the highest-observed hexanoate production from *Ckl* without concurrent in-reactor capture. Cocultures of *Csac* and *Ckl* produce more hexanoate more quickly than any published coculture (14 g/L in 42 hours). This work also provides more insight into coculture biology, as both flow cytometry and fluorescent microscopy have provided evidence of cell contact and protein sharing between coculture members of different species. To quickly evaluate the impacts of different coculture parameters, a custom rig of eight miniature bioreactors was designed and constructed. Modular and easy to adapt, this system has been key in tuning various coculture conditions and providing for high-throughput strain testing.

Our lab earlier found that in *Cac* and *Clj* cocultures, the cells come together and share proteins and metabolites, leading to the emergence of coculture phenomena that are unobserved in monocultures. Isopropanol is produced when the acetone natively produced by *Cac* is converted into isopropanol by a secondary alcohol dehydrogenase in *Clj*. Direct cell contact in coculture leads to lower acetone titers in the media as well as a faster onset of isopropanol production. The presence of *Clj* also enables complete utilization of the carbon and electrons in the sugar substrate due to its use of the Wood-Ljungdahl pathway. Our new work incorporates these insights to maximize the production of isopropanol with a *Cac*/*Clj* coculture that utilizes an engineered *Cac* M5 strain that produces mostly acetone. This coculture method will fully utilize the sugar substrate, maximizing the culture productivity and production of isopropanol. This coculture utilizes the same custom miniature bioreactor setup as the *Ckl* cocultures, which leads to quick parameter determination and optimization. Altogether, this work shows how *Clostridia* cocultures can expand the metabolic space by unlocking and better utilizing the capabilities of all coculture members so that more sustainable means of chemical production can be deployed.

Acknowledgement: ARPA-E Award DE-AR0001505: "Bioenergy production based on an engineered mixotrophic consortium for enhanced CO_2 fixation"

632g Efficient Co-Utilization of Biomass-Derived Sugar Mixtures By Engineered *Escherichia coli* for Enhanced Production of Aromatic Biochemicals

Arren Liu, Michael Machas, Aditya Sarnaik, Arul Mozhy Varman, Xuan Wang, David Nielsen

A significant challenge in biomanufacturing continues to be the efficient co-utilization of mixed sugar feedstocks, which commonly result from biomass hydrolysis. As such, continued efforts to engineer microbes for efficient conversion of substrate mixtures remains critical. Glucose and xylose represent some of the most abundant carbon sources in nature, and are commonly yielded from lignocellulosic waste biomass. Unfortunately, as a result of carbon catabolite repression microbes such as *Escherichia coli* first catabolize only glucose, after which xylose may then be consumed. However, in addition to reducing productivity, for certain applications, this inherent constraint misses a potential opportunity to realize synergistic benefits associated with the preferential synthesis of unique precursors from glucose vs. xylose, and vice versa. For instance, through glycolysis, glucose catabolism promotes the availability of phosphoenolpyruvate (PEP) whereas xylose catabolism via the pentose phosphate pathway promotes erythrose-4-phosphate (E4P). Since PEP and E4P are the two main precursors to the shikimic acid pathway, we hypothesized that, if efficiently co-utilized, glucose-xylose mixtures might represent and advantaged feedstock for the microbial production of aromatic chemicals. To investigate this prospect and improve our understanding of which precursor may be a limiting, flux balance analysis was performed, through which it was determined that increasing E4P availability would most significantly enhance total carbon-flux through the shikimate pathway. Then, to provide proof-of-concept, a phenylalanine over-producing strain of *E. coli* was further engineered by introducing a recently discovered mutant copy of the xylose-specific activator XylR (i.e, XylR*). Additional mutations aimed at increasing PEP availability and tuning the glucose consumption rates were also introduced. As a result of these modifications, 100% of the sugars initially provided in a model mixture of 2:1 glucose:xylose (2% (w/v) total sugars) were consumed within 96 h, resulting the production of up to 3.5 g/L L-phenylalanine at a yield of 156 mg/g total sugar in shake flask cultures. Finally, to investigate the utility of this approach with respect to real biomass-derived sugars, the same strain was grown using corn stover hydrolysate (also 2:1 glucose:xylose, here using 1% (w/v) total sugars). In this case, the strain carrying XylR* produced up to 1.2 g/L L-phenylalanine in 25 h (73 mg/g total sugar); representing a ~20% increase over the parent strain. Notably, with the XylR* mutation, maximum rates of xylose consumption were increased by 3-fold (from 0.04 vs. 0.12 g/L-h) without impeding the maximum rate of total sugar consumption (0.44 vs. 0.40 g/L-h). Overall, this study demonstrates a new strategy for improving glucose-xylose co-utilization in aerobic *E. coli* cultures, and provides evidence to suggest that, when used to target the correct product, substrate mixtures can provide synergistic benefits over single substrates. In particular, we expect that these findings will be readily translated to improve the production of other aromatic chemicals derived from the shikimate pathway.

632h Utilising a Genome Scale Metabolic Model to Design High-Producing CHO Cell Lines through Bi-Level Optimisation

Athanasios Antonakoudis, Alexandros Kiparissides, Cleo Kontoravdi

Chinese hamster ovary (CHO) cells are the dominant expression platform to produce therapeutic monoclonal antibodies (mAbs). However, despite increasing global demand and persistent societal needs, mammalian cell-based production systems suffer from low product yields as specific productivity is inversely proportional to cellular growth rates. In this work, bilevel optimization techniques are applied to the CHO-K1 cell genome-scale metabolic model (GEM) [1], to detect genetic regulations that will increase specific mAb productivity during the growth phase of the cell culture. Initially, the GEM was manually curated by removing the blocked reactions, limiting the cellular uptake and secretion rates based on four different experimental exometabolomic datasets and applying elemental balance algorithms to obtain realistic bounds for the intracellular reactions. Then, based on the OptReg algorithm [2], a bi-level optimization algorithm was formulated to maximise both the antibody production and cellular growth of the CHO-K1 GEM for the four experimental datasets. The optimisation resulted in a plethora of scenarios, each containing a different combination of regulations on the reaction bounds corresponding to enzyme knock-out and/or overexpression strategies that would theoritacally increase the antibody specific productivity during the exponential growth phase of the cell culture. The resulting scenarios of genetic regulations were then analysed using sensitivity analysis to identify a minimum set of genetic interventions that could achieve our goal and a select few were screened experimentally.

[1] Hefzi H et al. A Consensus Genome-scale Reconstruction of Chinese Hamster Ovary Cell Metabolism. Cell Systems. 2016; 3:434-43. e8.

[2] Pharkya P and Maranas C. An optimization framework for identifying reaction activation/inhibition or elimination candidates for overproduction in microbial systems. Metabolic Engineering. 2006; 8:1-13

251a Engineering Dually Adjuvanted Sars-CoV2 mRNA Vaccines

Allen Jiang, Bowen Li, Idris Raji, Caroline Atyeo, Theresa M. Raimondo, Akiva Gordon, Luke Rhym, Galit Alter, Robert Langer, Daniel G. Anderson

The successful development of mRNA-based vaccines against SARS-CoV-2 has provoked broad interest in RNA-based technologies and demonstrated the promising potential of mRNA as a new vaccine class for combating infectious diseases. Although two mRNA vaccines have been authorized for use against SARS-CoV2 and many more mRNA vaccines for other infectious diseases are in clinical trials, the widespread clinical translation of this technology requires overcoming certain limitations including eliciting a sufficiently potent immune response to the expressed antigen at a safely tolerable dose. To further improve the safety and efficacy of mRNA vaccines, we here develop a dually self-adjuvanted mRNA vaccine system whereby both the encapsulating lipid nanoparticle as well as the mRNA are designed to potentiate the immune response to an encoded antigen. We first screened a combinatorial library of 480 biodegradable ionizable lipids as part of LNP formulations for vaccine potential, identifying a top ionizable lipid capable of enhanced immune stimulation leading to a potentiated immune response when compared to the FDA-approved ionizable lipid MC3. We then designed an adjuvanted mRNA to encode a fusion protein consisting of a SARS-CoV2 antigen and a bio-inspired adjuvanting peptide (AP), within a single transcript. Compared to the mRNA encoding viral antigens alone, the inclusion of the AP in the mRNA transcript increased the magnitude of antigen-specific antibody titers by at least ten-fold in mouse sera for both wild-type SARS-CoV2 and Delta variant antigens. Vaccination with LNPs combining the dually adjuvanted ionizable lipid and mRNA transcript resulted in synergistic immune responses to the Delta variant of SARS-CoV2 for both intramuscular and intranasal administrations. In addition to significantly improving total antigen-specific IgG titers, vaccination with the dually-adjuvanting system also resulted in antibodies with higher levels of Fc-receptor binding which has recently been shown to be a predictor of resolution of severe COVID-19. Our results demonstrate that a dually-adjuvanted system holds great potential in improving the efficacy, safety, and ease of administration of mRNA vaccines.

251b Optimizing Combinations of Immune Inhibitors and Agonists to Generate Tolerogenic Antigen Presenting Cells

Peter Deak, Jeremiah Kim, Matthew Rosenberger, Aaron Esser-Kahn

Natural immune tolerance to self-antigens is maintained via T regulatory cells (Tregs), which actively surveil peripheral tissues or instruct naïve T cells in the thymus to remove or promote anergy of antigen specific T and B cells. This makes Tregs attractive targets for autoimmune therapies, but endogenous differentiation of *de novo* Tregs or *ex vivo* differentiation and reintroduction of Tregs has not been successful to date. A new strategy involves generating tolerogenic antigen presenting cells (APCs), which can instruct naïve T cells to differentiate into Tregs; however the types of immunosuppressant drugs used to generate tolerogenic APCs often triggers cell death or has off target effects on unrelated immune interactions. We have recently shown that combinations of both immunosuppressing and immune activating signals can induce longer lived tolerogenic APC populations that can effectively induce Tregs. This presentation will describe the efforts to optimize this combination through mining the data of a large screening library. RAW cells (a mouse macrophage cell line) were screened against a library of over 40,000 combinations of immunomodulators and immune agonists and their NF-kB and IRF3 activity measured. A smaller 10,000 combination library was also screened against THP-1 cells (a human monocyte cell line) for cytokine release. A subset of these combinations with low NF-kB and IRF3 activity as well as low inflammatory cytokine secretion, indicating more tolerogenic phenotypes, were further tested for tolerogenic cytokine release profiles and tolerance markers, such as high IL-10 and PD-L1. Finally, these combinations were tested on primary DCs isolated from mice to determine optimal tolerogenic APC generating formulations. These optimized drug combinations have translational potential for autoimmune diseases.

251c Engineered Multispecific Antibodies to Interrogate and Manipulate Immune Checkpoint Protein Trafficking

Seth Ludwig, Jamie Spangler

SIGNIFICANCE

Groundbreaking advances in immunotherapy, the mobilization of a patient's own immune system against disease, have revolutionized the cancer therapy landscape. A particularly transformative advance is the development of antibody drugs that block immunosuppressive networks known as immune checkpoint pathways, thereby unleashing the power of immune cells to destroy tumors. These drugs have been remarkably successful in multiple cancer types. However, the majority of patients (>70%) do not respond to treatment, and most tumor regressions are only partial. The determinants of clinical response are not well understood, and certain cancer types resist immune checkpoint blockade altogether. Moreover, the role of immune checkpoint protein trafficking in immune regulation and disease pathogenesis remains unexplored. There is thus an urgent demand for innovative tools to interrogate immune checkpoint pathways, and for new interventions to therapeutically modulate their activity.

OBJECTIVE

We addressed current limitations by engineering novel multispecific antibodies targeting immune checkpoint proteins that elucidate immune biology and also form the basis for a new class of cancer therapeutics. Whereas all current clinical antibodies targeting immune checkpoint proteins act through a common mechanism of inhibiting pathway activation by blocking ligand/receptor interaction, we pursued an alternative strategy designed to down-regulate the surface presentation of immune checkpoint proteins and thereby prevent the possibility of pathway activation. This approach employs a concept known as multiparatopic antibody-mediated down-regulation, wherein a multispecific antibody targets two or more epitopes on a single protein to bias protein trafficking and drastically down-regulate surface presentation. We hypothesized that multiparatopic antibodies that engage multiple epitopes on the immune checkpoint protein programmed death-ligand 1 (PD-L1) would abolish immunosuppressive signaling to stimulate robust anti-cancer immunity.

RESULTS

We used a combination of immunization and directed evolution techniques to discover a panel of anti-PD-L1 antibodies that are non-competitive with the FDA-approved PD-L1-targeted neutralizing antibody atezolizumab. We then incorporated the variable domains of these novel antibodies together with atezolizumab into engineered biparatopic and triparatopic antibodies in a range of orientations. Trafficking studies revealed that these multiparatopic antibodies induced rapid and robust clustering, internalization, and lysosomal degradation of PD-L1 in an epitope- and topology-dependent manner. T cell activation assays demonstrated that the complementary mechanisms of ligand blockade and receptor downregulation employed by multiparatopic antibodies led to more durable immune stimulation, and in vivo studies in tumor-bearing mice showed that our lead multiparatopic antibody dramatically reduced availability of PD-L1 at the site of disease.

CONCLUSIONS AND IMPACT

We established a panel of multiparatopic antibodies against PD-L1 that offer new insight into immune checkpoint pathways and demonstrate how we may manipulate these pathways for targeted immune activation. The universality of our immunostimulatory approach enables efficacy across cancer types, and our unique trafficking-focused strategy allows for synergy with other immunotherapeutic modalities, such as vaccines, cytokine therapy, and adoptive cell transfer. Moreover, our versatile multiparatopic antibody designs can be extended to other classes of transmembrane proteins to interrogate and manipulate molecular trafficking for a broad range of research and medical applications.

251d Phospho-Proteomic Analysis of CAR-T Cell Signaling Following Activation By Antigen-Presenting Cancer Cells

Melanie MacMullan, Zachary Dunn, Yun Qu, Xianhui Chen, Pin Wang, Nicholas Graham

Chimeric antigen receptors (CARs) are synthetic biomolecules comprised of an extracellular antigen recognition domain and intracellular signaling domains. Immune cells engineered to express CARs are one of the most promising forms of immunotherapy because CARs direct their host cells to kill diseased cells expressing the antigen recognized by the CAR. Although signaling pathways downstream of CAR activation control the cytotoxic function of CAR-expressing cells, studies of the phosphorylation-based signaling pathways initiated by CARs have been limited. For comprehensive, unbiased, and quantitative characterization of protein phosphorylation, liquid chromatography-mass spectrometry (LC-MS)-based phospho-proteomics is the technology of choice. However, phospho-proteomic studies of CAR signaling have been limited by the challenge of stimulating CAR-expressing cells with their *in vivo* stimulus, antigen-presenting cancer cells. Here, we demonstrate an efficient and cost-effective label-free phospho-proteomic approach to analyze CAR signaling in immune cells stimulated with antigen-presenting cancer cells (Fig. 1). Following co-culture of CAR-T cells with cancer cells, we first preserve phospho-signaling by cross-linking proteins with formalin. Then, we use magnet-activated cell sorting (MACS) to isolate CAR-T cells from the co-culture. Validation experiments demonstrated that formalin fixation did not alter the phospho-proteome and that MACS achieved >90% CAR-T cell purity. Next, we compared the phospho-proteome in CAR-T cells stimulated with either CD19-expressing or non-CD19-expressing SKOV3 ovarian cancer cells. This analysis revealed that CAR signaling activated known pathways including the mitogen-activated protein kinases (MAPKs) ERK1/2. Bioinformatic approaches further showed that CAR activation induced other signaling pathways including the MAPK p38α, protein kinase A, and checkpoint kinase 1 (CHK1). Taken together, our work represents the first label-free phospho-proteomic analysis of CAR signaling in CAR-T cells stimulated with antigen-presenting cancer cells. Our results provide insight into the signaling mechanisms downstream of CAR activation and suggest how CARs direct T cells to kill target cells. By coupling phospho-proteomic profiling with functional analysis of CAR-T cell cytotoxicity, we envision that an improved understanding of CAR signaling will enable engineering of next-generation CARs with increased potency, improved persistence, and reduced toxic side effects.

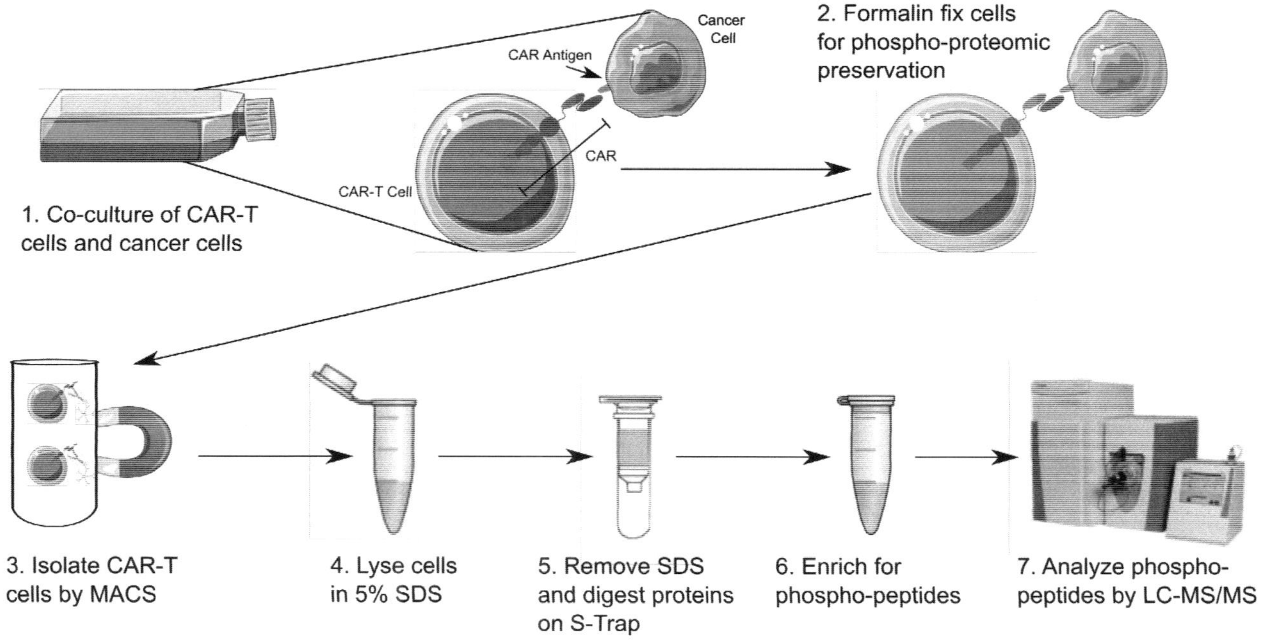

251e Targeting IL-10 to Atherosclerotic Plaques Reduces Vascular Inflammation

Jeffrey A. Hubbell, Yun Fang, Lisa Volpatti

Introduction: Atherosclerosis is a chronic inflammatory disease characterized by the accumulation of cholesterol on vascular walls and subsequent formation of plaques. Current preventative measures such as statins and dietary changes focus on reducing circulating "bad cholesterol" or low-density lipoprotein (LDL) and often lead to surgical interventions associated with high cost and morbidity. As an orthogonal approach, modulating the immune response to LDL can limit plaque formation. For example, overexpression of the anti-inflammatory cytokine IL-10 results in reduced lesion sizes. However, IL-10 has a poor pharmacokinetic profile and potential to cause systemic immunosuppression, which limit its therapeutic potential.

Methods: To overcome these challenges, we engineered constructs that target atherosclerotic plaques by fusing IL-10 to an antibody fragment (Fab) that binds LDL. We expressed six different Fab-IL-10 constructs: five with Fabs specific for LDL and one control Fab with irrelevant specificity (Fig. 1a).

Results: We validated the binding affinity of the constructs to LDL with surface plasmon resonance (Fig. 1b) and confirmed that IL-10 retains its bioactivity in vitro. We then conducted a screen of these constructs in hypercholesterolemic apolipoprotein $E^{-/-}$ mice. Our results showed that the total number of immune cells in the aorta was significantly reduced in the group treated with 2D03-IL-10 (Fig. 1c). The number of macrophages which contribute to plaque formation was similarly reduced. Importantly, neither wild type (WT) IL-10 nor IL-10 fused to the irrelevant Fab (IRR) showed any reduction of aortic immune cell infiltration.

Conclusions: We used biomolecular engineering techniques to target the cytokine IL-10 to atherosclerotic plaques. The 2D03-IL-10 fusion protein resulted in reduced vascular inflammation and has the potential to improve atherosclerosis disease outcomes.

a)

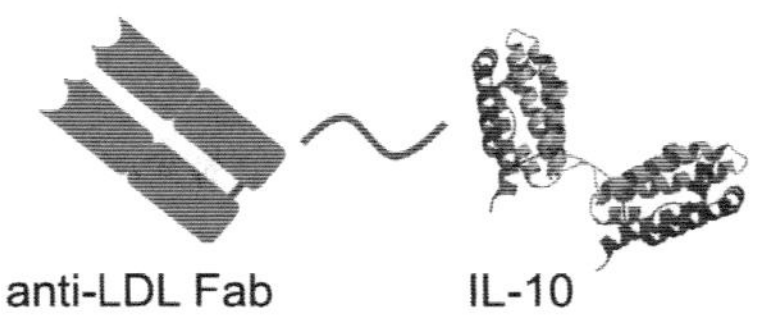

b)

Construct	K_D (nM)
2D03-IL-10	14.3
KTT-IL-10	33.4
E3-IL-10	1330
D8-IL-10	227
LDO-IL-10	55.4
IRR-IL-10	n.d.

c)

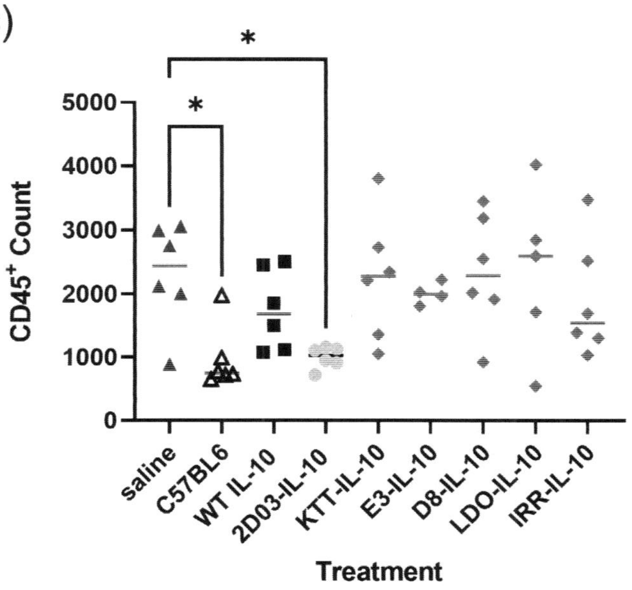

Fig. 1. In vivo reduction of vascular inflammation with LDL-targeted IL-10. a) Schematic of Fab-IL-10 constructs. b) Equilibrium dissociation constants (K_D) of Fab-IL-10 constructs and LDL by surface plasmon resonance. c) Number of $CD45^+$ cells in aortas of $apoE^{-/-}$ mice after 12 weeks of high fat diet and 4 weekly injections of saline or treatment compared to C57BL6 mice. Comparisons were made using one-way ANOVA with Dunnett's post-test. *$p < 0.05$. Data are compiled from 2 independent experiments for a total of n=4-6 per group.

251f Apoptotic Bio-Inspired Materials for Targeting and Engineering Macrophage Function

Kidochukwu Atube, Candice Cheung, Michael Gower

Introduction. A major initiative factor of various pathological conditions such as type 1 diabetes, obesity, and other autoimmune diseases is chronic low-grade inflammation. The immune cells responsible for such chronic inflammation are macrophages[1–3]. Currently, there exists several small molecule drugs that bind cognate receptors on macrophages and promote an anti-inflammatory response in the presence of inflammatory stimuli[1,4,5]. However, most of these small molecule drugs suffer from off-target and off-site side effects[6,7]. Therefore, there is a need for targeted therapies. Macrophage recognize apoptotic cells through the "eat me" signal, phosphatidylserine (PS)[8]. Interestingly, the recognition of PS on apoptotic cells leads to the internalization of the expressing apoptotic cell and subsequent initiation of an anti-inflammatory and regenerative gene program by macrophages[9,10]. This study aims to 1) Synthesize micron-sized polymer drug delivery carriers that present PS on their surface like that of an apoptotic cell and 2) Test the ability of these carriers to target macrophages and potentially modulate macrophage inflammatory function.

Methods. PS-presenting polymer microparticles were synthesized using the single oil-in-water emulsion/solvent extraction method, with poly(lactide-co-glycolide) (PLG) as our polymer of interest[11]. An annexin V binding assay was used to investigate the presence of functional PS on particle surface. The PS-presenting PLG microparticles (PS:PLG) were loaded with coumarin 6 (C6), to track particle uptake by bone marrow-derived macrophages (BMDMs) using flow cytometry and live cell imaging. BMDMs were co-treated with an inflammatory stimuli, lipopolysaccharide (LPS) and PS:PLG particles to monitor macrophage inflammatory response.

Results and Discussion. We show that PS:PLG particles of different sizes can be synthesized using the single oil-in-water emulsion/solvent extraction method (Figure 1). Thus, indicating that particle size can be modified to fit the right application. For example, the utility of larger PS:PLG particles where extended drug release rates are desired[11]. An annexin V binding assay demonstrated that PS:PLG particles bound fluorescently labeled Annexin V at increasing levels with more addition of PS to the emulsion during particle synthesis, while particles lacking PS did not (PLG particles) (Figure 2).

PS functions as an "eat me" signal to macrophages that promotes binding and uptake of apoptotic bodies. Confocal microscopy showed increased PS:PLG-C6 particle interaction with BMDMs compared to PLG particles having no surface PS (Figure 3A). Congruent to this visual data, flow cytometry shows a higher percentage of macrophages positive for PS:PLG-C6 particles compared to PLG-C6 particles (Figure 3B). Hence indicating that these PS:PLG particles could be employed in future drug formulations to improve the delivery of small molecule drug payloads to macrophages.

In addition, a co-treatment of macrophages with PS:PLG particles and LPS, significantly promoted an anti-inflammatory response characterized by a decrease in TNF-α and an increase in IL-10 (Figure 4). This suggests that the PS:PLG particles could have physiological effects as a drug carrier by promoting the resolution of inflammation without any drug payload.

Conclusion and Future Work. In summary, we report the development of macrophage-targeted drug delivery carriers that could be applied in therapeutic interventions to actively get drugs or bioactive agents to macrophages for better clinical outcomes and at the same time potentially act as therapeutics to treat low-grade inflammation. These macrophage-targeted carriers could also be applied to repurpose and improve the therapeutic efficacy of some FDA approved drug release formulations made from PLG. Future studies aim to test these PS:PLG particles *in vivo*. This information will further inform development of optimized particle formulations best suited to be delivered *in vivo*.

References:

1. Ma, L. *et al.* Efficient Targeting of Adipose Tissue Macrophages in Obesity with Polysaccharide Nanocarriers. *ACS Nano* **10**, 6952–6962 (2016).
2. Toita, R., Kawano, T., Murata, M. andamp; Kang, J. H. Anti-obesity and anti-inflammatory effects of macrophage-targeted interleukin-10-conjugated liposomes in obese mice. *Biomaterials* **110**, 81–88 (2016).
3. Oishi, Y. andamp; Manabe, I. Macrophages in inflammation, repair and regeneration. *Int. Immunol.* **30**, 511–528 (2018).
4. Xu, N. *et al.* Apoptotic cell-mimicking gold nanocages loaded with LXR agonist for attenuating the progression of murine systemic lupus erythematosus. *Biomaterials* **197**, 380–392 (2019).
5. Wang, J. *et al.* Enhancement of Anti-Inflammatory Activity of Curcumin Using Phosphatidylserine-Containing Nanoparticles in Cultured Macrophages. **6**, 1–19 (2016).
6. Zhao, Z., Ukidve, A., Kim, J. andamp; Mitragotri, S. Targeting Strategies for Tissue-Specific Drug Delivery. *Cell***181**, 151–167 (2020).
7. He, W., Kapate, N., Shields, C. W. andamp; Mitragotri, S. Drug delivery to macrophages: A review of targeting drugs and drug carriers to macrophages for inflammatory diseases. *Adv. Drug Deliv. Rev.* **165–166**, 15–40 (2020).
8. Budai, Z. *et al.* Macrophages engulf apoptotic and primary necrotic thymocytes through similar phosphatidylserine-dependent mechanisms. *FEBS Open Bio* **9**, 446–456 (2019).
9. Szondy, Z., Sarang, Z., Kiss, B., Garabuczi, É. andamp; Köröskényi, K. Anti-inflammatory mechanisms triggered by apoptotic cells during their clearance. *Front. Immunol.* **8**, (2017).
10. Szondy, Z., Garabuczi, É., Joós, G., Tsay, G. J. andamp; Sarang, Z. Impaired clearance of apoptotic cells in chronic inflammatory diseases: Therapeutic implications. *Front. Immunol.* **5**, 1–8 (2014).
11. Isely, C. *et al.* Development of microparticles for controlled release of resveratrol to adipose tissue and the impact of drug loading on particle morphology and drug release. *Int. J. Pharm.* **568**, (2019).

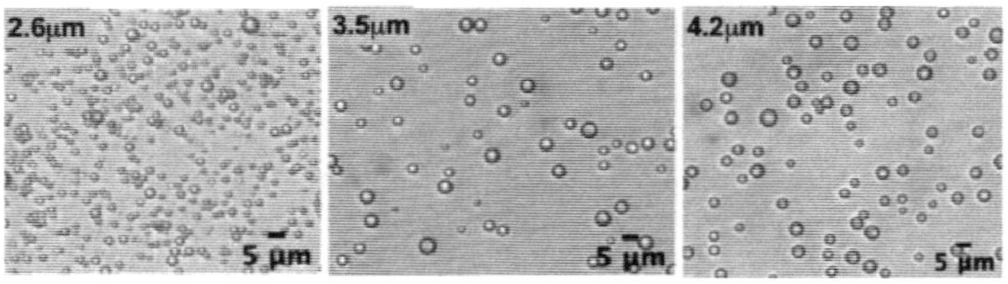

Figure 1: Images of PS:PLG particles of different sizes.

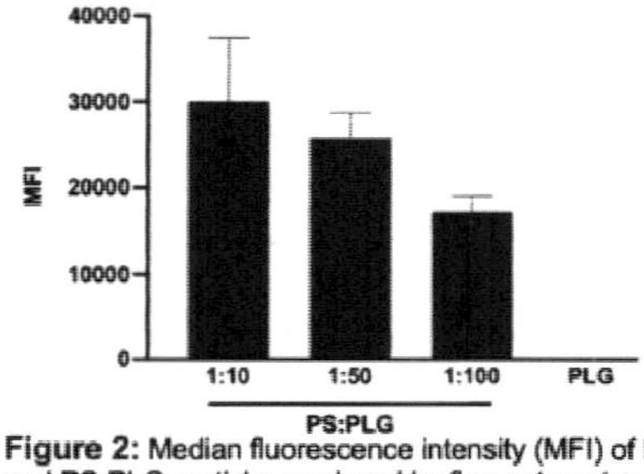

Figure 2: Median fluorescence intensity (MFI) of PLG and PS:PLG particles analyzed by flow cytometry.

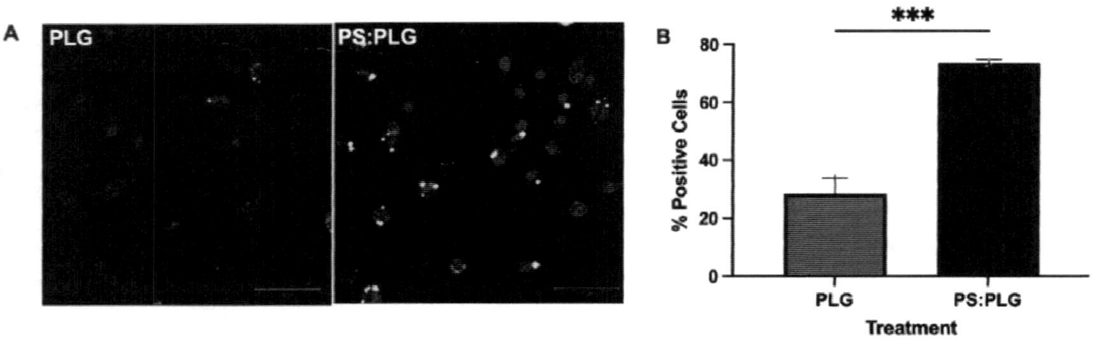

Figure 3: **A)** Representative confocal scanning laser microscopy imaging shows particle internalization by BMDMs 6 hrs following particle delivery. **B)** % of macrophages positive for C6-labeled particles

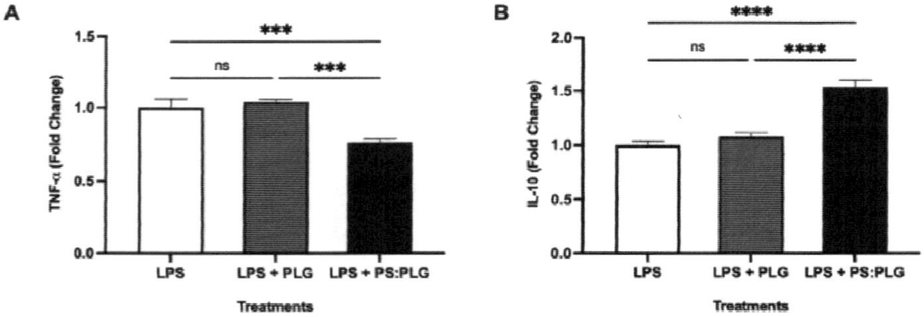

Figure 4: Levels of **A)** TNF-, **B)** IL-10 secreted by BMDMs co-treated with LPS and PS:PLG particles for 6hrs.

251g Calibrating Immunometabolism to Treat Diseases

Abhinav Acharya

Immunometabolism reprogramming is involved in progression, induction and therapy of several diseases such as cancer, infections, autoimmune disorders. Notably, modulation of immunometabolism can be performed by delivery of cell permeable metabolites, enzymatic inhibitors or through gene editing. For example, metabolites provided systemically may improve immune cell function. Importantly, these strategies can be targeted toward both innate and adaptive branch of the immune system to generate effector functions. In this presentation, I will introduce strategies for modifying immunometabolism of dendritic cells, macrophages, and T cells for supporting the induction of immune responses in rheumatoid arthritis, sepsis, multiple sclerosis, and melanoma (skin cancer), with special emphasis on melanoma treatment. Specifically, these projects are geared toward developing immunotherapies by directly modulating the metabolism of immune cells to develop robust immune responses in mouse models. Overall, these projects will demonstrate the importance of metabolite modulation in reprogramming immune responses.

58a Development of a Plant Generated Therapeutic Made in-Flight to Treat Microgravity Osteopenia

Kevin Yates, Imran Khan, Yongao Xiong, Nancy E. Lane, Abhaya M. Dandekar, Karen A. McDonald, Somen Nandi

Extended space missions in microgravity environments alter normal physiology in the human body. One critical issue in spaceflight is loss of bone mineral density. In the process of bone homeostasis, specialized cells facilitate bone remodeling with coupled bone resorption and formation. Reduced mechanical loading in microgravity causes resorption to outpace formation, which results in loss of bone mass. Medications such as parathyroid hormone (PTH) [amino acids 1-34], a peptide fragment of naturally occurring human PTH, stimulate bone formation and may be able to restore bone mass in microgravity. However, this requires a daily subcutaneous injection for a number of months, and in a limited resource environment, this is impractical. An alternative to transporting an injectable medication is to instead build capacity to produce it during the mission, which we propose to do in transgenic plants.

We are producing transgenic lettuce which expresses a fusion protein consisting of PTH [amino acids 1-34] linked with the fragment crystallizable (Fc) domain of human IgG1 via a flexible linker. The Fc component and sequestration of the PTH-Fc within the plant endoplasmic reticulum is intended to increase bioavailability via oral delivery. We constructed a binary vector for expression of this PTH-Fc protein and used it to constitutively produce PTH-Fc in two varieties of transgenic lettuce via *Agrobacterium tumefaciens* mediated transformation of wild type lettuce. We have successfully generated transgenic (T) plants and screened T_1 and T_2 generations using a selectable marker and PCR. An extraction protocol coupled with Western blotting has enabled the detection of PTH-Fc in plant extracts by targeting either the PTH or the Fc portions of the fusion protein. An optimized ELISA protocol has been developed to quantify expression levels of PTH-Fc. Preliminary results show an expression level on the order of 65 mg PTH-Fc per kilogram of fresh weight biomass.

58b Self-Remodeling Protein Complexes Inspired By Fungal Cellulosomes

Stephen Lillington, M. Scott Shell, Michelle O'Malley

Engineered protein systems that sense different environmental signals to produce specific outputs have broad applications towards developing smart biomaterials or for building programmable biotherapeutics. Parts with which to construct these "molecular computers" that function by altering sets of protein-protein interactions in response to stimuli are a key technology gap not well addressed by current sets of modular, interacting protein domains, which are limited in their generalizability to different protein cargo and their ability to be multiplexed. Fungal cellulosomes, the multi-enzyme complexes produced by anaerobic fungi to degrade plant matter in the guts of herbivores, are an attractive template for designer, stimuli-responsive protein complexes because they scaffold proteins of diverse structure and function using modular parts that are highly engineerable (Figure 1). Cellulosomes assemble via non-covalent interactions between enzyme-fused dockerin domains and cohesin domains repeated on a central scaffoldin protein. However, fungal cellulosome parts have never been explored for engineering applications because the sequence and structure of dockerin's binding partner, cohesin, remain unknown. Here, we combined molecular modeling with high throughput screening to develop the dockerin and a synthetic cohesin substitute, a nanobody, into a set of protein parts for constructing complexes that remodel in response to pH - a biologically useful signal. We first used modeling tools to predict the binding interface and constructed a yeast surface display screening library in which 22 positions were combinatorially mutated to histidine to impart pH-sensitive binding. Parallel screens produced switches with different binding and switch thermodynamics, providing a dataset with which to generate, via atomistic modeling, mechanistic hypotheses for switch function that will inform the design of new parts with tailored switch-like behavior. Towards synthesizing higher order complexes, we employed molecular dynamics to screen scaffolding protein architectures for designs capable of binding several partners simultaneously, accelerating the development of functional, self-remodeling complexes. Overall, this work showcases the potential for cellulosomes as synthetic biology tools and highlights the power of a combined modeling and high-throughput screening approach in protein engineering.

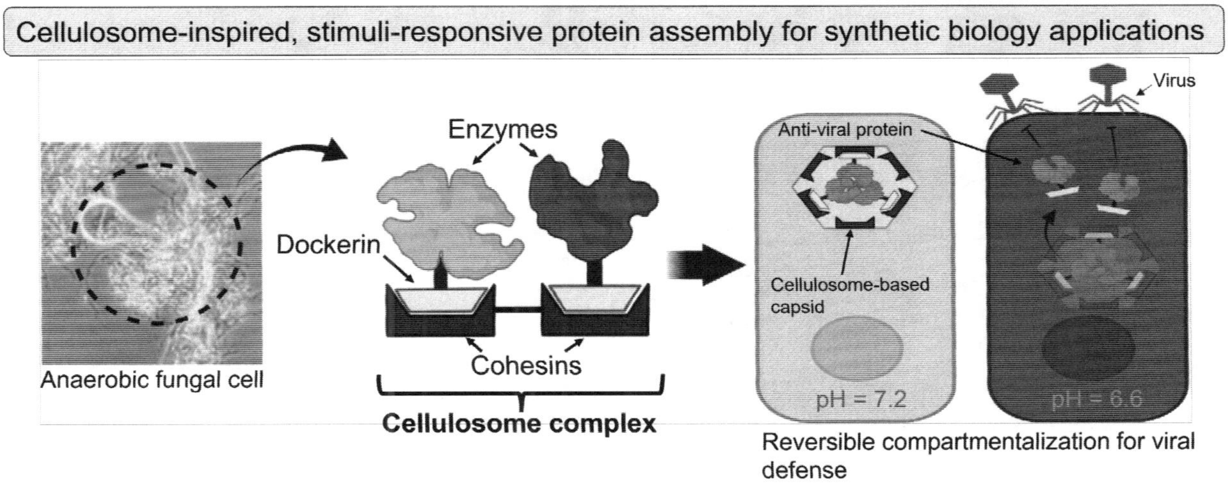

58c Use of Coiled-Coil Motifs to Assemble Membraneless Organelles for Tunable Control Liquid Liquid Phase Separation

Zhihui Su, Mikael Garabedian, Matthew C. Good, Daniel A. Hammer

In eukaryotic cells proteins can undergo self-assembly and phase separate to form membraneless organelles such as nucleoli, Cajal bodies, P-bodies, and stress granules. Because of their potential for subcellular compartmentalization and channel-less cargo recruitment and release, membraneless organelles made from liquid-liquid phase separation (LLPS) can be excellent bio-building blocks and hubs for cellular communication. Here, we combine intrinsically disordered RGG domains from Laf-1 with small coiled coil orthogonal peptide domains to facilitate targeted phase separation at room temperature. We engineered orthogonal coiled coil domains P1-P6, developed by the Jarala lab, on both N and C termini of the RGG domains and utilize fluorescence microscopy using RGG-GFP-RGG as a fluorescence reporter, and turbidity assays to measure coacervation. We find when two RGG domains are linked via P3-P4 and P5-P6 pairs, coacervation is facilitated. However, the incorporation of the P1-P2 pairs did not facilitate coacervation. In general, we find N terminal attachment of coiled coil linkers enhance coacervation but C terminus attachment inhibit coacervation. We then demonstrate phase separation induced through coiled coil linkers is reversible and could be thermally cycled and reconstituted. With these system, we are now exploring client recruitment into membraneless organelles assembled in yeast. This dynamic, synthetic RGG coiled coil system offers a tunable approach and generalizable platform for controlling coacervation, and has broad applications such as cargo recruitment and cellular enzymatic control in both bio-engineered cells and protocells.

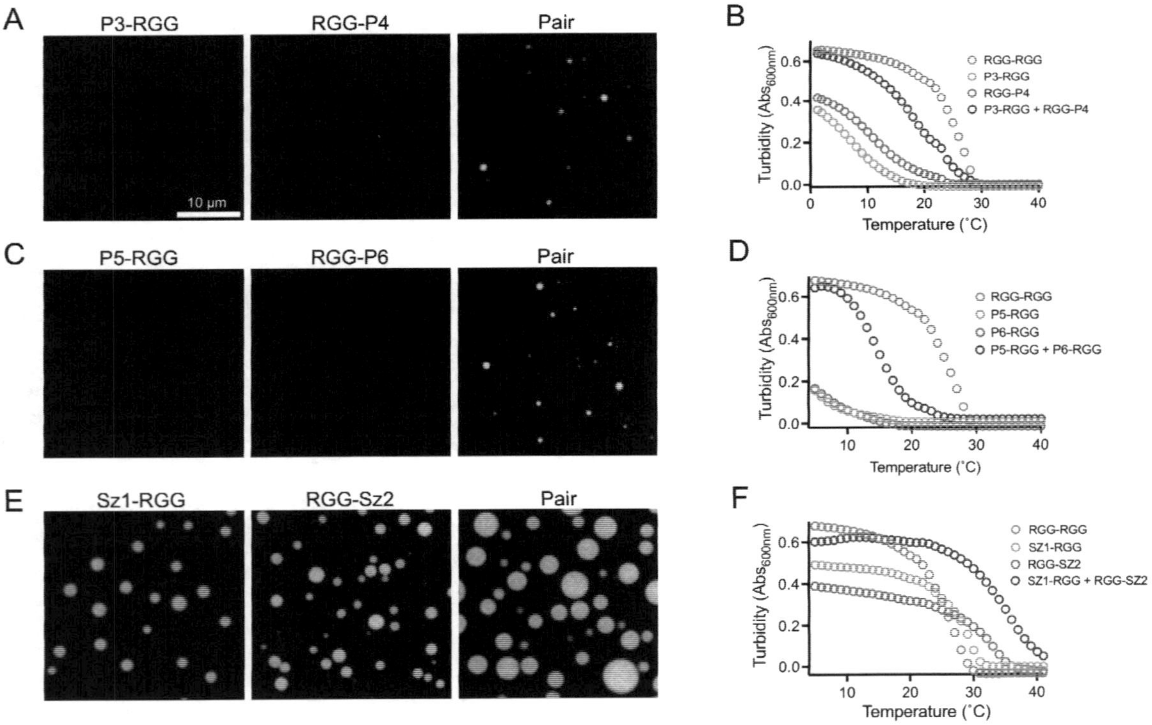

58d Pro-Apoptotic Stapled Peptides Discovered Using Bacterial Cell Surface Display Demonstrate Improved Affinity, Specificity, and Efficacy

Marshall Case, Greg Thurber

There is currently a wealth of disease-associated proteins that are "undruggable" by modern therapeutic formats. These targets reside inside the cell, where they are inaccessible to large biologics, but lack sites where a small molecule drug could bind. Peptide therapeutics, comprised of a short chain of amino acids, are an emerging therapeutic framework that attempt to fill this gap. These peptides are small enough to enter cells while being large enough to bind and inhibit protein-protein interactions. Unfortunately, peptide therapeutics for intracellular targets have had limited success in the clinic arising from several issues including low target affinity and low proteolytic stability. To overcome these challenges, a "stapled" peptide can be formed by covalently linking two amino acids, locking the peptide into its alpha-helical state. This single modification can improve target affinity and enhance proteolytic stability. However, a key challenge in the discovery of stapled peptide therapeutics is the reliance on solid-phase synthesis, where peptides are limited to low or moderate throughput and evaluation, which greatly limits the number of peptides that can be analyzed.

In this work, we employ a novel technique, developed in the Thurber Lab, known as **S**tabilized **P**eptide **E**ngineering by ***E**. coli* **D**isplay (SPEED) to rapidly characterize and develop $>10^9$ unique stapled peptide therapeutics against members of the Bcl-2 protein family, which are important regulators of apoptosis and play key roles in resistance to chemotherapy. In this work, libraries of DNA encoding unique peptides are computationally designed by mining genome and linear peptide library data, transformed into bacteria, and presented as peptides on the cell surface. To enable characterization of stapled peptides, azide containing residues are incorporated in place of methionine through use of methionine auxotrophic bacteria. Then, copper catalyzed click chemistry (CuAAC) is used to form stapled peptides directly on the cell surface through reaction with bisalkyne staples. Finally, cells are sorted for given molecular properties by incubating with fluorescently labeled target protein. This technique was previously applied to discover an inhibitor with eightfold improved affinity against mdm2, an important oncoprotein in the p53 pathway.

Here, we improve on this technique and further demonstrate its utility in discovering potent stapled peptide inhibitors by applying it to the Bcl-2 protein family, which is comprised of five highly homologous protein targets. Because of their homology, methods that can simultaneously optimize specificity and affinity are greatly needed. First, we utilized on-cell staple scanning to identify locations for the linker that maintain target binding. The next improvement in the technique is incorporating Next Generation Sequencing (NGS) into the workflow, which enables quantitative measurement of affinity and specificity based on high-throughput fluorescent-activated cell sorting data. To identify lead molecules, computational models were constructed based on affinity and specificity profiles of $>10^5$ peptides from cell sorting. Interestingly, these models generated peptides with improved specificity and affinity compared to those that emerged directly from the sorting campaign. Finally, we translated these molecules off the cell surface and evaluated their mechanisms of action and efficacy, proving they demonstrate improved efficacy *in vitro* over their linear counterparts. The ability to improve affinity, specificity, stability, and *in vitro* efficacy demonstrates SPEED's multifaceted capability.

In summary, we apply SPEED to design tight binding, highly specific, and protease resistant stapled peptide inhibitors of Bcl-2 proteins. To our knowledge, this work represents the first high-throughput study of stapled peptide Bcl-2 inhibitors. Peptides with these improved drug-like properties promise better selective treatment for cancer with *in vitro* and *in vivo* validation of efficacy. This approach provides a framework to discover potent stapled peptide inhibitors towards other important protein targets.

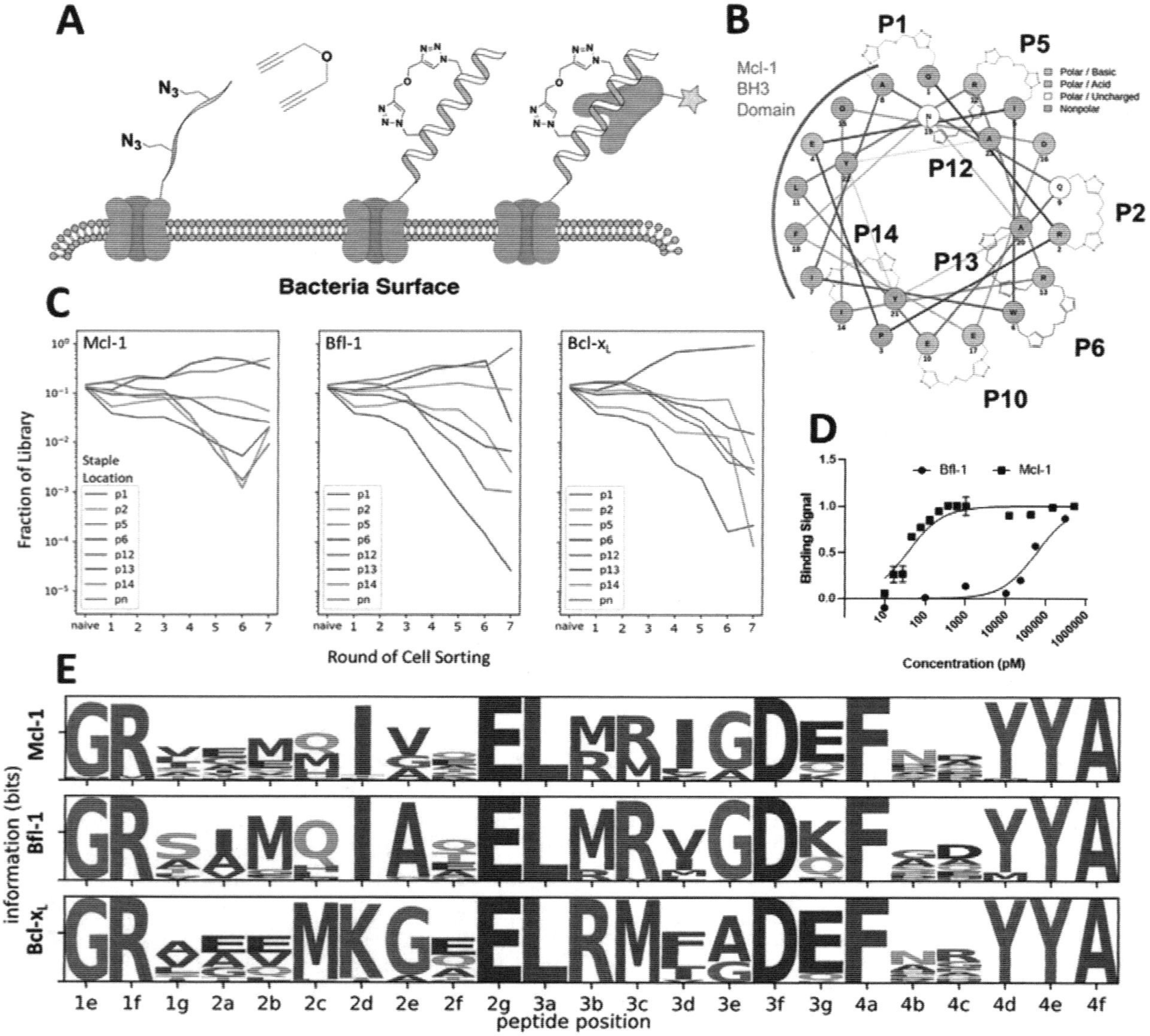

Figure 1: Discovery of stabilized peptides on the bacterial cell surface. (A) Peptides are displayed on the surface of bacteria and form stapled peptides using copper catalyzed click chemistry. After treatment with fluorescently labeled Bcl-2 protein, cells of varying signal are sorted by activity using fluorescent activated cell sorting and quantified using next generation sequencing. **(B)** Helix-wheel diagram of WT (BIM) staple scan, where the location of staple in context of the protein interaction is varied along the length of the peptide. High affinity peptides are shown with their staples in chemical structure. **(C)** Different Bcl-2 proteins develop specificity profiles based on staple location. **(D)** Surface titration of Mcl-1 versus Bfl-1 demonstrates >1000x specificity for stapled peptides. **(E)** Logoplots for Mcl-1 versus Bfl-1 versus Bcl-x$_L$ binding peptides after cell sorting demonstrate key sequence features that delineate specificity.

58e *Lon* Deletion Impairs Persister Cell Resuscitation in *Escherichia coli*

Sayed Golam Mohiuddin, Aslan Massahi, Mehmet Orman

Abstract

Bacterial persisters are a small subpopulation of phenotypic variants in isogenic cell populations that are transiently tolerant to high concentrations of antibiotics. Persisters aggravate the crisis of antibiotic failure as they can evolve to antibiotic-resistant mutant strains (1). The deletion of *lon* has been frequently shown to reduce fluoroquinolone persistence (2, 3), and the mechanism underlying this phenomenon is not clear. Lon was initially thought to induce persistence by degrading antitoxin molecules through a ppGpp/polyphosphate-dependent mechanism (4). Another suggested mechanism for the role of Lon in bacterial persistence depends on the cell division inhibitor protein SulA and is based on the observation that deletion of the *sulA* gene restores fluoroquinolone persister levels in *lon*-deficient strains (2, 5). However, it is unclear from these studies whether targeting Lon protease chemically or genetically eradicates persister cells or converts them to a viable but non-culturable (VBNC) state. In this study, using vector constructs that allow fine-tuning of recombinant protein expression, we have demonstrated that deletion of *lon* impairs the ability of persister cells to form colonies during recovery (**Fig. 1AB**), without killing these cells, through a *sulA*- and *ftsZ*-dependent mechanism. We have also shown that the reduced persister levels in Δ*lon* strain can be transient, and starvation can restore their culturability (**Fig. 1C**). Starvation-induced SulA degradation or expression of Lon during recovery facilitates Z-ring formation in Δ*lon* persisters. When we calculated the ratio of the cell length (L in μm) to the number of Z-rings (Z) for each ofloxacin-treated intact cell analyzed, we found a correlation between persister resuscitation and calculated L/Z values. Both WT and Δ*lon* persisters having L/Z ≈1 with L>5 μm and Z>5 were mostly resuscitated under the conditions studied here (**Fig. 1DE**). Overall, our findings here suggest that Z-ring formation may be a key biomarker for persisters.

Note: This study has been published in *mBio* (6).

References

1. Levin-Reisman I, Ronin I, Gefen O, Braniss I, Shoresh N, Balaban NQ. 2017. Antibiotic tolerance facilitates the evolution of resistance. Science. 355:826–830.
2. Matange N. 2020. Highly contingent phenotypes of lon protease deficiency in Escherichia coli upon antibiotic challenge. J Bacteriol 202:561–580.
3. Theodore A, Lewis K, Vulić M. 2013. Tolerance of Escherichia coli to fluoroquinolone antibiotics depends on specific components of the SOS response pathway. Genetics 195:1265–1276.
4. Harms A, Fino C, Sørensen MA, Semsey S, Gerdes K. 2017. Prophages and growth dynamics confound experimental results with antibiotic-tolerant persister cells. MBio 8.
5. Shan Y, Gandt AB, Rowe SE, Deisinger JP, Conlon BP, Lewis K. 2017. ATP-Dependent persister formation in Escherichia coli. MBio 8.
6. Mohiuddin SG, Massahi A, Orman MA. 2022. lon Deletion Impairs Persister Cell Resuscitation in Escherichia coli. MBio 13:e02187-21.

Figure Legend

Fig. 1. Lon overexpression during recovery or PBS starvation rescue Δ*lon* persisters. (**A-B**) *E. coli* WT and Δ*lon* cells containing an inducible *lon* overexpression plasmid (pUA66-*lon*) or empty vector control (pUA66-EV) were treated with ofloxacin (5 μg/ml) for 6 h and transferred to LB agar plates supplemented with isopropyl β-D-1-thiogalactopyranoside (IPTG) at the indicated concentrations (0 and 1 mM). Colony formation units (CFU) were measured for each strain before and after ofloxacin treatment. The number of independent biological replicates, N = 4. (**C**) WT and Δ*lon* cells with or without pUA66-*lon* or pUA66-EV were treated with ofloxacin for 6 h, transferred to sterile PBS solution with or without 1 mM IPTG, and incubated at 37°C in a shaker for 7 days. At the designated time points starved cells were collected and spotted onto LB agar plates to enumerate surviving cells. n = 4. (**D-E**) WT and Δ*lon* strains harboring the IPTG-inducible FtsZ-GFP expression plasmid (pUA66-*ftsZ-gfp*) were transferred to PBS solution after 6 h of ofloxacin treatment and cultured at 37°C with shaking. IPTG (1 mM) was added to cultures before and during ofloxacin treatment to express FtsZ-GFP. After 2 days of incubation in PBS solution, cells were then collected and spread on LB agarose pads containing 1 mM IPTG. Fluorescence microscope with an onstage incubator was used to monitor the cellular growth on the pads for 24 h. Cell lengths, as well as Z-ring structures and numbers, were determined for both resuscitating (persisters) and nonresuscitating (VBNC) cells to generate plots of the number of Z-rings versus cell length N=4. One-way analysis of variance (ANOVA) with Dunnett's post hoc test was used to determine statistical significance. *, $P<0.05$; **, $P < 0.01$; ***, $P < 0.001$; ****, $P < 0.0001$; *ns*, nonsignificant.

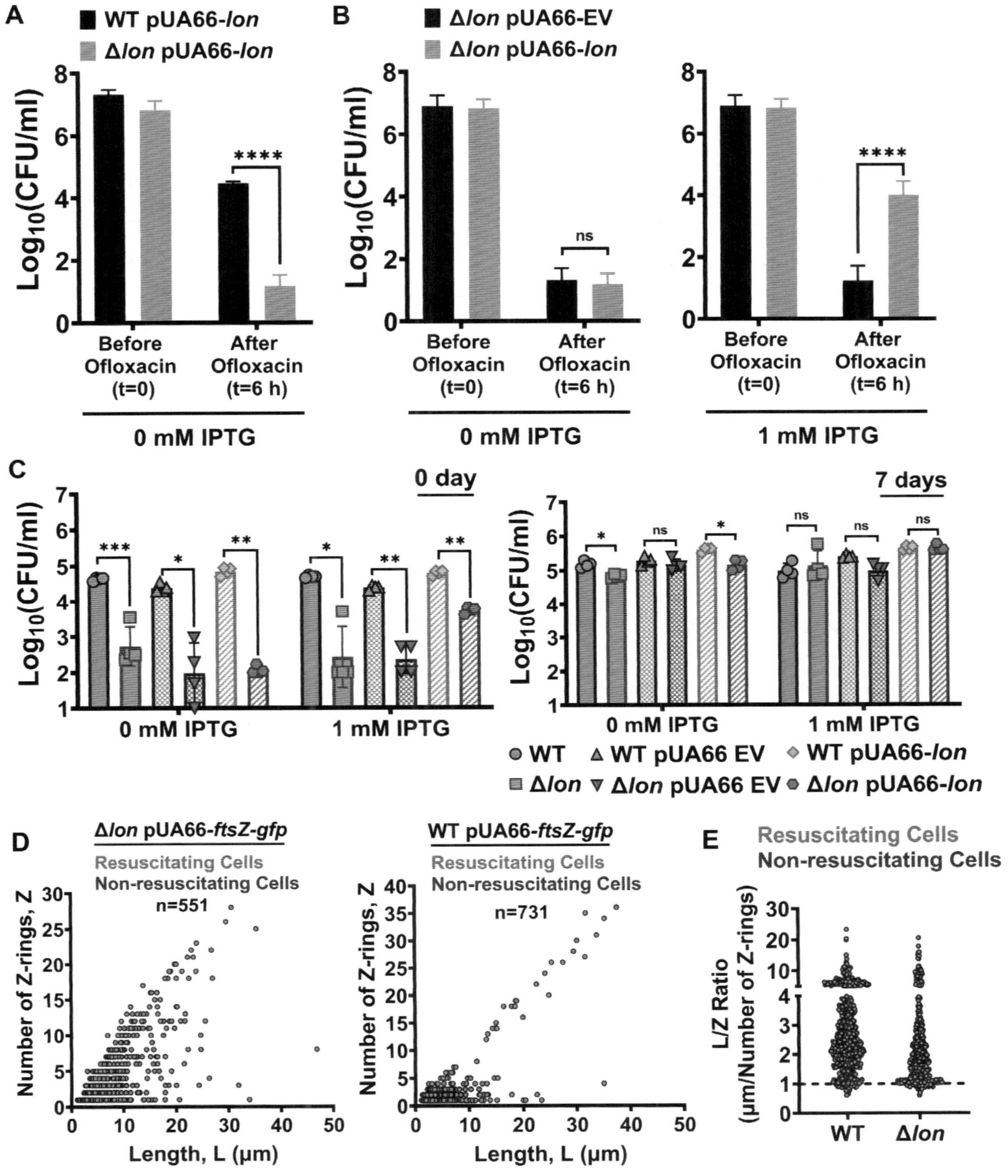

A
WT pUA66-lon
Δlon pUA66-lon
Log10(CFU/ml)
****
Before Ofloxacin (t=0)
After Ofloxacin (t=6 h)
0 mM IPTG
B
Δlon pUA66-EV
Δlon pUA66-lon
ns
0 mM IPTG
1 mM IPTG
C
0 day
7 days
0 mM IPTG
1 mM IPTG
WT
WT pUA66 EV
WT pUA66-lon
Δlon
Δlon pUA66 EV
Δlon pUA66-lon
D
Δlon pUA66-ftsZ-gfp
WT pUA66-ftsZ-gfp
Resuscitating Cells
Non-resuscitating Cells
n=551
n=731
Number of Z-rings, Z
Length, L (μm)
E
Resuscitating Cells
Non-resuscitating Cells
L/Z Ratio (μm/Number of Z-rings)
WT
Δlon

58f Investigating Bacterial Biofilm Infection in a 3D Bovine Airway Model

Neeti Gandhi, Padmavathy Rajagopalan

Introduction

Bovine Respiratory Disease (BRD) is a multifactorial condition where different genera of bacteria, such as Mannheimia haemolytica (M. Haemolytica), Histophilus somni (H. Somni), Pasteurella multocida, Mycoplasma bovis, and viruses, like bovine respiratory syncytial virus, bovine viral diarrhea virus, and bovine herpes virus-1, infect the respiratory system of cattle. This condition causes severe illness or death within livestock, resulting in a loss of approximately $1 billion to the cattle industry annually. Because of the multiple pathogens involved in this disease, antimicrobials and vaccines are difficult to administer.

Current models of the bovine airway system are primary submerged tissue cultures (STC) and air-liquid interface (ALI) models. STCs involve a monolayer, most commonly bovine epithelial cells, submerged in a bacterial concentrated cultured supernatant (CCS). ALIs are seeded on Transwell® plates where cells are exposed to the environment on the apical side, which allows cells to differentiate into goblet cells and ciliated epithelial cells. However, this process can take anywhere from 14 to 42 days. In order to recapitulate the bovine respiratory tract, we have designed a novel 3D co-culture model that incorporates bovine epithelial (B-Epis) and airway endothelial cells (BAECs). We focus our analysis on cultures exposed to CCS from H. Somni to understand signaling factors that could contribute to the pathogenesis of BRD.

Methods

The assembly of the co-culture 3D models began on day 1 and finished on day 2. In general, on day 1, B-Epis were encapsulated in a 1.1 mg/ mL type 1 collagen gel and seeded on a well-plate. On day 2, BAECs were encapsulated at a ratio of 3 B-Epis : 1 BAEC in type 1 collagen and then seeded on top of the B-Epis layer. B-Epis and BAEC monocultures served as controls. Three media conditions were then administered to all cultures on day 3, 24h after co-culture assembly, which included (1) cell media (or optimized media for co-cultures) (2) 50% cell media and 50% H. Somni CCS (50/50) and (3) H. Somni CCS. Cultures were ended for DNA analysis at 24h (day 4), 72h (day 6), and 108h (day 8) after CCS administration. For the cultures ended on day 6 and day 8, five media conditions were continued, which included (1) cell media, (2) 50/50 revert, where cell media was administered from day 4 onwards, (3) 50/50 (continuous), (4) H. Somni CCS revert, where cell media was also administered from day 4 onwards, and (5) H. Somni CCS (continuous).

Results

DNA analysis indicates that on day 8, B-Epi monocultures with cell media exhibited a 58% increase in cell proliferation compared to B-Epi monoculture controls at day 4. Co-cultures with cell media displayed a 73% increase in DNA on day 8 compared to co-cultures on day 4. However, both B-Epi monocultures and co-cultures with continuous H. Somni CCS administration do not exhibit significant changes in DNA from day 4 to day 8. B-Epi monocultures show a 67% increase in cell proliferation in 50/ 50 revert cultures on day 8 compared to 50/50 cultures on Day 4. Co-cultures also displayed a 73% increase in cell proliferation in H. Somni CCS revert cultures on day 8 compared to H. Somni CCS cultures on day 4. However, BAECs do not exhibit statistically significant changes in DNA under 50/50 and

H. Somni CCS revert and continuous conditions from day 4 to day 8.

Conclusions and Future Work

H. Somni CCS results in cell death in B-Epi only and co-culture models. However, B-Epi monocultures and co-cultures exhibit cell proliferation in revert 50/50 media and revert H. Somni CCS cultures, potentially indicating cell recovery after initial bacterial media administration. However, BAECs do not display cell death upon H. Somni CCS administration. Further characterization of the monoculture and co-culture models will include examining changes in the epithelial barrier and tight junctions, as well as cytokine secretion with H. Somni CCS administration. Future investigations of bacterial biofilm interactions will expand to M. Haemolytica administration.

58g ENGINEERED POLYSACCHARIDE-BASED NANOMATERIALS FOR TISSUE SEALING, REPAIR, AND ENDOSCOPIC IMAGING

Kaushal Rege

Early detection of abnormalities and effective repair of tissues is critical for improving health outcomes of patients. Amine-containing small-molecules and polysaccharides have been investigated in therapeutic applications and drug delivery because of their biological activity and biocompatibility. We have extensively used small-molecule aminoglycoside antibiotics as substrates for generating polymers for therapeutic delivery, hydrogels for 3D organoid culture, and microparticles for applications in nucleic acid biotechnology. In this presentation, I will discuss our more recent work on engineering polysaccharide-derived injectable formulations for endoscopic marking of gastrointestinal lesions. Endoscopic tattooing is commonly used to mark suspect regions, including polyps, in colorectal tissue in vivo. The one product currently approved by the FDA for endoscopic marking diffuses significantly away from the injection site and demonstrates poor localization, can cause fibrosis, and is not amenable to multimodal imaging. We have developed a set of injectable nano- and micro-scale formulations based on engineered polysaccharides for endoscopic tattooing with application in detection and marking of colorectal pathologies. These formulations demonstrated excellent spot localization, high contrast under endoscopic light, high biocompatibility, and potential for multi-modal imaging. Several cases of GI trauma and disease necessitate surgical intervention, but leakage after surgery is a common occurrence and results in high morbidity, need for additional interventions and increased hospitalizations. We developed an intestinal leakage model in live rats and our results demonstrate that laser sealing using polysaccharide-based laser-activated sealants (LASEs) significantly enhanced biomechanical recovery of intestinal tissues ex vivo and survival in vivo. Reinforcement using LASEs is therefore an attractive strategy for minimizing leakage-associated adverse events in GI trauma. Our results demonstrate that engineered polysaccharide-derived nanomaterials have significant promise for clinical translation in endoscopic marking and for tissue sealing and repair in the GI tract.

19a Discovery and Characterization of a Novel Lasso Peptide Using a Guilt By Association Approach

Drew Carson, A. James Link

With the rise in resistant bacterial infections jeopardizing the efficacy of the antibiotics developed in the 21st century, new compounds are needed to combat this potential loss of life that was once treatable. Natural products present us with a plethora of novel compounds that can be harnessed for medicinal purposes. Lasso peptides are a member of the ribosomally synthesized and post-translationally modified peptide class of natural products, and much of the interest on these natural products has focused on their unique lariat-knot shape and narrow-spectrum antimicrobial activity against clinically relevant pathogens.[1,2] Genome mining approaches can be focused on organisms that are "pathogen adjacent" in hopes of discovering lasso peptides with antimicrobial activity against a pathogen of interest.[3-5] This talk will focus on an example of a successful genome mining campaign of this type that resulted in a novel lasso peptide with potent antimicrobial activity. This talk will focus on the solution structure of this novel lasso peptide as determined by 2D NMR experiments as well as a structure-function analysis carried out via mutagenesis of the lasso peptide. These mutagenesis experiments reveal interesting behavior regarding the stability of the lasso peptide. The breadth of the antimicrobial spectrum of activity of this peptide will be discussed as well as its potential in treating nosocomial infections.

1. Montalbán-López M, Scott TA, Ramesh S, Rahman IR, van Heel A, Viel JH, Bandarian V, Dittmann E, Genilloud O, Goto Y, Grande Burgos MJ, Hill C, Kim S, Koehnke J, Latham JA, Link AJ, Martínez B, Nair SK, Nicolet Y, Rebuffat S, Sahl HG, Sareen D, Schmidt EW, Schmitt L, Severinov K, Süssmuth RD, Truman AW, Wang H, Weng JK, van Wezel GP, Zhang Q, Zhong J, Piel J, Mitchell DA, Kuipers OP, van der Donk, WA: New developments in RiPP discovery, enzymology and engineering. *Natural Product Reports* 2021, **38**: 130-239.

2. Tan S, Moore G, Nodwell J: Put a Bow on It: Knotted Antibiotics Take Center Stage. *Antibiotics (Basel, Switzerland)* 2019, **8**(3): 117

1. Cheung-Lee WL, Link AJ: Genome mining for lasso peptides: past, present, and future. *Journal of Industrial Microbiology andamp; Biotechnology* 2019, **46**: 1371-1379.
2. Cheung-Lee WL, Parry ME, Zong C, Jaramillo Cartagena A, Darst SA, Connell ND, Russo R, Link AJ: Discovery of ubonodin, an antimicrobial lasso peptide active against members of the *Burkholderia cepacia* complex. *ChemBioChem* 2020, **21**: 1335-1340.

5. Metelev M, Arseniev A, Bushin L, Kuznedelov K, Artamonova TO, Kondratenko R, Khodorkovskii M, Seyedsayamdost MR, Severinov K: Acinetodin and Klebsidin, RNA Polymerase Targeting Lasso Peptides Produced by Human Isolates of *Acinetobacter gyllenbergii* and *Klebsiella pneumoniae. ACS Chem. Biol.* 2017, **12**: 814-824

19b Rational Design of an Adenosine-Degrading Enzyme for Therapeutic Administration in Tumor Mouse Models

Maria Rain Jennings, John Blazeck

Adenosine serves important functions in the body such as inducing sleep and calming immune cells to allow healing in damaged tissues. When cancer cells in tumors upregulate the machinery to produce adenosine, local concentrations of the molecule can reach a 1000-fold increase relative to normal tissue. Such high adenosine concentrations deactivate immune cells that may otherwise recognize and destroy the cancerous cells. An adenosine-degrading enzyme, ADOase, breaks down adenosine with a high catalytic efficiency on the order of 10^6 $M^{-1}s^{-1}$. We have shown that administration of ADOase to a colorectal cancer mouse model slowed tumor growth (Fig 1A) and delayed mortality significantly (Fig 1B). However, ADOase loses ~50% of its activity within eight hours at physiological conditions (Fig 1C). We solved the crystal structure of our enzyme (Fig 1D), which allowed us to identify structural motifs that could contribute to its instability through comparison to enzymes that do not deactivate at physiological conditions. Structural and sequence alignment of ADOase with similar enzymes demonstrated regions conserved amongst more resilient proteins that differed from the associated ADOase sequences. We thus rationally determined particular amino acid positions within ADOase's makeup for mutagenesis that could potentially curb the enzyme's tendency to deactivate. To streamline and hasten evaluation of each variant for activity retention, we developed an *Escherichia coli* lysate-based assay to allow characterization of the stability of enzyme variants in a more high throughput manner, allowing us to isolate variants with improved stability after 8 hours (up to ~90% activity retained) at physiological conditions (Fig 1E), and serendipitously, with 10-fold improved yield retained from affinity chromatography purifications relative to the wild-type enzyme (Fig 1F). Armed with a mutant form of ADOase less likely to degrade at physiological conditions, we now plan to determine the therapeutic efficacy and pharmacokinetic/pharmacodynamic profile of the mutant compared to that of the original enzyme in tumor mouse models. In summary, we detail an engineering workflow to identify enzyme mutants with reduced tendency to deactivate at a given temperature that can be customized to meet the needs of other protein engineers.

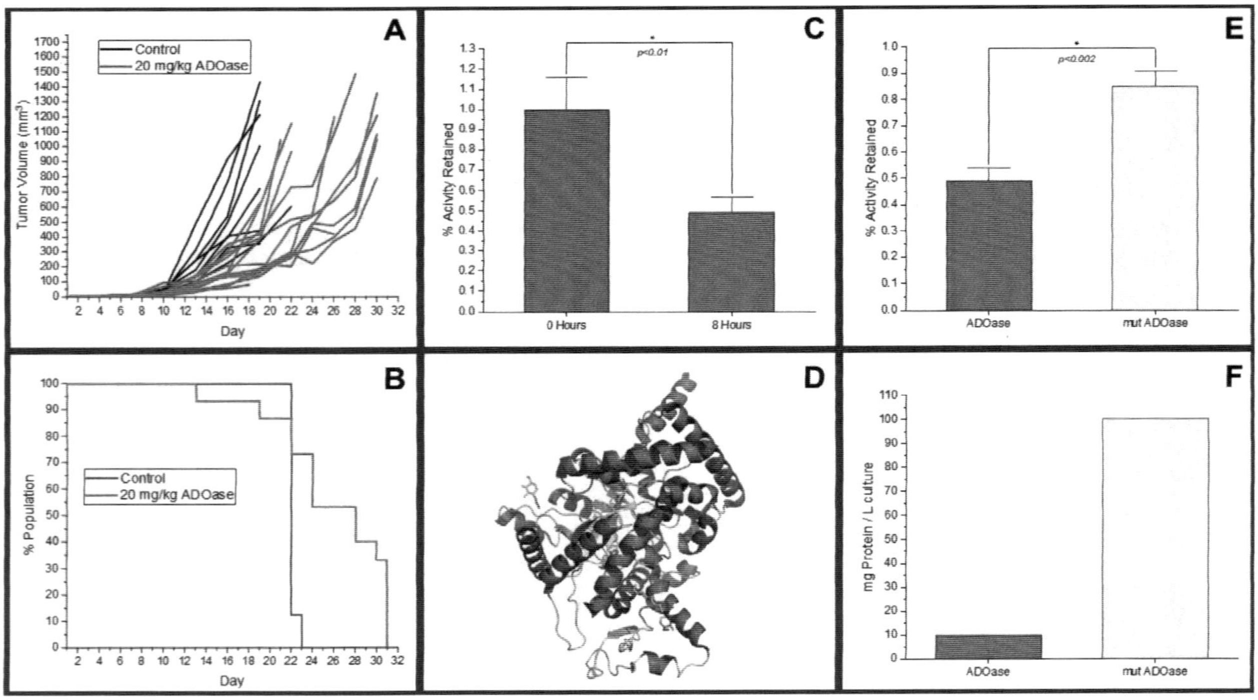

19c Peptide Aggregation Induced Immunogenic Rupture (PAIIR)

Gokhan Gunay, Seren Hamsici, Handan Acar

Peptide Aggregation Induced Immunogenic Rupture (PAIIR)

Gokhan Gunay M.Sc., PhD Candidate: Gokhan.Gunay-1@ou.edu

Immunogenic cell death (ICD) involves damage to the cell membrane and release of damage associated molecular patterns (DAMPs), molecules that can engage innate immune cells and are natural adjuvants for inducing immune response. In the presence of an antigen, released DAMPs can be a decisive factor regarding the type and magnitude of the immune response, and therefore the longevity and efficacy of an antigen-specific immunity. The role of ICD and DAMP release has been studied in diverse set of applications such as vaccine design and immunotherapies. Introducing a co-assembling oppositely charged peptides (CoOP) strategy[1] we have shown a controllable peptide aggregation for the induction of ICD. We have used this strategy to design a vaccine against influenza hemagglutinin antigens[2] as well as model antigen ovalbumin (OVA). Our results show that aggregation induced ICD and DAMP release is comparable and further activates bone marrow derived dendritic cells into a pro-inflammatory state through the release of DAMPs. This system will further be used in an in vivo tumor model and would represent a strategy to engage ICD and DAMP release for a diverse set of applications.

Figure 1 Peptide aggregation induced immunogenic rupture (PAIIR) is a platform engaging damage-associated molecular patterns for robust antigen-specific immunity.

References

1. Hamsici, S., White, A. D. andamp; Acar, H. Peptide framework for screening the effects of amino acids on assembly. *Science Advances* (2022) doi:10.1126/sciadv.abj0305.
2. Gunay, G. *et al.* Peptide Aggregation Induced Immunogenic Rupture (PAIIR). *BioRxiv* **472230**, (2021).

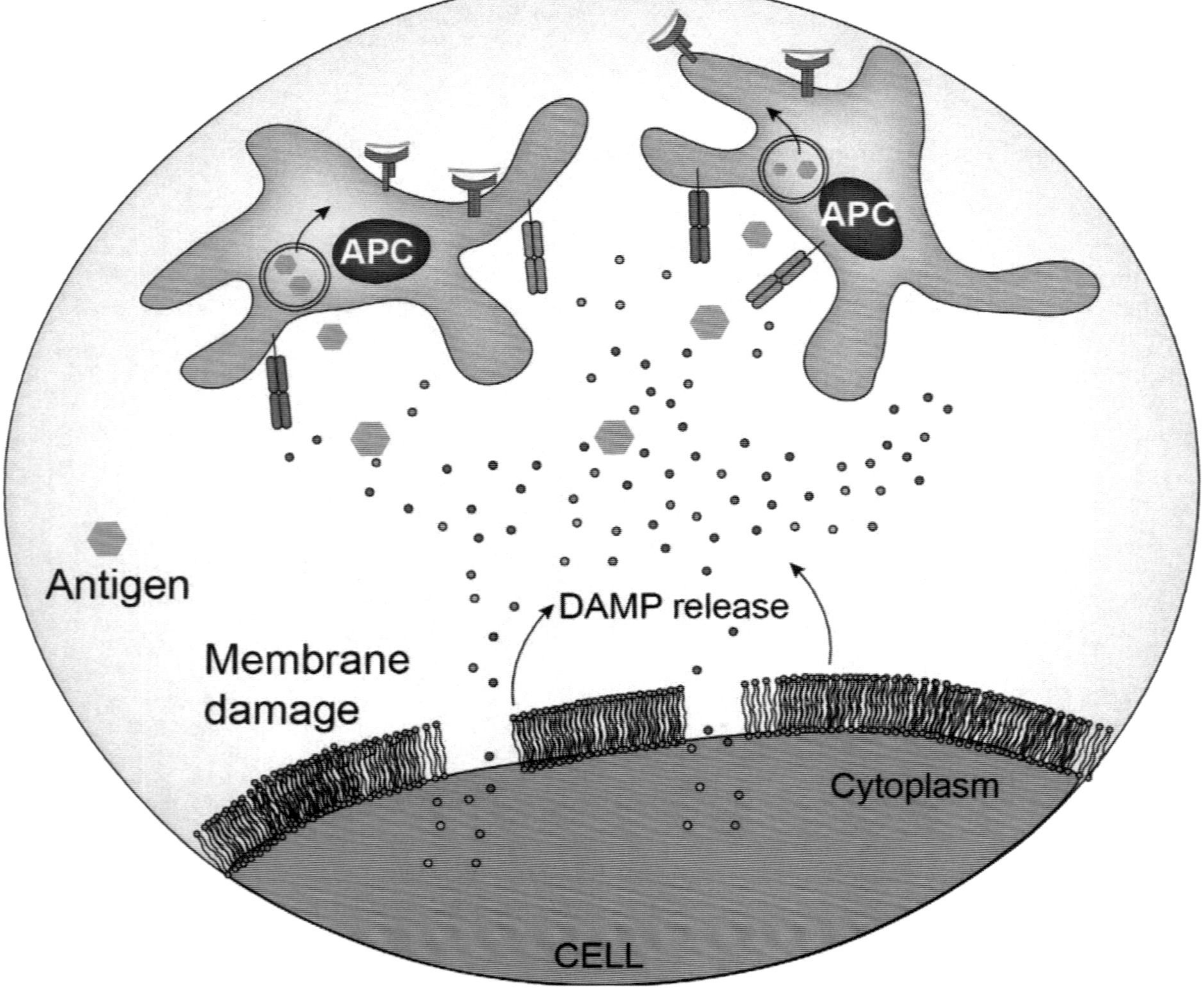
APC
APC
Antigen
DAMP release
Membrane
damage
Cytoplasm
CELL

19d Recombinant Serine Protease Inhibitors (serpins) for Tissue Repair

Jordan Yaron, Jayda Hylton-Pelaia, Jordan Roberts, Holly Gildar, Jacquelyn Kilbourne, Liqiang Zhang, Alexandra Lucas, Kaushal Rege

Purpose/Objectives: Cutaneous wound repair is a major global economic and medical burden characterized by several continuous, overlapping phases. Delay in the onset or resolution of any phase leads to impaired healing, as occurs with complex comorbidities such as diabetes. Bioactive treatments to modulate these phases offer the potential to improve healing and to enhance the performance of biomaterial solutions for wound repair. Re-epithelialization and appropriate immune response in excisional and incisional wounds are critical early steps in healing. Re-epithelialization describes the process of epidermal keratinocytes undergoing a partial epithelial-to-mesenchymal transition (EMT) to mobilize and is a well-recognized characteristic of non-healing diabetic wounds and a goal for enhanced tissue regeneration. Sustained and inappropriate immune response to wound healing is also known to convert chronic wounds to acute wounds. There is growing interest in precisely defining the cellular and molecular mechanisms of repair and treatments to specifically augment re-epithelialization and immune response to wounds are under investigation. Serp-1 is a *Myxoma virus*-derived serine protease inhibitor (serpin) deployed as an immune evasion protein which demonstrates potent immune modulating activities when utilized as a recombinant protein therapeutic. Serpinb3a (paralog of human SerpinB3/B4) is a cross-class serpin canonically associated with cancer EMT, but which has not yet been investigated in wound healing. Here, we investigate the ability for Serp-1 and Serpinb3a to therapeutically augment healing in two different models of wound repair: granulation-mediated (second intention) excisional wound repair and light-activated sealant (LASE) biomaterial-mediated incisional wound closure.

Methodology: For Serpinb3a studies: NCBI Gene Expression Omnibus (GEO) was queried for relevant publicly available datasets and GEO2R was used to evaluate gene expression levels of Serpinb3a in acute and diabetic wounds. Western blot for Serpinb3a and immunohistochemistry for E-cadherin and Slug was performed on healing skin of immunocompetent Balb/c mice with 5-mm full-thickness splinted wounds at 6 hrs and 2, 4 and 7 days post-wounding. Full-length Serpinb3a with a C-terminal 6x-Histag was expressed in E. coli BL21(DE3) and purified using Ni-NTA IMAC and polished by Triton X-114 to endotoxin-free levels. Serpinb3a identity was confirmed by immunoblot and serpin inhibitory function was validated against purified human neutrophil Cathepsin G. *In vitro* closure was measured by HaCaT scratch assay and compared to EGF. Cell proliferation was measured by PrestoBlue assay. Therapeutic efficacy of Serpinb3a was evaluated in 5-mm full-thickness splinted wounds in C57BL6/J and diabetic and obese BKS.Cg-Dock7^{m} +/+ Lepr^{db}/J (db/db) mice. Wound closure rate was evaluated by planimetry. For Serp-1 studies: Light-activated sealants (LASE) were generated by solvent evaporation of aqueous solutions of *Bombyx mori* cocoon silk fibroin impregnated with indocyanine green dye (ICG). Serp-1 was recombinantly produced in CHO-K1 cells. Incisional wounds were treated with sutures or LASE activated by a handheld 808 nm laser and treated with saline or with recombinant Serp-1. Barrier function recovery was measured by trans-epidermal water loss (TEWL) with a VapoMeter.

Results: NCBI GEO analysis indicated evidence for skin-specific expression of Serpinb3a after dermal wounding, but not mucosal (tongue) wounding, and further suggested that Serpinb3a gene expression was lower in migrating keratinocytes at the wound edge of db/db but not C57BL6/J mice. Immunoblot analysis of acute wounds in Balb/c mice demonstrated high levels of Serpinb3a within 6 hours of wounding, peaking at 2-4 days and resolving by day 7 post-wounding. Importantly, proteins levels of Serpinb3a correlated tightly with immunohistochemical markers of EMT (E-cadherin loss and nuclear localization of Slug) in keratinocytes at the healing wound edge. Treatment of HaCaT cells with Serpinb3a dose-dependently reduced proliferation and significantly accelerated wound closure compared to controls and at levels comparable to EGF. Functional recombinant Serpinb3a was topically applied to the wound bed of healing acute and diabetic wounds. Significant acceleration of early closure, during the epidermal EMT phase of healing, was observed with exogenous Serpinb3a in both C57BL6/J and db/db mice. When co-delivered during laser tissue welding of incisional wounds with ICG-loaded LASE sealants, MYXV Serp-1 enhanced barrier function recovery nearing equivalence to intact tissue as early as 1-day post-incision.

Conclusion/Significance: Serpins represent a novel class of recombinant therapeutic modulators for wound healing and tissue repair. Serpinb3a is a novel endogenous mediator of wound healing in the skin and therapeutic administration accelerates poorly healing wounds. Serp-1 is a potent virus-derived immune modulator which enhances functional recovery of biomaterial-sealed skin injuries. Topical or controlled delivery of these recombinant serpins may provide new treatment approaches for tissue repair and regeneration.

19e Generating Extended Grnas with Hairpins for Increased CRISPR Editing Specificity

Hillary Dimig, Ashley Herring-Nicholas, Eric Josephs, Miranda Roesing

CRISPR, or clustered regularly interspaced short palindromic repeats, systems have gained notoriety for their gene editing ability. Clinical trials using CRISPR systems are currently underway for several blood diseases that are difficult to treat otherwise including sickle cell disease, inherited eye disease, and other conditions. However, off-target editing is still a concern. For example, for *EMX1*, a well-characterized clinically relevant gene, on-target editing accounted for 4,521 events, while the most common off-target accounted for 3,123 editing events. Extensions of the 5' sgRNA with secondary structure have been shown to increase specificity by on average 55-fold, but there is no good method for designing these extensions. We developed a method for generating guide RNAs with 5' extensions that result in high on-target efficiency and low off-target editing. For *EMX1*, this method was used to generate sgRNAs that were able to reduce off-target cleavage at the top four most common off-targets while maintaining on-target cleavage efficiency of the wild type sgRNA. In most cases no cleavage was observed at these off-target sites.

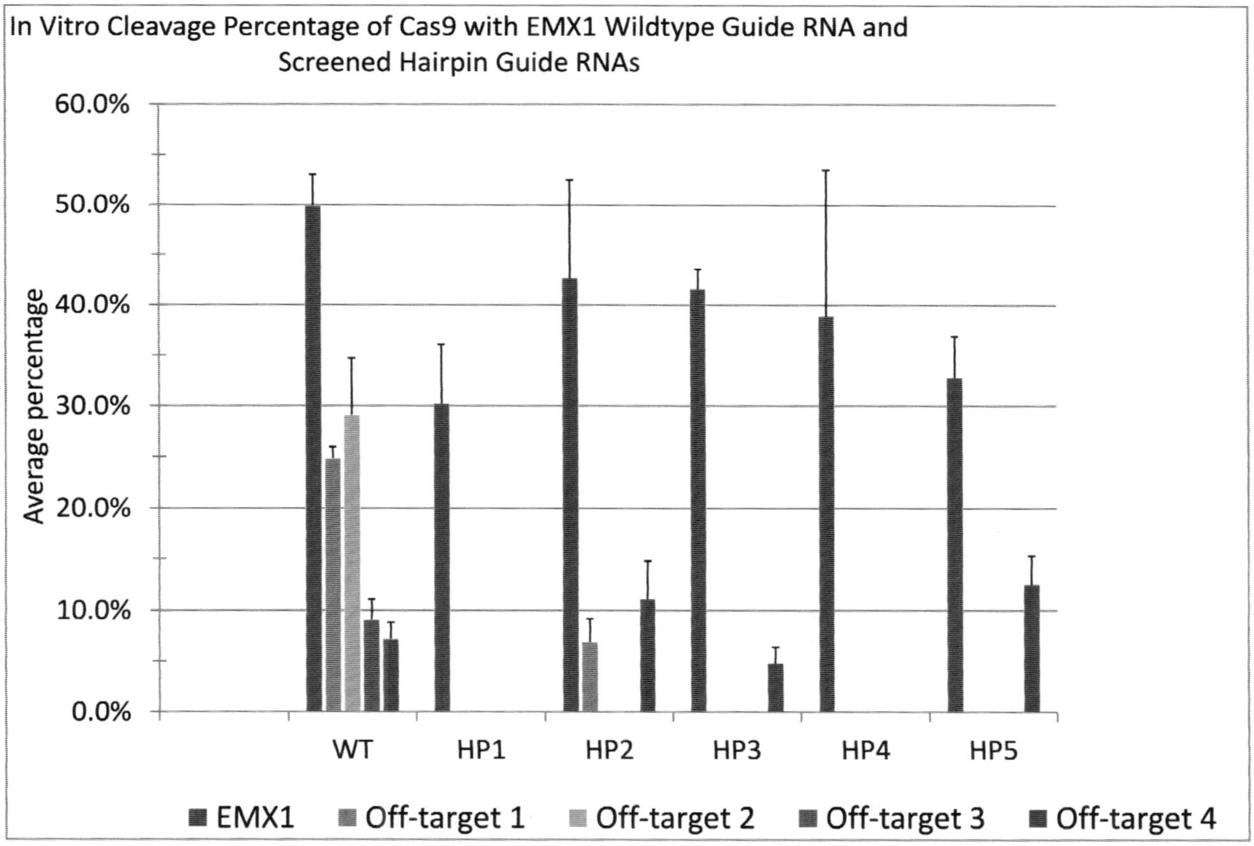

19f Engineering Improved Crispr Repressors for Targeted Mammalian Gene Regulation

Andrew Kristof, John Blazeck

The ability to precisely silence endogenous gene expression is vital for performing robust whole-genome screens and discovering non-coding transcriptional regulatory motifs in mammalian cells. CRISPR interference (CRISPRi) technology has emerged as a powerful tool enabling site-specific transcriptional repression for these applications. The CRISPRi platform employs the RNA-guided, catalytically inactive endonuclease, dCas9, to deliver conjoined transcriptional repressor domains near to the transcription start site of target genes. Almost all CRISPRi systems implement direct fusions of dCas9 with the Krüppel-associated box (KRAB) domain from the human protein KOX1, often referred to only as KRAB, but here called KOX1-KRAB to prevent ambiguity. Previous studies have demonstrated improved CRISPRi transcriptional repression by combining KOX1-KRAB with additional repressor domains or by utilizing KRAB domains from other human proteins. Despite this progress, the CRISPRi platform still suffers from incomplete repression and high performance variability across cell lines, as well as a general overreliance on KOX1-KRAB.

In order to improve the CRISPRi platform in mammalian cells, we sought to engineer more efficient repressors by discovering novel dCas9-repressor fusions, characterizing their transcriptional mechanisms of action, and incorporating this insight to construct highly efficient repressors that incorporate multiple modalities to silence local gene expression. We first screened single putative transcriptional repressor domains using a reporter assay in HEK293T cells and identified eleven novel protein domains compatible with the CRISPRi system, including one domain that outperformed KOX1-KRAB (**Figure 1**).

Next, we computationally quantified gene expression patterns of regulatory co-factors that interact with each novel repressor domain (as well as KOX1-KRAB) and discovered that numerous interaction partners for repressor domains have highly variable expression levels across human cell lines. To further investigate the generalizability of our novel repressors, we compared performance of individual repressors across three diverse human cell lines, and we are further assembling and screening bipartite repressor domains to evaluate potential synergistic activity when two domains, with diverse interaction co-factors, are directly fused together in conjunction with dCas9. We envision these efforts will mark a significant step forward in applying the CRISPRi platform to regulate transcription in human cells from diverse lineages.

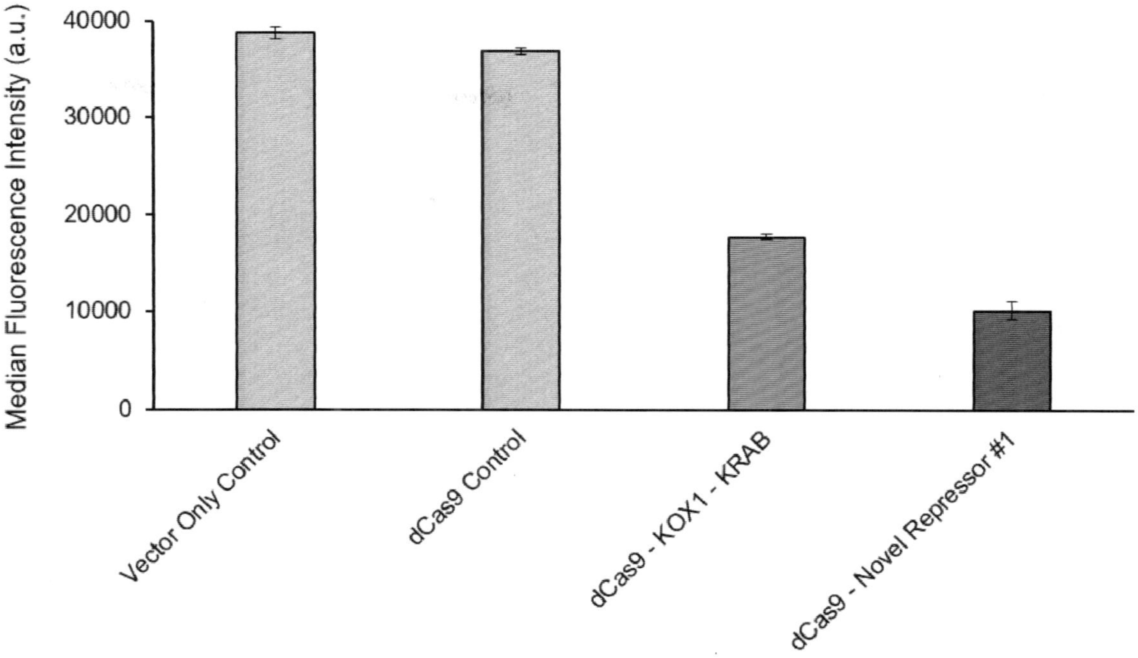

Figure 1: Characterization of novel dCas9-repressor fusions for CRISPRi applications. Putative repressor domains (blue bar) were screened by transfecting dCas9-repressor fusion constructs into HEK293T-EGFP reporter cell lines and measuring EGFP fluorescence levels 72 hours post-transfection using flow cytometry. Widely adopted repressor domain KOX1-KRAB is included (orange bar) for reference. In addition, 10 novel repressors performed as well or better than the commonly utilized MeCP2 protein (data not shown). All repressor domains were fused to the C-terminus of dCas9. n = 3 biologically independent samples (cell cultures).

19g Invited Talk: Engineering Specificity in Next-Generation Immunotherapies

Larry Stern

Engineered proteins have had profound impacts in the clinical setting. This class of molecules is vast, including a wide variety of functionalities, such as 1) synthetic receptors that turn engineered inputs into augmented natural outputs, 2) inhibitory proteins that block cell-cell and cell-ligand interactions from affecting cell behavior, and 3) molecular imaging agents capable of detecting disease with drastically improved sensitivity and specificity compared to conventional means. Although these advances have been great, many challenges remain in the pipeline of engineering new proteins to meet clinical demands. Chief among these challenges is the engineering of new proteins with specific binding function, such as selectivity for individual protein family members or the ability to inhibit protein-protein interactions. In this talk, I will discuss efforts to overcome each of these challenges and applications of the described technologies. First, I will describe the engineering of new chimeric antigen receptors (CARs) that use mutated IL13 variants to selectively target cancer cells overexpressing the cancer-testis antigen IL13Rα2 while sparing tissue expressing the ubiquitous receptor IL13Rα1. Through stringent in vitro and in vivo experiments, we studied the levels of binding affinity required to achieve this selectivity, and validated these results using in vivo CAR T cell trafficking studies to understand how CAR binding affinity influences CAR T cell localization. Second, I will discuss my lab's recent efforts in developing a novel yeast surface display platform that uses inhibition rather than target binding as a selective pressure for ligand selection. We have demonstrated the ability to display a natural receptor extracellular domain or natural ligand simultaneously with a candidate inhibitory protein on the yeast surface. Application of an EGFR-Erbitux model system leads to robust detection of the exclusion of a titrated soluble competitor. Recent progress in further optimizing this system for affinity ranges and construct architecture will be discussed.

316b N-Terminal Engineering Improves the Quality of Recombinant Proteins Secreted from Yeast

Neil C. Dalvie, Christopher Naranjo, Sergio Rodriguez Aponte, Ryan Johnston, J. Christopher Love

There is a global need to increase the capacity for manufacturing recombinant proteins for food, industrial enzymes, and therapeutics. The cost and complexity of protein manufacturing can be reduced by secretion of the recombinant product into the extracellular medium during fermentation. Secreting cells can be retained in the fermentation to enable continuous upstream processes, and downstream purification of the product from culture supernatant is preferable to cell lysates. Manufacturing cell lines, however, may secrete recombinant products with unwanted modifications or variations in primary, secondary, or tertiary structure. Product-related variants are especially problematic for therapeutic proteins because they may impact the function or safety of a medical intervention. Specifically, product molecules with *N*-terminal truncation or extension, proteolytic cleavage, or N- or O-linked glycosylation may be non-functional, immunogenic, prone to aggregation, or unstable in formulation. These product-related variants pose challenges in downstream purification because they may not be removed by affinity chromatography and typically require additional unit operations like ion exchange or hydrophobic interaction chromatography. Development and manufacturing costs can be reduced, therefore, by design of upstream processes that produce homogenous, properly modified recombinant proteins.

The yeast *Komagataella phaffii* (*Pichia pastoris*) is a common alternative host with high potential for low-cost manufacturing of therapeutic proteins like vaccine antigens and monoclonal antibodies. *K. phaffii* has a productive, highly developed secretory pathway, in addition to fast growth to high cell densities, and simple genetic manipulation. In this host, secretion of a recombinant protein requires genetic attachment of a signal peptide to facilitate translocation into the endoplasmic reticulum, post-translational modification, and ultimately packing into secretory vesicles. The most commonly used signal peptide in recombinant *K. phaffii* is the α-mating factor signal peptide (αSP) from *Saccharomyces cerevisiae*, because it typically yields the highest secreted titer of the protein of interest. Processing of the αSP, however, may depend on the recombinant protein, and improper cleavage of the αSP can result in product-related variants.

Here, we elucidate the impact of signal peptide processing on the quality and secreted titer of several recombinant vaccine antigens. We observed that an aggregated product-related variant of the SARS-CoV-2 receptor binding domain (RBD) is due to *N*-terminal extension from incomplete processing of the αSP. We eliminated this product variant and increased secreted titers by steric extension of the *N*-terminus of the RBD by a functional peptide or addition of one or more *N*-terminal amino acid residues. We also demonstrated that this strategy improves the quality of three other subunit antigens from a trivalent vaccine for rotavirus. Finally, we applied *N*-terminal engineering to a monoclonal antibody and observed the same improvement in product quality.

These results together suggest that processing of the signal peptide is a critical determinant of the quality of secreted recombinant proteins. To date, wide adoption of alternative manufacturing hosts like *K. phaffii* has been limited by product-related variants of unique proteins like subunit vaccine antigens, and by improper processing and modification of complex intravenously delivered products like monoclonal antibodies. The framework presented here will enable reliable secretion of a wide range of recombinant proteins with a single signal peptide, which should accelerate the development of new, low-cost manufacturing processes.

316c Antioxidant-Immobilized Backpack-Carrying Monocytes for TBI Therapy

Rick Liao, Samir Mitragotri

Introduction: Traumatic brain injury (TBI) afflicts around 3 million people in the United States annually, with around 55,000 fatal cases. Furthermore, moderate or severe TBI can cause long-term disabilities including difficulty in sensorimotor function, memory issues, depression, and dementia. There are no clinically approved therapeutics to reduce TBI damage from the initial insult or secondary injury. Due to the lifelong impact of TBI, there is a strong clinical need to develop therapeutics that mitigate TBI damage. TBI damage encompasses the initial impact trauma and secondary brain injury. Elevated oxidative stress contributes a pivotal role in secondary brain injury, exacerbating excitotoxicity, blood-brain barrier (BBB) and blood-cerebrospinal fluid-barrier (BCSFB) dysfunction, neuroinflammation, and ultimately cell death in TBI. Thus, the antioxidant enzyme catalase is a promising therapeutic for mitigating TBI secondary injury. However, effective enzyme delivery is hindered by poor penetration and accumulation in diseased brain parenchyma and enzyme degradation from extracellular proteases. To overcome these obstacles, we have implemented a novel approach of immobilizing catalase into discoidal hydrogel scaffolds termed backpacks and attaching these backpacks onto monocytes to leverage chemotactically targeted therapeutic delivery to the diseased brain after TBI.

Materials and Methods: We formulated hydrogel backpacks composed of 250kDa hyaluronic acid-methacrylate (HA-MA), 1kDa 2-arm poly(ethylene glycol) methacrylate (PEGDMA), Irgacure 2959 (I2959) photoinitiator, and catalase. We converted free primary amines on catalase to sulfhydryl groups with Traut's reagent to form immobilization reaction sites. For backpack synthesis, we spin-coated our hydrogel solution onto polydimethylsiloxane (PDMS) elastomer molds with 8 μm wells. Upon 365 nm irradiation with I2959 photo-initiation, HAMA and PEGDMA pendant MA groups covalently conjugate together via free radical polymerization to form HAMA/PEGDMA hydrogels, while MA groups also react with catalase sulfhydryl groups via the thiol-ene Michael addition-based click reaction for immobilization. Subsequent O_2 plasma etching to remove the thin hydrogel film and printing yields uniform 8 μm hydrogel backpacks. With hydrogen peroxide (H_2O_2) as the substrate for catalase, we measured catalase activity with kinetic 240nm spectrophotometric detection of H_2O_2 decomposition and evaluated catalase activity in the presence of degradative proteases over time. We isolated bone marrow cells from the tibias and femurs of C57BL/6J mice and cultured them with macrophage colony stimulating factor (MCSF)-containing media on ultra-low adhesion plates to obtain monocytes. After backpack incubation with our primary mouse monocytes, we obtained ABC-monocytes.

Results: Catalase modification with Traut's reagent was non-denaturing, resulting in enzyme activity retention of 89.53%. Starting at 100% at time 0h, upon incubation with degradative proteases, freely suspended catalase rapidly degraded to 4.7% activity in 2h. Unmodified catalase loaded within backpacks exhibited 24.6% activity at 2h and decreased to 2.5% activity at 24h. On the other hand, Traut's reagent-modified catalase immobilized within backpacks exhibited 42.3% activity at 2h, 39.6% activity at 24h, and 19.8% activity at 21 days before reducing to 0.0% activity at 28 days. We achieved backpack binding on 56.2% of mouse monocytes with a 3:1 catalase-immobilized backpack to mouse monocyte ratio.

Conclusions: Our work demonstrates that by immobilizing catalase within hydrogel backpacks to prevent release, the hydrogel backpack scaffold sterically protects catalase from environmental protease degradation while permitting H O diffusion for retained enzyme activity. This antioxidant

2 2

immobilization strategy extends catalase activity from less than 1 day to over 21 days. Furthermore, our hydrogel backpacks bind onto the surface of monocytes without phagocytosis, enabling continuous extracellular H_2O_2 scavenging. Antioxidant-immobilized backpack-carrying monocytes are a promising neurotherapeutic strategy for reducing oxidative stress to ultimately mitigate TBI secondary injury.

316d A Genetic Toolkit for Anaerobic Gut Fungi

Tejas Navaratna, Michelle O'Malley

Anaerobic fungi, found in the rumen of herbivores, are powerful degraders of lignocellulosic biomass and are equipped with a broad suite of enzymes to accomplish this goal. These organisms are found in diverse multi-kingdom communities and have been shown to produce a range of biotechnologically valuable products, including short- and medium-chain fatty acids and bioactive small molecules. However, a major bottleneck remains in the translation of anaerobic fungi to the industrial setting as there is a lack of robust genetic engineering strategies to direct metabolic flux(es) towards desired products, generate protein diversity, and introduce heterologous genes.

We have built and characterized a diverse series of vectors with promoters driving the expression of anaerobe-compatible flavin-based fluorescent proteins in a transient expression system and are advancing the most promising candidates for genomic integration using CRISPR and transposon-based engineering. By mining the transcriptomes across many genera of anaerobic fungi, we identified putatively active mobile genetic elements, consisting of long terminal repeats (LTRs) that flank protein machinery for reverse transcription and genomic integration. We have co-opted this system for heterologous gene insertion by constructing a library of promoter-optimized antibiotic selection markers flanked by LTRs and found evidence of long-term persistence of resistance to hygromycin and expression of a suite of flavin-based fluorescent reporters.

Careful consideration of the life cycle of anaerobic fungi is important to the genetic engineering of these complex organisms. Anaerobic fungal sporangia produce motile, unicellular zoospores that due to their thinner cell wall, are more amenable to genetic manipulation. We have found that fungal zoospores can be sorted by FACS while retaining viability, enabling high throughput sorting for downstream applications. We also show zoospores are capable of the uptake of fluorescently-tagged oligonucleotides and have optimized transformation conditions for high levels of DNA entry and cellular viability. Additionally, we also show that these zoospores can be analyzed by fluorescence-activated cell sorting (FACS) to gain insights into the production of cell-surface proteins involved in lignocellulose degradation.

316e Efflux of Cerebrospinal Fluid through Cervical Lymph Vessels Is Reduced in Aged Mice

Aditya Raghunandan, Ting Du, Maiken Nedergaard, Douglas Kelley

Introduction: Incidences of Alzheimer's disease (AD) and other dementias increase with age and are linked to the reduced clearance of the toxic waste products (such as amyloid-β (Aβ)) accumulating in the brain [1]. Cerebrospinal fluid (CSF) entering and leaving the brain is integral to this clearance. CSF is propelled into the brain along perivascular spaces, mixes with the brain's waste-laden interstitial fluid (ISF) and then drains these toxins predominantly via the cervical lymph vessels of the neck. This transport of fluid deteriorates with age, presumably because lymphatic function is compromised [1]. However, the dominant physiological mechanisms that drive cervical lymphatic flow (or fail in aging) remain unclear as most studies assess fluid transport using isolated lymph vessels. These conditions do not accurately represent physiological transport, where a network of vessels undergoes coordinated pulsations to transport fluid. In this talk, we present measurements of fluid flow in the cervical lymph vessels of mice in vivo. We use these first-of-their-kind measurements to explore the failure mechanisms impacting fluid clearance in mouse models of aging.

Methods: Intravital two-photon microscopy was used to visualize and quantify CSF flow in the cervical lymph vessels of both young (2-month) and aged mice (18 – 22-month, n>5). The mice were anaesthetized using ketamine/xylazine, and the cervical lymph vessels were surgically exposed. We simultaneously visualized vessel diameter fluctuations and the motion of 1 μm fluorescent microspheres (injected in the cisterna magna) flowing in the vessels. Detailed velocity measurements from particle motion and correlations to the changes in vessel diameter were performed using custom particle tracking and vessel wall detection algorithms in MATLAB [2]. All experiments were approved and conducted in accordance with the relevant guidelines and regulations stipulated by the University Committee on Animal Resources of the University of Rochester Medical Center, certified by the Association and Accreditation of Laboratory Animal Care.

Results: Particle tracking revealed significantly impaired clearance in the cervical lymph vessels of aged mice. The reduced speeds and lower volumes cleared were strongly correlated to the decrease in the contraction frequency (intrinsic rate) of the lymphatic vessels – the dominant driving force of fluid transport. Additionally, valve function is compromised in aged vessels, leading to stronger retrograde flows that reduce the net fluid volumes cleared.

Discussion: The techniques and results presented here allow for the first in vivo investigation of fluid dynamics and vessel function in the cervical lymph vessels – a dominant exit pathway draining CSF from the brain. With simultaneous particle tracking and vessel contraction measurements, we present mechanistic evidence that the severely reduced volumes of CSF cleared through these vessels is strongly correlated to the decrease in vessel function with age. Our measurements bolster the hypothesis that the cervical lymph vessels are a bottleneck in the aged brain's fluid transport and restoring their function could be a powerful therapeutic target.

Acknowledgements: This work was supported by the NIH/National Institute on Aging (grant RF1AG057575).

References:[1] Nedergaard, M., et al., *Science*. 2020, 370(6512);[2] Kelley, D.H., et al., *Am. J. Phys.* 2011, 69, 267

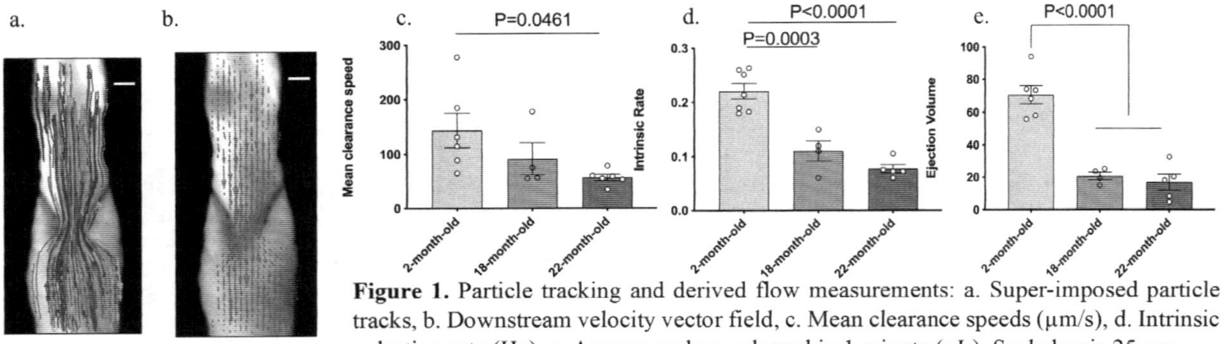

Figure 1. Particle tracking and derived flow measurements: a. Super-imposed particle tracks, b. Downstream velocity vector field, c. Mean clearance speeds (μm/s), d. Intrinsic pulsation rate (Hz), e. Average volume cleared in 1 minute (μL). Scale bar is 25 μm.

316f Self-Powered, Programmable, and Bioresorbable Drug Delivery Devices

Yamin Zhang, John A. Rogers

Implantable drug delivery devices are powerful for localized and controlled pharmacological treatments. The emergence of bioresorbable materials in recent years presents an opportunity to develop bioresorbable drug delivery devices, which can naturally degrade and disappear in the human body after a period of operation. We will introduce an approach that allows programmable control over the self-powered drug release process in the form of a fully bioresorbable drug delivery device that combines electrochemically degradable valves and bioresorbable polymer reservoirs. Fully bioresorbable construction eliminates the need for secondary surgical removal processes after treatment. Lidocaine, a local anesthetics medicine, has been successfully released to the sciatic nerve in the rat animal model.

316g Engineering Functional Vascularization By Synthetic Regulation of Paracrine Signaling

Mai T. Ngo, Ahmad S. Khalil, Christopher S. Chen

The ever-increasing demand for tissue or organ transplants, coupled with a limited supply of suitable donors, motivates the development of engineered tissue replacements for transplantation into patients. However, the translational utility of engineered tissues is currently limited by the inability to create functional vasculature that can support the perfusion, survival, and function of clinically-relevant sizes of highly-metabolic tissues during in vitro culture or post-transplantation. While numerous strategies have delivered angiogenic growth factors or genes alongside engineered tissues during transplantation, these approaches have largely failed to establish perfused, long-lasting vasculature.

During development, the human body utilizes a process called vascular morphogenesis to create vasculature. Vascular morphogenesis is a multi-stage process, during which distinct phases of paracrine signaling between endothelial and stromal cells direct the assembly of endothelial cells into premature cords that eventually acquire lumen and organize into a perfused vascular network. Although paracrine factors that are characteristic of each stage of vascular morphogenesis have been identified, it is unknown how the magnitude and timing of signaling within each stage dictates vascular architecture and functionality.

In this work, we combined synthetic biology and tissue engineering tools to investigate how the magnitude and timing of paracrine signaling during vascular morphogenesis impacts the formation of functional vasculature. We engineered endothelial and stromal cell populations to express characteristic paracrine factors (e.g. VEGF, bFGF, etc.) from each stage of vascular morphogenesis. Each paracrine factor is installed under the control of an orthogonal synthetic zinc finger transcription regulator (synZiFTR), such that the expression of each factor is controllably induced by a different small molecule (e.g. grazoprevir, doxycycline, etc.). Through small molecule induction, we demonstrated independent and tunable control over the kinetics, concentration, and identity of paracrine factor production. We encapsulated these engineered cells in a microfluidic model of vascular morphogenesis, in which vascular architecture and functionality were evaluated by immunofluorescent staining and perfusion with fluorescent dextran respectively. Using different programs of paracrine signaling by varying the concentration, timing, and duration of each small molecule, we found that vascular density can be controlled by varying the magnitude of paracrine signaling in the early stages of vascular morphogenesis, while induction of late-stage signaling drives morphogenetic changes required for perfusion of the vascular network. Importantly, we identify a critical window of time for inducing late-stage signaling in order to maximize perfusion and vascular network formation.

Overall, these results provide insight into how paracrine signaling is regulated during vascular morphogenesis to give rise to functional vascular networks with varied architecture, and point to an approach to usurp these controls to control the vascularization of tissues. More broadly, the toolbox developed in this work for controlling paracrine signaling offers a new generalizable approach to investigate and uncover a wide variety of tissue morphogenetic programs that could be used to direct the assembly of artificial tissues.

AUTHOR INDEX